Engineering Fluid Dynamics 2019–2020

Engineering Fluid Dynamics 2019–2020

Volume 1

Editor

Bjørn H. Hjertager

MDPI • Basel • Beijing • Wuhan • Barcelona • Belgrade • Manchester • Tokyo • Cluj • Tianjin

Editor
Bjørn H. Hjertager
University of Stavanger
Norway

Editorial Office
MDPI
St. Alban-Anlage 66
4052 Basel, Switzerland

This is a reprint of articles from the Special Issue published online in the open access journal *Energies* (ISSN 1996-1073) (available at: https://www.mdpi.com/journal/energies/special_issues/eng_fluid_dyn2019-2020).

For citation purposes, cite each article independently as indicated on the article page online and as indicated below:

LastName, A.A.; LastName, B.B.; LastName, C.C. Article Title. *Journal Name* **Year**, *Volume Number*, Page Range.

Volume 1
ISBN 978-3-0365-0214-4 (Hbk)
ISBN 978-3-0365-0215-1 (PDF)

Volume 1-2
ISBN 978-3-0365-0252-6 (Hbk)
ISBN 978-3-0365-0253-3 (PDF)

Contents

About the Editor

Bjørn H. Hjertager received his Ph.D. from the University of Trondheim, Norway (now the Norwegian University of Science and Technology, NTNU), in 1979. His thesis examined combustion, heat transfer, and fluid flow. Thereafter, he stayed for almost 10 years at Chr. Michelsen Institute. From the late 1980s to the late 1990s, he worked at the Telemark Institute of Technology (HiT-TF), and at the research institute Tel-Tek. He then spent 11 years at Aalborg University in Denmark. Since 2008, he has returned to Norway, as Professor in Fluid Dynamics at the University of Stavanger. He has published more than 180 papers on fluid flow, heat transfer, combustion, gas explosions, and chemical reactors, and has supervised 21 Ph.D. candidates in Norway and Denmark.

Preface to "Engineering Fluid Dynamics 2019–2020"

This book contains the successful submissions to a Special Issue of Energies entitled"Engineering Fluid Dynamics 2019–2020". The topic of engineering fluid dynamics includes both experimental and computational studies. Of special interest were submissions from the fields of mechanical, chemical, marine, safety, and energy engineering. We welcomed original research articles and review articles. After one-and-a-half years, 59 papers were submitted and 31 were accepted for publication. The average processing time was about 41 days. The authors had the following geographical distribution: China (15); Korea (7); Japan (3); Norway (2); Sweden (2); Vietnam (2); Australia (1); Denmark (1); Germany (1); Mexico (1); Poland (1); Saudi Arabia (1); USA (1); Serbia (1). Papers covered a wide range of topics including analysis of free-surface waves, bridge girders, gear boxes, hills, radiation heat transfer, spillways, turbulent flames, pipe flow, open channels, jets, combustion chambers, welding, sprinkler, slug flow, turbines, thermoelectric power generation, airfoils, bed formation, fires in tunnels, shell-and-tube heat exchangers, and pumps. I found the task of editing and selecting papers for this collection to be both stimulating and rewarding, and I would like to thank the staff and reviewers for their efforts and input.

Bjørn H. Hjertager
Editor

energies

MDPI

Article

CFD Simulations of the Propagation of Free-Surface Waves Past Two Side-By-Side Fixed Squares with a Narrow Gap

Haoyu Zhang [1], Shengnan Liu [2], Muk Chen Ong [2] and Renqing Zhu [1,*]

1 School of Naval Architecture and Ocean Engineering, Jiangsu University of Science and Technology, Zhenjiang 212003, China
2 Department of Mechanical and Structural Engineering and Materials Science, University of Stavanger, 4036 Stavanger, Norway
* Correspondence: zhurq@just.edu.cn; Tel.: +86-130-3359-2343

Received: 1 June 2019; Accepted: 4 July 2019; Published: 11 July 2019

Abstract: Two-dimensional computational fluid dynamics (CFD) simulations are carried out to investigate the gap resonance phenomenon that occurs when free-surface waves travel past twin squares in tandem. The volume of fluid method is used to capture the free surface. Validation studies of the present numerical model are conducted for different incident wave frequencies. The numerical results agree well with the published experimental data in terms of the free surface elevations in the gap. The hydrodynamic characteristics of the water column in the gap are investigated at different incident wave frequencies and gap widths. It is found that the free surface elevation in the gap increases and then decreases with the increasing incident wave frequency. The horizontal force on the weather side square structure (the structure in front of the gap) reaches the peak value at a larger frequency than the gap resonance frequency, whereas the variation of the horizontal force on the lee side structure (the structure behind the gap) is in-phase with the free surface elevation in the gap. Moreover, the added mass of the water column in the gap increases with the increasing gap width, which results in the decrease of resonance amplitude and frequency in the gap. However, this does not necessarily reduce the peak value of the horizontal forces on the structures.

Keywords: CFD; gap resonance; hydrodynamic forces; free surface waves

1. Introduction

In a multiple floating body system, such as the ship-by-ship offloading system and the assembly of very large floating structures (VLFS), the free surface elevations in the narrow gap between structures can be much higher than the incident wave heights. This phenomenon is called gap resonance, and it is caused by the proximity of incident wave frequencies to the natural frequency of the fluid vibrations in the gap. The gap resonance may cause higher wave forces, lead to violent body motions, and influence the stability of the system.

The sketch of the gap resonance motion is shown in Figure 1. The water column in the gap can have more than five times higher free surface elevations as compared to the incident waves, according to the experiments of Saitoh et al. [1]. The water column in the gap can be simplified as a rigid body (Molin, [2]). The motion of this rigid body then can be solved by considering the gravity force G, buoyancy force F_b, wave excitation force F_w, and radiation force F_r. Therefore, the oscillation of the water column in the gap can be influenced by these forces, which are relevant to the dimension of the gap and the incident wave conditions.

Extensive theoretical analyses, numerical simulations, and experiments have been carried out to investigate the generation mechanism and hydrodynamic characters of this phenomenon. The potential flow theory is an efficient method to investigate gap resonance. Theoretical explanations have been made to figure out the generation mechanism of the narrow gap, and simplified solutions have been proposed to estimate the natural frequency of the gap when the gap width is relatively small (Saitoh et al., [1]; Molin, [2]; Liu and Li, [3]; Tan et al., [4]). The estimated two-dimensional (2-D) gap resonance frequencies agree well with experiments (Saitoh et al., [1]; Iwata et al., [5]; Cong et al., [6]). However, since the linear potential flow theory over-predicts experimentally determined resonant responses in the gap under the excitation of sinusoidal waves (Faltinsen et al., [7]; Kristiansen et al., [8]), most of the potential flow models are semi-analytical or combined with empirical terms (Huijsmans et al., [9]; Newman et al., [10]; Chen, [11]).

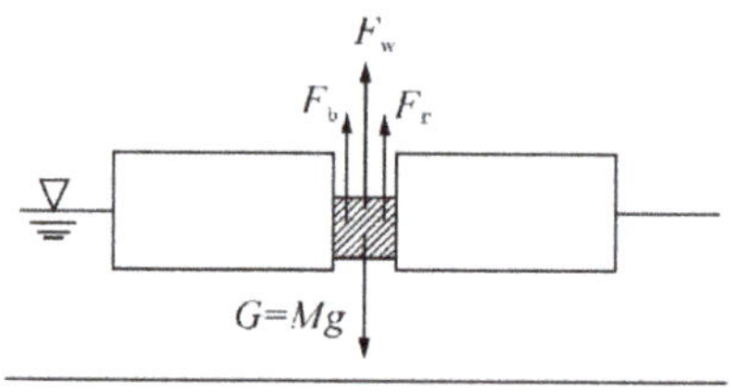

Figure 1. The sketch of the water column in the gap (reproduced from Cong et al. [6]).

As one of the sources of the damping terms in the potential flow model, the viscous effect is inevitable while solving the hydrodynamic forces and the resonance amplitudes at the gap between structures. Faltinsen and Timokha [7] and Kristiansen and Faltinsen [12] systematically investigated the damping in the gap both by experimental measurement and the potential flow method. The damping effects of the free surface nonlinearity, the flow separation, and the boundary layer on the hull were evaluated. It was concluded that the flow separation from the entrance of the gap is the main source of the discrepancy between the potential solver and their experimental results. Furthermore, Kristiansen and Faltinsen [13] investigated the coupled ship motion and the piston-mode flow in a gap. The flow separation at the ship bilge was modeled by an inviscid vortex tracking method. In this manner, the potential flow results were effectively suppressed. They further put forward a combined potential and viscous solver to deal with the flow separation from the barge bilge (Kristiansen and Faltinsen, [13]). Feng and Bai [14] investigated the gap resonance under nonlinear waves by a mixed Eulerian–Lagrangian scheme by the high-order boundary element method (HOBEM). They demonstrated that the main source of the overestimated response is viscosity instead of the free surface nonlinearity. The effect of free surface nonlinearity is negligible except for slightly increasing the resonance frequency. Another truth that proves the importance of viscosity is the influence of the hull bilge shape on the resonance amplitude. Moradi et al. [15] investigated a 2-D gap resonance through numerical simulations including viscosity. They found that the resonance amplitude became smaller with the increasing bilge radius. Tan et al. [16] carried out large quantities of experiments to investigate the effects of the bilge shape. Their results suggested that the energy dissipations of sharp bilge cases are larger than round bilge cases. This was likely due to the different flow separation behavior at the bilge.

Several computational fluid dynamics (CFD) simulations have been carried out to evaluate the viscous dissipation of narrow gap resonance problem. Lu et al. [17,18] investigated the 2-D gap resonance both by the modified potential model [11] and viscous numerical methods. By optimizing the artificial damping of the potential solver, the numerical results based on the potential theory agreed well with the turbulent viscous solver in terms of the resonance amplitude and the hydrodynamic forces. Tan et al. [16] figured out the relationship between the damping coefficients of a theoretical dynamic model (ε) and that used in the modified potential model [11,17] (μ_p): $\mu_p = 3\pi\varepsilon\omega_n/8$, where ω_n is the natural frequency, and the damping coefficient ε could be calculated according to the surface elevation results of the numerical viscous solver. Zhao et al. [19] experimentally investigated the 3-D gap resonance under new wave-type transient waves. They found that each gap resonance mode had a characteristic damping which was somewhat larger than the damping calculated using the linear potential flow theory alone. Later, Wang et al. [20] employed OpenFOAM-based Navier–Stokes (N–S) equations to reproduce the experiment results of Zhao et al. [19]. The grid resolution, mesh topology, domain size, and boundary conditions were systematically optimized. The transient wave group was considered to be a better choice for investigating the 3-D gap resonance phenomenon as compared to the regular incident waves. Based on a CFD solver, Chua et al. [21,22] developed a framework to evaluate the damping coefficients of a 3-D gap resonance problem. The energy dissipations regarding wave scattering, frictional force, flow separation, appendages, and hull motions were investigated thoroughly. It could be concluded that the modified potential model considering the damping effects is able to predict well regarding the gap resonance problem. The value of damping, however, still relies on CFD simulations, which are less expensive compared to experiments.

The previous research concerns on the gap resonance problem were usually the viscous effects and the free surface elevations in the gap. There is still lack of research on the characteristics of hydrodynamic loads on structures with a narrow gap. In the present study, the hydrodynamic loads, viscous effects, and free surface elevations related to the narrow gap resonance were investigated thoroughly. The forces on the floating structures are more important as compared to the surface elevations in the gap, while the horizontal wave loads are more sensitive to the gap resonance, as compared to the vertical wave loads [18]. The present study is organized as follows. Firstly, a mesh and time-step refinement study is conducted for the case with the smallest gap width, which is considered to be the most severe case. Secondly, the free surface elevations in the gap, the viscous dissipation, and the horizontal wave loads on the structures are studied at different incident wave frequencies and different gap widths between structures. Meanwhile, viscous dissipation and horizontal wave loads versus the gap width are also discussed. The present research can provide a reasonable reference to the engineering design process in term of nap gap damping.

2. Numerical Modeling and Setup

2.1. Governing Equations

The open source CFD software OpenFOAM [23] was used to solve the N–S equations numerically. The toolbox waves2Foam (Jacobsen et al., [23]) was applied for the numerical wave tank. The governing equations for the incompressible viscous flow are given as follows [24]:

$$\begin{cases} \nabla \cdot \boldsymbol{v} = 0 \\ \frac{\partial(\rho \boldsymbol{v})}{\partial t} + \nabla \cdot (\rho \boldsymbol{v} \boldsymbol{v}) = \nabla \cdot \mu \nabla \boldsymbol{v} - \nabla p_{rgh} - \boldsymbol{g} \cdot (\boldsymbol{x} - \boldsymbol{x}_r) \nabla \rho \end{cases} \tag{1}$$

where ∇ denotes the Hamiltonian operator, $\boldsymbol{v}$ is the velocity field, ρ is the fluid density, μ is the dynamic viscosity coefficient, t is the time, p_{rgh} is the pressure in excess of the hydrostatic pressure $p_{rgh} = p - \rho \boldsymbol{g} \cdot (\boldsymbol{x} - \boldsymbol{x}_r)$, $\boldsymbol{g}$ is the gravity vector, $\boldsymbol{x}$ is the Cartesian coordinate vector, and $\boldsymbol{x}_r$ is the reference coordinate vector defined at sea level. The contribution of surface tension effect is less than 1% of the inertial force when T > 0.35 s, H > 0.02 m [24], where T is the deep-water wave period and H is the deep-water wave height. Therefore, the surface tension effect is considered negligible in the present study (see the numerical set-up in Section 3).

2.2. Free Surface Capturing

The volume of fluid (VOF) method was applied to capture the free surface. The water volume fraction is defined as follows:

$$\alpha(\boldsymbol{x}, t) = \begin{cases} 0 & air \\ 0 < \alpha_s < 1 & free\ surface \\ 1 & water \end{cases} \tag{2}$$

where α is the volume fraction of the water phase. The density ρ and the dynamic viscosity μ are calculated as follows:

$$\begin{cases} \rho = \alpha\rho_{water} + (1-\alpha)\rho_{air} \\ \mu = \alpha\mu_{water} + (1-\alpha)\mu_{air} \end{cases} \tag{3}$$

The transport equation of the water volume fraction is:

$$\frac{\partial\alpha}{\partial t} + \nabla\cdot(\alpha \boldsymbol{v}) + \nabla\cdot[\alpha(1-\alpha)\boldsymbol{v}_r] = 0 \tag{4}$$

where t is the time; $\alpha = \alpha(\boldsymbol{x}, t)$ denotes the cell-based water volume fraction at coordinate $\boldsymbol{x}$ in time t; $\boldsymbol{v}$ is the local fluid velocity; and $\boldsymbol{v}_r$ is the compress velocity at the interface [25]:

$$\boldsymbol{v}_r = \boldsymbol{v}_{water} - \boldsymbol{v}_{air} \tag{5}$$

2.3. The Numerical Wave Flume

The relaxation zone technique [24] was applied for making and absorbing waves. In each time-step, this technique corrects fluid fields in relaxation zones by Equation (6). As a consequence, fluid fields in these zones change gradually from the computed fields (that are obtained by theories) to the target fields (according to the selected wave theory):

$$\Phi = (1-\omega_R)\Phi_{target} + \omega_R\Phi_{computed} \tag{6}$$

where Φ denotes the corrected fields; Φ_{target} and $\Phi_{computed}$ are the target and the computed fields, respectively; and $\omega_R = \omega_R(\sigma) \in [0,\ 1]$ denotes the weighting function which varies from 1 to 0 as the relaxation zone local coordinate σ increases from 0 to 1. The default weighting functions used in waves2Foam is exponential:

$$\omega_R = 1 - \frac{\exp(\sigma^{3.5}) - 1}{\exp(1) - 1} \tag{7}$$

2.4. The Post-Processing

The nondimensionalized surface elevations inside the gap and the forces on the floating structures are defined in Equations (8)–(13), which will be used in the post-processing. The free surface elevations at the wave gauges in the wave flume are normalized by the incident wave height H_i:

$$\eta^* = \frac{\eta(t)}{H_i} \tag{8}$$

where $\eta(t)$ is the instantaneous free surface elevation relative to the still water level and H_i is the incident wave height. Correspondingly, the normalized surface elevation amplitude is:

$$\eta_a^* = \frac{\eta_a}{A_i} \tag{9}$$

where η_a is the average value of the elevation amplitudes of several steady-state periods and A_i is the incident wave amplitude.

The wave force f is calculated by integrating over the surface Ω of the squares:

$$f = \int_\Omega \left(-\mathbf{n}p_{rgh}(\mathbf{r},t) + \mathbf{n}\cdot\boldsymbol{\tau}(\mathbf{r},t)\right)d\Omega \tag{10}$$

where $\mathbf{n}$ is the unit normal vector pointing into the fluid, $\mathbf{r}$ is the coordinate vector, $p_{rgh}(\mathbf{r},t)$ is the in excess of the static pressure (see Equation (1)), and $\boldsymbol{\tau}(\mathbf{r},t) = \mu\nabla\boldsymbol{v}$ is the shear stress tensor.

As a consequence, the instantaneous magnitude f of the wave force f contains both the hydrodynamic components and the hydrostatic components:

$$f = f_{inertial+viscous} + f_{hydrostatic} \quad (11)$$

where $f_{hydrostatic}$ is the hydrodynamic component—the hydrostatic component $f_{hydrostatic}$ is the variation of the transient hydrostatic force relative to the initial hydrostatic force in still water.

The normalized parameter f^* is introduced to normalize the hydrodynamic force on the squares:

$$f^* = \frac{f}{\rho g B A_i} \quad (12)$$

where B is the breadth of the square and d is the draft of the square;

Furthermore, the subscript x was introduced for horizontal force components, e.g., f_x^* is the horizontal normalized force and f_x is the horizontal hydrodynamic force.

The normalized horizontal force amplitude is defined as:

$$F_x^* = \frac{F_x}{\rho g B A_i} \quad (13)$$

where F_x is the average value of the force amplitude of steady-state periods.

3. Simulation Cases

Twin square bodies are fixed side-by-side in regular incident waves—see SQ1,2 in Figure 2. B is the width of the square, h is the water depth. Several normalized parameters are introduced here: $d^* = \frac{d}{B}$; $h^* = \frac{h}{d}$; $\omega^* = \omega/\sqrt{\frac{g}{B}}$, where ω is the incident wave frequency; $B_g^* = \frac{B_g}{B}$, where B_g is the gap width; the Keulegan–Carpenter number (KC number) $KC = \frac{2\pi A_i}{B_g}$. From hereon, the parameter ω^* indicates the nondimensional incident wave frequency of the cases, while the parameter B_g^* indicates the nondimensional gap width of the cases.

Two variables were considered in the present study, i.e., the incident wave frequency and the gap width between two squares, which are relevant to the resonance of the water column in the gap. The variations of the gap width can result in different natural frequencies of the water column in the gap. The incident wave frequency can influence the behavior of the gap resonance. Both the characteristics of the hydrodynamic forces and the gap resonance were investigated in the present study. Here B, d^*, and h^* are constants which are same as the existing experiments (Saitoh et al., [1]; Tan et al., [16]): $B = 0.5$ m; $d^* = 0.5$; and $h^* = 2$. The KC numbers of the present cases were small, i.e., less than 3 for all the simulation cases. Therefore, the flows were considered to be inertial dominated [26].

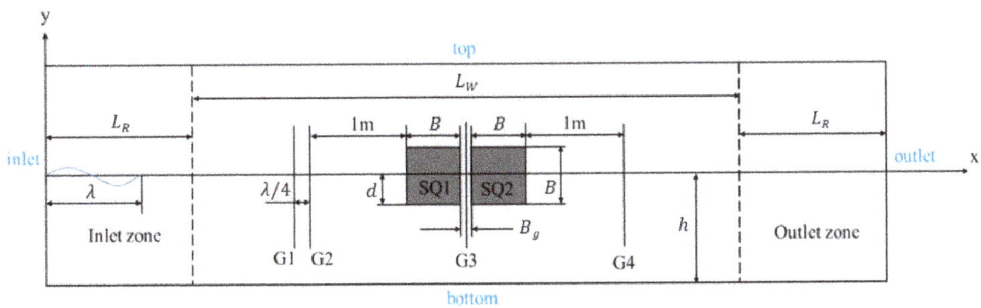

Figure 2. Geometrical parameters of the numerical wave flume.

3.1. Numerical Wave Flumes

The reference length for the wave flume design is the resonance wave length λ_n, which is estimated by the first order wave theory:

$$\begin{cases} \lambda_n = 2\pi/k \\ \omega_n^2 = gk\,tanh(kh) \end{cases} \tag{14}$$

where g is the gravitational acceleration, k is the wave number, h is the water depth (see Figure 2); and the resonance frequency ω_n is estimated according to [1]:

$$\omega_n = \sqrt{\frac{g}{B_g B/(h-d)+d}} \tag{15}$$

Jacobsen et al. [24] recommended that the length of the relaxation zones should be larger than the incident wave length. Therefore, L_R (Figure 2) was set to $2\lambda_n$, L_W was $5\lambda_n$, and the total length L of the numerical wave tank was $9\lambda_n$. The B_g^* was fixed as 0.1 to investigate the influence of incident wave frequencies, which is the same as the experimental set-up of Saitoh et al. [1] and Tan et al. [4].

3.2. Simulation Cases

(i) The Influences of the Incident Wave Frequencies

In the present simulation cases, different incident wave frequencies were set at the input boundary, which covers the resonance frequency region. However, waves with higher steepness result in larger viscous dissipation in the process of wave propagation [27,28], and the grid resolution needs to vary correspondingly to obtain sufficient numerical accuracy. Therefore, the largest wave frequency was $1.1\omega_n$, i.e., the shortest wave length was no smaller than 80% of λ_n. On the other hand, the waves with the smallest frequency pertain to the second-order Stokes wave theory. The incident wave frequencies are listed in Table 1.

Table 1. Summary of the present numerical simulation cases.

Index	B_g^*	ω_n^*	KC	ω^*									
C1~C9	0.10	1.20	1.51	C1 0.99	C2 1.07	C3 1.11	C4 1.16	C5 1.17	C6 1.20	C7 1.21	C8 1.24	C9 1.28	
C10~C17	0.17	1.09	0.89	C10 0.89	C11 1.03	C12 1.08	C13 1.09	C14 1.10	C15 1.13	C16 1.14	C17 1.20		
C18~C27	0.25	1.00	0.60	C18 0.77	C19 0.92	C20 0.98	C21 1.00	C22 1.01	C23 1.04	C24 1.06	C25 1.07	C26 1.09	C27 1.11

The incident wave amplitude A_i was fixed to 0.012 m in accordance with existing experiments (Saitoh et al., [1]; Tan et al., [4]). Besides, four wave gauges (G1–G4) were applied for the free surface elevations and the wave energy dissipation ratio, $R_d = 1 - {R_r}^2 - {R_t}^2$, where $R_r = \frac{A_r}{A_i}$ is the reflection ratio, A_r is the reflected wave amplitude obtained by the two point method [29], $R_t = \frac{A_t}{A_i}$ is the transmission ratio, and A_t is the transmitted wave amplitude measured at G4.

(ii) The Influences of the Gap Width

The B_g^* changed from 0.1 to 0.25 to include the general range of gap width between two squares in published studies. It should be noted that B_g^* is usually small (e.g., $B_g^* \approx 0.1$ [1,4,5]). For each B_g^*, ω_n was calculated by Equation (15) and normalized to ω_n^*. The selections of incident wave frequencies are shown in Table 1 in detail.

3.3. Boundary Conditions

As shown in Figure 2, the 2-D rectangular computational domain was built around the square structures with inlet, outlet, top, and bottom boundaries. The boundary conditions used in the present study are summarized as follows:

(i) At the inlet boundary, the velocity for the water was given according to the linear wave theory while the air velocity is zero. The normal zero-gradient condition was applied for the pressure.

(ii) At the outlet boundary, the pressure was specified as normal zero-gradient, and the velocities for both water and air were set to zero.

(iii) The top boundary of the computational domain was set as an atmospheric condition, which allowed the air to flow in and out of the domain. The pressure at the top boundary was calculated by $p_T = p_0 - 0.5u^2$. The velocity u was obtained from the flux at the patch for the inflow and a normal zero-gradient condition for the outflow.

(iv) The no-slip wall boundary condition was used at the bottom and the structures' surface, where the velocity was zero. The pressure was set as the normal zero-gradient condition.

(v) The "empty" boundary condition was employed at the front and back boundaries due to the present 2-D simulations in OpenFOAM.

4. Mesh and Time-Step Refinement Studies

Mesh and time-step refinement studies were carried out for case C6, which was the case with the smallest gap. The numerical results with different meshes and time-steps were compared to each other in terms of the normalized free surface elevations at G2 (in front of the squares), G3 (in the gap), G4 (behind the squares; see Figure 2), and the hydrodynamic forces on SQ1. As B_g^* becomes larger, the resonance amplitude in the gap decreases (Moradi et al., [15]). Therefore, the verified grid settings could be utilized for the other cases with larger B_g^*. Figure 3 shows the zoom-in view of the mesh around the twin squares of case C6. The vertical grid size over the free surface region is noted as Δy_f, the horizontal grid size in the free surface region is Δx_f, and the grid size in the gap and in the vicinity of the square structure boundaries is Δy_{bn}. The grid size changes gradually at different locations, e.g., between the free surface and the bottom region the grids distribute in the form of hyperbolic cosine function vertically according to the water wave velocity. Details of the grid resolutions are shown in Table 2. An adaptive time-step scheme was used for simulations with a maximum Courant number of 0.25.

Table 2. The three different grid resolutions used in the present study.

Mesh Index	Δx_f	Δy_f	Δy_{bn}	Cell Number
A	$\lambda_n/80$	$H_i/8$	$B_g/40$	139,270
B	$\lambda_n/200$	$H_i/20$	$B_g/100$	723,362
C	$\lambda_n/250$	$H_i/25$	$B_g/125$	1,100,640

Figure 4 shows the results of the different grid resolutions over five wave periods after the results repeat their cycles. Figure 4a is the normalized surface elevation η^* at G2, G3, and G4 versus the normalized time $t^* = \frac{t-t_0}{T_n}$, where t is the simulation time and t_0 is the start time of the captured steady-state periods. The relative error of η^* between different grid resolutions is noted as ε, and the numerical model is considered converged when $\varepsilon < 5\%$. In Figure 4a, the largest ε (the ε at G4) between the numerical results of mesh A and mesh B is 19.79%, and the largest ε between the numerical results of mesh B and mesh C is 4.95%. Therefore, mesh B is considered to give the sufficient accuracy in terms of the prediction of η^*. Figure 4b illustrates F_{x1}^* versus t^* over the same duration as in Figure 4a, where the largest ε between the numerical results of mesh A and mesh B is 2.55%, and the largest ε between the numerical results of mesh B and mesh C is 0.40%. The results show that mesh B is sufficiently fine for predicting both the surface elevations and the horizontal wave forces. Therefore, mesh B was employed for the numerical simulations in the present study.

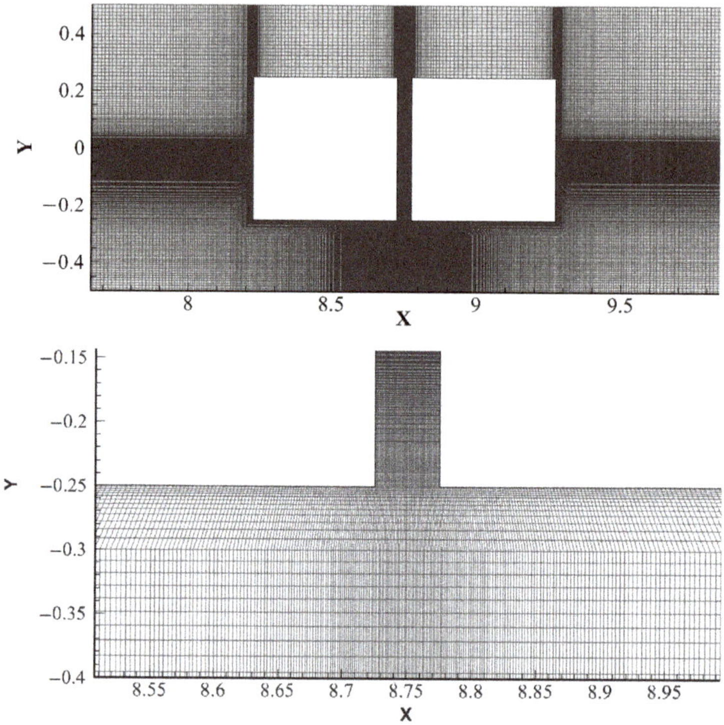

Figure 3. The zoom-in view of mesh around the twin squares for case C6.

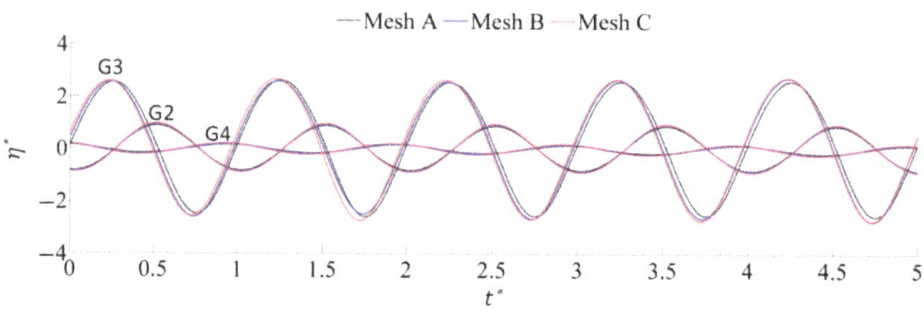

(**a**)

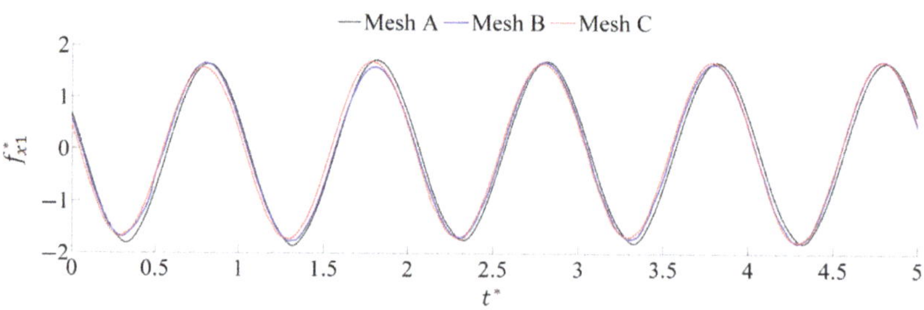

(**b**)

Figure 4. Time history curves of C6 using different mesh settings over five steady-state periods: (**a**) η^* versus t^* at G2, G3, and G4; (**b**) f^*_{x1} versus t^*.

A time-step refinement study was carried out based on mesh B. A simulation with a maximum Courant number of 0.15 was performed to verify the convergence of the time-step settings. Figure 5 shows the results of C6 using different time-steps over five steady-state periods. The maximum ε between η^* with two time-steps was 1.91%; the ε between F^*_{x1} with two different time-steps was 2.83%. Therefore, a maximum Courant number of 0.25 was employed for the present numerical simulations.

The verified grid resolution and time-step settings were further applied to the cases with $B^*_g = 0.17$ and the cases with $B^*_g = 0.25$. The mesh for cases with $B^*_g = 0.17$ (cases C10~C17) contained 730,410 cells, while the mesh for the cases with $B^*_g = 0.25$ (cases C18~C27) contained 899,190 cells; see Table 2 for more information.

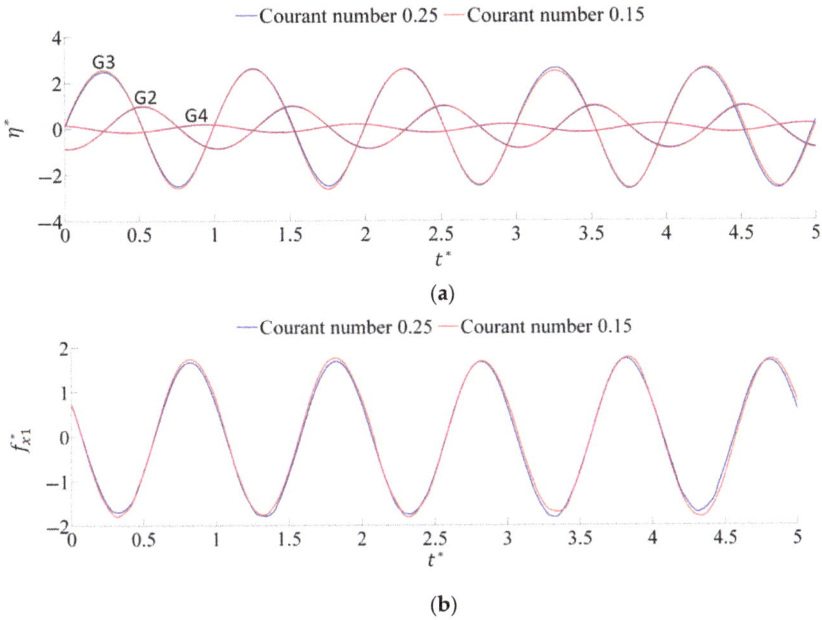

Figure 5. Time history curves of C6 using different time-steps over five steady-state periods: (**a**) η^* versus t^* at G2, G3, and G4; (**b**) f^*_{x1} versus t^*.

5. Results and Discussions

5.1. The Influences of Incident Wave Frequencies

Simulations were performed for cases C1~C9 ($B^*_g = 0.1$) to investigate the influence of incident wave frequencies. Discussions were carried out in terms of the surface elevation amplitudes in the gap, the viscous dissipation in the gap, the phase-frequency characters of the wave field, and the horizontal wave forces on squares. The subscripts 1 and 2 were introduced to distinguish the forces on the SQ1 and the SQ2, e.g., f_{x2} denotes the horizontal force on SQ2. The present numerical results were compared with the existing experimental and numerical results.

Figure 6 is the comparison of elevation amplitudes inside the gap (η^*_a) versus the incident wave frequency ω^* between the present study and the experiment by Saitoh [1]. The relative difference ε here was the relative error between the interpolation points of the simulations and the experiment data points. For case C1~C7, ε was less than 5%, and ε was slightly larger, i.e., less than 15% for cases C7~C9. The good agreement shows that the present numerical model is able to predict the elevation amplitudes in the gap with reasonable accuracy. As the incident wave frequency ω^* became larger, the elevation amplitude inside the gap η^*_a increased rapidly and reached its peak at $\omega^* = 1.195$, before decreasing. This tendency is in accordance with the resonance phenomenon. The mechanism of the resonance can be explained by considering the motion of fluid in the gap as rigid body motion. The motion equation of the fluid vibration in the gap can be expressed as:

$$(m + m_a)\ddot{\eta} + c\dot{\eta} + k_s\eta = f_{excitation} \tag{16}$$

where η is the fluid displacement or the surface elevation inside the gap, $f_{excitation}$ is the excitation force, m is the mass of fluid inside the gap, m_a is the added mass, c is the damping coefficient, k_s is the stiffness. Based on Equation (16), the natural frequency of the fluid vibration in the gap is:

$$\omega_n = \sqrt{\frac{k_s}{m + m_a}} \tag{17}$$

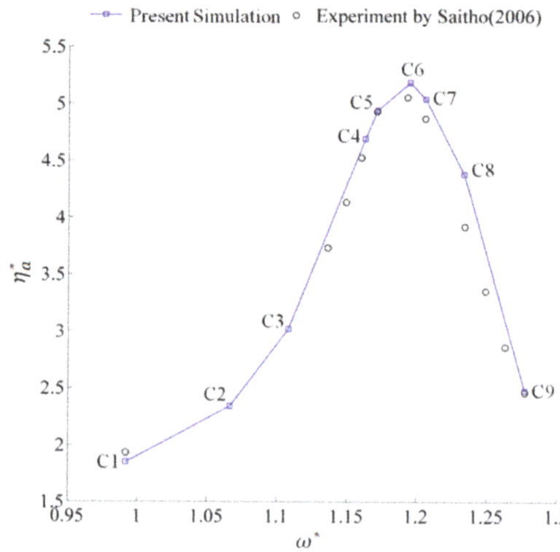

Figure 6. The comparison of the present numerical results and the experimental data by Saitoh [1] in terms of the free surface elevations in the gap (η_a^*) ($B_g^* = 0.1$).

Resonance occurs when the incident wave frequency is equal or nearly equal to ω_n. To be specific, the stiffness k_s is the coefficient of y in the buoyancy variation term as the fluid moves, $k_s = \rho g B_g$. The first order mass of the fluid inside the gap is the fluid mass inside the gap of still-water state, $m = \rho B_g d$. Equation (17) can further be simplified according to these formulas:

$$\omega_n = \sqrt{\frac{g}{\frac{m_a}{\rho B_g} + d}} \tag{18}$$

Equations (16)–(18) are the basis of follow-up discussions in the present study. It should be noted that these equations are not limited to the rigid body assumption. It is a general motion equation of the vertical flow motion in the gap. The coefficients m_a and c change significantly between different theoretical models. For example, Equation (15) (derived by Saitoh, [1] using the energy method) is equivalent to take m_a as $\frac{\rho B B_g^2}{h-d}$. In Tan et al. [4], m_a was $\frac{\rho B^* B_g^2}{4(h-d)}$, where B^* is the artificial coefficient.

The viscous dissipation influences the flow field around the squares pronouncedly. In analytical and potential methods, damping is usually obtained from CFD simulations or experimental data. It is generally agreed that the vorticity near the entrance of the gap is the main source of the viscous dissipation. Figure 7 shows the variation of the vorticity contour in the water phase for case C6 (ω^* equals to ω_n^*) at $t^* = 0$, 0.17, 0.33, 0.50, 0.67, 1. When $t^* = 0$, the water level in the gap was at the lowest position, and the average flow velocity in the gap was nearly zero. As the wave gradually travels past the square bodies at $t^* = 0 \sim 0.25$, part of the waves reflected back and overlapped with the incident waves, part of the wave energy was absorbed by the movement of the elevation in the gap, and the rest was the transmitted waves. Due to the large velocity gradient at the corners of the SQ1 and SQ2, two vortices began to grow symmetrically in the gap to dissipate the energy. From $t^* = 0.25 \sim 0.5$, the decreasing flux in the gap resulted in the hydrodynamic pressure gradient, which was at the same direction as the flow direction in the gap, and pushed the vortex back to the gap entrance. From $t^* = 0.5 \sim 1$, the vortices gradually spread to the bottom region below the square structures as the water level in the gap returned back to the lowest position.

Figures 8 and 9 present the variations of the vorticity contour of case C2 (ω^* smaller than ω_n^*) and C9 (ω^* larger than ω_n^*). In these two cases, the flow in the gap produced smaller vortices as compared to case C6. Furthermore, Figure 10 shows the square of transmission ratio ${R_t}^2$, the square of the reflection ratio ${R_r}^2$ and the dissipation ratio R_d versus the incident wave frequency ω^*. The present simulation results were compared with the experimental data from Tan et al. [16]. The linear interpolation method was applied for the numerical results according to the sample points from the experimental data. The root mean square errors (RMSE) of the interpolated numerical results relative to the experimental data were calculated. For the mth wave frequency, R_{1m} and R_{2m} represent the interpolated numerical results and the experiment data, and the RMSE was calculated by: $\varepsilon = \sqrt{\sum_{m=1}^{M}[(R_{1m} - R_{2m})^2/M]}$, where M is the total number of frequencies of the experiments. In Figure 10, the RMSE of R_r^2, R_t^2, and R_d are 0.061, 0.04, and 0.053, respectively, which are relatively small values. There was a slight difference of peak frequency in the present simulation and experiments—i.e., R_t^2 was approximately 1.17 in the present numerical results, and in the experiments, the peak frequency for R_t^2 was 1.15. Though the difference was only around 1.7%, it was able to result in a significant difference when the resonance interval was narrow and R_t^2 was nearly zero out of the resonance interval. Another reason for the discrepancies was the uncertainties of the experiments. For example, when the incident wave frequency ω^* was approximately 1.02, the R_d in the experiment was negative, Tan et al. (2016) believed that this unphysical phenomenon was caused by the measurement uncertainties. In general, the present numerical results are in good agreement with the experiments.

When gap resonance happens, the wave reflection ratio decreases a lot, which is mainly caused by the increased dissipation ratio. Assuming that the energy loss in the gap is equal to the energy loss of the whole wave field in each period, the value of damping coefficient c (Equation (16)) can be estimated as:

$$\frac{\rho g \omega}{4k}\left[1 + \frac{2kh}{\sinh(2kh)}\right]A_i^2 R_d = \frac{1}{T}\int_0^T c\dot{\eta}^2 dt \Rightarrow c = \frac{\rho g}{2k\omega}\left[1 + \frac{2kh}{\sinh(2kh)}\right]\frac{R_d}{(\eta_a^*)^2} \tag{19}$$

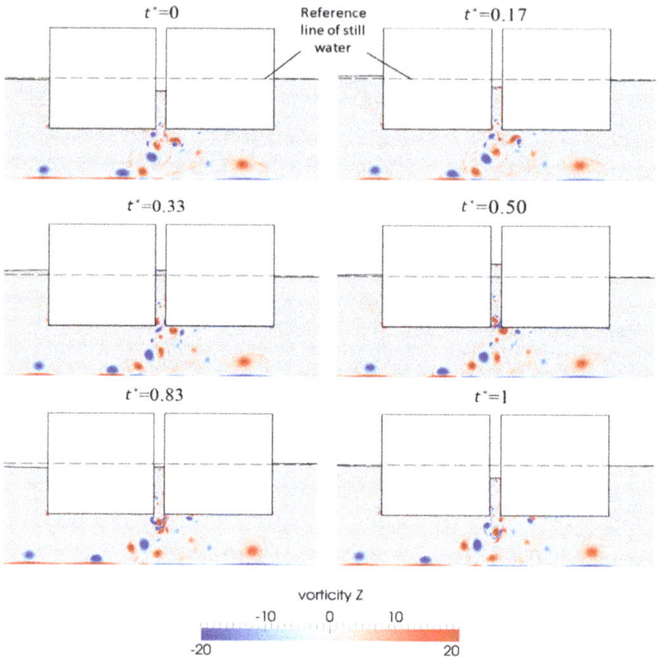

Figure 7. The variations of the vorticity contour in the water phase for Case C6 (incident wave frequency approaching the resonance) at different time instants.

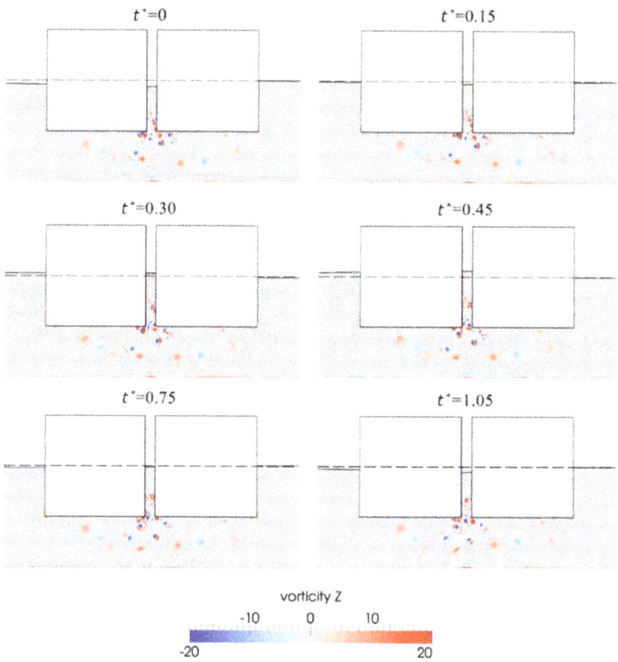

Figure 8. The variations of the vorticity contour in the water phase for Case C2 (incident wave frequency smaller than the resonance frequency) at different time instants.

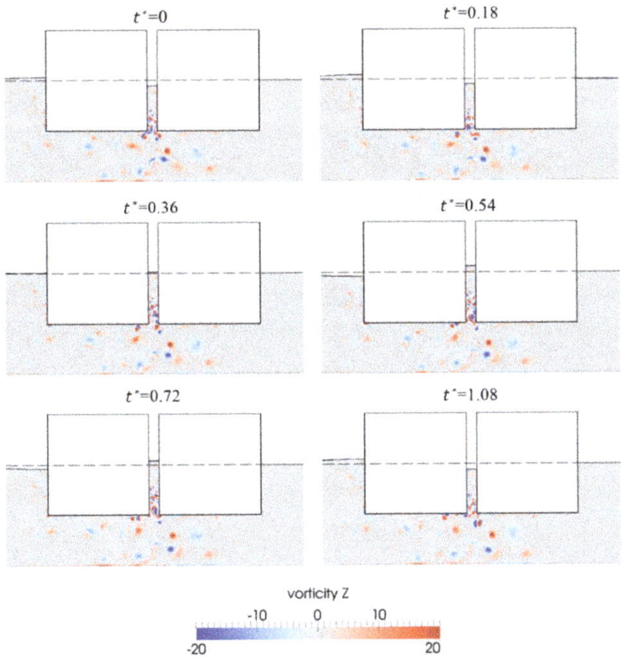

Figure 9. The variations of the vorticity contour in the water phase for Case C9 (incident wave frequency larger than the resonance frequency) at different time instants.

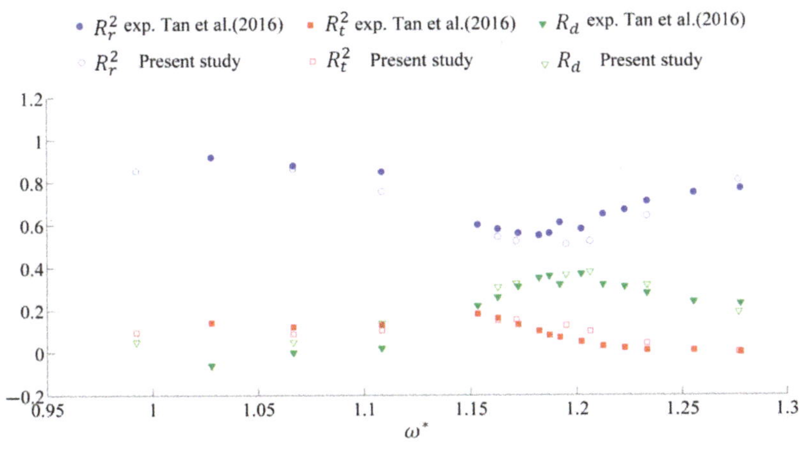

Figure 10. The square of transmission ratio $R_t{}^2$, the square of the reflection ratio $R_r{}^2$, and the dissipation ratio R_d versus the incident wave frequency ω^* in the present study and in the experiments conducted by Tan et al. [4] ($B_g^* = 0.1$).

The damping coefficient c versus ω^* is illustrated in Figure 11. As ω^* approaches ω_n^*, the damping coefficient c gradually converges to a constant value. Tan et al. [16] suggested that the linearized damping coefficient is proportional to the excitation frequency ω: $c = \varepsilon\omega M$, where ε is a non-dimensional damping coefficient obtained from the CFD results of the resonance case. In order to make comparisons, the damping coefficient c in present simulations was reformulated as:

$$c = c_{\omega}\omega \tag{20}$$

where c_{ω} is the damping coefficient c of the resonance case divided by the resonance frequency ω_n. In cases with the gap width $B_g^* = 0.1$ (cases C1~C9) the value of $c_{\omega} = \frac{5.858}{\omega_n} = 1.107$, which agrees well with that of the cases with the same gap width B_g^* of Tan et al. [16] ($c_{\omega} = 1.010$). It should be noted that their results were obtained from CFD calculations including turbulence modeling. This demonstrates that the present laminar model has enough accuracy for predicting the viscous dissipation in the gap, as mentioned in Section 3.

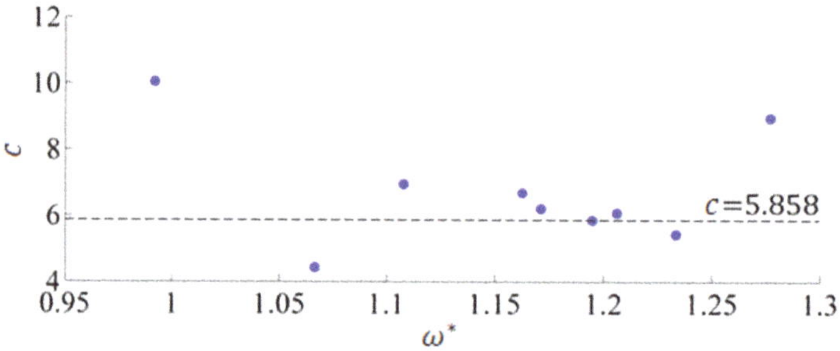

Figure 11. The damping coefficient c versus the incident wave frequency ω^* in the present numerical study ($B_g^* = 0.1$).

To simplify the problem, in later discussions, it was assumed that the only influence of viscosity is to determine the surface elevation inside the gap, while the flow outside the gap is irrotational. Firstly, as shown in Figures 7–9, the vorticity mostly gathered in the bottom area of the gap and had little influence on the flow tendency of the entire flow field. Secondly, several studies [9–11,17,18] indicated that the linear potential flow theory based results are reasonable after taking into account the damping inside the gap.

From Figures 7–9, the phase retardation of the elevation inside the gap relative to the wave profile in front of SQ1 varied visibly. This interesting character was further investigated. The phase of the elevation inside the gap (Φ_1) was evaluated from the elevation time history inside the gap ($\eta(t)$):

$$\Phi_1 = \omega t_{crest} \tag{21}$$

where t_{crest} is the time corresponding to the peak of $\eta(t)$ and ω is the incident wave frequency.

In the present simulations the incident wave phase in front of SQ1 (Φ_2) was:

$$\Phi_2 = kx_{front} + 2\pi n_p \tag{22}$$

where k is the wave number, x_{front} is the x coordinate in front of the SQ1, and n_p is the number of the wave period.

The wave phase in front of the SQ1 may deviate slightly from the result of Equation (22) as a consequence of the superposition between the incident wave and the reflected wave. However, the effect is negligible. It was assumed that the incident wave completely reflected back at the left side of SQ1. Based on the wave superposition theory, this generated a standing wave with the following profile function:

$$\begin{aligned} \eta_{sw} &= A_i \cos(\omega t - kx) + A_i \cos\left(\omega t + kx - 2x_{front}\right) \\ &= 2\mathrm{A_i} \cos\left(\omega t - kx_{front}\right) \cos\left[k\left(x - x_{front}\right)\right] \end{aligned} \tag{23}$$

Figure 12 is the comparison between the surface elevations in front of SQ1 in the present simulations and of the presumed standing waves in front of SQ1. The time history curves of the surface elevation in front of SQ1 were almost coincident with the elevation time histories of the presumed standing wave in front of SQ1, which were calculated based on Equation (22). This indicates that most parts of the incident wave reflected back because of the strong shielding effect of the twin square structures. By comparing Equations (22) and (23), it could readily be drawn that the wave profile in front of SQ1 was almost in-phase with the incident wave profile at the left side surface of SQ1.

Therefore, the phase retardation of the elevation inside the gap relative to the wave phase in front of SQ1 (Hereinafter referred to as the phase leg Φ) was:

$$\Phi = \Phi_1 - \Phi_2 \tag{24}$$

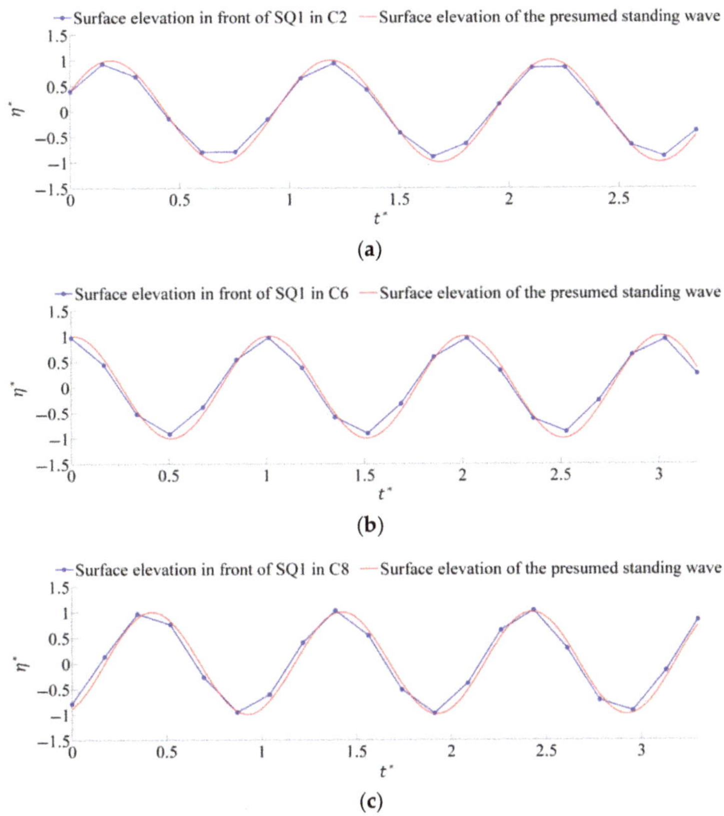

Figure 12. The comparison of the surface elevations in front of SQ1 with the presumed standing wave in front of SQ1: (**a**) for cases C2; (**b**) for case C6; (**c**) for case C8.

In order to clarify the character of the phase leg Φ, a further investigation on the excitation force ($f_{excitation}$) of the liquid inside the gap was made. As a matter of investigating the excitation force, the liquid inside the gap was considered stationary. The twin squares and the liquid inside the gap could be regarded as an entire rectangular obstacle with a width of $B_t = 2B + B_g$. In the control volume below the obstacle (see region CV in Figure 13), the velocity potential ψ followed the impermeable boundary condition at the boundaries *ad* and *bc*; it could therefore be written as (Cong et al., [6]):

$$\psi = A_0 x\, e^{-j\omega t} + \sum_{n=1}^{\infty}\left[A_n e^{\mu_n x} + B_n e^{\mu_n (B_t - x)}\right]\cos\mu_n(y+h)e^{-j\omega t} \tag{25}$$

where $\mu_n = \frac{n\pi}{h-d}$ is the eigenvalue ($n = 1,2,\ldots$); A_0, A_n, and B_n are complex constants; and j is the imaginary unit.

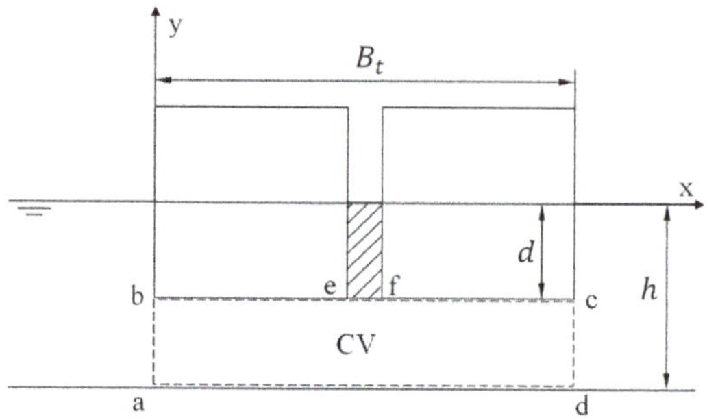

Figure 13. The sketch of the control volume below the twin squares.

Equation (25) indicates that the flow in the region CV was composed of two parts, i.e., the first part is the horizontal uniform oscillation with the velocity potential of $A_0 x\, e^{-j\omega t}$, and the second part is the disturbance of velocity with the velocity potential of $\sum_{n=1}^{\infty}\left[A_n e^{\mu_n x} + B_n e^{\mu_n (B_t - x)}\right]\cos\mu_n(y+h)e^{-j\omega t}$. In fact, the free surface wave velocity on the left side of the boundary *ab* decays with the water depth. When the relative draft d/λ is sufficiently large, the disturbance of velocity at the boundary *ab* becomes small quantity. The fluid flow in the control volume is mainly the horizontal uniform oscillation. The pressure distribution in the uniform flow field is the same everywhere. This indicates that the pressure inside the whole control volume CV was in-phase with the pressure at the boundary *ab*, which is the pressure of the wave field on the left side of boundary *ab*. When the wave profile in front of SQ1 rose up, the pressure inside the control volume CV increased and vice versa. Therefore, the excitation force ($f_{excitation}$) on the liquid inside the gap synchronized with the wave elevation in front of SQ1. The phase leg Φ was the same as the phase retardation of the response relative to the excitation of the vibration system.

Liu [30] carried out a series of experiments to investigate the phase leg Φ and concluded that the change of the phase leg Φ versus the incident wave frequency satisfied the phase-frequency character of the vibration system under harmonic excitation. For verification of the present simulation results, the phase leg Φ was theoretically estimated based on Equations (16)–(18) and the structural dynamics theory (relevant theories could be found in textbooks like Humar, [31]):

$$\Phi = \arctan\frac{2\zeta\gamma}{1-\gamma^2} \tag{26a}$$

$$\zeta = \frac{c}{2\left(\rho B_g d + m_a\right)\omega_n} \tag{26b}$$

$$\gamma = \frac{\omega}{\omega_n} \tag{26c}$$

At present, the parameters $\left(\rho B_g d + m_a\right)$, ω_n in Equation (26b,c) are known constants. The damping coefficient c is proportional to the incident wave frequency ω (Equation (20)). Therefore, the phase leg Φ versus the incident wave frequency ω could be plotted immediately based on Equation (26a–c). On the other hand, the phase leg Φ in the present simulations was evaluated based on Equation (24). Figure 14 shows the comparison of the phase leg Φ of the theoretical estimation (based on Equation (26a–c) and the simulation results (based on Equation (24)). The theoretical estimation curve and the simulation results are in good agreement with each other. A brief summary could be made from the above results. Firstly, in the present simulations, the excitation force ($f_{excitation}$) on the liquid inside the gap synchronized with the wave elevation in front of SQ1, which was almost in-phase with the incident wave profile in front of SQ1. Secondly, the phase leg Φ could be estimated by referring to the phase-character of the vibration system, i.e., Equation (26a–c).

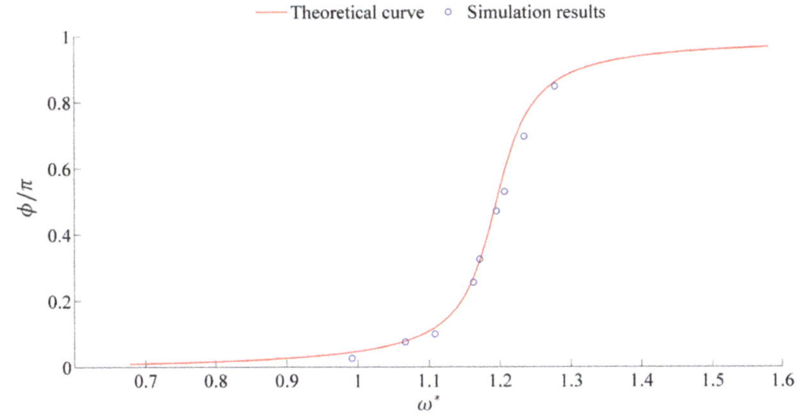

Figure 14. The comparison of the dimensionless phase leg (Φ/π) of the elevation inside the gap in the present simulations (evaluated based on Equation (24)) and the theoretical curve based on Equation (26a–c) ($B_g^* = 0.1$).

Figure 15 shows the horizontal force amplitudes on the twin squares. The results of the present simulations were compared with the results of Lu et al. [18] using a viscous flow solver based on the three-step Taylor–Galerkin finite element method. The normalized horizontal force amplitude F_x^* in resonance interval of Lu et al. [18] was smaller than that of the present simulation, whereas the variation tendencies of force amplitudes were similar in both the present solver and Lu et al. [18]'s solvers. It was found that the horizontal force amplitude on SQ2 (F_{x2}^*) has the same phase with the elevation amplitude in the gap (η_a^*). However, the peak frequency of the horizontal force amplitude on SQ1 (F_{x1}^*) was considerably larger than that of SQ2 (F_{x2}^*).

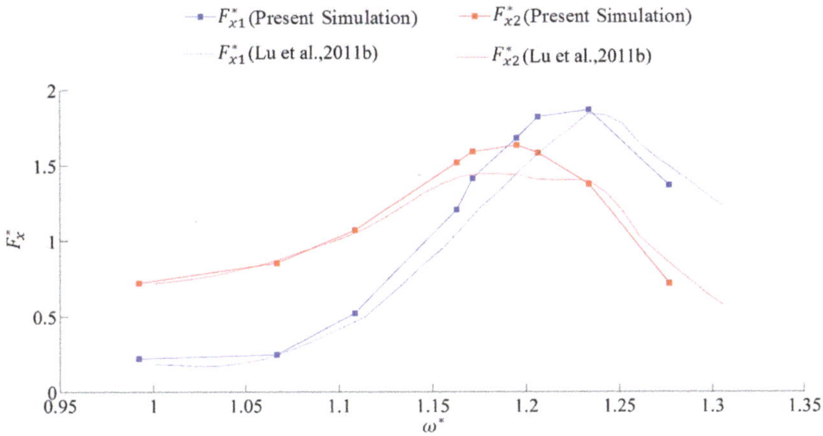

Figure 15. The horizontal force amplitude on SQ1 (F_{x1}^*) and SQ2 (F_{x2}^*) versus the incident wave frequency ω^* ($B_g^* = 0.1$) in both the present numerical simulation and numerical results from Lu et al. [18].

To explain the generation mechanics of the horizontal forces, first order theoretical estimations of the horizontal force amplitudes were conducted. The first order free surface elevation in the gap was harmonic and can be expressed as:

$$\eta = \eta_a \cos(\omega t - \Phi) \tag{27}$$

Assuming that the flow in the gap is uniform, the horizontal force produced by the vibration in the gap can be estimated using the Lagrange integral from the free surface to the target location:

$$f_{xg}^{(1)} = \int_{-d}^{0} (-\rho \frac{\partial \psi}{\partial t})\, dy + \int_{-d}^{0} \rho g \eta dy = (\rho g d \eta_a - \frac{\omega^2 d^2}{2} \eta_a) \cos(\omega t - \Phi) \tag{28}$$

where $\psi = \dot{\eta} y$ is the velocity potential of the uniform flow inside the gap.

Using the standing wave theory, the first order wave force on the left side of SQ1 could be estimated by:

$$f_{xw}^{(1)} = \int_{-\mathrm{d}}^{0} \rho g \xi_a \frac{\cos \mathrm{h} k(y+h)}{\cos \mathrm{h}(kh)} \cos(\omega t)\, dy = \rho g \xi_a \frac{\sinh kh - \sinh k(h-d)}{k \cos \mathrm{h}(kh)} \cos(\omega t) \tag{29}$$

where ξ_a is the equivalent standing wave amplitude in front of SQ1. According to the energy conservation law, ξ_a is limited between two times the reflection wave amplitude and is two times the incident wave amplitude. Therefore, it is estimated as $\frac{2A_r + 2A_i}{2}$.

The horizontal force on SQ1 includes the effects of both $f_{xg}^{(1)}$ and $f_{xw}^{(1)}$:

$$f_{x1}^{(1)} = f_{xw}^{(1)} - f_{xg}^{(1)} = \sqrt{\left(F_{xw} - F_{xg} \cos \Phi\right)^2 + \left(F_{xg} \sin \Phi\right)^2} \cos(\omega t - \Phi_1) \tag{30}$$

where Φ_1 is the phase difference between $f_{x1}^{(1)}$ and $f_{xw}^{(1)}$.

The horizontal force on SQ2 is mainly induced by the vibration in the gap:

$$f_{x2}^{(1)} = f_{xg}^{(1)} \tag{31}$$

Figure 16 shows the comparison between the horizontal force amplitudes that were estimated by Equations (30) and (31) and in the present simulation results. F_{x1}^* and F_{x2}^* (F^* and F respectively denote the normalized amplitude and amplitude of corresponded force term (see Section 2.4) estimated by Equations (30) and (31) agree well with the results from the present CFD simulations.

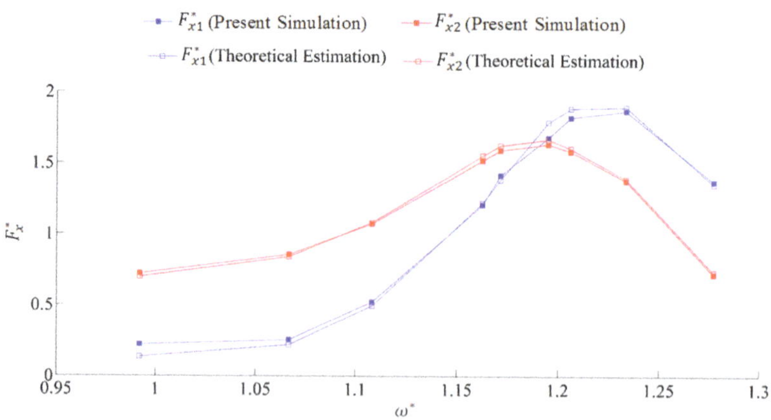

Figure 16. The horizontal forces F_x^* estimated based on Equations (30) and (31) are in good agreements with the simulation results ($B_g^* = 0.1$).

As a summary, firstly, the phase difference between f_{xw} and f_{xg} influenced F_{x1} significantly. To be specific, f_{xw} and f_{xg} counteracted each other when ω^* was smaller than ω_n^*, resulting in a smaller F_{x1} than F_{xg}; f_{xw} and f_{xg} mutually promoted when ω^* was larger than ω_n^*; the peak frequency of F_{x1} therefore fell behind the resonance frequency. Secondly, due to the shielding effect of the up-stream structures, the wave forces from the transmitted waves were negligible compared with the hydrodynamic forces in the gap f_{xg}. Therefore, the force amplitude on SQ2 F_{x2} was almost equal to the hydrodynamic forces in the gap F_{xg}. Aforementioned discussions reveal the mechanism of the horizontal forces on the square structures.

5.2. The Influences of Gap Width

Figure 17 shows the elevation amplitude inside the gap η_a^* versus the incident wave frequency ω^* for the cases with different gap width B_g^*. The resonance frequency and the resonance amplitude became smaller as the gap width increased. A similar relationship has also been reported in the experiments conducted by Saitoh et al. [1] and the numerical simulations of Lu et al. [17] and Moradi et al. [15]. As the gap width B_g^* increased, the resonance frequency ω_n^* in the present simulations gradually became smaller than the theoretical estimations based on Equation (17) (Saitoh et al., [1]), which is due to the stronger transverse flow in the gap. However, Equations (18)–(20) could still be applied to describe the mean vertical flow in the gap, as described in Section 5.1. For the cases with larger gap width B_g^*, the value of the resonance frequency ω_n^* could be estimated based on the characteristics of the phase leg Φ of the elevation inside the gap relative to the wave phase in front of SQ1. According to Equation (26a), the phase leg Φ approached $\pi/2$ when the incident wave frequency approached to the gap resonance frequency ω_n. For example, the phase leg Φ of cases C23 and C24 were most close to $\pi/2$. among the cases with the gap width $B_g^* = 0.25$. Note the value of the phase leg Φ of case C23 as $\Phi_{\pi/2-}$, which was slightly smaller than $\pi/2$; the phase leg Φ of case C24 as $\Phi_{\pi/2+}$, which was slightly larger than $\pi/2$. The resonance frequency ω_n^* of the cases with the gap width $B_g^* = 0.25$ could be estimated by interpolation:

$$\omega_n^* = \omega_{C23}^* + \frac{\pi/2 - \Phi_{\pi/2-}}{\Phi_{\pi/2+} - \Phi_{\pi/2-}}\left(\omega_{C24}^* - \omega_{C23}^*\right) \tag{32}$$

where ω_{C23}^*, ω_{C24}^* are, respectively, the incident wave frequencies of case C23 and case C24.

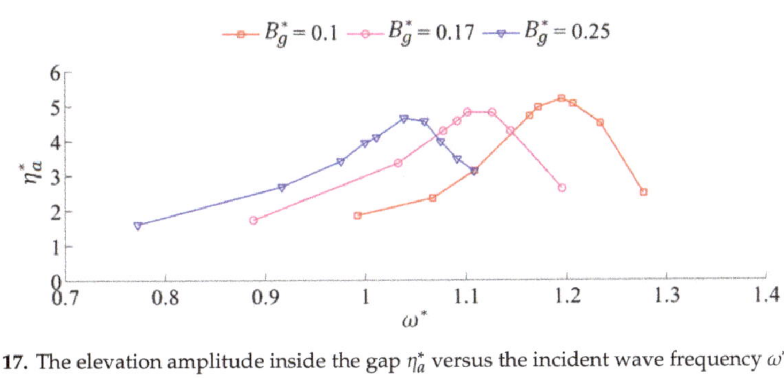

Figure 17. The elevation amplitude inside the gap η_a^* versus the incident wave frequency ω^* of cases with different gap width B_g^* in the present simulations.

The resonance frequency ω_n^* of the cases with different gap width B_g^* and the dimensionless added mass $m_a/\left[\rho B B_g^2/(h-d)\right]$ that were deduced from the resonance frequency ω_n^* based on Equation (18) are illustrated in Figure 18. The dissipation ratio R_d and the damping coefficient c of the cases with different B_g^* were calculated and shown in Figure 19, and the phase leg Φ is illustrated in Figure 20. The resonance frequency ω_n^* and the corresponding dimensionless added mass $m_a/\left[\rho B B_g^2/(h-d)\right]$

decreased with the increase of B_g^*. Meanwhile, the damping coefficient c increased with the increase of the gap width B_g^*, which indicates that the viscous dissipation in the gap became stronger when the gap width B_g^* became larger. However, in Figure 20, the phase leg Φ estimated based on Equation (26a–c) shows a big difference from the present simulation results when B_g^* was large. To be specific, as the incident wave frequency increasing, the phase leg Φ estimated based on Equation (26a–c) increased faster than the data points of the present simulations, which indicates that the damping ratio ζ and the damping coefficient c that are used in Equation (26a–c) were smaller than the corresponding values that the simulations indicate. The reasonable explanation is that when the gap width B_g^* was relatively large, the reduction of the piston flow in the gap was not only caused by the viscous dissipation—it was also caused by the sloshing mode flow in the gap. For example, Figure 21 shows the comparison of the flow structures in the gap between cases C6 ($B_g^* = 0.1$) and C21 ($B_g^* = 0.25$). For case C6, the flow in the gap was mostly in the vertical direction. However, for case C21, the sloshing mode of the flow became much stronger and extracted a large part of the mechanical energy from the vertical flow in the gap. The damping coefficient c in Equations (16) and (26) could not simply be estimated according to the dissipation ratio R_d (Equation (19)). The coupling between the piston and the sloshing types of flow motion in the gap was very complex, which could be captured well using the present numerical method.

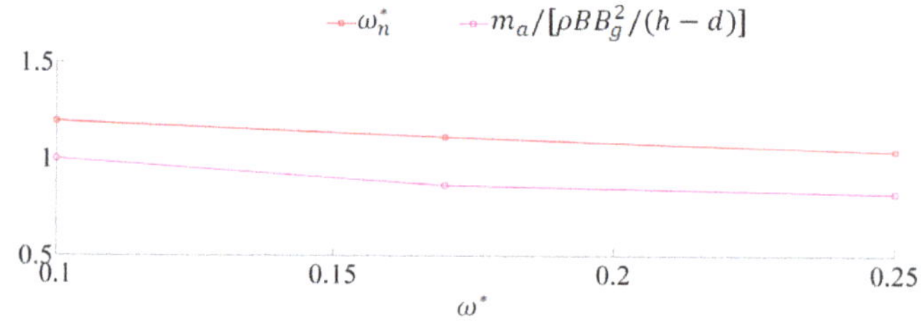

Figure 18. The resonance frequency ω_n^* and the dimensionless added mass $m_a/\left[\rho B B_g^2/(h-d)\right]$ decreases with the increase of the gap width B_g^*.

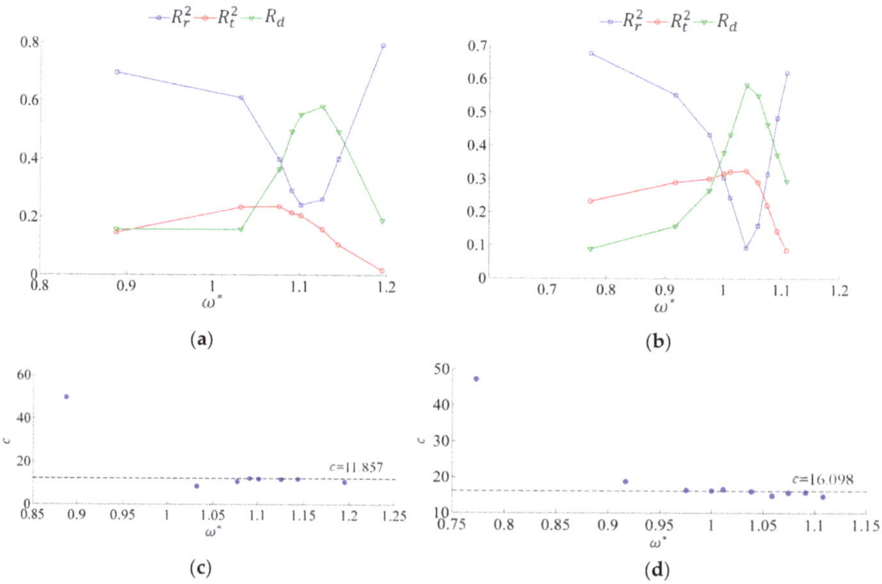

Figure 19. The dissipation ratio R_d and the damping coefficient c versus the incident wave frequency ω^* in the present simulations: (**a**,**c**) the gap width $B_g^* = 0.17$; (**b**,**d**) the gap width $B_g^* = 0.25$.

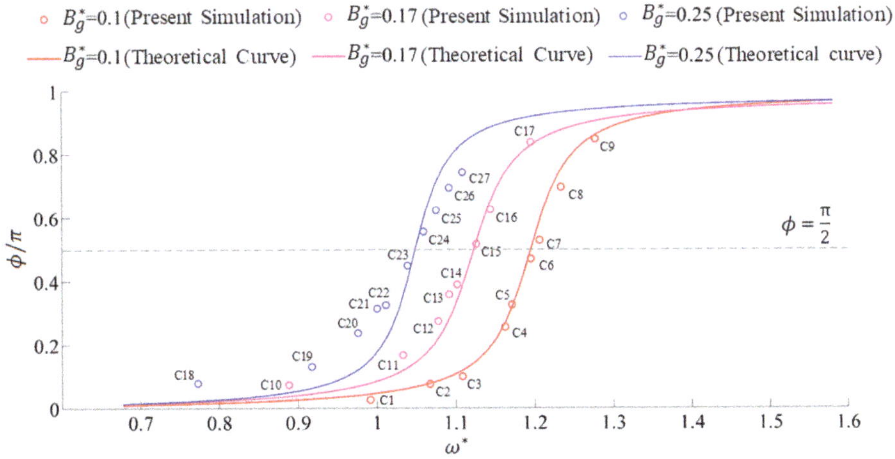

Figure 20. The dimensionless phase retardation (Φ/π) of the elevation inside the gap relative to the wave phase in front of SQ1 of the present numerical simulations (evaluated based on Equation (24)) and the theoretical estimation based on Equation (26a–c).

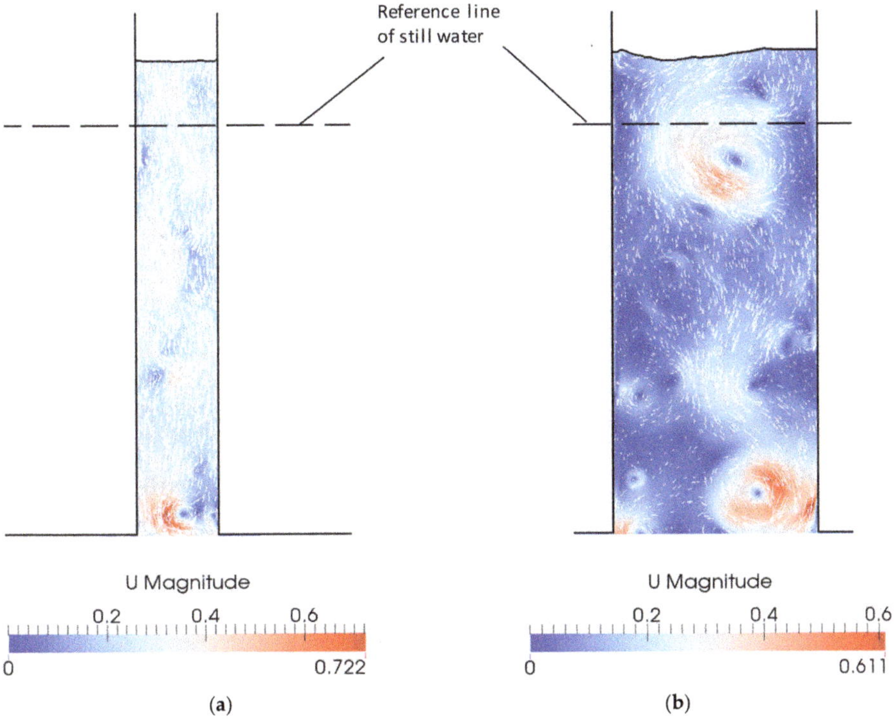

Figure 21. The transverse flow in the gap becomes violent as the gap width increases. (**a**) the velocity profile of case C6; (**b**) the velocity profile of case C21.

The horizontal force amplitude versus the incident wave frequency of the cases with $B_g^* = 0.17$ and $B_g^* = 0.25$ are presented in Figures 22 and 23, respectively. The variation tendencies of the horizontal force amplitudes were well predicted by the theoretical estimations based on Equations (30) and (31).

As the gap width B_g^* increased, the decreasing resonance amplitude in the gap did not necessarily reduce the peak value of the horizontal force amplitude F_x^*. For example, the maximum F_{x1}^* increased from 1.85 to 1.92 as B_g^* increased from 0.17 to 0.25. The maximum F_{x2}^* increased from 1.63 to 1.68 as B_g^* increased from 0.1 to 0.17. The above results are reasonable. In Equation (28), the horizontal force amplitude induced by the fluid motion inside the gap (F_{xg}) includes two parts, namely the hydrostatic part $\rho g d\eta_a$ and the hydrodynamic part $-\frac{\omega^2 d^2}{2}\eta_a$. As the resonance amplitude decreases, the hydrostatic part decreases proportionally. However, the hydrodynamic part may increase due to the rapid decrease of the resonance frequency ω_n. As a consequence, the horizontal force amplitude on the SQ2 (F_{x2}) was likely to increase. On the other hand, due to the effect of the wave field at the left side of SQ1, the probability for a horizontal force amplitude on the SQ1 F_{x1} increased even greater. The aforementioned discussions indicate that the gap width influences the hydrodynamic forces from various aspects. Enlarging gap width can reduce resonance amplitude; however, it does not directly result in the decrease of the maximum horizontal force amplitude on the floating structures.

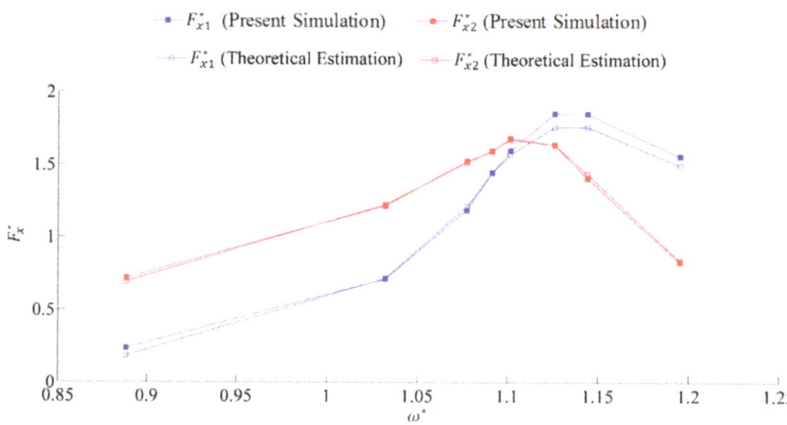

Figure 22. The variation tendencies of the horizontal force amplitudes are well predicted by the theoretical estimations based on Equations (30) and (31) ($B_g^* = 0.17$).

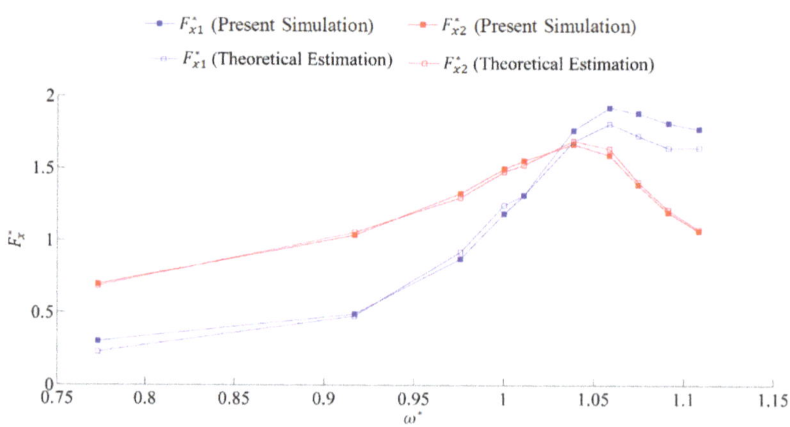

Figure 23. The variation tendencies of the horizontal force amplitudes are well predicted by the theoretical estimations based on Equations (30) and (31) ($B_g^* = 0.25$).

6. Conclusions

Two-dimensional computational fluid dynamics (CFD) simulations were carried out to investigate the gap resonance phenomenon that occurs when free-surface waves propagate past twin alongside placed squares. The laminar model was used to solve the governing equations, and the volume of fluid method was used for free surface capturing. Both mesh and time-step convergence studies were carried out to ensure sufficient numerical accuracy. The influences of the incident wave frequency, as well as the gap width, were investigated. Detailed discussions were carried out based on the motion equation of the fluid inside the gap in terms of the damping coefficient, the phase retardation of the elevation inside the gap relative to the wave phase in front of the weather side structure, and the horizontal force amplitudes on the square structures. The following conclusions can be drawn from the discussions:

1. For the cases of free surface waves past through twin alongside placed 2-D identical rectangular squares, the laminar model with the volume of fluid method was able to predict the gap resonance amplitudes, the viscous dissipation effects, and wave forces on the structures accurately when the KC number was small (e.g., $KC = \frac{2\pi A_i}{B_g} < 3$, where A_i is the incident wave amplitude and B_g is the gap width).
2. As the incident wave frequency increased, the surface elevation amplitude inside the gap first increased and then decreased, a tendency which is in accordance with resonance phenomena. The horizontal force amplitude on the lee side square structure changed in-phase with the elevation amplitude inside the gap, while the horizontal force amplitude on the weather side structure reached the peak value at a larger frequency than the gap resonance frequency.
3. In present simulations, the phase retardation of the elevation inside the gap relative to the wave phase in front of the weather side structure satisfied the phase-frequency characteristics of the vibration system and determined the change tendency of the horizontal force amplitude on the weather side structure. To be specific, the horizontal wave force on the left side of the weather side structure and the horizontal force on the right side of the structure (induced by the fluid inside the gap) counteracted each other when the wave frequency was smaller than the gap resonance frequency. This resulted in a smaller horizontal force amplitude on the structure than the horizontal fluid force inside the gap. The forces on each side of the structure mutually promoted when the incident wave frequency was larger than gap resonance frequency, making the peak frequency of the horizontal force on the structure fall behind the gap resonance frequency.
4. The increase of the gap width resulted in the increase of the added mass and reduced the resonance frequency. Moreover, as the gap width increased, the decrease of resonance amplitude in the gap did not necessarily reduce the peak value of the horizontal forces on the squares.
5. As the gap width increased, the viscous dissipation and the sloshing mode flow inside the gap both became stronger.

Author Contributions: The author contributions are listed as following: Conceptualization, M.C.O. and S.L.; methodology, S.L. and M.C.O.; software, S.L. and H.Z.; validation, H.Z. and S.L.; formal analysis, H.Z.; investigation, H.Z. and R.Z.; writing—original draft preparation, H.Z.; writing—review and editing, S.L., M.C.O., and R.Z.; visualization, H.Z.; supervision, S.L., M.C.O., and R.Z.

Acknowledgments: Grateful acknowledgments are given to the financially support from the National Science Foundation of China (Grant No. 51579120) and the computational resources provided by the Norwegian Metacenter for Computational Science (NOTUR), under Project No: NN9372K.

Conflicts of Interest: The authors declare no conflict of interest.

Nomenclature

A nomenclature table used in the present study is provided as follows.

The subscripts

i	The parameter of the incident wave
r	The parameter of the reflected wave
t	The parameter of the transmitted wave
d	The subscript for the wave energy dissipation ratio
n	The resonance parameter of the surface elevation in the gap
x	The horizontal physical quantity
xw	The horizontal physical quantity produced by waves
xg	The horizontal physical quantity produced by the fluid vibration in the gap
1	The hydrodynamic parameter for the square structure 1
2	The hydrodynamic parameter for the square structure 2

The flow parameters

ρ	The fluid density
μ	The dynamic viscosity coefficient
t	The time
$\boldsymbol{v}$	The flow velocity
p	The fluid pressure
p_{excess}	The excessive pressure
α	The water volume fraction
Φ	The general symbol of the fluid variables $\boldsymbol{v}$, p, p_{excess} and α
$\boldsymbol{g}$	The gravity vector
g	The gravitational acceleration

The geometrical parameters

h	The water depth
B	The breadth of the square structures
B_g	The gap width between the square structures
d	The draft of the square structures

The hydrodynamic parameters

A	The wave amplitude
H	The wave height
ω	The wave frequency
ω^*	The dimensionless wave frequency, $\omega^* = \omega/\sqrt{g/B}$
T	The wave period
t^*	The dimensionless time coordinate, $t^* = (t - t_0)/T_i$, where t_0 is the start time
λ	The wave length
k	The wave number
η	The instantaneous surface elevation in the gap
η^*	The dimensionless (normalized) surface elevation in the gap, $\eta^* = \eta/H_i$
η_a	The surface elevation amplitude in the gap
η_a^*	The dimensionless surface elevation amplitude in the gap, $\eta_a^* = \eta_a/A_i$
$\boldsymbol{f}$	The instantaneous force on the square structures
f	The instantaneous force on the square structures
f^*	The dimensionless force magnitude, $f^* = f/\rho g B A_i$
F	The force amplitude on the square structures
F^*	The dimensionless force amplitude, $F^* = F/\rho g B A_i$
m	The mass of the fluid in the gap
m_a	The added mass of the fluid motion in the gap
c	The damping coefficient of the fluid motion in the gap
k_s	The stiffness of the fluid motion in the gap
R_r	The wave reflection ratio
R_t	The wave transmission ratio
R_d	The wave dissipation ratio
Φ_1	The wave phase of the surface elevation in the gap
Φ_2	The wave phase in front of the square structure 1
Φ	The phase leg of the surface elevation inside the gap, $\Phi = \Phi_1 - \Phi_2$

References

1. Saitoh, T.; Miao, G.P.; Ishida, H. Theoretical analysis on appearance condition of fluid resonance in a narrow gap between two modules of very large floating structure. In Proceedings of the 3rd Asia-Pacific Workshop on Marine Hydrodynamics, Beijing, China, 27 June 2006; pp. 170–175.
2. Molin, B. On the piston and sloshing modes in moonpools. *J. Fluid Mech.* **2001**, *430*, 27–50. [CrossRef]
3. Liu, Y.; Li, H.J. A new semi-analytical solution for gap resonance between twin rectangular boxes. *Proc. Inst. Mech. Eng. Part M J. Eng. Marit. Environ.* **2014**, *228*, 3–16. [CrossRef]
4. Tan, L.; Tang, G.Q.; Zhou, Z.B.; Cheng, L.; Chen, X.; Lu, L. Theoretical and numerical investigations of wave resonance between two floating bodies in close proximity. *J. Hydrodyn.* **2017**, *29*, 805–816. [CrossRef]
5. Iwata, H.; Saitoh, T.; Miao, G.P. Fluid resonance in narrow gap of very large floating structure composed of regular modules. In Proceedings of the Fourth International Conference on Asian and Pacific Coasts, Nanjing, China, 21–24 September 2007; pp. 815–826.
6. Cong, L.; Teng, B.; Gou, Y. Analytical analysis of the fundamental frequency of 2D gap resonance. *J. Harbin Eng. Univ.* **2015**, *36*, 882–887.
7. Faltinsen, O.M.; Rognebakke, O.F.; Timokha, A.N. Two-dimensional resonant piston-like sloshing in a moonpool. *J. Fluid Mech.* **2007**, *575*, 359–397. [CrossRef]
8. Kristiansen, T.; Faltinsen, O.M. A two-dimensional numerical and experimental study of resonant coupled ship and piston-mode motion. *Appl. Ocean Res.* **2010**, *32*, 158–176. [CrossRef]
9. Huijsmans, R.H.M.; Pinkster, J.A.; De Wilde, J.J. Diffraction and radiation of waves around side-by-side moored vessels. In Proceedings of the Eleventh International Offshore and Polar Engineering Conference, Stavanger, Norway, 17–22 June 2001.
10. Newman, J.N. Wave effects on multiple bodies. *Hydrodyn. Ship Ocean Eng.* **2001**, *3*, 3–26.
11. Chen, X.B. Hydrodynamic analysis for offshore LNG terminals. In Proceedings of the 2nd International Workshop on Applied Offshore Hydrodynamics, Rio de Janeiro, Brazil, 14–15 April 2005.
12. Kristiansen, T.; Faltinsen, O.M. Application of a vortex tracking method to the piston-like behaviour in a semi-entrained vertical gap. *Appl. Ocean Res.* **2008**, *30*, 1–16. [CrossRef]
13. Kristiansen, T.; Faltinsen, O.M. Gap resonance analyzed by a new domain-decomposition method combining potential and viscous flow DRAFT. *Appl. Ocean Res.* **2012**, *34*, 198–208. [CrossRef]
14. Fen, X.; Bai, W. Wave resonances in a narrow gap between two barges using fully nonlinear numerical simulation. *Appl. Ocean Res.* **2015**, *50*, 119–129.
15. Moradi, N.; Zhou, T.; Cheng, L. Effect of inlet configuration on wave resonance in the narrow gap of two fixed bodies in close proximity. *Ocean Eng.* **2015**, *103*, 88–102. [CrossRef]
16. Tan, L.; Lu, L.; Tang, G.Q.; Cheng, L. Experimental Study of Wave Resonance in a Narrow Gap with Various Edge Shapes. In Proceedings of the 20th Australasian Fluid Mechanics Conference, Perth, Australia, 5–8 December 2016.
17. Lu, L.; Teng, B.; Cheng, L.; Sun, L.; Chen, X. Modelling of multibodies in close proximity under water waves-fluid resonance in narrow gaps. *Sci. China Phys. Mech. Astron.* **2011**, *54*, 16–25. [CrossRef]
18. Lu, L.; Teng, B.; Sun, L.; Chen, B. Modelling of multibodies in close proximity under water waves-fluid forces on floating bodies. *Ocean Eng.* **2011**, *38*, 1403–1416. [CrossRef]
19. Zhao, W.H.; Wolgamot, H.A.; Taylor, P.H.; Eatock Taylor, R. Gap resonance and higher harmonics driven by focused transient wave groups. *J. Fluid Mech.* **2017**, *812*, 905–939. [CrossRef]
20. Wang, H.C. Development of a CFD Model to Simulate Three-Dimensional Gap Resonance Applicable to FLNG Side-By-Side Offloading. In Proceedings of the ASME 2017 36th International Conference on Ocean, Offshore and Arctic Engineering OMAE, Trondheim, Norway, 25–30 June 2017.
21. Chua, K.H.; Eatock Taylor, R.; Choo, Y.S. Hydrodynamics of side-by-side fixed floating bodies. In Proceedings of the 35th International Conference on Ocean, Offshore and Arctic Engineering, Busan, South Korea, 19–24 June 2016; pp. 19–24.
22. Chua, K.H.; Eatock Taylor, R.; Choo, Y.S. Hydrodynamic interaction of side-by-side floating bodies part I: Development of CFD-based numerical analysis framework and modified potential flow model. *Ocean Eng.* **2017**, *166*, 404–415. [CrossRef]
23. Jasak, H. Openfoam: Open source CFD in research and industry. *Int. J. Nav. Archit. Ocean Eng.* **2009**, *1*, 89–94.

24. Jacobsen, N.G.; Fuhrman, D.R.; Fredsøe, J. A wave generation toolbox for the open-source CFD library: OpenFoam®. *Int. J. Numer. Meth. Fluids.* **2011**, *70*, 1073–1088. [CrossRef]
25. Rusche, H. Computational Fluid Dynamics of Dispersed Two-Phase Flows at High Phase Fractions. Ph.D. Thesis, Imperial College of Science, Technology and Medicine, London, UK, 2002.
26. Liu, S.; Ong, M.C.; Obhrai, C.; Seng, S. CFD Simulations of Regular and Irregular Waves Past a Horizontal Semi-submerged Cylinder. *J. Offshore Mech. Arctic Eng.* **2017**, *140*. [CrossRef]
27. Lamb, H. *Hydrodynamics*, 6th ed.; The Syndics of Cambridge University Press: London, UK, 1932; pp. 623–631.
28. Yang, Y.; Grétar, T. Dissipation of energy by finite-amplitude surface waves. *Comput. Fluids.* **1998**, *27*, 829–845. [CrossRef]
29. Goda, Y.; Suzuki, Y. Estimation of incident and reflected waves in random waves. In Proceedings of the Fifteenth Conference on Coastal Engineering, ASCE, New York, NY, USA, 11–17 July 1976; pp. 828–865.
30. Liu, C.Y. Experimental and numerical study of hydrodynamic response of side-by-side floating bodies under wave action. Master Thesis, Dalian University of Technology, Dalian, China, 2017.
31. Humar, J.L. *Dynamics of Structures*, 2nd ed.; A.A. Balkema Publishers: Lisse, The Netherlands; Abingdon, UK; Exton, PA, USA; Tokyo, Japan, 2002; pp. 225–226.

energies

Article

Unsteady RANS Simulations of Flow around a Twin-Box Bridge Girder Cross Section

Wonmin Jeong, Shengnan Liu *, Jasna Bogunovic Jakobsen and Muk Chen Ong

Department of Mechanical and Structural Engineering and Materials Science, University of Stavanger, Postboks 8600 Forus, 4036 Stavanger, Norway

* Correspondence: shengnan.liu@uis.no; Tel.: +47-90822195

Received: 21 May 2019; Accepted: 8 July 2019; Published: 11 July 2019

Abstract: The aerodynamic performance of bridge deck girders requires a thorough assessment and optimization in the design of long-span bridges. The present paper describes a numerical investigation of the aerodynamic characteristics of a twin-box bridge girder cross section in the range of angles of attack between −10.0° and +10.2°. The simulations are performed by solving 2D unsteady Reynolds-averaged Navier–Stokes (URANS) equations together with the k–ω shear stress transport (SST) turbulence model. The investigated Reynolds number (Re) based on the free stream velocity (U_∞) and the height of the deck (D) is 31,000. The predicted aerodynamic characteristics such as the mean drag, lift and moment coefficients, are generally in good agreement with the results from the wind tunnel tests. Changes of flow patterns and aerodynamic forces with different angles of attack are investigated. Flow characteristics during one vortex shedding period are highlighted. Relative contributions of each of the two bridge decks to the overall drag and lift coefficients, with respect to the angle of attack, are also discussed.

Keywords: URANS; CFD; twin-box deck; aerodynamics; vortex shedding

1. Introduction

In wind-resistant bridge design, many factors should be considered, such as static wind load, wind forces due to turbulence, aerodynamic instability and vortex-shedding excitation. For bridge spans around 1 km and longer, twin-box girders are being increasingly used, due to their favorable aerodynamic properties in term of flutter stability. This is due to a favorable effect of the central gap on the surface pressure distribution, and a higher ratio between the twisting and the heaving eigen-frequencies [1,2]. On the other hand, this configuration is prone to vortex-induced vibration, since vortices shed from the windward box can cause a significant excitation of the leeward box and thereby important oscillations of the bridge girder [3,4]. In the early design stage of long-span bridges, several bridge deck design alternatives are usually considered, and their aerodynamic performance must be assessed to ensure the most feasible design for the specific project.

Both experimental and numerical approaches are available for the assessment of the bridge deck aerodynamic performance [4,5]. Extensive wind tunnel experiments are normally performed to optimize the design of long-span bridges. However, wind tunnel investigations have limitations in terms of time, cost, applicable Reynolds numbers and the representation of the details such as railings and stabilizers at small scales. For an improved bridge design, computational fluid dynamics (CFD) based simulations can facilitate a deeper understanding of complex flow conditions around a bridge girder, as demonstrated by [6–13].

The application of a 2D unsteady Reynolds-averaged Navier–Stokes (URANS) model for flow around simple rectangular geometry has been validated thoroughly in previous research [14–18]. Mannini et al. [16] applied 2D URANS equations with advanced turbulence modeling closures for predicting flow around a 5:1 rectangular cylinder at $Re = 1 \times 10^5$. They found that the predicted force coefficients are in reasonable agreement with the experimental data reported by Schewe [19,20]. Ong [17] conducted CFD simulation for the same structure at high Re numbers (5×10^5, 1×10^6, 1.5×10^6 and 2×10^6) using the 2D URANS together with the standard high Reynolds number k–ω model. For engineering design purposes, it gave reasonably good agreements with the published experimental data in terms of time-averaged drag coefficients, Strouhal numbers and time-averaged base pressure coefficient. Patruno et al. [18] studied the flow field around the 5:1 rectangular cylinder with an angle of attack of 0 to 4 degrees using both Large Eddy Simulation (LES) and URANS (i.e., k–ω shear stress transport (SST)) turbulence models. They found that LES and URANS approaches appeared to provide comparable results in terms of accuracy of first and second order pressure statistics measured at the central section of the considered prism. The asymmetric characteristics of the flow field induced by small angles of attack were under-estimated by LES as compared to URANS simulations, and the experimental data were between them generally.

The main goal of the present study is to evaluate the validity of 2D URANS simulations with k–ω SST turbulence model [21] for the flow around a twin-box bridge girder cross-section. One of the preliminary bare deck designs of the Halsafjord suspension bridge in Norway will be considered in the present study, see Figure 1. This bare deck design was chosen due to the availability of experimental data for validation studies. Several CFD studies of flow characteristics around the sharp-edged twin-boxes, particularly with asymmetric geometry as in the present case, have been carried out [22–24]. These studies have mainly dealt with the overall flow conditions and their variations with the relative gap size, Reynolds number and turbulence models used in the simulations. To the authors' knowledge, there is a limited number of studies addressing the flow characteristics around the bridge girder subject to flow at different angles of attack. The present study will provide a thorough 2D CFD investigation of the flow conditions around the twin-deck inclined to the approaching flow at angles of attack (AoA) ranging from −10.0 to +10.2 degrees (°).

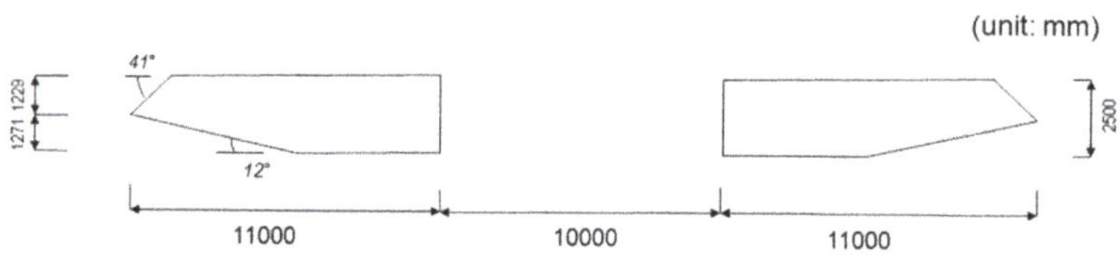

Figure 1. Cross section of the twin-box bridge decks in full scale, reproduced from [25]. Model scale is 1:40.

The study utilizes data from a wind tunnel investigation for validation. The tests were performed with a 1:40 section model at Re = 31,000, as reported in [25]. The cross-section geometry and the Reynolds number in the simulations are thus adopted as for the section model, which was exposed to a uniform flow. The results are discussed in terms of the time-averaged drag coefficient ($\overline{C_D}$), the time-averaged lift coefficient ($\overline{C_L}$) and the time-averaged moment coefficient ($\overline{C_M}$) in the range of AoA from −10.0° to +10.2°, respectively. The time histories of the force coefficients as well as the flow physics changes with different AoA are investigated. The vortex shedding process and the associated variation of the instantaneous drag and lift coefficients for each of the two decks with different AoA are also presented and discussed in detail.

2. Mathematical Formulation

2.1. Flow Model

The Reynolds-averaged equations for conservation of mass and momentum are given by:

$$\frac{\partial u_j}{\partial x_j} = 0 \tag{1}$$

$$\frac{\partial u_i}{\partial t}+u_j\frac{\partial u_i}{\partial x_j}=-\frac{1}{\rho}\frac{\partial p}{\partial x_i}+v\frac{\partial^2 u_i}{\partial x_j \partial x_j}-\frac{\partial \overline{u'_i u'_j}}{\partial x_j} \tag{2}$$

where $i, j = 1, 2$. Here x_1 and x_2 denote the streamwise and cross-stream directions respectively; u_1 and u_2 are the corresponding mean velocity components; $\overline{u'_i u'_j}$ is the Reynolds stress component, where u'_i denotes the fluctuating part of the velocity; p is the pressure; ρ is the density of the fluid and t is the time.

The k–ω SST turbulence model [26] used in the present study is a blending of the k–ω and the k–ε models. The original k–ω model of [27] is implemented in the near-wall region and the standard k–ε model of [28] in the outer wake region and in the free shear layers. Following [26], the equations for the k–ω SST turbulence model is taken as:

$$\frac{D(\rho k)}{Dt}=\widetilde{P_k}-\beta^*\rho\omega k+\frac{\partial}{\partial x_j}\left[(\mu+\sigma_k\mu_t)\frac{\partial k}{\partial x_j})\right] \tag{3}$$

$$\frac{D(\rho\omega)}{Dt}=\alpha\rho S^2-\beta\rho\omega^2+\frac{\partial}{\partial x_j}\left[(\mu+\sigma_\omega\mu_t)\frac{\partial \omega}{\partial x_j})\right]+2(1-F_1)\rho\sigma_{\omega 2}\frac{1}{\omega}\frac{\partial k}{\partial x_j}\frac{\partial \omega}{\partial x_j} \tag{4}$$

where $\widetilde{P_k}$ is given by:

$$\widetilde{P_k}=\min\left[\mu_t\frac{\partial u_i}{\partial x_j}\left(\frac{\partial u_i}{\partial x_j}+\frac{\partial u_j}{\partial x_i}\right),\ 10\beta^*\rho k\omega\right] \tag{5}$$

Here ϕ_1 represents any constant in the original k–ω turbulence model (i.e., $\sigma_{k1}, \beta_1, \sigma_{\omega 1}$) and ϕ_2 represents any constant in the original k–ε turbulence model (i.e., $\sigma_{k2}, \beta_2, \sigma_{\omega 2}).\phi$, the corresponding constant of the new $k-\omega$ SST turbulence model given by Equations (3) and (4), is

$$\phi=F_1\phi_1+(1-F_1)\phi_2 \tag{6}$$

$$F_1=\tanh(\arg_1{}^4) \tag{7}$$

$$\arg_1=\min\left[\max\left(\frac{\sqrt{k}}{0.09\omega y},\frac{500v}{y^2\omega}\right),\frac{4\rho\sigma_{\omega 2}k}{CD_{k\omega}y^2}\right] \tag{8}$$

$$CD_{k\omega}=\max\left(2\rho\sigma_{\omega 2}\frac{1}{\omega}\frac{\partial k}{\partial x_j}\frac{\partial \omega}{\partial x_j},10^{-10}\right) \tag{9}$$

Here y is the distance to the nearest wall and $CD_{k\omega}$ is the positive portion of the cross-diffusion term of Equation (4).

The turbulence viscosity can be estimated by:

$$v_t=\frac{a_1 k}{\max(a_1\omega, SF_2)} \tag{10}$$

where S is the invariant measure of the strain rate and F_2 is given by:

$$F_2=\tanh(\arg_2{}^2),\ \arg_2=\max\left(\frac{2\sqrt{k}}{0.09\omega y},\frac{500v}{y^2\omega}\right) \tag{11}$$

The empirical constants of the k–ω SST model are [19]:

$\beta^* = 0.09$, $a_1 = 0.31$, $\alpha_1 = 0.5532$, $\alpha_2 = 0.4403$, $\beta_1 = 0.075$, $\beta_2 = 0.0828, \sigma_{k1} = 0.85034, \sigma_{k2} = 1.0$ and $\sigma_{\omega 1} = 0.5$, $\sigma_{\omega 2} = 0.85616$.

2.2. Numerical Simulation Scheme, Computational Domain and Boundary Conditions

The open source CFD code OpenFOAM [29] is used in the present study. The pressure implicit with splitting of operators (PISO) scheme is chosen for the solver. The spatial schemes for gradient, Laplacian and divergence are Gauss linear, Gauss linear corrected and Gauss linear schemes, respectively. All these schemes are in second order accuracy. The second order Crank–Nicolson scheme is used for the time integration.

Figure 2 shows the investigated twin-box bridge girder cross section and the corresponding computational domain. The size of the entire computational domain is 6 m by 1.55 m, which is equivalent to 91.0*D* by 24.8*D*, where *D* is the height of the girder cross section. The origin of the coordinate system is located at the left lower corner of the domain. The flow inlet boundary is 18.8*D* upstream from the center of the gap between the decks and the flow outlet boundary is 77.2*D* downstream from the center of the gap. The top and bottom boundaries are positioned at a distance of 12.4*D* from the center of the gap between the decks. Although the shear center of the bridge deck section does not coincide with the mid height of the deck, the rotation center is adopted in the middle of the deck height. This facilitates a direct comparison with the experimental results which also refer to such a rotational center. The deck position and the domain size in the Y direction are same as in the wind tunnel working section. The boundary conditions used for the present numerical simulations are as follows:

1. A uniform flow, $u_1 = U_\infty$, $u_2 = 0$, is set at the inlet boundary; the pressure is specified as zero normal gradient at the inlet boundary. k and ω at the inlet boundary are set equal to:

$$k_{inlet} = 0.5(U_\infty I_u + U_\infty I_v + U_\infty I_w)^2 \tag{12}$$

where I_u and I_v are the turbulence intensities in X and Y directions, respectively. $I_u = 12\%$ and $I_v = 8\%$ are taken from [25].

$$\omega_{inlet} = k_{inlet}{}^{0.5} / \left(C_\mu{}^{0.25} l\right) \tag{13}$$

where $C_\mu = 0.09$ is the empirical constant specified in the turbulent model and turbulence length $l = 0.07$D [30–32]. Effects of l on the calculated results have been studied by using a much lower value $l = 0.04D$, and small variations (less than 0.27%) are observed in the aerodynamic quantities, i.e., mean drag coefficient, root mean square (r.m.s.) of the lift coefficient and Strouhal number.

2. Along the outlet boundary, u_1, u_2, k and ω are specified as zero normal gradient. The pressure is specified as zero. The zero pressure outlet boundary condition has been widely used to calculate the unsteady flow around bluff bodies [33–35]. The distance used in the present study for the downstream 77.2*D* is considerably longer than the value of 50*D* previously used by [10]. It is considered that the effect of the outlet boundary condition on the numerical results is negligible.
3. On the deck surfaces, no-slip boundary condition is specified (i.e., $u_1 = u_2 = 0$). The pressure is set as zero normal gradient. k is fixed at 0 and ω is calculated as follows [26]:

$$\omega_{deck} = 10 \times \frac{6\nu}{\beta_1 (\Delta y_1)^2} \tag{14}$$

4. For the top and bottom boundaries, $u_1 = u_2 = 0$, the pressure is set as zero normal gradient; k is fixed at 0. ω is specified as zero normal gradient.

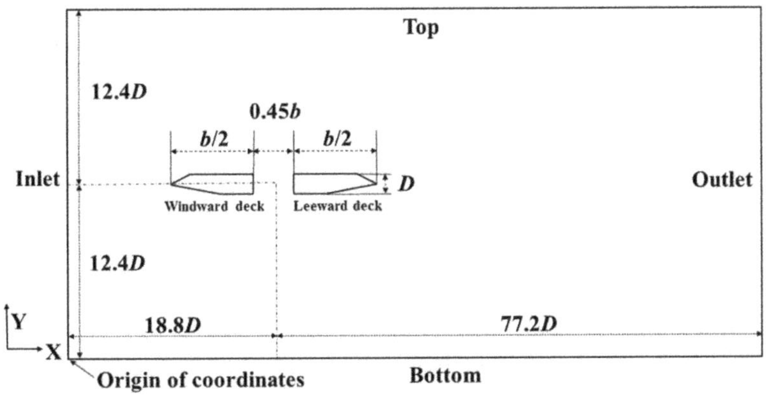

Figure 2. Computational domain and boundary conditions for the case of the angle of attack +0.1°, D = 0.0625 m in the model test.

2.3. Grid and Time Resolution Tests

Convergence studies considering different grid sizes and the time steps have been carried out for AoA ranging from −10.0° to +10.2°. Time-averaged drag coefficient ($\overline{C_D}$), time-averaged lift coefficient ($\overline{C_L}$), time-averaged moment coefficient ($\overline{C_M}$) as well as Strouhal number (St) are considered in the tests. The time-averaged values are obtained with a duration of 30 vortex shedding cycles after the numerical results have repeated their cycles. Here the drag coefficient (C_D), lift coefficient (C_L), moment coefficient (C_M) and Strouhal number (St) are defined as follows:

$$C_D = \frac{F_{drag}}{\frac{1}{2}\rho U_\infty{}^2 D} \tag{15}$$

$$C_L = \frac{F_{lift}}{\frac{1}{2}\rho U_\infty{}^2 b} \tag{16}$$

$$C_M = \frac{M}{\frac{1}{2}\rho U_\infty{}^2 b^2} \tag{17}$$

$$St = \frac{fD}{U_\infty} \tag{18}$$

where F_{drag}, F_{lift} are the along-wind and the cross-wind force components acting on the decks per unit length and b is the width of the two decks, i.e., of the solid part of the cross-section. M is the overturning moment about the aforementioned rotational center per unit length, and is positive in the clockwise direction. The frequency of vortex shedding f is obtained from C_L. Figure 3 shows the sign convention. Like the overturning moment, AoA is defined positive in the clock-wise (nose-up) direction in the cases of wind coming from left to right.

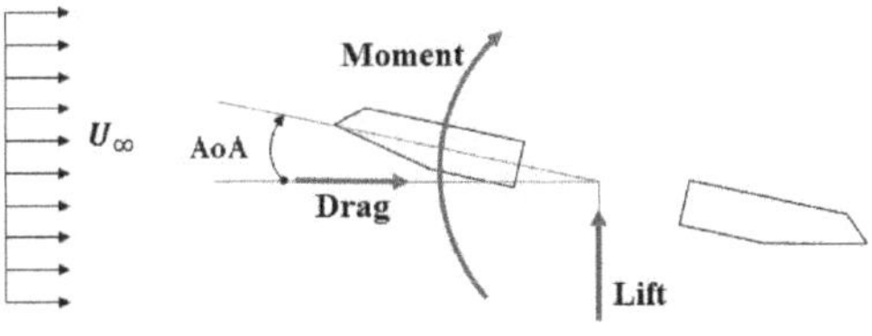

Figure 3. Sign convention.

The results of both the grid size and the time step convergence studies in terms of the force coefficients are shown in Tables 1 and 2, where Δt. represents the time step in seconds. The relative change of results between two consecutive meshes or time steps is investigated in terms of $\overline{C_D}$, $\overline{C_L}$, $\overline{C_M}$ as well as St. The cases with the converged solution in terms of grid size and time-step are marked with '*' and, adopted for the further investigations.

For most of the cases, the grid resolution tests have been carried out with 3 sets of meshes with 138,354 (M1), 198,834 (M2) and 288,804 (M3) elements and two different time-steps (i.e., M3 with time-step of 5.00×10^{-5} s, and M3T1 with time-step of 2.50×10^{-5} s), see Tables 1 and 2. For example, in the AoA = +0.1° case, the relative change of the $\overline{C_D}$ values between M1 and M2 meshes is 14.0%. The difference decreases to 1.7% when the mesh is further refined from M2 to M3, i.e., a negligible change is observed. A similar trend is observed for the other cases. For the cases of AoA = +4.4°, +6.3° and +10.2°, 138,354 to 545,954 (or 362,514) elements were used for the grid resolution tests. The convergence is considered to be achieved when the relative variation of the aerodynamic quantities between two cases is lower than 5% or if the parameters change by less than 0.01 in magnitude. For instance, in the AoA = +0.1° case, the relative change of $\overline{C_M}$ values between M3 and M3T1 cases is 8.2%, the absolute change of the $\overline{C_M}$ values is less than 0.01, see Table 1. An example of the mesh structure with 288,804 elements, in the case of AoA = +0.1°, is presented in Figure 4.

For the AoA = +0.1° case, the values of y^+ over the entire wall boundaries of two decks range from 0 to 7.8 with an average value from 1.0 to 1.5, where $y^+ = u_* \Delta y_1 / \nu = \sqrt{\tau_w / \rho} \cdot \Delta y_1 / \nu$, where u_* and τ_w denote the wall friction velocity and wall shear stress, respectively. The maximum y^+ value is found at the windward deck at the leading edge lower corner, where high flow acceleration exists. A similar procedure is taken to calculate the averaged $\overline{y^+}$ for other AoA cases, and the averaged values of $\overline{y^+}$ give a range of 1.0 to 2.1. All refinement studies of the flow at different AoA are carried out with the fixed y^+ values in order to minimize influence of the boundary layer calculation.

Overall, it is concluded that the present simulations marked with '*' in Tables 1 and 2 can provide satisfactory spatial and time resolutions for all the different AoA cases at $Re = 31{,}000$.

Table 1. Aerodynamic properties for different grid and time resolutions for AoA > 0°.

AoA. (°)	Case	Elements	Δt (s)	$\overline{C_D}$	$\overline{C_L}$	$\overline{C_M}$	St	$\frac{\Delta}{C_D}$ (%)	$\frac{\Delta}{C_L}$ (%)	$\frac{\Delta}{C_M}$ (%)	$\frac{\Delta}{St}$ (%)
0.1	M1	138,354	5.00×10^{-5}	1.164	−0.154	0.077	0.224	-	-	-	-
0.1	M2	198,834	5.00×10^{-5}	1.129	−0.176	0.078	0.216	−3.0	−14.0	1.3	−3.7
0.1	M3 *	288,034	5.00×10^{-5}	1.125	−0.178	0.075	0.216	−0.3	−1.7	−2.8	0.0
0.1	M3T1	288,034	2.50×10^{-5}	1.142	−0.180	0.082	0.225	1.5	−0.6	8.2	4.3
1.5	M1	138,354	5.00×10^{-5}	1.088	−0.051	0.127	0.203	-	-	-	-
1.5	M2	198,834	5.00×10^{-5}	1.096	−0.057	0.125	0.201	0.7	−12.3	−1.8	−0.7
1.5	M3 *	288,034	5.00×10^{-5}	1.099	−0.057	0.123	0.205	0.2	−0.7	−1.8	2.0
1.5	M3T1	288,034	2.50×10^{-5}	1.114	−0.047	0.126	0.196	1.4	18.2	2.2	−4.4
3.2	M1	138,354	5.00×10^{-5}	1.080	0.141	0.182	0.201	-	-	-	-
3.2	M2	198,834	5.00×10^{-5}	1.087	0.160	0.181	0.191	0.7	14.0	−0.5	−5.1
3.2	M3 *	288,034	5.00×10^{-5}	1.091	0.154	0.180	0.192	0.3	−3.8	−0.8	0.4
3.2	M3T1	288,034	2.50×10^{-5}	1.101	0.147	0.180	0.200	1.0	−5.0	−0.2	4.4
4.4	M1	138,354	5.00×10^{-5}	1.106	0.245	0.218	0.186	-	-	-	-
4.4	M2 *	288,034	5.00×10^{-5}	1.106	0.228	0.210	0.192	0.0	−6.8	−3.8	3.4
4.4	M3	545,954	5.00×10^{-5}	1.083	0.238	0.211	0.200	−2.1	4.1	0.7	4.0
4.4	M2T1	545,954	2.50×10^{-5}	1.112	0.241	0.211	0.196	2.7	1.5	−0.1	−2.2
6.3	M1	138,354	2.50×10^{-5}	1.429	0.260	0.177	0.261	-	-	-	-
6.3	M2 *	288,034	2.50×10^{-5}	1.111	0.313	0.235	0.250	−22.3	20.4	32.7	−4.1
6.3	M3	545,954	2.50×10^{-5}	1.096	0.313	0.239	0.240	−1.4	0.0	1.5	−3.9
6.3	M2T1	545,954	$1.25 \times 10^{-}$	1.107	0.325	0.240	0.244	1.0	3.7	0.5	1.6
8.1	M1	138,354	2.50×10^{-5}	1.820	0.338	0.207	0.126	-	-	-	-
8.1	M2	198,834	2.50×10^{-5}	1.654	0.371	0.202	0.136	−9.1	9.8	−2.4	8.5
8.1	M3 *	288,034	2.50×10^{-5}	1.628	0.387	0.206	0.142	−1.6	4.4	2.0	4.2
8.1	M3T1	288,034	$1.25 \times 10^{-}$	1.636	0.395	0.207	0.137	0.5	2.0	0.7	−3.3
10.2	M1	198,834	2.50×10^{-5}	2.045	0.263	0.148	0.125	-	-	-	-
10.2	M2	272,914	2.50×10^{-5}	2.037	0.210	0.141	0.107	−0.4	−20.1	−5.0	−14.6
10.2	M3 *	362,514	2.50×10^{-5}	2.039	0.208	0.132	0.110	0.1	−1.0	−5.8	2.8
10.2	M3T1	362,514	1.25×10^{-5}	2.057	0.200	0.130	0.101	0.9	−4.0	−1.8	−7.9

In the columns 9–12, Δ indicates the variation of the aerodynamic parameter relative to the previous case.

Table 2. Aerodynamic parameters for different grid and time resolutions for AoA < 0°.

AoA (°)	Case	Elements	Δt (s)	$\overline{C_D}$	$\overline{C_L}$	$\overline{C_M}$	St	$\frac{\Delta}{C_D}$(%)	$\frac{\Delta}{C_L}$(%)	$\frac{\Delta}{C_M}$(%)	$\frac{\Delta}{St}$(%)
−1.4	M1	138,354	5.00×10^{-5}	1.277	−0.190	0.024	0.228	-	-	-	-
−1.4	M2	198,834	5.00×10^{-5}	1.230	−0.208	0.020	0.214	−3.7	−9.2	13.6	−6.3
−1.4	M3 *	288,034	5.00×10^{-5}	1.230	−0.209	0.019	0.217	0.0	−0.7	4.9	1.5
−1.4	M3T1	288,034	2.50×10^{-5}	1.205	−0.210	0.024	0.211	−2.0	−0.3	−22.2	−2.7
−2.9	M1	138,354	5.00×10^{-5}	1.345	−0.208	−0.053	0.226	-	-	-	-
−2.9	M2	198,834	5.00×10^{-5}	1.339	−0.207	−0.038	0.218	−0.5	0.0	27.2	−3.8
−2.9	M3 *	288,034	5.00×10^{-5}	1.340	−0.200	−0.038	0.220	0.1	3.7	1.0	1.0
−2.9	M3T1	288,034	2.50×10^{-5}	1.318	−0.191	−0.039	0.226	−1.6	4.6	−2.6	2.6
−4.0	M1	138,354	5.00×10^{-5}	1.512	−0.251	−0.080	0.212	-	-	-	-
−4.0	M2	198,834	5.00×10^{-5}	1.482	−0.219	−0.066	0.215	−2.0	12.9	17.2	1.3
−4.0	M3 *	288,034	5.00×10^{-5}	1.487	−0.217	−0.066	0.212	0.4	0.9	0.6	−1.5
−4.0	M3T1	288,034	2.50×10^{-5}	1.485	−0.216	−0.065	0.213	−0.1	0.2	0.9	0.4
−6.2	M1	138,354	5.00×10^{-5}	1.512	−0.354	−0.165	0.203	-	-	-	-
−6.2	M2	198,834	5.00×10^{-5}	1.517	−0.362	−0.163	0.196	0.3	−2.3	−1.5	−3.6
−6.2	M3 *	288,034	5.00×10^{-5}	1.519	−0.364	−0.163	0.198	0.1	−0.6	0.1	1.2
−6.2	M3T1	288,034	2.50×10^{-5}	1.512	−0.381	−0.161	0.203	−0.5	−4.6	−0.8	2.1
−8.1	M1	138,354	1.25×10^{-5}	2.094	−0.681	−0.181	0.160	-	-	-	-
−8.1	M2	198,834	1.25×10^{-5}	1.963	−0.653	−0.188	0.180	−6.3	4.2	3.5	12.8
−8.1	M3 *	288,034	1.25×10^{-5}	2.023	−0.664	−0.182	0.182	3.1	−1.8	−3.3	1.1
−8.1	M3T1	288,034	6.25×10^{-6}	2.064	−0.649	−0.191	0.176	2.0	2.2	5.3	−3.2
−10.0	M1	138,354	2.50×10^{-5}	2.567	−0.793	−0.135	0.182	-	-	-	-
−10.0	M2	198,834	2.50×10^{-5}	2.473	−0.733	−0.185	0.100	3.6	7.6	37.1	−45.1
−10.0	M3 *	288,034	2.50×10^{-5}	2.526	−0.733	−0.176	0.104	−2.2	−0.1	−5.1	4.0
−10.0	M3T1	288,034	1.25×10^{-5}	2.507	−0.727	−0.173	0.095	0.7	0.9	−1.8	−8.9

In the columns 9–12, Δ indicates the variation of the aerodynamic parameter relative to the previous case.

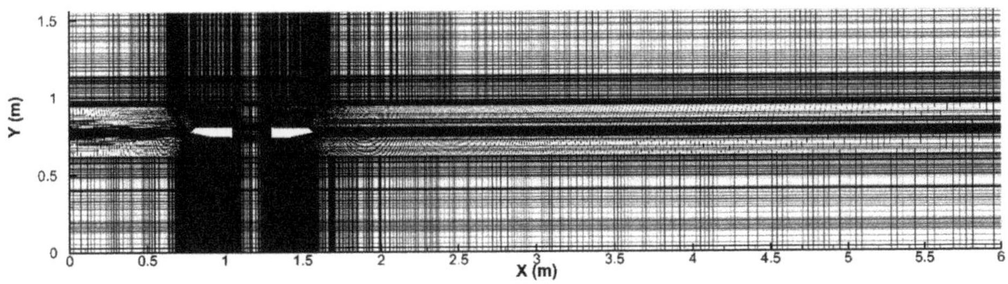

(a)

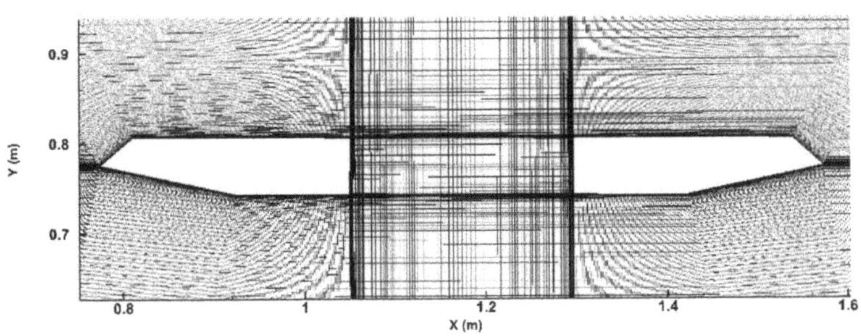

(b)

Figure 4. *Cont.*

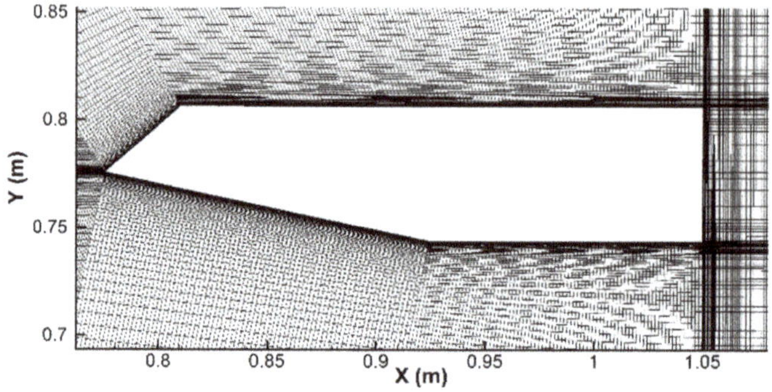

(c)

Figure 4. An example of the mesh for the case of AoA = +0.1° with 288,034 elements: (**a**) entire computational domain, (**b**) grids around the decks and (**c**) grids near the windward deck.

3. Results and Discussion

3.1. Validation of Numerical Model

To assess the validity of URANS with the k–ω SST turbulence model for flow around twin-box girder bridge cross section, a comparison has been carried out between the present simulations and the wind tunnel test results from [18]. The time-averaged aerodynamic quantities such as $\overline{C_D}$, $\overline{C_L}$ and $\overline{C_M}$, in the range of AoA from −10.0° to +10.2° at $Re = 31{,}000$ are studied, see Figure 5. A good overall agreement between the present numerical simulations results and the experimental data is observed. Generally, the presently predicted $\overline{C_D}$ are in good agreement with the experimental data for all AoA. However, a relatively larger deviation from the experimental data is found at AoA = +6.3° and +8.1°. This may be due to a change in flow physics, which will be discussed in detail in Section 3.3.

The numerically based $\overline{C_L}$ and $\overline{C_M}$ values are generally in good agreement with the experimental counterparts. However, significant reductions of $\overline{C_L}$ and $\overline{C_M}$ at AoA = +10.2° are observed, where $\overline{C_L}$ is under-predicted by 57% and $\overline{C_M}$ by 31% as compared to the experimental data. Such a reduction in magnitude of $\overline{C_L}$ and $\overline{C_M}$ at large AoA are also reported in Nieto et al. [8], who performed the 2D simulations of the Stonecutters bridge deck cross section with the k–ω SST turbulence model. A similar over- or under-prediction of $\overline{C_L}$ and $\overline{C_M}$ at high AoA by the 2D URANS model can be found in several previous studies [3,6–9,13]. This might be attributed to the fact that the present 2D numerical model cannot account for the fluctuating flow in the axial direction (Z-direction), while the 3D flow effects become significant with the increase of AoA. The higher energy dissipation in 3D cannot be correctly evaluated in 2D, i.e., larger pressure fluctuations and stronger vortex generation on the body surface are observed in 2D analysis when compared to 3D analysis, see also [36]. In the normal operation condition, the bridge is mainly exposed to a relatively small AoA (< 3°), associated with the bridge motion and large-scale turbulence. For this range of angles, the 2D simulation can provide good and quick results, and can thus be an efficient evaluation tool for the bridge design, particularly as a screening tool for the bridge girder designs.

The predicted St, shown in Figure 5d, does not vary significantly (from 0.18 to 0.22) for AoA ranging from −8.1° to +4.4°. For the case of AoA = +6.3°, St increases to 0.250. This may be due to the flow structure change associated with an interaction of the vortices on the different parts of the deck surface, which will be discussed in detail in Section 3.3. For the cases of AoA = +8.1° and +10.2°, St gradually decreases to 0.142 and 0.110, respectively. The decrease in St at large AoA is consistent with an increase of the projected area in the cross-flow direction.

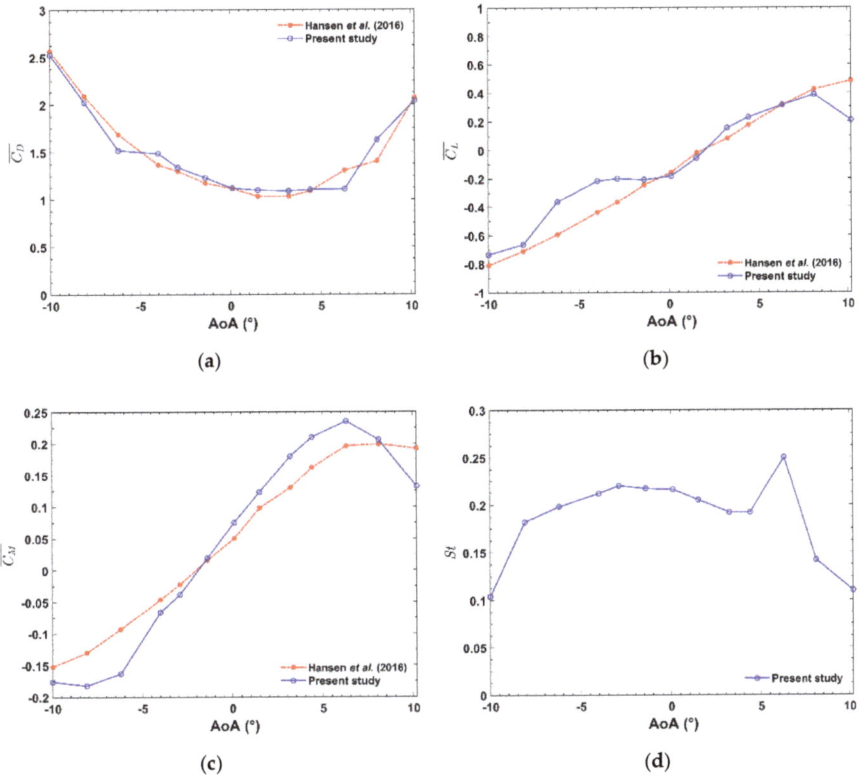

Figure 5. Variation of aerodynamic quantities with respect to AoA: (**a**) time-averaged drag coefficient $\overline{C_D}$, (**b**) time-averaged lift coefficient $\overline{C_L}$, (**c**) time-averaged moment coefficient $\overline{C_M}$ and (**d**) Strouhal number S_t.

Overall, it appears that the present 2D RANS simulations with the k–ω SST turbulence model are in satisfactory agreement with the experimental data under the same flow conditions, especially at small AoA. The discrepancies of the aerodynamic coefficients between numerical and experimental results at large AoA (i.e., +10.2°) may be due to the three-dimensional effects of the flow structure along the spanwise direction.

3.2. Vortex Formation around the Decks in One Vortex Shedding Period

One of the main concerns in bridge design is the vortex-induced vibration (VIV) behavior of bridge girders. In order to investigate the vortex formation around the decks, eight time instants during one cycle of the C_L fluctuation at AoA = +0.1° are studied, as defined in Figure 6. It is observed that the values of the amplitudes of the maximum and minimum C_L are not the same, and this is mainly due to the asymmetric geometry of the decks.

Figure 7 shows the vorticity contour plots for the eight time instants for AoA = +0.1°. During one C_L fluctuating period, it is observed that the vortices shed from the downstream upper and lower corners of the windward deck. Then the vortices propagate through the gap between the two decks and attack the leeward deck periodically. At the time instant 1, where C_L is the maximum, the lower vortex dominates the flow at the downstream edge of the windward deck what generates the lift force. Until the time instant 5, the upper vortex at the windward deck increases in size, and reduces the lift force gradually. As the vortices move to the leeward deck, the new vortices are generated from the downstream vertical edge of the windward deck. The upper and the lower vortices are different in size due to the asymmetric geometry of the decks. Vortices shed from the leeward edge of the downstream deck are very limited in size and strength due to a more streamlined form of this part of the cross-section.

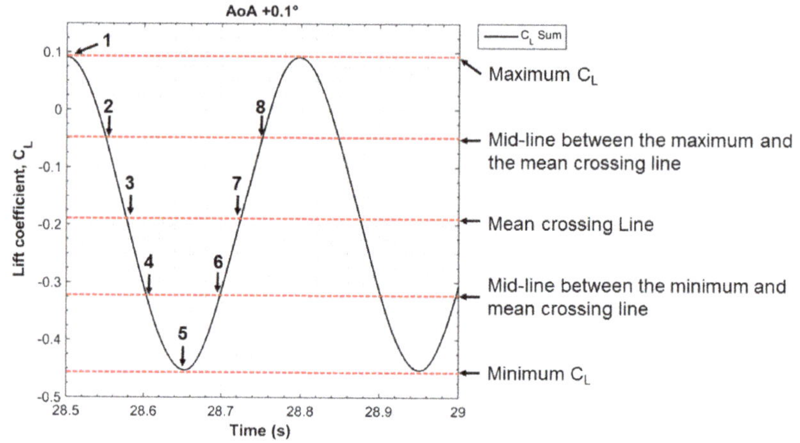

Figure 6. Eight time instances during one vortex shedding cycle which are investigated in Figure 7 (AoA = +0.1°).

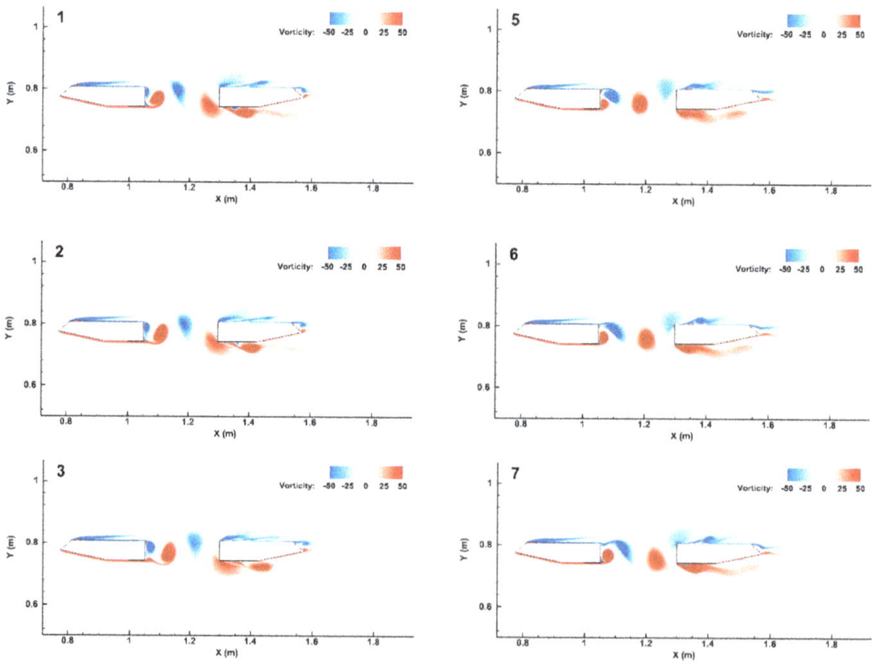

Figure 7. *Cont.*

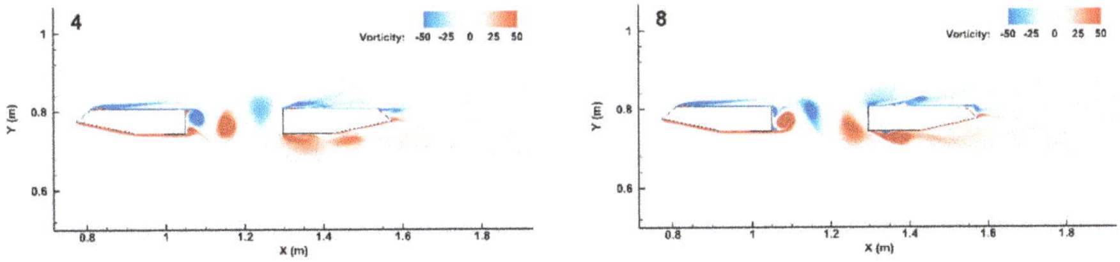

Figure 7. Vorticity contour plots of the eight time instances in one vortex shedding period for the case of AoA = +0.1°.

3.3. Flow Characteristics at Different Angles of Attack (AoA)

Figure 8 shows the time histories of C_D, C_L and C_M of the AoA = +0.1° case, respectively. The normalized forces on each deck, as well as the sum of the normalized forces on the two decks are shown. It is observed that C_L and C_M vary at a more or less single frequency. The total drag coefficient C_D also includes a frequency component at twice the frequency of the vortex shedding. The values of C_D from the individual boxes do not follow a pure harmonic function. This is mainly attributed to the generation and the downstream action of the asymmetric vortices, see Figure 7. Figure 9 shows the time history of C_D for different AoA, i.e., +3.2°, +6.3°, −6.2° and −8.1°. By comparing Figures 8a and 9a, it can be observed that the time history of C_D for AoA = +0.1° has different trend from that for AoA = +3.2°. This is due to the different vortex formation between these two cases. However, the time histories of C_L and C_M show similar behavior in both cases. By comparing Figure 9a–d, it is found that the time history of C_D shows significant changes for different AoA. The phase differences between the force coefficients on the two decks can be explained by the vortex shedding variations in Figure 7, i.e., the vortex shedding is generated first at the windward deck, and is reattached at the leeward deck.

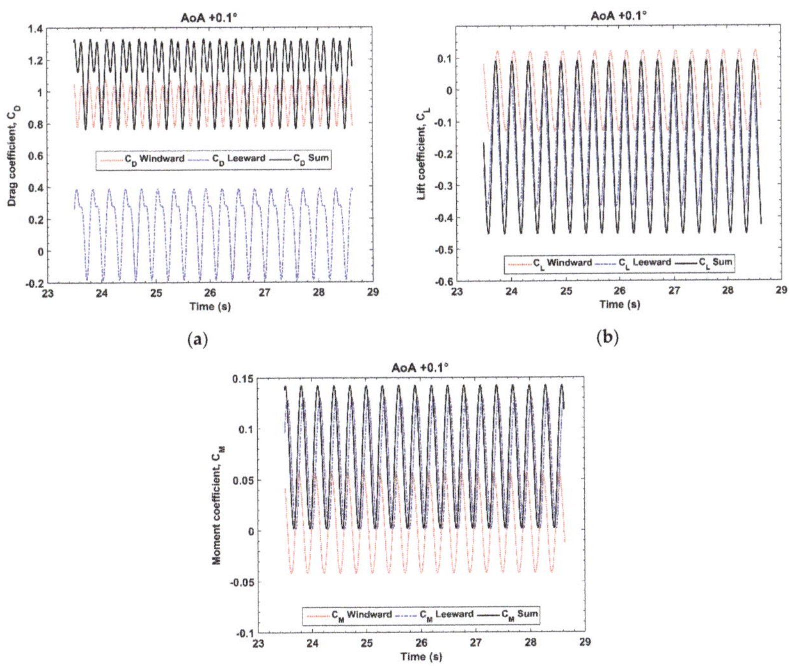

(a) (b) (c)

Figure 8. Time histories of force coefficients (AoA = +0.1°): (**a**) drag coefficient, (**b**) lift coefficient and (**c**) moment coefficient.

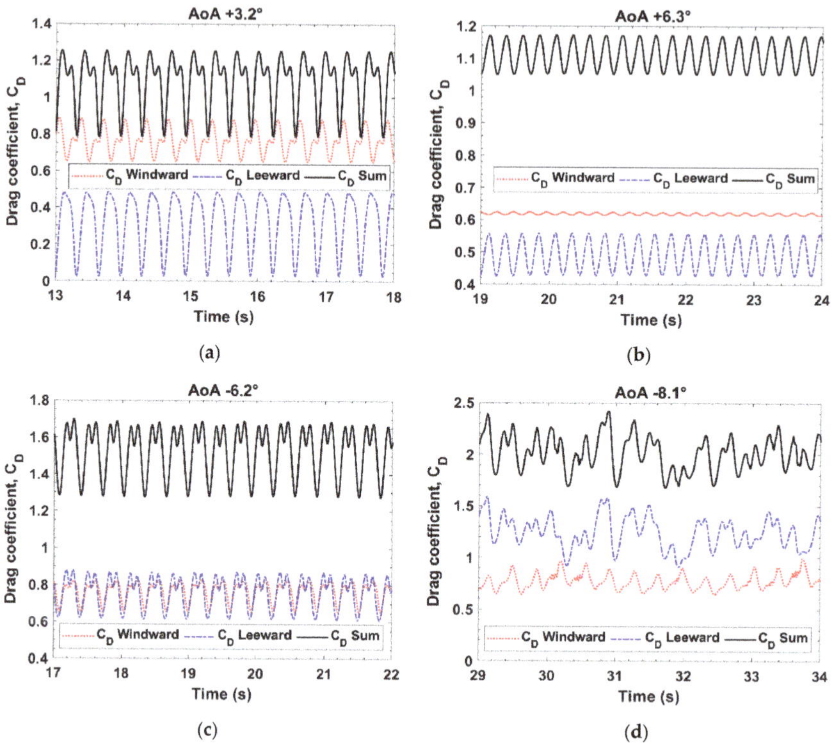

Figure 9. Time history of drag coefficient for different angles of attack: (**a**) AoA = +3.2°, (**b**) AoA = +6.3°, (**c**) AoA = −6.2° and (**d**) AoA = −8.1°.

Figure 9b shows time history of C_D for the AoA = +6.3° case. At this AoA, the flow characteristics begin to change. The second peak of C_D in one period becomes smaller as AoA increases. At AoA = +6.3°, the time history of C_D becomes a single harmonic, see Figure 9b. As the AoA increases, the projected area in the cross-flow direction increases and the drag force fluctuation of the windward deck decreases. The standard deviation (RSD) of C_D relative to its mean value, is less than 0.6% at the windward deck at AoA = +6.3°. It means that the drag force at the windward deck is almost stable, see Figure 9b. To study flow pattern change in detail, the C_L time histories for different AoA are looked into, as shown in Figure 10. For the AoA = +6.3° case, there are more vortex cycles than the other cases in the same time interval, corresponding to a higher St number value at this angle, see Figure 10b. Unlike the other cases, C_L at the windward deck is stable with little fluctuations during period at AoA = +6.3°. Figure 11a,b show instantaneous streamlines of the AoA = +6.3° case at the time instants of the maximum C_L at $tU_\infty/D = 304$ and the minimum C_L at $tU_\infty/D = 368$, respectively. It appears that the flow structure at the windward deck does not change significantly at these two time instants by comparing Figure 11a,b. However, the flow patterns at the leeward deck change significantly between these time instants. In particular, the vortices at the upper surface of the leeward deck and at the trailing edge of the leeward deck are observed to have strong interaction shown in Figure 11a,b. At the leeward deck, the interaction of the vortices on the upper deck surfaces seems to result in a flow separation over the entire surface.

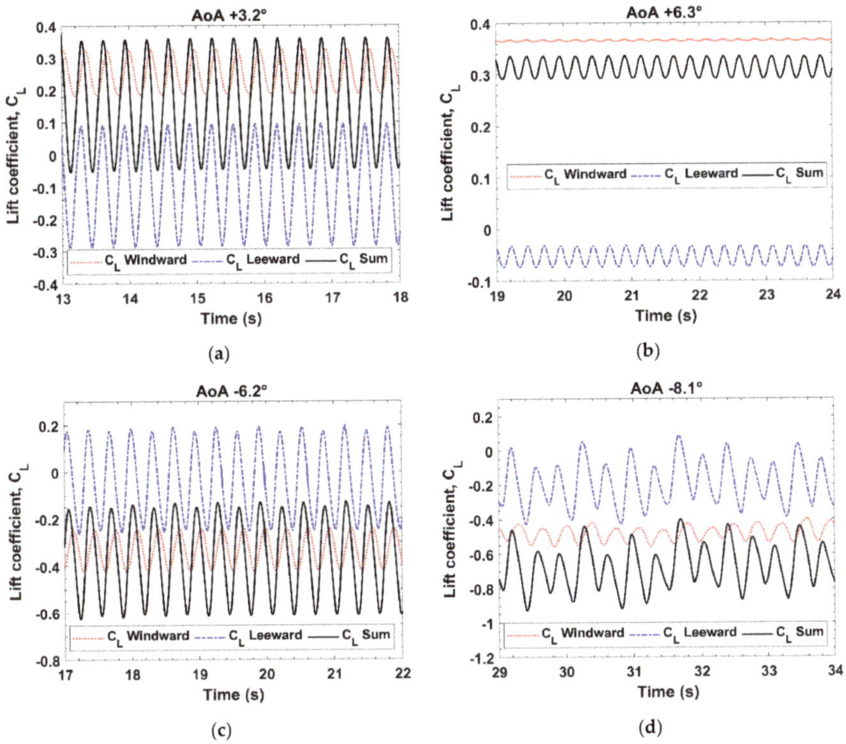

Figure 10. Time history of lift coefficient for different angles of attack: (**a**) AoA = +3.2°, (**b**) AoA = +6.3°, (**c**) AoA = −6.2° and (**d**) AoA = −8.1°.

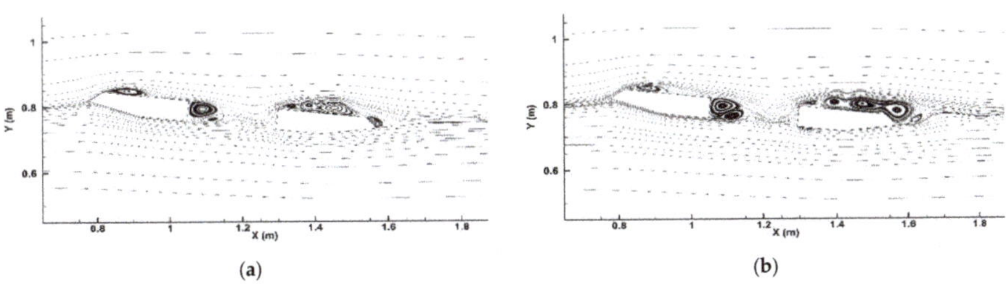

Figure 11. Instantaneous streamlines of AoA = +6.3° at (**a**) $tU_\infty/D = 304$ and $2tU_\infty/b = 69$, and (**b**) $tU_\infty/D = 368$ and $2tU_\infty/b = 84$.

Figure 12 show instantaneous streamlines of the AoA = +10.2°case at $tU_\infty/D = 504$. At AoA = +10.2°, the vortices at the upper part of both the windward and the leeward decks encompass the entire surfaces. The flow separation leads to the loss of lift as indicated by a reduced C_L value, see Figure 5b.

In the range of negative AoA from −1.5° to −6.2°, the time histories of C_D, C_L and C_M have similar characteristics as those in the range of AoA from +0.1° to +4.4°. Figure 9c shows the time history of C_D for AoA = −6.2°, where a double peak within the main C_D cycle can be observed. Figure 9d shows the time history of C_D for the case of AoA = −8.1°. For the AoA = −8.1° case, the variation in the force coefficients is more random than that for the AoA = −6.2° case, see Figures 9 and 10. Unlike the other cases depicted above, the time histories of aerodynamic quantities are no longer periodic. Generally, the time history of C_D shows a more random behavior than C_L by comparing Figures 9 and 10. For the cases of AoA = −6.2° and AoA = +3.2°, the total C_D has a double peak due to the vortices generation at the windward deck, see Figure 9a,c. For the AoA = +6.3° case, the total C_D has one peak per period because the flow structure at the windward deck in one vortex shedding cycle changes significantly less than at the leeward deck. This explains the small variation of C_D at the windward deck observed in Figure 9b.

Figure 13a–c show instantaneous streamlines for the AoA = −10.0° case. For the AoA = −10.0° case, the flow both at the windward deck and the leeward deck changes significantly compared to the AoA = +0.1° case (small angle) shown in Figure 7. Thus, the different time-histories of the force coefficients reflect the differences in the flow pattern. For the AoA = +10.2° case in Figure 12, the big vortex at the upper surface of the leeward deck is nearly stationary and only small changes of the vortex structure are observed during the period. However, for the AoA = −10.0° case shown in Figure 13a–c, the vortices generated at lower surface of the leeward deck are not stationary. The vortex formation and the flow pattern change significantly. The lower inclined surface of the windward deck for the AoA = −10.0° case has a smaller angle to the flow. Thus, a weaker vortex generation mechanism is observed as compared to the conditions at the upper deck surface of the AoA = +10.2° case shown in Figure 12.

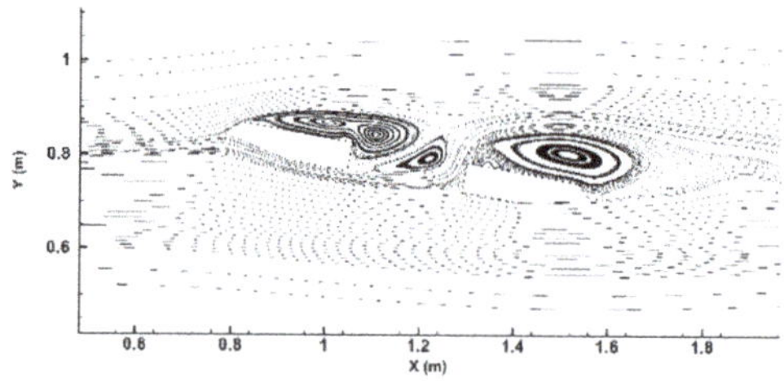

Figure 12. Instantaneous streamlines of AoA = +10.2° at tU_∞/D = 504 and $2tU_\infty/b$ = 115.

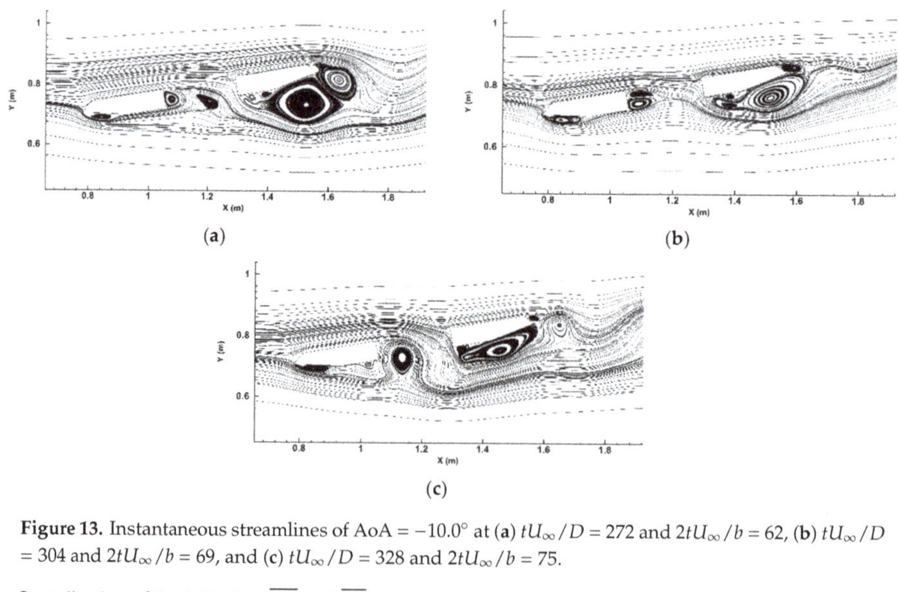

Figure 13. Instantaneous streamlines of AoA = −10.0° at (**a**) tU_∞/D = 272 and $2tU_\infty/b$ = 62, (**b**) tU_∞/D = 304 and $2tU_\infty/b$ = 69, and (**c**) tU_∞/D = 328 and $2tU_\infty/b$ = 75.

3.4. Contribution of Each Deck to $\overline{C_D}$ and $\overline{C_L}$

The relative contribution of the windward and the leeward deck, to the overall $\overline{C_D}$ and $\overline{C_L}$ is further presented in Figure 14. The $\overline{C_L}$ contributions are calculated based on the absolute mean values of each deck.

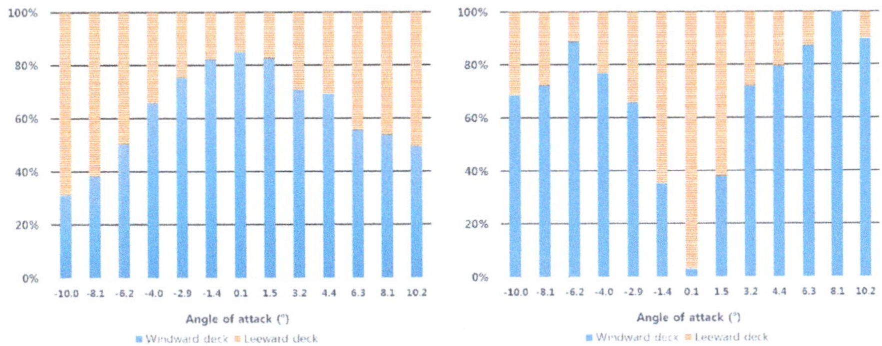

(**a**) $\overline{C_D}$ (**b**) $\overline{C_L}$

Figure 14. Relative contribution of each deck to the mean force coefficients ((**a**) $\overline{C_D}$ and (**b**) $\overline{C_L}$) versus AoA.

At AoA = +0.1°, the contribution of the windward deck to C_D is over 80%, and is less than 20% for the leeward deck, i.e., the along-wind force is mainly applied to the windward deck. As AoA increases, the contribution of the leeward deck to $\overline{C_D}$ increases. This is because the projected area of the leeward deck in the cross-flow direction increases as AoA increases, and the shielding effect of the upstream deck is reduced. The contribution of each deck to $\overline{C_D}$ is different for all the cases. For instance, the contributions of the two decks to $\overline{C_D}$ at AoA = +10.2° are nearly equal, i.e., 50% from the windward deck and 50% from the leeward deck. However, for the case of AoA = −10.0°, 30% contribution to $\overline{C_D}$ is from the windward deck, and 70% is from the leeward deck. The cross-sections of the bridge girders have asymmetric geometry, which result in different flow structures in the cases with same magnitude of AoA but in opposite signs.

Figure 12 shows the streamlines of the case of AoA = +10.2°, where the accelerated flow at the upper surface of the windward deck does not attack the leeward deck. Figure 13a–c show the streamlines of the AoA = −10.0° case, the accelerated flow at the upper surface of the windward directly hits the vertical edge of the leeward deck. Thus, even though two cases have same magnitude of AoA, the flow pattern and the contributions of each deck to the aerodynamic quantities are different. At all the negative AoA the leeward deck contributes more $\overline{C_D}$ than at the corresponding positive AoA. This is due to the accelerated flow at the windward deck acting on the leeward deck. For the positive AoA cases, the accelerated flow at the windward deck passes by the leeward deck and does not hit the leeward deck. The vortex at the upper surface of the windward deck prevents the interaction, see Figure 12.

For the contribution of each deck to $\overline{C_L}$, it is observed that the contribution of the windward deck to $\overline{C_L}$ increases as AoA increases. At AoA = +8.1°, nearly 100% lift force is generated by the windward deck and almost no contribution from the leeward deck. Then, at AoA = +10.2° the $\overline{C_L}$ contribution from the windward deck drops to around 90%, where the abrupt lift drop is observed. A similar trend can be observed in the negative AoA. From AoA = −1.4° to −6.2°, the contribution of the windward deck to $\overline{C_L}$ increases. For the cases of AoA = −8.1° and −10.0°, the contribution of the windward deck to $\overline{C_L}$ drops. The latter two cases have shown that the flow pattern begins to change and becomes chaotic. Thus, this C_D and C_L contribution study can offer a general overview of the aerodynamic responses and flow physics of the bridge. In particular, it is a good qualitative tool to figure out where the flow pattern begins to change.

Figures 15 and 16 show the variations of the relative standard deviation (RSD) of C_D and C_L to each deck versus AoA, respectively. The RSD of C_D and C_L are obtained based on the same period (30 vortex shedding cycles) used for the mesh and time-step refinement study in Table 1. This parameter shows the average force fluctuations at the two decks with respect to the average total value, and helps to understand the flow physics as a function of AoA. In Figure 15, the windward deck has quite stable RSD values within the investigated range of AoA, i.e., RSD values of 5% to 10%. For the cases of AoA = +6.3° and +8.1°, RSD values are less than 1%, see red line plot in Figure 10. The leeward deck has much larger RSD values within the investigated range of AoA as compared to the windward deck. The values show that the RSD values decrease as AoA increases. In Figure 16, positive lift forces on the windward deck can be observed when AoA ≥ 1.5°, whereas the lift force on the leeward deck is negative at all AoA. The smallest RSD magnitude of C_L on windward deck occurs at AoA = 6.3°, and the smallest RSD magnitude of C_L on leeward deck occurs at AoA = 8.1°, which indicate the change of the flow physics at these two AoA. The RSD of C_L on the windward deck at AoA = 0.1° and the RSD of C_L on the leeward deck at AoA = 6.3° show extremely large magnitude (i.e., 1627% and 977% respectively) due to the small value of $\overline{C_L}$ at these two AoA (see also Figures 8 and 10).

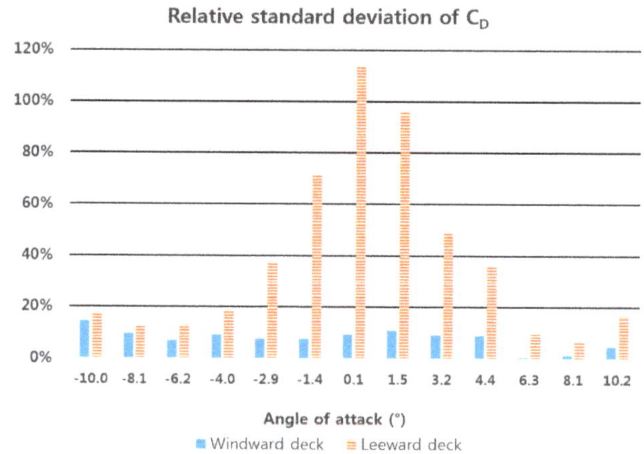

Figure 15. Variation of the C_D relative standard deviation of each deck with respect to AoA.

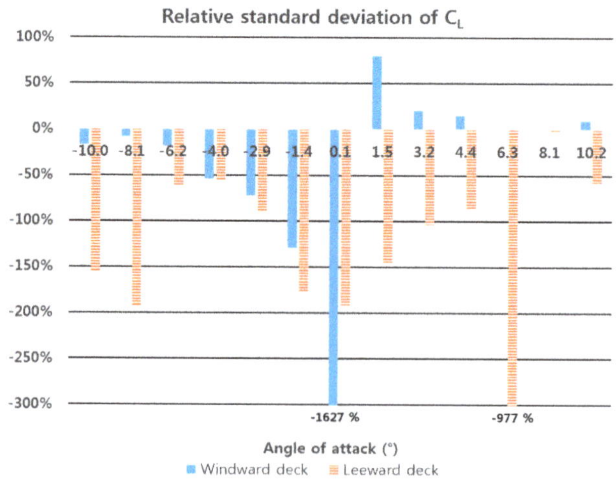

Figure 16. Variation of the C_L relative standard deviation of each deck with respect to AoA.

4. Conclusions

The aerodynamic force coefficients and the Strouhal number for a preliminary design of the long-span bridge girder cross-section have been obtained numerically, using the k–ω SST turbulence model at *Re*=31,000. The objective of this study is to examine the validity of the 2D URANS simulations for the flow around a sharp-edged twin-box cross section for the quick and robust evaluation of the bridge design. The aerodynamic coefficients have been calculated and compared with the wind tunnel test results. The vortex formation mechanisms as a function of AoA have been studied. The relative contribution of each deck to the resultant force has also been investigated. The main conclusions are summarized as follows:

1. The numerically predicted time-averaged force coefficients $\overline{C_D}$, $\overline{C_L}$ and $\overline{C_M}$ are in a good agreement with the wind tunnel experiment results. In particular, the drag coefficient has shown a good agreement with the experimental data at different AoA. The lift and moment coefficients show a good agreement with the experimental measurement in the low AoA range. A large reduction of $\overline{C_L}$ and $\overline{C_M}$ is observed at AoA = +10.2°. This indicates a premature stalling of the bridge decks simulated by the turbulence model, compared to the experimental observations in this high AoA region. The discrepancy may be due to the three-dimensional effects along the spanwise direction, which cannot be captured using the present 2D numerical model. Such high angles of attack are not expected during the normal bridge operation. Thus, the present 2D simulations are generally able to provide efficient and reliable assessment of the bridge girder aerodynamic performance under normal operating conditions.
2. The flow structure shows a different pattern from AoA = +6.3°, as the vortices of the upstream deck become rather steady, while the vortices traveling along the upper surface of the downstream deck begin to merge towards a complete flow separation at even higher angles of attack. The flow pattern variations at AoA = +10.2°, −8.1° and −10.0° also have been discussed.
3. Relative contributions of each deck to C_D and C_L varies with the AoA. This is also a good assessment tool, aiding the understanding of the flow physics and the screening purpose of the bridge design.

Overall, the present 2D URANS simulations with k–ω SST turbulence model give satisfactory results compared with the experimental results. The investigation of the flow pattern in the set of CFD studies can provide key information and insights to judge feasibility of aerodynamic performance of the bridge in a relatively cost-efficient manner.

Author Contributions: W.J. conducted CFD and data analysis. M.C.O. and J.B.J. conceived and designed the research. S.L. provided assistance in carrying out the CFD analysis and provided the MATLAB code for analyzing Strouhal number. W.J., S.L., J.B.J. and M.C.O. wrote the paper.

Funding: The authors are grateful to the Norwegian Public Road Administration (NPRA) for a partial financial support for the present study, as well as for sharing the reference data from the wind tunnel tests, performed by Svend Ole Hansen ApS. This study was also supported in part with computational resources provided by the Norwegian Metacenter for Computational Science (NOTUR), under Project No: NN9372K.

Conflicts of Interest: The authors declare no conflict of interest.

References

1. Fok, C.H.; Kwok, K.C.S.; Qin, X.R.; Hitchcock, P.A. Sectional pressure tests of a twin-deck bridge: Part 1: Experimental techniques and effects of angle of wind incidence. In Proceedings of the 11 AWES Workshop, Darwin, Australia, 28–29 June 2004.
2. Fok, C.H.; Kwok, K.C.S.; Qin, X.R.; Hitchcock, P.A. Sectional pressure tests of a twin-deck bridge: Part 2: Effects of gap-width on a twin-deck configuration. In Proceedings of the 11 AWES Workshop, Darwin, Australia, 28–29 June 2004.
3. Sánchez, R.; Nieto, F.; Kwok, K.C.S.; Hernández, S. CFD analysis of the aerodynamic response of a twin-box deck considering different gap widths. In Proceedings of the Congresso de Métodos Numéricos em Engenharia 2015, Lisboa, Portugal, 29 June–2 July 2015.
4. Kwok, K.C.S.; Qin, X.R.; Fok, C.H.; Hitchcock, P.A. Wind-induced pressures around a sectional twin-deck bridge model: Effects of gap-width on the aerodynamic forces and vortex shedding mechanisms. *J. Wind Eng. Ind. Aerodyn.* **2012**, *110*, 50–61. [CrossRef]

5. Diana, G.; Fiammenghi, G.; Belloli, M.; Rocchi, D. Wind tunnel tests and numerical approach for long span bridges: The Messina bridge. *J. Wind Eng. Ind. Aerodyn.* **2013**, *122*, 38–49. [CrossRef]
6. Nieto, F.; Zasso, A.; Rocchi, D.; Hernandez, S. CFD verification of aerodynamic devices performance for the Messina Strait bridge. In Proceedings of the Bluff Bodies Aerodynamics and Applications sixth international colloquium, Milano, Italia, 20–24 July 2008.
7. Nieto, F.; Hernández, S.; Jurado, J.Á.; Baldomir, A. CFD practical application in conceptual design of a 425 m cable-stayed bridge. *Wind. Struct.* **2010**, *13*, 309–326. [CrossRef]
8. Nieto, F.; Kusano, I.; Hernandez, S.; Jurado, J.Á. CFD analysis of the vortex-shedding response of a twin-box deck cable-stayed bridge. In Proceedings of the fourth international symposium on Computational Wind Engineering, Chapel Hill, NC, USA, 23 May 2010.
9. Nieto, F.; Hernández, S.; Kusano, I.; Jurado, J.Á. CFD aerodynamic assessment of deck alternatives for a cable-stayed bridge. In Proceedings of the bluff Bodies Aerodynamics and Applications seventh international colloquium, Shanghai, China, 2 September 2012.
10. Fumoto, K.; Watanabe, S. An estimation of aerodynamics of slotted one-box girder section using computational fluid dynamics. In Proceedings of the Fourth International Symposium on Computational Wind Engineering, Yokohama, Japan, 16–19 July 2006.
11. Maruoka, A.; Takou, M.; Sasaki, H. Detached eddy simulation of flow around a box girder bridge section. In Proceedings of the Fourth International Symposium on Computational Wind Engineering, Yokohama, Japan, 16–19 July 2006.
12. Larsen, A. Computation of aerodynamic derivatives by various CFD techniques. In Proceedings of the Fourth International Symposium on Computational Wind Engineering, Yokohama, Japan, 16–19 July 2006.
13. Shirai, S.; Ueda, T. Aerodynamic simulation by CFD on flat box girder of super-long-span suspension bridge. *J. Wind Eng. Ind. Aerodyn.* **2003**, *91*, 279–290. [CrossRef]
14. Bruno, L.; Salvetti, M.V.; Ricciardelli, F. Benchmark on the aerodynamics of a rectangular 5: 1 cylinder: An overview after the first four years of activity. *J. Wind Eng. Ind. Aerodyn.* **2014**, *126*, 87–106. [CrossRef]
15. Mannini, C.; Marra, A.M.; Pigolotti, L.; Bartoli, G. The effects of free-stream turbulence and angle of attack on the aerodynamics of a cylinder with rectangular 5: 1 cross section. *J. Wind Eng. Ind. Aerodyn.* **2017**, *161*, 42–58. [CrossRef]
16. Mannini, C.; Soda, A.; Schewe, G. Numerical investigation on the three-dimensional unsteady flow past a 5: 1 rectangular cylinder. *J. Wind Eng. Ind. Aerodyn.* **2011**, *99*, 469–482. [CrossRef]
17. Ong, M.C. Unsteady rans simulation of flow around a 5: 1 rectangular cylinder at high reynolds numbers. In Proceedings of the ASME 2012 31st International Conference on Ocean, Offshore and Arctic Engineering, American Society of Mechanical Engineers, Rio de Janeiro, Brazil, 1–6 July 2012; pp. 841–846.
18. Patruno, L.; Ricci, M.; de Miranda, S.; Ubertini, F. Numerical simulation of a 5: 1 rectangular cylinder at non-null angles of attack. *J. Wind Eng. Ind. Aerodyn.* **2016**, *151*, 146–157. [CrossRef]
19. Schewe, G. Influence of the Reynolds-number on flow-induced vibrations of generic bridge sections. In Proceedings of the International Conference on Bridges, Dubrovnik, Croatia, 21–24 May 2006; pp. 351–358.
20. Schewe, G. Reynolds-number-effects in flow around a rectangular cylinder with aspect ratio 1:5. *J. Fluids Struct.* **2013**, *39*, 15–26. [CrossRef]
21. Menter, F.R. Two-equation eddy-viscosity turbulence models for engineering applications. *Am. Insti. Aero. Astro. J.* **1994**, *32*, 1598–1605. [CrossRef]
22. Laima, S.; Li, H. Effects of gap width on flow motions around twin-box girders and vortex-induced vibrations. *J. Wind Eng. Ind. Aerodyn.* **2015**, *139*, 37–49. [CrossRef]
23. Laima, S.; Jiang, C.; Li, H.; Chen, W.; Ou, J.A. numerical investigation of Reynolds number sensitivity of flow characteristics around a twin-box girder. *J. Wind Eng. Ind. Aerodyn.* **2018**, *172*, 298–316. [CrossRef]
24. Miranda, S.; Patruno, L.; Ricci, M.; Ubertini, F. Numerical study of a twin box bridge deck with increasing gap ratio by using RANS and LES approaches. *Eng. Struct.* **2015**, *99*, 546–558. [CrossRef]
25. Hansen, S.O.; Srouji, R.G.; Rasmussen, J.T. *Julsundet and Halsafjorden Bridges Part II: Suspension Bridge Crossing Halsafjorden, Flutter and Vortex-Induced Vibrations of a Sharp-Edged Cross Section*; Project Report; Kvisvik, Norway, 2016.
26. Menter, F.R.; Kuntz, M.; Langtry, R. Ten years of industrial experience with the SST turbulence model. *Turbulence Heat. Mass. Transf.* **2003**, *4*, 625–632.
27. Wilcox, D.C. *Turbulence Modeling for CFD*, 2nd ed.; DCW Ind., Inc.: La Canada, California, USA, 1998.

28. Launder, B.; Spalding, D. The numerical computation of turbulent flows. *Comput. Methods Appl. Mech. Eng.* **1974**, *3*, 269–289. [CrossRef]
29. OpenFoam. *The Open Source CFD Toolbox, Programmer's Guide*; Version 3.0.1.; OpenCFD Limited: Bracknell, UK, 2009.
30. Tian, X.; Ong, M.C.; Yang, J.; Myrhaug, D. Unsteady RANS simulations of flow around rectangular cylinder with different aspect ratios. *J. Ocean Eng.* **2013**, *58*, 208–216. [CrossRef]
31. Rahman, M.M.; Karim, M.M.; Alim, M.A. Numerical investigation of unsteady flow past a circular cylinder using 2-D finite volume method. *J. Naval. Arch. Marine. Eng.* **2007**, *4*, 27–42. [CrossRef]
32. Gao, Y.; Chow, W.K. Numerical studies on air flow around a cube. *J. Wind Eng. Ind. Aerodyn.* **2005**, *93*, 115–135. [CrossRef]
33. Shimada, K.; Ishihara, T. Application of a modified $k-\varepsilon$ model to the prediction of aerodynamic characteristics of rectangular cross-section cylinders. *J. Fluid. Struct.* **2002**, *16*, 465–485. [CrossRef]
34. Sandham, N.D.; Yao, Y.F.; Lawal, A.A. Large-eddy simulation of transonic turbulent flow over a bump. *Int. J. Heat. Fluid. Flow.* **2003**, *24*, 584–595. [CrossRef]
35. Narasimhamurthy, V.D.; Andersson, H.I.; Pettersen, B.R. Cellular vortex shedding in the wake of a tapered plate. *J. Fluid. Mech.* **2010**, *617*, 355–379. [CrossRef]
36. Hirano, H.; Watanabe, S.; Maruoka, A.; Ikenouchi, M. Aerodynamic characteristics of rectangular cylinders. *Int. J. Comp. Fluid. Dyn.* **1999**, *12*, 151–163. [CrossRef]

energies

MDPI

Article

Influence of the Dynamic Motion of a Splash-Lubricated Gearbox on Churning Power Losses

Xiaozhou Hu *, Pingping Li and Minggui Wu

School of Mechanical and Electrical Engineering, State Key Laboratory of High Performance Complex Manufacturing, Central South University, Changsha 410083, China

* Correspondence: huxiaozhou@csu.edu.cn

Received: 22 July 2019; Accepted: 19 August 2019; Published: 21 August 2019

Abstract: To investigate the influence of the dynamic motion of a gearbox on lubricating flow field and churning power losses under splash lubrication, a computational fluid dynamics (CFD) method based on a combination of the fluid of volume (VOF) method and turbulence model is presented in this paper. A non-inertial coordinate system was employed to simulate the motion of the gearbox, and the feasibility and accuracy of the method was validated by the available experimental results. Numerical models of the gearbox with a spur gear pair under no load operation condition were established, and sinusoidal motions with different frequencies and amplitudes were implemented in the gearbox. The effects of the rotational speed of gears, oil immersion depth, and the frequency and amplitude of sinusoidal motions were studied. The results showed that the dynamic motion of the gearbox can exert a significant influence on churning losses and the oil supply of the gear contacting zone, and the gear pair may be in a loss-of-lubrication state.

Keywords: splash lubrication; dynamic motion; gearbox; churning power losses; non-inertial coordinate system; CFD

1. Introduction

The efficiency of gear transmissions is an important study object in the field of energy-saving. The total energy losses in splash-lubricated gearboxes are mainly categorized in two ways the first are load dependent losses, which are a result of the inter-tooth meshing friction of gears and inter-surface friction of bearings and seals; the second category are independent of load and are known as churning power losses, which are losses due to the fluid resistance torque between the gear and the surrounding oil. When the rotational speeds of the gear speed is low, churning losses represent a tiny proportion of the total energy losses and can generally be ignored; however, several researchers have pointed out that when the gears rotate at higher speeds, the influence of the churning losses increases significantly. For example, Concli and Gorla [1] quoted Strasser's investigations to show that in a parallel-shaft gearbox with splash lubrication, 70% of the total losses of gears is meshing friction loss and 30% is oil churning loss; Michaelis et al. [2] suggested that the no-load loss of rotating parts accounted for the majority of the overall loss of a gearbox under high-speed and light-load conditions; calculations by Neurouth et al. [3] showed that under high speed operating conditions (32,000 rpm), the gear churning loss is 1.28 times that of the meshing friction loss. Therefore, reducing churning losses can significantly improve the transmission efficiency of the gear transmission and effectively save energy; thus it is essential to accurately calculate the churning power losses of gear transmission.

Oil churning losses are generated by the combination of interactions between rotating gears and the surrounding oil, and the oil pumping effect between the meshing gear pair [4,5]. It is difficult to predict oil churning loss directly by theoretical methods, because its determination is affected by many factors, such as the complex inner shape of housing of gearboxes, flow field characteristics inside gearboxes, etc. [6–9].

In previous works, several researchers have performed experimental studies on oil churning losses inside gearboxes and presented analytical formulations under different assumptions [10–12]. Changenet and Velex [13] conducted several experiments on a specific test setup to study the influences of rotational speed, oil immersion depth, oil parameters and gear geometrical parameters, and derived a formula for calculating churning losses of spur gears. Höhn et al. [14] found that churning power losses can be significantly reduced with oil immersion depth. Seetharaman and Kahraman et al. [4,5] carried out experiments under different rotational speeds, temperatures, oil immersion depths and geometrical parameters of gears, and found that the static oil level and tooth width had important effects on the churning losses. Neurouth et al. [15] performed experiments to study the influence of air aeration phenomenon on churning losses and discovered that properly inserting movable walls inside the gearbox can significantly reduce air aeration and churning losses. Polly et al. [16] developed an experimental setup to study churning power losses with different operating parameters and geometrical parameters of gears and gear pair.

Continuous improvements in computer technology and CFD methods make it more reliable and effective to employ numerical methods to calculate churning losses of gears. Gorla et al. [17] built a CFD model to numerically investigate churning losses of power transmissions. Concli et al. [18], Concli and Gorla [19,20] used different and improved CFD modeling or simulation methods to predict no-load power losses of geared transmissions under varying operating and geometric parameters. Liu et al. [21] numerically calculated churning losses in a gearbox with a spur gear pair based on finite volume method, and observed good agreement between simulation results and experimental measures. Then, Liu et al. [22] also applied a smoothed particle hydrodynamics (SPH) method for the prediction of churning losses, but found that this method was not able to precisely calculate churning losses. Peng et al. [23] performed CFD simulations with a flank-moving method and obtained churning losses of a splash-lubricated hypoid gearbox. Jiang et al. [9] investigated the effect of an oil guide device on oil supply for bearings, and found churning power losses were not affected.

The above-mentioned studies reveal that churning power losses is significantly affected by oil flow behavior, and they pay most attention to on churning power losses of gearboxes in a static state, that is, the motion state of gearboxes was not often considered. However, in actual applications, a gearbox may be a part of movable objects, such as aircrafts, helicopters or vehicles, etc. Velocity, motion attitude and the direction of movement of those objects may change at any time, thus the motion state of the gearboxes are correspondingly dynamic changes, which cause the flow field inside gearboxes to be more complex and also to have an effect on churning power losses. Therefore, it is necessary to investigate the impact of the motion state of gearboxes on churning power losses. On this research topic, Lemfeld et al. [24] firstly performed CFD simulations to study the flow field of gearboxes under different inclined conditions, which resulted in varying angles of oil sump surface, and thus, obvious changes in splashing oil flow. Chen and Matsumoto et al. [25] developed a test setup to experimentally investigate churning losses of a spur gearbox considering the gearbox inclination and found that the inclination angle of gearboxes can exert an important influence on churning losses. Hu et al. [26] investigated the influence of the static tilt angle of the helicopter on the flow field and the churning losses, which showed that the flight attitude had a remarkable influence on churning power losses.

All of the previous studies have focused on the investigation of the internal oil distribution and churning losses of gearboxes in a static or quasi-stationary state, without considering the dynamic change of motion state, such as tilt angle, velocity, etc. However, in real working conditions, the motion state of gearboxes may change all the time, and the movement of gearboxes has an important influence on the internal flow field, and thus, churning power losses. Therefore, there is a strong demand to clearly reveal the influence mechanism of the motion state of gearboxes. In the present work, instead of employing a traditional CFD method, we establish CFD models based on the combination of a Fractional Area/Volume Obstacle Representation (FAVOR) method and a non-inertial coordinate system for realization of the dynamic motion of a gearbox. Second, the numerical approach is validated with available experimental results [25]. The effects of the rotational speed of gears, oil immersion depth and motion parameters of the gearbox on churning power losses are studied by CFD simulations.

2. Numerical Method

To study the effect of the dynamic motion of gearboxes on churning losses by numerical simulations, it is necessary to simultaneously simulate dynamic motion of gearboxes and gear pairs and precisely track the time-varying oil free surface inside gearboxes. To achieve this goal, a non-inertial coordinate system was used to simulate the motion of gearboxes, and the FAVOR method [27,28] was employed to model complex surfaces (for example, the gear meshing zone) in a size-fixed grid with two important parameters, namely, area fractions A_i and volume fractions V_F, which describe the rotating gears at every time step and reconstruct the geometry according to the rotation of gears. Therefore, this method needs no body-fitted mesh, which can significantly improve calculation efficiency, and when in combination with the Volume of Fluid (VOF) method [29], it can precisely describe complex geometric shapes and capture oil free surfaces, which exactly meets the demand to solve problems in the present work.

In this paper, the Cartesian coordinate system was used to simulate the dynamic motion of the gearbox, and the lubricating oil was supposed to be a viscous incompressible fluid. The continuity equation can be given as:

$$\frac{\partial(u_i A_i)}{\partial x_i} = 0 \tag{1}$$

where $i = 1, 2, 3$ represents x, y, and z directions, and A_i is the area fraction.

The momentum equations for three-dimensional flows are Navier-Stokes (N-S) equations with some additional terms, which are given as follows:

$$\begin{aligned}
\frac{\partial u}{\partial t} + \frac{1}{V_F}\left(uA_x\frac{\partial u}{\partial x} + vA_y\frac{\partial u}{\partial y} + wA_z\frac{\partial u}{\partial z}\right) &= -\frac{1}{\rho}\frac{\partial p}{\partial x} + G_x + f_x \\
\frac{\partial v}{\partial t} + \frac{1}{V_F}\left(uA_x\frac{\partial v}{\partial x} + vA_y\frac{\partial v}{\partial y} + wA_z\frac{\partial v}{\partial z}\right) &= -\frac{1}{\rho}\frac{\partial p}{\partial y} + G_y + f_y \\
\frac{\partial w}{\partial t} + \frac{1}{V_F}\left(uA_x\frac{\partial w}{\partial x} + vA_y\frac{\partial w}{\partial y} + wA_z\frac{\partial w}{\partial z}\right) &= -\frac{1}{\rho}\frac{\partial p}{\partial z} + G_z + f_z
\end{aligned} \tag{2}$$

where u, v and w are fluid velocity components, ρ is the oil density, p is the average pressure, G_i is the body acceleration, and f_i is the viscous acceleration ($i = x, y, z$).

For a dynamic viscosity μ, the viscous acceleration can be calculated as:

$$\begin{aligned}
\rho V_F f_x &= -\left[\frac{\partial}{\partial x}(A_x\tau_{xx}) + \frac{\partial}{\partial y}\left(A_y\tau_{yx}\right) + \frac{\partial}{\partial z}(A_z\tau_{zx})\right] \\
\rho V_F f_y &= -\left[\frac{\partial}{\partial x}\left(A_x\tau_{xy}\right) + \frac{\partial}{\partial y}\left(A_y\tau_{yy}\right) + \frac{\partial}{\partial z}\left(A_z\tau_{zy}\right)\right] \\
\rho V_F f_z &= -\left[\frac{\partial}{\partial x}(A_x\tau_{xz}) + \frac{\partial}{\partial y}\left(A_y\tau_{yz}\right) + \frac{\partial}{\partial z}(A_z\tau_{zz})\right]
\end{aligned} \tag{3}$$

where τ_{ij} is the shear stress ($i, j = x, y, z$):

$$\begin{aligned}
\tau_{xx} &= -2\mu\left[\frac{\partial u}{\partial x} - \frac{1}{3}\left(\frac{\partial u}{\partial x} + \frac{\partial v}{\partial y} + \frac{\partial w}{\partial z}\right)\right] \\
\tau_{yy} &= -2\mu\left[\frac{\partial v}{\partial y} - \frac{1}{3}\left(\frac{\partial u}{\partial x} + \frac{\partial v}{\partial y} + \frac{\partial w}{\partial z}\right)\right] \\
\tau_{zz} &= -2\mu\left[\frac{\partial w}{\partial z} - \frac{1}{3}\left(\frac{\partial u}{\partial x} + \frac{\partial v}{\partial y} + \frac{\partial w}{\partial z}\right)\right] \\
\tau_{xy} = \tau_{yx} &= -\mu\left(\frac{\partial u}{\partial x} + \frac{\partial v}{\partial y}\right) \\
\tau_{xz} = \tau_{xx} &= -\mu\left(\frac{\partial u}{\partial x} + \frac{\partial v}{\partial z}\right) \\
\tau_{zy} = \tau_{yz} &= -\mu\left(\frac{\partial w}{\partial z} + \frac{\partial v}{\partial y}\right)
\end{aligned} \tag{4}$$

There is a dramatic change of oil free surface inside gearboxes with splash lubrication, where large time-average strain rates and strong shear regions also exist. Therefore, to complete turbulence closures for N-S equations, the renormalization group (RNG) k-ε model is employed to simulate oil flow field of gearboxes. Transport equations for the turbulent kinematic energy k_T and its dissipation rate ε_T equations are as follows

$$\frac{\partial k_T}{\partial t} + \frac{1}{V_F}\left(uA_x\frac{\partial k_T}{\partial x} + vA_y\frac{\partial k_T}{\partial y} + wA_x\frac{\partial k_T}{\partial z}\right) = P_T + G_T + Diff_{k_T} - \varepsilon_T \tag{5}$$

$$\frac{\partial \varepsilon_T}{\partial t} + \frac{1}{V_F}\left(uA_x\frac{\partial \varepsilon_T}{\partial x} + vA_y\frac{\partial \varepsilon_T}{\partial y} + wA_z\frac{\partial \varepsilon_T}{\partial z}\right) = \frac{CDIS1\cdot\varepsilon_T}{k_T}(P_T + CDIS3\cdot G_T) + Diff_\varepsilon - CDIS2\frac{\varepsilon_T^2}{k_T} \tag{6}$$

where P_T is turbulent production; G_T is the buoyancy production term; $Diff_{k_T}$ and $Diff_\varepsilon$ are the diffusion of turbulent kinematic energy and its dissipation rate, respectively; $CDIS1$, and $CDIS3$ are dimensionless user-defined parameters, the defaults of which are 1.42 and 0.2, respectively; and $CDIS2$ is calculated by the turbulent kinetic energy k_T and turbulent production P_T.

The air-oil two-phase flow inside gearboxes is complex, and the lubricating oil surface changes dramatically at all times. As mentioned above, the VOF method is used to capture the complex and time-varying free surface. However, different to the typical VOF method, the one used in the present work takes no account of the mass of the air adjacent to the oil, and a void space with uniform pressure and temperature is used to replace the air-occupied volume, thus it is also known as the TruVOF method and more suitable for two-fluid flows than typical VOF method [28].

By using the TruVOF method, a function $F(x, y, z)$ is defined [29] to denote the volume of fluid per unit volume, which can satisfy the following formula,

$$\frac{\partial F}{\partial t} + \frac{1}{V_F}\left[\frac{\partial}{\partial x}(FA_xu) + \frac{\partial}{\partial y}\left(FA_yv\right) + \frac{\partial}{\partial z}(FA_sw)\right] = F_{\mathrm{DIF}} \tag{7}$$

where

$$F_{\mathrm{DIF}} = \frac{1}{V_F}\left\{\frac{\partial}{\partial x}\left(v_FA_x\frac{\partial F}{\partial x}\right) + \frac{\partial}{\partial x}\left(v_FA_y\frac{\partial F}{\partial y}\right) + \frac{\partial}{\partial z}\left(v_FA_z\frac{\partial F}{\partial z}\right)\right\} \tag{8}$$

where the diffusion coefficient is given by $v_F = c_F\mu/\rho$, and c_F is a constant which can sometimes be determined as the reciprocal of a turbulent Schmidt number.

For a single fluid, F represents the fluid-occupied volume fraction. Therefore, $F = 1$ means that fluid exists, while locations where $F = 0$ indicate that there is no fluid mass and a uniform pressure, thus these are also called "void" regions. Physically, they are not void but denote regions full of air or gas with much lower density compared to the fluid.

3. Verification of the Numerical Method

In order to evaluate the feasibility of the present numerical method, a model of a rectangular parallelepiped gearbox with a length of 224 mm, height of 144 mm, and width of 49 mm was established with reference to the experimental model available in [25]. A spur gear pair with splash lubrication under no-load condition was mounted in the rectangular gearbox.

The simulation model of the gearbox was placed at different angles of inclination β, as shown in Figure 1, which means the angle for oil moved from the vertical direction to the central line of the gear pair according to the rotation direction of the gears. In the simulations, six different inclination angles 120°, 150°, 180°, 210°, 240°, and 270° were used. The initial oil surface immersion depth of the lubricating oil was 4 times the tooth depth and the temperature of oil in the tank was set as 25 °C, which is consistent with that of the experiment available in [25]; besides, for the convenience of the investigation of dynamic motion parameters of the gearbox, oil for all the simulations below was also set as 25 °C. The basic parameters of gears and lubricating oil are shown in Tables 1 and 2.

Figure 1. Schematic of the simulation model and the definition of β.

Table 1. Geometrical parameters of gears.

Number of Teeth (Z)	28
Module (m)	2.0 mm
Face width of gear (b)	20 mm
Tip diameter (d_a)	60 mm
Tooth depth (h)	4.5 mm
Center distance (a)	56 mm

Table 2. Physical and chemical properties of lubricating oil.

Oil	Gear Oil ISO VG320
Kinetic viscosity ν (40 °C)	305.3 cst
Kinetic viscosity ν (100 °C)	25.57 cst
Density ρ_{oil}	0.8873 g/cm^3

Mesh is a significant factor that can influence the precision and efficiency of simulations. In the present simulations, the grid size was 1.4 mm and the grid type was hexahedron, which can improve the calculation efficiency and maintain the calculation accuracy. The grid boundary conditions were set as symmetric boundaries. The model grid number is 602784, and the mesh details are shown in Figure 2.

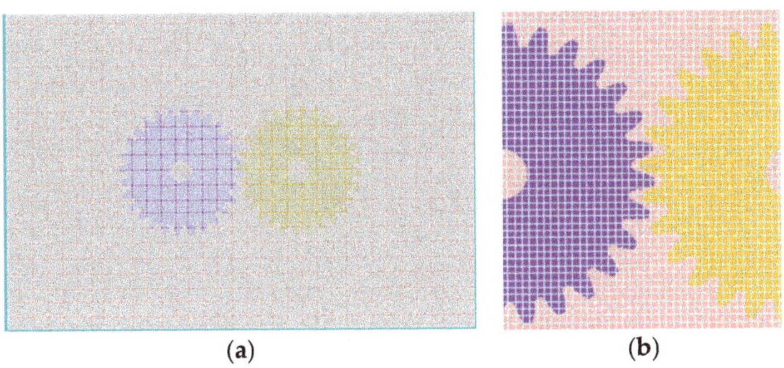

Figure 2. Mesh. (**a**) Entire mesh; and (**b**) enlarged view of mesh around gears.

The gears rotate by using the general moving objects (GMO) model [28], in which the gear was defined as a moving part with the rotation speed of 500 rpm, and the gear center was set as the rotation center. The gravity acceleration was taken into account during calculations and its value was set to 9.8 m/s^2, and the viscosity and turbulence model was activated where the RNG model is chosen.

Figure 3 shows the ratios of the churning loss torque of simulations to the corresponding experimental ones, from which it can be observed that the simulation results are in good quantitative agreement with the ones obtained by experiments. Both the experimental results and the simulation data show the same trend, and data error is less than 10%, except those of 90°and 120°. As can be seen from Figure 4, the larger volume of lubricating oil concentrated in the meshing area of simulations or experiments may be the cause of the bigger data error of 90° and 120°.

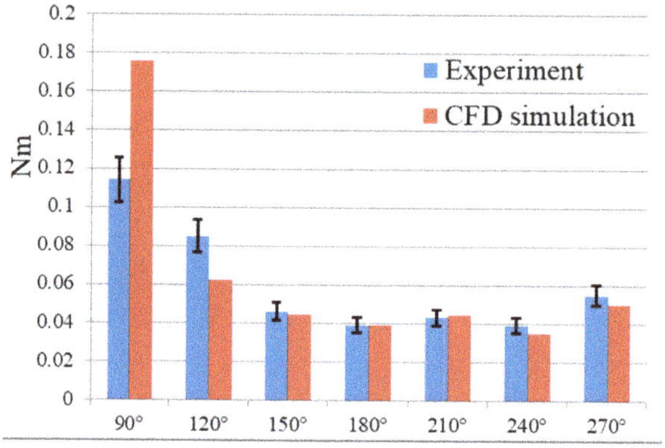

Figure 3. Churning loss torque from experiments and simulations with different inclination angles.

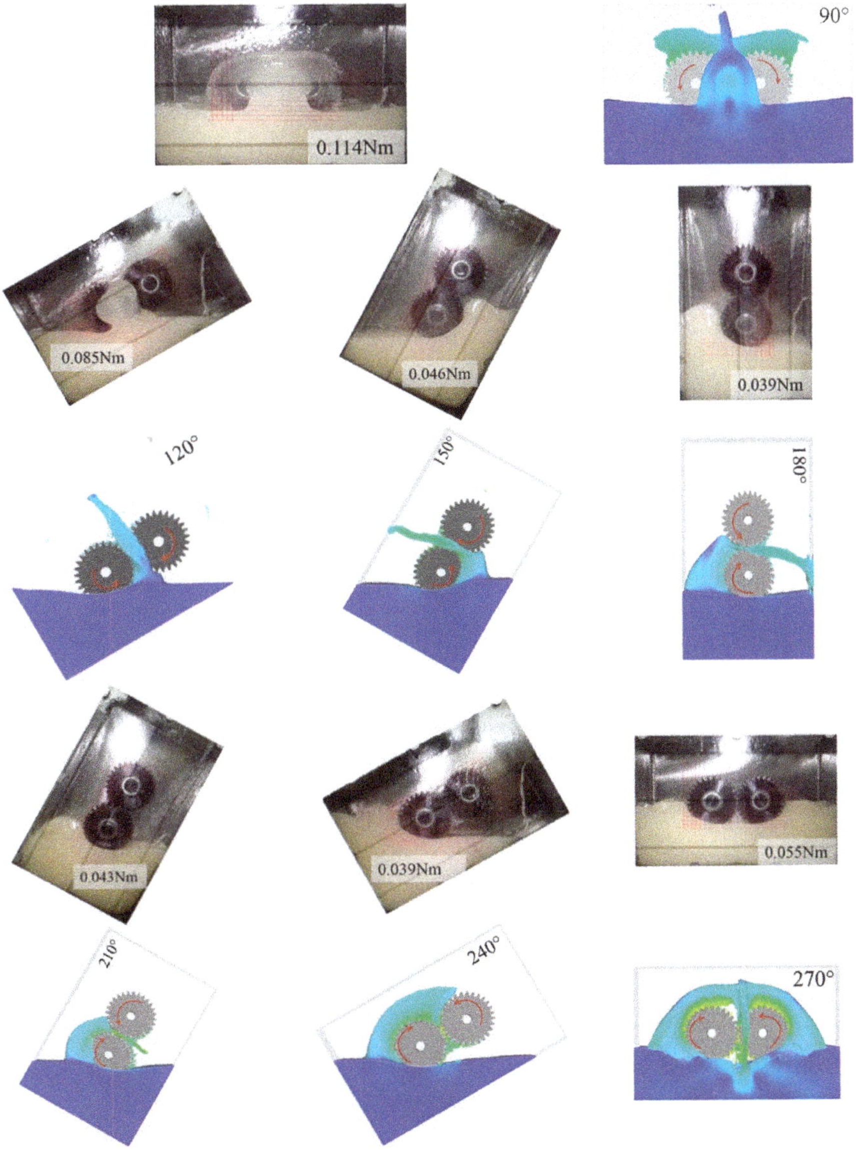

Figure 4. Comparison between numerical and experimental results for the oil free surface with different inclination angles.

Figure 4 shows the comparison of the oil free surface inside the gearbox with different inclination angles obtained by the experiments and the corresponding simulations, from which a good qualitative agreement can be observed. Also, it is worth noting that the pictures and drawings from 180° to 270° are reversed in Figure 4.

From Figures 3 and 4, it can be inferred that the present numerical method demonstrates a good ability to predict churning power losses and the oil flow field inside gearboxes with inclination angles, thus, the feasibility and accuracy of the method can be verified.

4. Simulation Results and Analysis

The above analyses indicate that the inclination angle of the initial oil surface has an important influence on the internal flow field and churning losses inside the gearbox. In most real industry cases, such as transmissions of helicopters or automobiles, gearboxes move together with these mechanical devices, therefore, it is necessary to study the effect of dynamic motion of gearboxes on flow field and churning losses.

To simultaneously simulate the motion or dynamic attitude of gearboxes and the movement of oil inside gearboxes, the non-inertial reference frame method was employed; this method inserts the computational mesh in a non-inertial reference frame, which moves with the moving gearboxes and applies fictitious forces to fluid for considering the non-inertial effects.

As shown in Figure 5, supposing that a gearbox moves towards the negative x-axis direction sinusoidally, its acceleration can be given by

$$a = -\omega^2 \times d \times \sin(\omega \times t - \phi) \tag{9}$$

where ω is the angular frequency (rad/s), d is the magnitude and ϕ is the phase angle (rad).

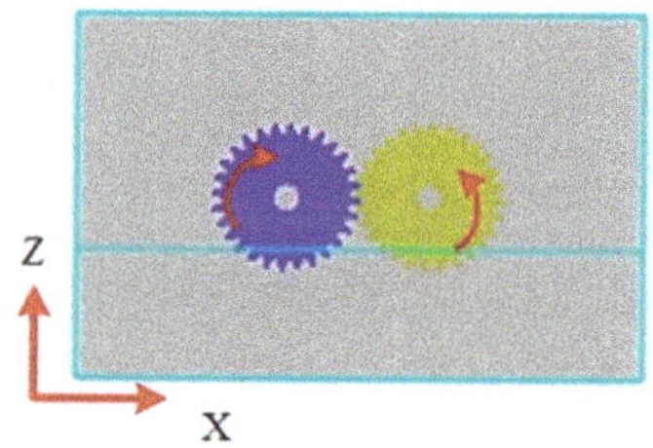

Figure 5. Model of a gearbox in sinusoidal motion.

To investigate the effect of gear speed n, oil immersion depth H, magnitude d and frequency f of the sinusoidal motion on churning power losses inside the moving gearbox, five different rotational speeds, four different oil immersion depths, two frequencies and magnitude of the sinusoidal motion of the gearboxes were employed to form eighteen parameter combinations, which are shown in Table 3. It is noted that the parameter combination No. 0 is also called "static simulation" below. In addition, for real industrial applications, the magnitude 60 cm may be too large a value because it would cause the maximum acceleration to reach 0.6 g (gravity acceleration), which is more valid for aircraft than for automotive applications, however, to clearly observe the influence of the dynamic motion of the gearbox on churning power losses, the magnitude 60 cm was used.

A comparison was made between the churning power losses of gearboxes with and without motion of gearboxes to reveal the influence mechanism. Figure 6 shows the churning loss torque of the gearbox with parameter combination No. 0 and No. 1, and it indicates that the internal lubricating oil sloshing caused by the motion of the gearbox has a significant influence on the churning losses. Because of the sinusoidal motion of the gearbox, the oil free surface inside the gearbox dynamic changes at all times, which causes the oil immersion depth to change correspondingly. Previous researchers have pointed out that an increase in the oil immersion depth would result in an increase in churning losses [3,5,14,28]. Figure 6 also reveals that the overall trend in the curves of the churning loss torque fluctuates greatly with the motion of the gearboxes, and the fluctuation amplitude is much larger than that at rest. Moreover, as shown in Figure 7, the average value of the churning power torque of the former is much larger than the latter, which indicates that the dynamic motion of the gearbox can exert a significant influence on churning power losses.

Table 3. Parameter combinations for simulations.

No.	Speed, n (rpm)	Immersion Depth, H (mm)	Frequency, f	Magnitude, d (cm)	Oil Temperature (°C)	β (°)
0	500	9 (2h)	0	0	25	270
1	500	9 (2h)	π	60	25	270
2	1000	9 (2h)	π	60	25	270
3	1500	9 (2h)	π	60	25	270
4	2000	9 (2h)	π	60	25	270
5	2500	9 (2h)	π	60	25	270
6	500	4.5 (h)	π	60	25	270
7	500	13.5 (3h)	π	60	25	270
8	500	18 (4h)	π	60	25	270
9	500	9 (2h)	2π	60	25	270
10	1000	9 (2h)	2π	60	25	270
11	1500	9 (2h)	2π	60	25	270
12	2000	9 (2h)	2π	60	25	270
13	2500	9 (2h)	2π	60	25	270
14	500	9 (2h)	π	30	25	270
15	1000	9 (2h)	π	30	25	270
16	1500	9 (2h)	π	30	25	270
17	2000	9 (2h)	π	30	25	270
18	2500	9 (2h)	π	30	25	270

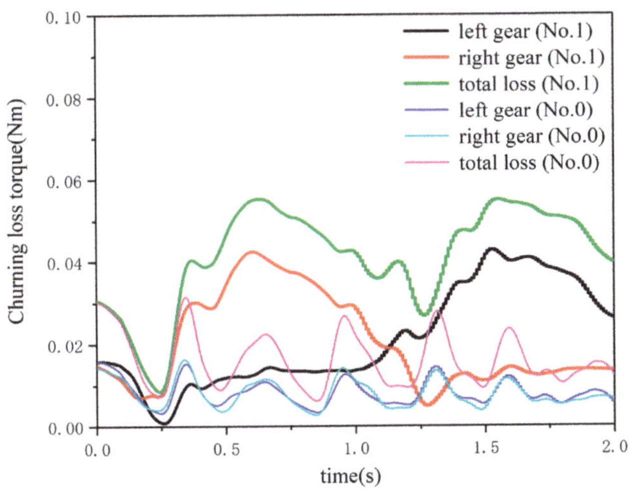

Figure 6. Time history curves of churning loss torques of parameter combination No. 0 and No. 1.

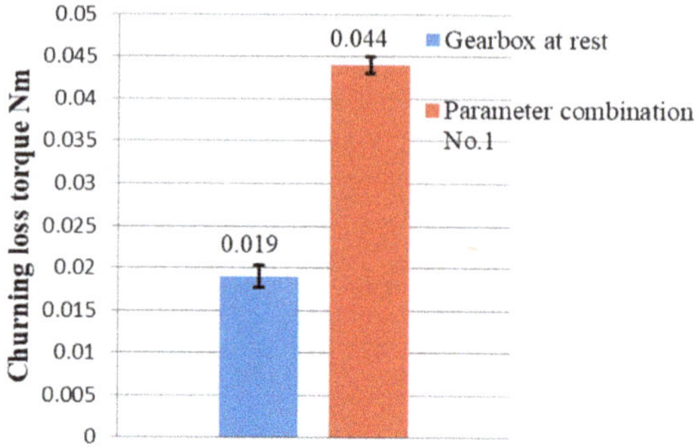

Figure 7. Average churning loss torque with and without motion of the gearbox.

Figure 8 shows the oil free surface and velocity distribution of the flow field inside the gearbox at four time points. At the initial point, the oil level is in the horizontal position (the inclination angle β of the gearbox is 270°). At time = 0.54 s, with the moving of the gearbox, oil moves towards the right housing wall of the gearbox, which causes the oil immersion depth of the left gear to decrease significantly while that of the right gear increases, it even becomes fully immersed; therefore, at this moment, the churning loss torque of the left gear is smaller than that of the right gear. At time = 1.5 s, the reverse phenomenon can be observed.

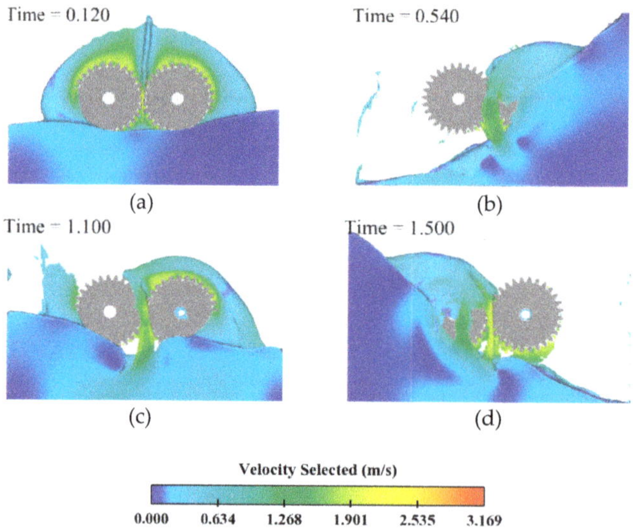

Figure 8. Oil free surface and velocity distribution at different time points. (**a**) Time = 0.120 s; (**b**) Time = 0.540 s; (**c**) Time = 1.100 s; (**d**) Time = 1.500 s.

Figures 9 and 10 show the time history curves and average values of churning power torques with five parameter combinations, No. 1, 2, 3, 4, and 5, of which only the rotational speeds are different. As can been seen in Figure 8, the curves of the churning power torque show a trend of fluctuations, and the curve fluctuation becomes more severe as the gear speed increases. It is noted that there is an obvious minimum value at about 1.1 s–1.25 s; this phenomenon can be interpreted in terms of the internal flow field inside the gearbox. Taking the case with a rotational speed of 1000 rpm as an example, as shown in Figure 11, it can be seen that both gears are in an un-immersed state with little oil lubrication when the time is about 1.1 s, and the lubricating oil is concentrated at the bottom and side walls of the gearbox. If the gears are in a loss-of-lubrication operating condition for a long time, their heat dissipation performance would deteriorate, and thus result in a dramatic increase in tooth temperature, even scuffing failure.

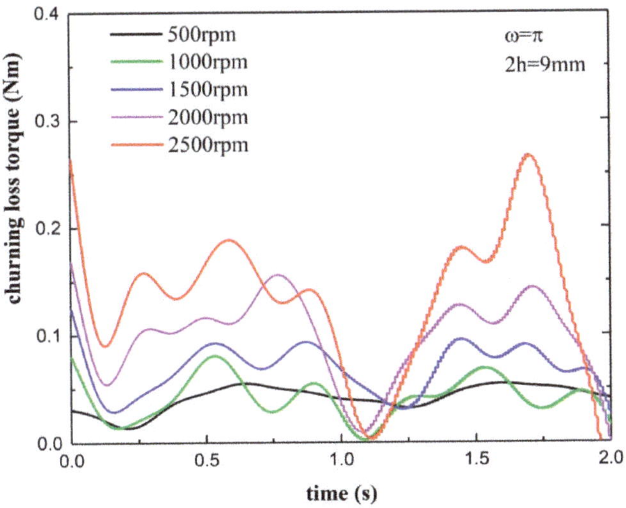

Figure 9. Time history curve of churning loss torque with rotational speeds.

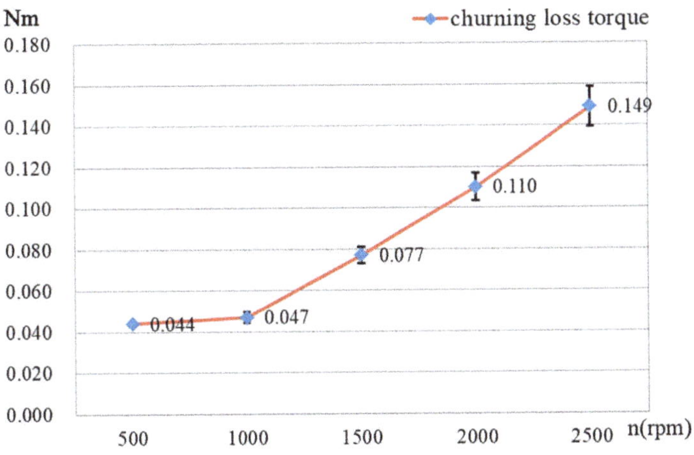

Figure 10. Average churning loss torques with rotational speeds.

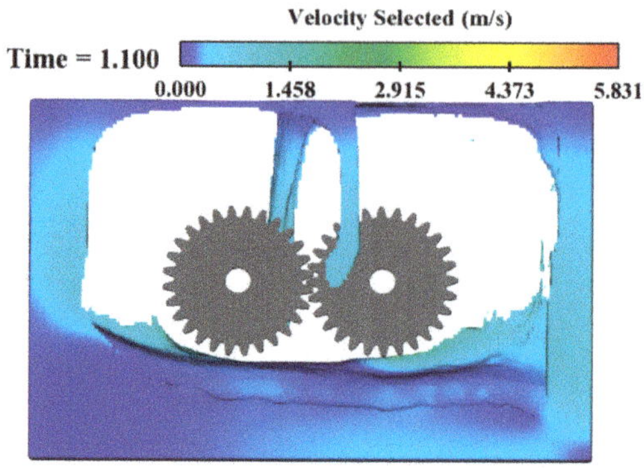

Figure 11. Oil flow field and velocity distribution inside the gearbox (n = 1000 rpm).

To quantitively indicate the starved-oil degree, a probe was placed at the gear meshing zone to monitor the oil fraction. Figure 12 shows that the oil fraction at the probe does not exceed 40% for most of the time, and there is no lubricating oil in the meshing zone at some moments, which is unfavorable to the lubrication of the gears. Combined with Figure 11, it can be inferred that when churning losses are approximate to zero, gears are almost in a loss-of-lubrication operating condition with no lubricating oil in the meshing zone.

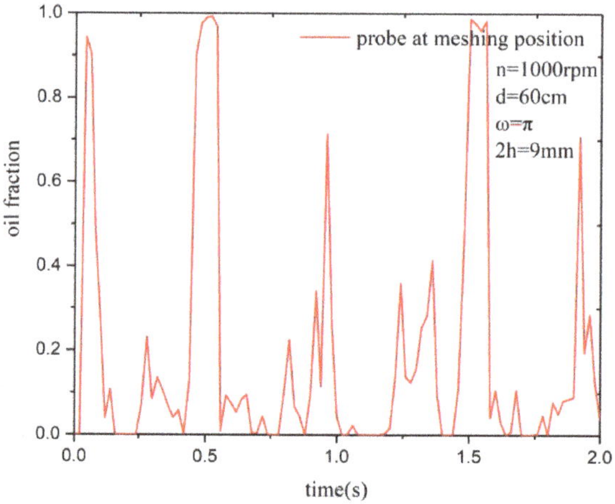

Figure 12. Oil fraction of the probe at the gear meshing zone.

To study the effect of the oil viscosity on churning power losses, a different lubricating oil (VG32), the viscosity of which is a much lower than that of lubricating oil VG320, was used for working condition No. 14 and No. 15. The basic parameters of the lubricating oil are shown in Table 4. Figure 13 reveals the churning loss torque and the oil fraction in the meshing zone with lubricating oil VG32 and VG320. From Figure 13a,b, it can be seen that the churning power losses with VG320 are more significant than with VG32 at the initial moment for both cases at 500 rpm or 1000 rpm. With a rotational speed of 500 rpm, in the 2 s operation process, the churning loss torque with VG32 is obviously less than that with VG320; however, when the rotational speed is 1000 rpm, no great difference between the curves for VG32 and VG320 can be observed. Therefore, oil viscosity can affect the churning power losses at a lower speed, while it has less influence at a higher speed, and thus, the influence of rotational speed is obviously greater than that of oil viscosity.

Table 4. Physical and chemical properties of lubricating oil VG32.

Oil	Gear Oil ISO VG32
Kinetic viscosity v (40 °C)	33.03 cst
Kinetic viscosity v (100 °C)	5.718 cst
Density ρ_{oil}	0.8727 g/cm^3

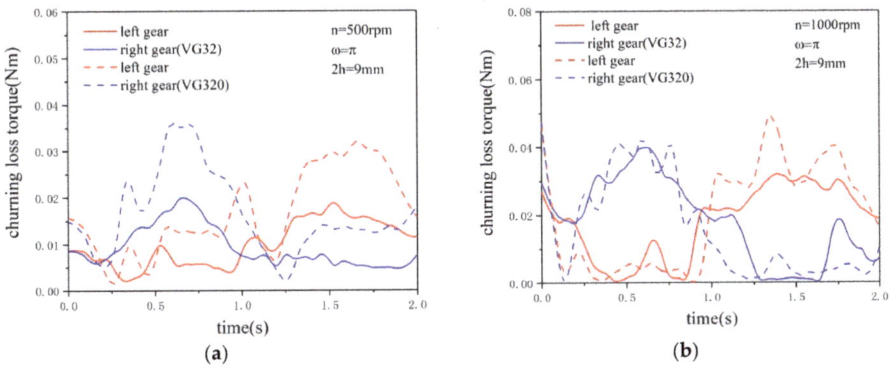

Figure 13. Influence of oil viscosity. (**a**) Cases with n = 500 rpm; (**b**) Cases with n = 1000 rpm.

Figure 14 shows the time history curves of churning loss torques with different oil immersion depth. The overall trend is that with increasing oil immersion depth, the churning loss torque increases, while the fluctuation amplitude of its curves decreases. The greatest fluctuation occurs when the oil immersion depth is h, and the fluctuation is gentle when the oil immersion depth is $4h$, which indicates that gearboxes with a low oil immersion depth are more susceptible to their own motion.

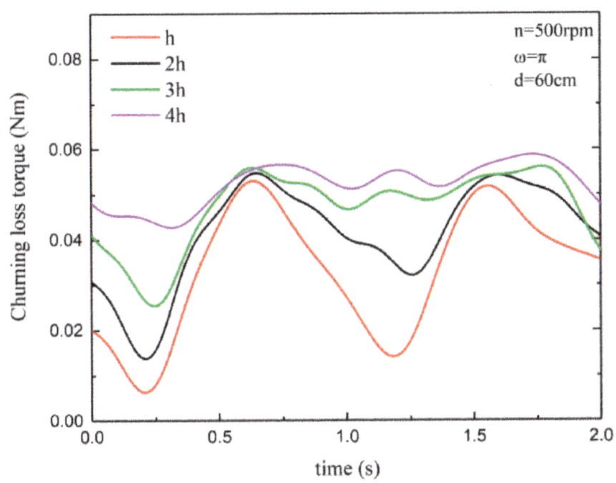

Figure 14. Churning loss torque of gears with different oil immersion depth.

To study the effect of the frequency of the sinusoidal motion of the gearbox on churning power losses, simulations of gears at four different rotational speeds with two different frequencies were carried out, the results of which are shown in Figure 15. On one hand, when the frequency of the sinusoidal motion of the gearbox increases from π to 2π, the number of peaks of the curves of churning loss torques also increases correspondingly. On the other hand, as the rotational speed of the gears increases, the peaks of curves with a frequency 2π are larger than those with a frequency π, and at higher rotational speeds, peaks of the churning loss curves of the two gears occur simultaneously. It is worth noting that the increase in frequency reduces the time interval in which the gears are in a loss-of-lubrication state.

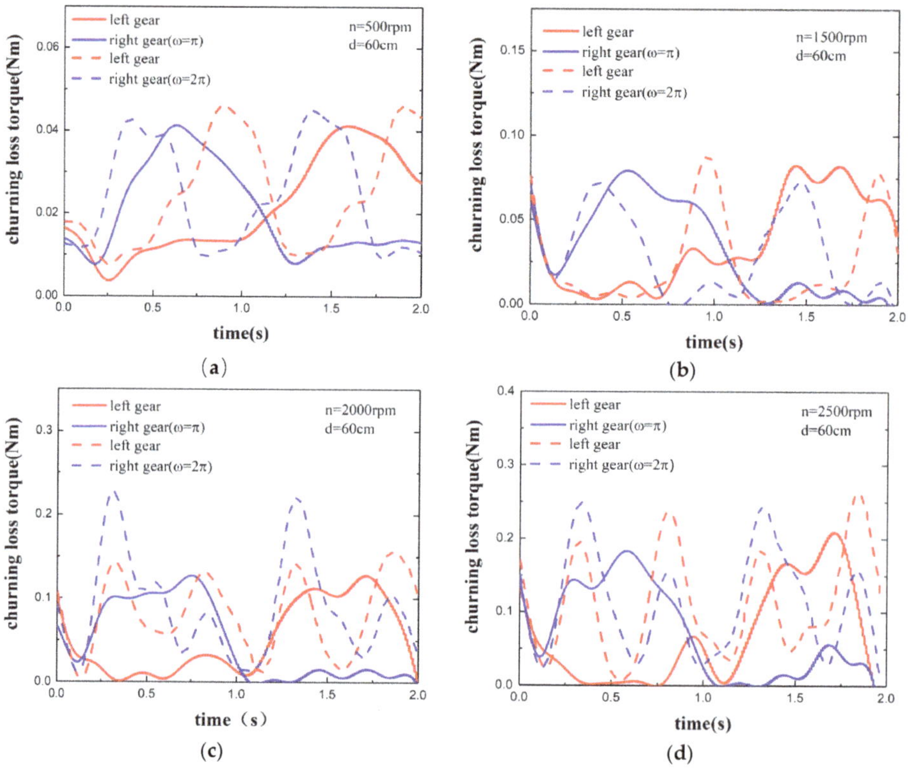

Figure 15. Influence of the frequency of the sinusoidal motion at different rotating speeds. (**a**) Cases with $n = 500$ rpm; (**b**) Cases with $n = 1500$ rpm; (**c**) Cases with $n = 2000$ rpm; (**d**) Cases with $n = 2500$ rpm.

Figure 16 shows the churning loss torque of gears with different magnitudes of the sinusoidal motion of gears at four different speeds. The churning loss torque curves of gears inside the gearbox with different motion magnitudes have a similar shape and trend. Specifically, the value of the churning loss torque with a motion magnitude of 30 cm is slightly smaller than that with 60 cm; the peaks of curves occur at the same time, and the peak value of the churning loss torque decreases with the decrease of magnitude.

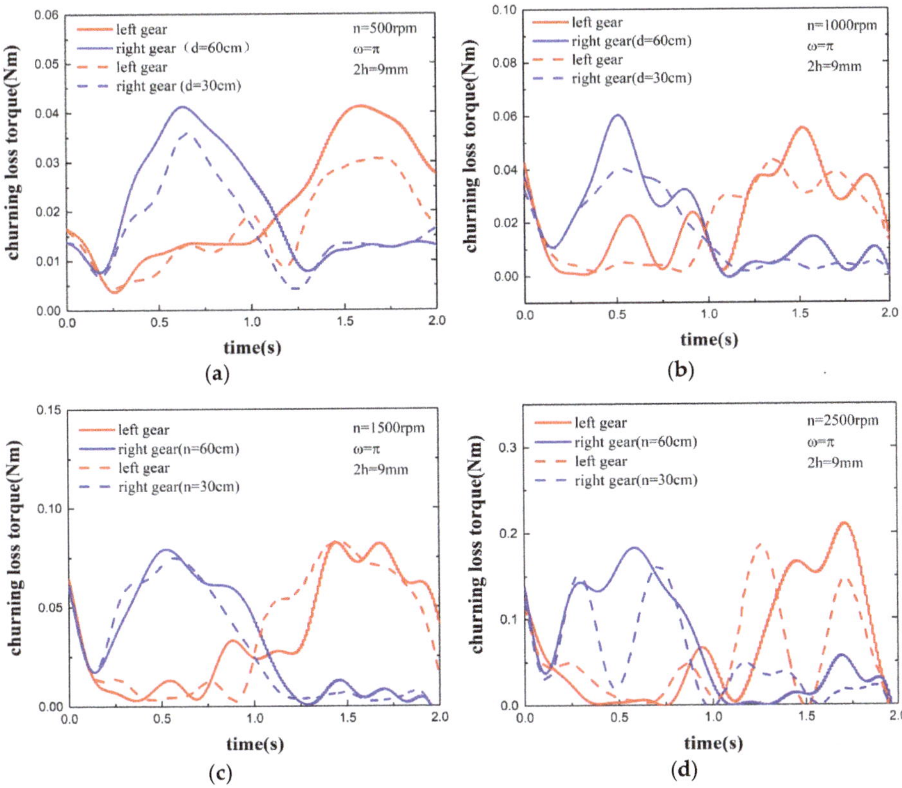

Figure 16. Influence of the magnitude of the sinusoidal motion at different rotating speeds. (**a**) Cases with $n = 500$ rpm; (**b**) Cases with $n = 1000$ rpm; (**c**) Cases with $n = 1500$ rpm; (**d**) Cases with $n = 2500$ rpm.

Figure 17 shows the churning loss torque value under different simulation conditions. An obvious upward trend of churning loss torque can be observed with the increase in the rotational speed of gears. When the frequency is π, the churning loss torque slightly decreases as the magnitude of the sinusoidal motion increases, and is close to that of static simulation. However, when the frequency is 2π, the churning loss torque significantly increases compared with other conditions at the same rotational speed. Therefore, Figure 17 also reveals that the frequency of the sinusoidal motion of the gearbox and rotational speed of gears have a great influence on churning losses, while the magnitude of the sinusoidal motion of the gearbox and oil immersion depth of gears have less influence.

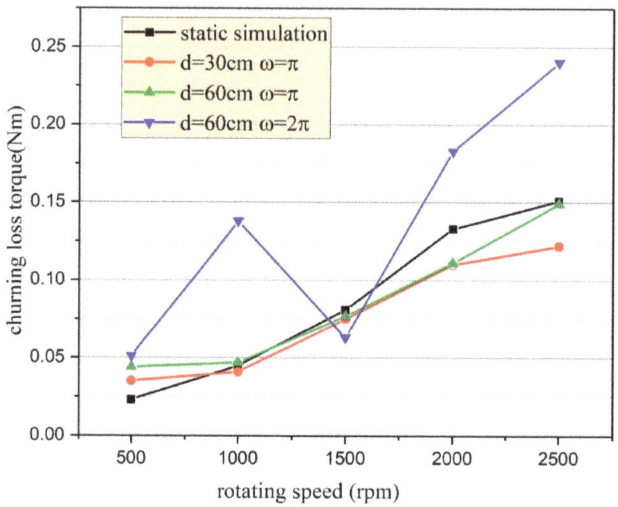

Figure 17. Churning loss torques under different simulation conditions.

5. Conclusions

In this study, a FAVOR method combined with an on-inertial coordinate system were applied to CFD simulations to calculate churning losses inside a gearbox with sinusoidal motion. The feasibility and accuracy of the present numerical method was verified by available test results. The effects of oil immersion depth, rotational speed, the frequency and magnitude of the sinusoidal motion of the gearbox were investigated, and the following conclusions are drawn:

(1) The sinusoidal motion of the gearbox causes the lubricating oil to oscillate, and the curves of the churning losses of the gears fluctuates with the fluctuation of the lubricating oil. When the lubricating oil tilts to the right side, the churning loss torque of the right gear is larger than that of the left gear loss, and vice versa. Due to the sloshing of lubricating oil, the gear pair may briefly be in a loss-of-lubrication state.

(2) The churning loss torque increases as the rotational speed of the gears increases, and the fluctuation amplitude increases as well. The churning loss torque increases as the oil immersion depth increases, while the fluctuation becomes gentle as the immersion depth increases.

(3) An increase in frequency reduces the time interval during which gears are in an un-immersed state, and the peak value of the churning loss torque and the number of the peak value increases. The peak value of churning loss torque decreases slightly when the magnitude decreases, and the peak for different magnitudes occurs at almost the same time.

Author Contributions: All authors contributed to this work. Conceptualization, X.H.; Methodology, P.L. and X.H.; Modelling and simulations, P.L. and M.W.; Analysis and discussion, P.L. and X.H.; Writing—Original Draft Preparation, P.L. and M.W.; Writing & Editing, X.H. and M.W.

Funding: This research was funded by the National Key Research and Development Project of China grant number 2018YFB2001500.

Acknowledgments: The authors are grateful to the National Key Research and Development Project of China (Grant No. 2018YFB2001500) for financial support of the work presented in this article.

Conflicts of Interest: The authors declare that there are no conflicts of interest regarding the publication of this article.

References

1. Concli, F.; Gorla, C. Numerical modeling of the power losses in geared transmissions: Windage, churning and cavitation simulations with a new integrated approach that drastically reduces the computational effort. *Tribol. Int.* **2016**, *103*, 58–68. [CrossRef]
2. Michaelis, K.; Höhn, B.R.; Hinterstoißer, M. Influence factors on gearbox power loss. *Ind. Lubr. Tribol.* **2011**, *63*, 46–55. [CrossRef]
3. Neurouth, A.; Changenet, C.; Ville, F.; Octrue, M. Is splash lubrication compatible with efficient gear units for high-speed applications? In Proceedings of the 2014 International Gear Conference, Lyon, France, 24–28 August 2014; pp. 1060–1068.
4. Seetharaman, S.; Kahraman, A.; Moorhead, M.D.; Petry-Johnson, T.T. Oil churning power losses of a gear pair: Experiments and model validation. *J. Tribol.* **2009**, *131*, 022202. [CrossRef]
5. Seetharaman, S.; Kahraman, A. Load-independent spin power losses of a spur gear pair: Model formulation. *J. Tribol.* **2009**, *131*, 022201. [CrossRef]
6. Changenet, C.; Velex, P. Housing influence on churning losses in geared transmissions. *J. Mech. Des. N. Y.* **2008**, *130*, 062603. [CrossRef]
7. Kolekar, A.S.; Olver, A.V.; Sworski, A.E.; Lockwood, F.E. Windage and churning effects in dipped lubrication. *J. Tribol.* **2014**, *136*, 021801. [CrossRef]
8. Liu, H.; Jurkschat, T.; Lohner, T.; Stahl, K. Detailed investigations on the oil flow in dip-lubricated gearboxes by the finite volume CFD Method. *Lubr.* **2018**, *6*, 47. [CrossRef]
9. Jiang, Y.Y.; Hu, X.Z.; Hong, S.J.; Li, P.P.; Wu, M.G. Influences of an oil guide device on splash lubrication performance in a spiral bevel gearbox. *Tribol. Int.* **2019**, *136*, 155–164. [CrossRef]
10. Terekhov, A.S. Hydraulic losses in gearboxes with oil immersion. *Russ. Eng. J.* **1975**, *55*, 7–11.
11. Boness, R.J. Churning losses of discs and gears running partially submerged in oil. In Proceedings of the ASME International Power Transmission Gearing Conference, Chicago, IL, USA, 24–27 April 1989; Volume 1, pp. 355–359.
12. Luke, P.; Olver, A.V. A study of churning losses in dip-lubricated spur gears. *Proc. Inst. Mech. Eng. Part G J. Aerosp. Eng.* **1999**, *213*, 337–346. [CrossRef]
13. Changenet, C.; Velex, P. A model for the prediction of churning losses in geared transmissions—Preliminary results. *J. Mech. Des. N. Y.* **2007**, *129*, 128–133. [CrossRef]
14. Höhn, B.R.; Michaelis, K.; Otto, H.P. Influence of immersion depth of dip lubricated gears on power loss, bulk temperature and scuffing load carrying capacity. In Proceedings of the International Conference on Mechanical Transmissions, Chongqing, China, 26–30 September 2006; Volume 1–2, pp. 837–845.
15. Neurouth, A.; Changenet, C.; Ville, F.; Octrue, M.; Tinguy, E. Experimental investigations to use splash lubrication for high-speed gears. *J. Tribol.* **2017**, *139*, 061104. [CrossRef]
16. Polly, J.; Talbot, D.; Kahraman, A.; Singh, A.; Xu, H. An Experimental Investigation of Churning Power Losses of a Gearbox. *J. Tribol.* **2018**, *140*, 031102. [CrossRef]
17. Gorla, C.; Concli, F.; Stahl, K.; Höhn, B.R.; Michaelis, K.; Schultheiss, H.; Stemplinger, J.P. Hydraulic losses of a gearbox: CFD analysis and experiments. *Tribol. Int.* **2013**, *66*, 337–344. [CrossRef]
18. Concli, F.; Della, T.A.; Gorla, C.; Montenegro, G. A new integrated approach for the prediction of the load independent power losses of gears: Development of a mesh-handling algorithm to reduce the CFD simulation time. *Adv. Tribol.* **2016**, *2016*, 2957151. [CrossRef]
19. Concli, F.; Gorla, C. CFD simulation of power losses and lubricant flow in gearboxes. In Proceedings of the American Gear Manufacturers Association Fall Technical Meeting (AGMA), Columbus, Ohio, USA, 22–24 October 2017.
20. Concli, F.; Gorla, C. Numerical modeling of the churning power losses in planetary gearboxes: An innovative partitioning-based meshing methodology for the application of a computational effort reduction strategy to complex gearbox configurations. *Lubr. Sci.* **2017**, *29*, 455–474. [CrossRef]
21. Liu, H.; Jurkschat, T.; Lohner, T.; Stahl, K. Determination of oil distribution and churning power losses of gearboxes by finite volume CFD method. *Tribol. Int.* **2017**, *109*, 346–354. [CrossRef]
22. Liu, H.; Arfaoui, G.; Stanic, M.; Montigny, L.; Jurkschat, T.; Lohner, T.; Stahl, K. Numerical modeling of oil distribution and churning gear power losses of gearboxes by smoothed particle hydrodynamics. *Proc. Inst. Mech. Eng. Part J* **2019**, *233*, 74–86. [CrossRef]

23. Peng, Q.L.; Gui, L.J.; Fan, Z.J. Numerical and experimental investigation of splashing oil flow in a hypoid gearbox. *Eng. Appl. Comput. Fluid Mech.* **2018**, *12*, 324–333. [CrossRef]
24. Lemfeld, F.; Fran, K.; Unger, J. Numerical simulations of unsteady oil flows in the gearboxes. *J. Appl. Sci. Thermodyn. Fluid Mech.* **2007**, *1*, 1–5.
25. Chen, S.W.; Matsumoto, S. Influence of Relative Position of Gears and Casing Wall Shape of Gear Box on Churning Loss under Splash Lubrication Condition-Some New Ideas. *Tribol. Trans.* **2016**, *59*, 993–1004. [CrossRef]
26. Hu, X.Z.; Jiang, Y.Y.; Luo, C.; Feng, L.F.; Dai, Y. Churning power losses of a gearbox with spiral bevel geared transmission. *Tribol. Int.* **2019**, *129*, 398–406. [CrossRef]
27. Hirt, C.W.; Sicilian, J.M. A porosity technique for the definition of obstacles in rectangular cell meshes. In Proceedings of the 4th International Conference on Numerical Ship Hydrodynamics, Washington, DC, USA, 24–27 September 1985; pp. 1–19.
28. Flow Science, Inc. *Flow 3D User Manual*; Version 11.2; Flow Science, Inc.: Leland, NC, USA, 2016.
29. Hirt, C.W.; Nichols, B.D. Volume of fluid (VOF) method for the dynamics of free boundaries. *J. Comput. Phys.* **1981**, *39*, 201–225. [CrossRef]

energies

MDPI

Article

Large Eddy Simulations of the Flow Fields over Simplified Hills with Different Roughness Conditions, Slopes, and Hill Shapes: A Systematical Study

Zhenqing Liu [1,*], Yiran Hu [1] and Wei Wang [2]

1 School of Civil Engineering & Mechanics, Huazhong University of Science and Technology, Wuhan 430074, China
2 Department of Architecture and Building Engineering, Tokyo Institute of Technology, Yokohama, Kanagawa 1528550, Japan
* Correspondence: liuzhenqing@hust.edu.cn

Received: 15 August 2019; Accepted: 30 August 2019; Published: 4 September 2019

Abstract: Turbulent flow fields over topographies are important in the area of wind energy. The roughness, slope, and shape of a hill are important parameters affecting the flow fields over topographies. However, these effects are always examined separately. The systematic investigations of these effects are limited, the coupling between these effects is still unrevealed, and the turbulence structures as a function of these effects are still unclear. Therefore, in the present study, the flow fields over twelve simplified isolated hills with different roughness conditions, slopes, and hill shapes are examined using large eddy simulations. The mean velocities, velocity fluctuations, fractional speed-up ratios, and visualizations of the turbulent flow fields are presented. It is found that as the hill slope increases, the roughness effects become weaker, and the roughness effects will further weaken as the hill changes from 3D to 2D. In addition, the fractional speed-up ratio at the summit of rough hills can even reach to three times as large as that over the corresponding smooth hills. Furthermore, the underestimation of the ratios of spanwise fluctuation to the streamwise fluctuation by International Electrotechnical Commission (IEC) 61400-1 is quite obvious when the hill shape is 3D. Finally, coherent turbulence structures can be identified for smooth hills, and as the hill slope increases, the coherent turbulence structures will experience clear evolutions. After introducing the ground roughness, the coherent turbulence structures break into small eddies.

Keywords: ground roughness; hill shape; hill slope; large-eddy simulations; turbulent flow fields

1. Introduction

Flow structures over complex topographies are featured by the flow separations and reattachments which strongly depend on the topographic aspects, including surface roughness conditions, slopes, and shapes. On the other hand, the turbulent flow fields over complex terrain are of a great interest for many applications, such as wind turbine sittings [1], pollution diffusions [2], estimation of aerodynamic loadings on structures [3], identifications of tree damage [4], and forest fire propagation [5]. As a starting point modeling flow fields over real complex terrains, great efforts have been made to clarify the turbulent boundary layer (TBL) over simplified isolated hills. Two simplified isolated hills, i.e., two-dimensional (2D) and three-dimensional (3D) hills have been extensively examined in wind-tunnel experiments and numerical simulations.

For the roughness effects, Britter et al. experimentally found that at the lee side of the hill the roughness significantly alters the flow [6]. Then Pearse et al. examined the flow over several triangular and bell-shaped hills [7]. Importantly, it was found that increasing the surface roughness results in an increasing of amplification factors of the velocities at the hill crests. Using large eddy simulations (LES), Brown et al. concluded that the critical slope for the separation will increase as the ground roughness increases [8]. Moreover, the flow over a series of sinusoidal hills was investigated in wind tunnel experiments by Athanassiadou and Castro [9], who concluded that the effective roughness length is much greater than the length of the individual roughness blocks covering the hill. Importantly, it was pointed out by Finnigan and Belcher that even the canopy density is very limited [10], the canopy continues to play an important role in the flow above the canopy. Ross and Vosper found that the linear analytical solutions for the flow over a low and wide forested hill are still applicable [11]. From the studies by Cao and Tamura [12,13], it was determined that the speed-up ratio above the crest of a rough hill is larger than that of a smooth hill, same as the study by Pearse et al. [7], and it was also found that adding or removing ground roughness will create completely different turbulence structure in the wake, consistent with the study by Britter et al. [6]. Adopting LES, a constant mixing-length assumption was found not to be strictly valid within the canopy by Ross [14], the same conclusion reached in a recent study by Okaze et al. [15], and the structure of the turbulence over a forested hill was found to be broadly similar to that over flat ground, with sweeps and ejections dominating. Then, Loureiro et al. developed a consistent theory on the flow over rough ground using water-tank experiments [5,16]. Furthermore, the surface roughness was found to have a great influence on the flow separation point which occurs earlier with rougher surface leading to a larger recirculation area, in the studies by Takahashi et al. [17], Cao and Tamura [12], Tamura et al. [18], Loureiro et al. [5], and Cao et al. [19]. Most recently, Liu et al. concluded that with intent to capture the turbulent characteristics accurately the horizontal gird size should be at least as large as the height of the roughness canopy [20].

For the hill slope effects, Finardi et al. numerically found that the Cartesian coordinates and a brick-like terrain mesh are effective for modeling the flow over steep topographies [21]. Adopting a Reynolds stress model, Ying et al. concluded that a domain containing steeply-sloped topography requires higher horizontal resolutions for improving computational stability [3], which was also confirmed in the recent studies by Hu et al. and Ma and Liu [22,23]. Ferreira et al. found that the size of the recirculating region is strongly dependent on the hill slope and a quite evident growth of the recirculation bubble can be identified as the hill slope increases [24]. The wind tunnel experiments by Neff and Meroney and Athanassiadou and Castro as well as the numerical simulations by Griffiths and Middleton and Cao et al. showed that the flow separation occurs in the steep hill while the flow remains attached over the low-slope hill [9,19,25,26]. Most recently, the hills with different slopes were examined by Pirooz and Flay [27], who found that in comparison with the more peaked hill crests, the flat-topped hills would have a lower speed-up.

For the hill shape effects, Ishihara et al. experimentally found that the cavity zone behind a 3D hill is smaller than that behind the 2D hill attributing to the convergence of the flow in the wake of the 3D hill [28]. It was also found that the lateral velocity variances behind the 3D hill provide a secondary local maximum in the wall layer. This phenomenon was not observed in the 2D hill. Lubitz and White measured the flow past the hills with different shapes [29], and it was observed that owing to the secondary flow around the 3D hill, the speed-up and the size of the wake region of the flow over a 3D hill decrease significantly when compared with that over a 2D hill. Liu et al. adopted LES and found that the spectra of the fluids in the wake are sensitive to the oncoming turbulence condition for the 2D hill [30], which is not true for the 3D hill. Most recently, different vortices were identified numerically by Ishihara and Qi [31], in which the roller vortices were found to be significant on the lee side of the 2D hill, while horseshoe vortices appear around the 3D hill.

However, the roughness effects, the slope effects, and the effects of the shape of the isolated hills are always examined separately. The systematic examinations of these effects are limited, the coupling of these effects remains unrevealed, and turbulence structures as a function of these effects remain unclear.

In the present LES, twelve isolated hills with different ground roughness conditions, hill slopes and hill shapes were chosen. The flow fields were systematically examined, including the mean velocities, root mean square of the velocity fluctuations, speed-up ratios, fluctuation ratios; in addition, different turbulence structures in the wake were analyzed through the visualization of Q-criteria. The effects of the ground roughness, hill slopes and hill shapes, as well as the coupling between them are revealed.

2. Numerical Model

2.1. Configurations

The numerical models are shown in Figure 1, where x, y, and z denote the streamwise, spanwise and vertical directions, respectively. The hills and the ridges are placed at (0, 0, 0). The shapes of the 3D hills are determined by:

$$\begin{cases} z_s(x,\ y) = h\cos^2\pi(x^2+y^2)^{1/2}/2L, & \text{if } (x^2+y^2)^{1/2} < L \\ z_s(x,\ y) = 0, & \text{if } (x^2+y^2)^{1/2} \geq L \end{cases} \tag{1}$$

where, L is a constant equaling to 100 mm, and h = 20 mm, 40 mm, and 80 mm. The shapes of the ridges are determined by:

$$\begin{cases} z_s(x,\ y) = h\cos^2\pi|x|/2L, & \text{if } |x| < L \\ z_s(x,\ y) = 0, & \text{if } |x| \geq L \end{cases} \tag{2}$$

where h = 20 mm, 40 mm, and 80 mm. Two vertical coordinates, z denoting the absolute height and $z' = z - z_s(x, y)$ denoting the height above the local surface, will be used for the convenience of discussion. Both the smooth and rough ground conditions are considered. For all of the rough cases, the height of the roughness canopy equals 5 mm. Varying the hill heights but constant radius is from the consideration that in a certain district, the higher the hill is, the steeper the hill is more likely to be in the real situation. And the constant height of roughness canopy is from the consideration that the height of the forest is nearly independent with the hill height in the real situation. The case settings are listed in Table 1 and the shape of the topographies are shown in Figure 2. Four of the cases are modelled to be in accordance with the Ishihara et al. [28] experiments for validation purposes. The wind tunnel experiments by Ishihara et al. [28] did not consider the effects of the hill slopes, which is one of the motivations of the present study.

Table 1. Case parameters.

Case Name.	Shape	Ground Condition	Height of Hill h (mm)	Validation Case
H2S	3D hill	smooth	20	
H4S	3D hill	smooth	40	√
H8S	3D hill	smooth	80	
H2R	3D hill	rough	20	
H4R	3D hill	rough	40	√
H8R	3D hill	rough	80	
R2S	2D ridge	smooth	20	
R4S	2D ridge	smooth	40	√
R8S	2D ridge	smooth	80	
R2R	2D ridge	rough	20	
R4R	2D ridge	rough	40	√
R8R	2D ridge	rough	80	

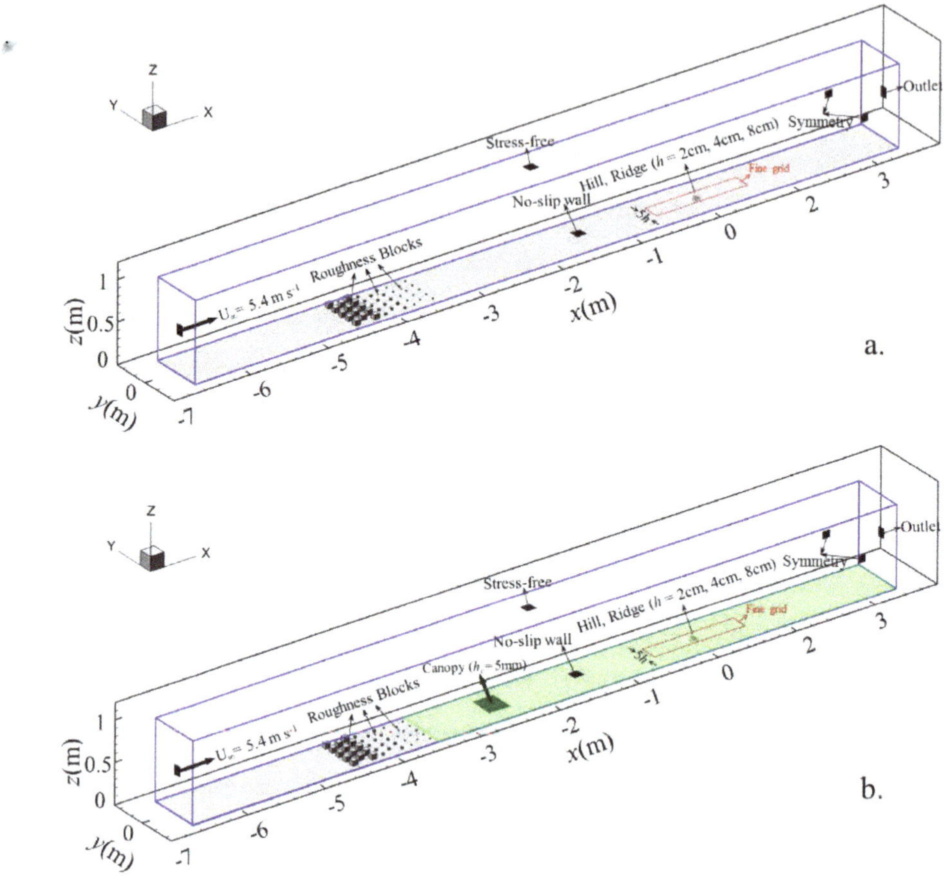

Figure 1. Sketch of the computational domain (**a**) smooth ground and (**b**) rough ground.

To best reproduce the experimental results, the configuration for our numerical model is set same as that in the wind tunnel experiments by Ishihara et al. [28], except the domain size in the spanwise direction, L_y and the upstream necking zone. In the experiment of a smooth 3D hill by Ishihara et al. [28], turbulent boundary layer (TBL) was simulated using two 60 mm high cubic arrays placed downstream of the contraction exit, followed by 20 mm and 10 mm cubic roughness elements, covering 1.2 m of the test-section floor. The remaining 5.8 m of the test section floor was covered with plywood, which was as smooth as the hill surface. The 3D hills or 2D ridges were mounted 4.6 m downstream of the contraction exit. In the simulation, the domain extends over (L_x, L_y, L_z) = (9, 0.65, 0.9) m^3 = (25, 1.8, 2.5) δ^3 = (90, 6.5, 9) L^3. Two nested domains (coarse and fine) are adopted, as illustrated by the red dashed lines in Figure 1. The fine-grid domain covers the range of (L_x', L_y', L_z') = (10, 2, 3) L^3 in the x-, y-, and z- directions, respectively. Both the upstream and the downstream fine-grid regions are 5L long.

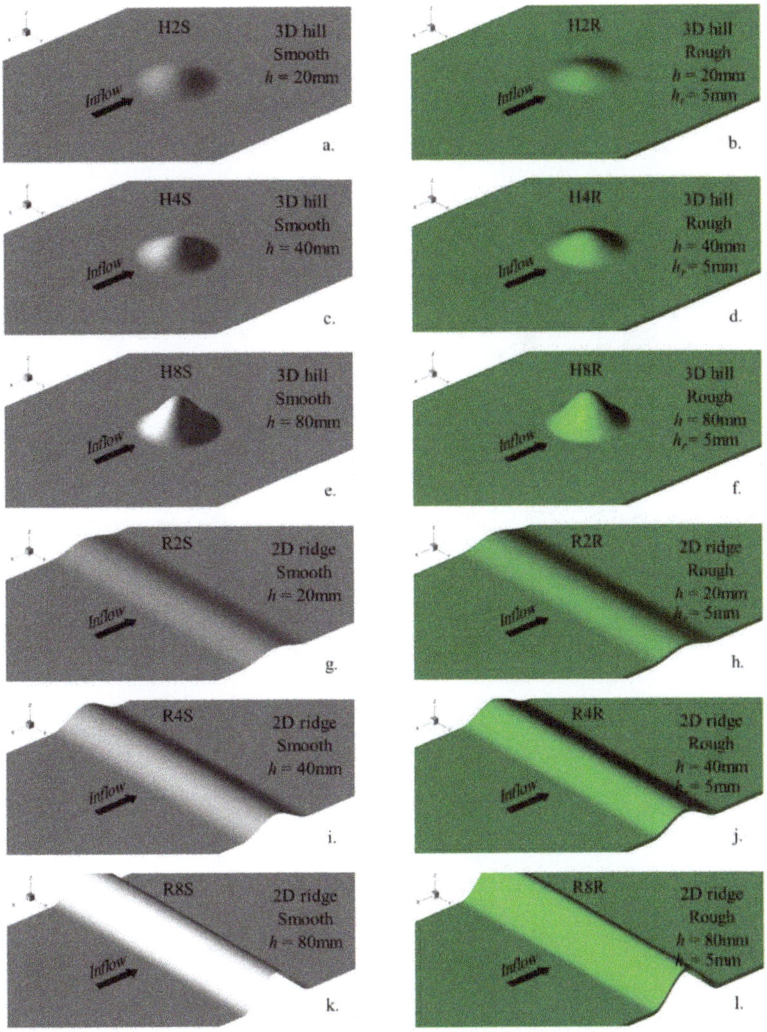

Figure 2. Configurations of the topographies. (**a**) 3D hill (smooth, h = 20 mm), (**b**) 3D hill (rough, h = 20 mm), (**c**) 3D hill (smooth, h = 40 mm), (**d**) 3D hill (rough, h = 40 mm), (**e**) 3D hill (smooth, h = 80 mm), (**f**) 3D hill (rough, h = 80 mm), (**g**) 2D ridge (smooth, h = 20 mm), (**h**) 2D ridge (rough, h = 20 mm), (**i**) 2D ridge (smooth, h = 40 mm), (**j**) 2D ridge (rough, h = 40 mm), (**k**) 2D ridge (smooth, h = 80 mm), (**l**) 2D ridge (rough, h = 80 mm).

2.2. Numerical Method

2.2.1. Governing Equations

In the LES strategy, large eddies are explicitly resolved, while the small eddies are parameterized by SGS models. The governing equations are usually obtained by filtering the time-dependent Navier-Stokes equations in Cartesian coordinates (x, y, z):

$$\frac{\partial \rho \tilde{u}_i}{\partial x_i} = 0 \tag{3}$$

$$\frac{\partial \rho \tilde{u}_i}{\partial t} + \frac{\partial \rho \tilde{u}_i \tilde{u}_j}{\partial x_j} = \frac{\partial}{\partial x_j}\left(\mu \frac{\partial \tilde{u}_i}{\partial x_j}\right) - \frac{\partial \tilde{p}}{\partial x_i} - \frac{\partial \tau_{ij}}{\partial x_j} \tag{4}$$

where $\tilde{u}_i$ and $\tilde{p}$ are the filtered velocity and pressure, respectively; μ is the viscosity; ρ is the density; and τ_{ij} is the SGS stress. To close the equations for the filtered velocities, a model for the anisotropic residual stress tensor τ_{ij} is needed, which is obtained as:

$$\tau_{ij} = -2\mu_t \tilde{S}_{ij} + \frac{1}{3}\tau_{kk}\delta_{ij} \tag{5}$$

$$\tilde{S}_{ij} = \frac{1}{2}\left(\frac{\partial \tilde{u}_i}{\partial x_j} + \frac{\partial \tilde{u}_j}{\partial x_i}\right) \tag{6}$$

where μ_t denotes the SGS turbulent viscosity; $\tilde{S}_{ij}$ refers to the rate-of-strain tensor for the resolved scale, and δ_{ij} is the Kronecker delta. The Smagorinsky-Lilly model is used to parameterize the SGS turbulent viscosity as:

$$\mu_t = \rho L_s^2 |\tilde{S}| = \rho L_s^2 \sqrt{2\tilde{S}_{ij}\tilde{S}_{ij}} \tag{7}$$

$$L_s = \min\left(\kappa d,\ C_s \Lambda^{1/3}\right) \tag{8}$$

where L_s stands for the SGS mixing length; κ represents the von Kármán constant (= 0.42); d is the distance to the closest wall, and Λ is the volume of a computational cell. Here, C_s denoting the Smagorinsky constant is set to a value of 0.1 as studied by Liu et al. [20]. Since z^+ is below 2 in all of the cases, the shear stresses are obtained from the viscous stress-strain relation:

$$\frac{\tilde{u}}{u_*} = \frac{\rho u_* \delta_n}{\mu} \tag{9}$$

where u_* is the friction velocity and δ_n is the distance between the cell centre and the wall.

2.2.2. Method Modeling Roughness

Accurate modeling of the inflow is essential for the success of simulating the flow in the 3D hill wake. In the present study, the arrangement of the upstream roughness blocks in the simulations is exactly same as that in the wind-tunnel experiments by Ishihara et al. [32]. In addition, the distance between the roughness blocks and the location of 3D hills placed is also set exactly same as that in the wind-tunnel experiment. In the volumes occupied by the roughness blocks, the drag force term is added:

$$\frac{\partial \rho \tilde{u}_i}{\partial t} + \frac{\partial \rho \tilde{u}_i \tilde{u}_j}{\partial x_j} = \frac{\partial}{\partial x_j}\left(\mu \frac{\partial \tilde{u}_i}{\partial x_j}\right) - \frac{\partial \tilde{p}}{\partial x_i} - \frac{\partial \tau_{ij}}{\partial x_j} + f_{u,i} \tag{10}$$

in which the drag force term $f_{u,i} = C_f A_f \rho \tilde{u}_i |\tilde{u}_i|$; since the velocity in the volume occupied by the roughness blocks should be zero, the drag coefficient C_f is set as 100 to model the drag effect from the solid roughness blocks; A_f is the frontal area of the roughness block.

In order to model the canopy, the drag force term, $f_{u,i}$ is added to the momentum equations in the same form as Equation (10). The drag force term is determined by $f_{u,i} = C_f A_f \rho \tilde{u}_i |\tilde{u}_i|$, where $C_f = 0.2$ is the drag force coefficient, $A_f = 0.6\ \text{m}^{-1}$ is the leaf-area density, and $|\tilde{u}_i|$ is the velocity magnitude. C_f and A_f are determined following the study by Liu et al. [20].

2.2.3. Boundary Conditions

With respect to the boundary conditions, a stress-free condition is applied at the top of the domain ($\partial\tilde{u}/\partial n = 0$, $\partial\tilde{v}/\partial n = 0$, $\tilde{w} = 0$) and the spanwise sides ($\partial\tilde{u}/\partial n = 0$, $\partial\tilde{w}/\partial n = 0$, $\tilde{v} = 0$). Uniform wind flow with a constant speed of 5.4 m·s^{-1} with time is set at the inlet ($\tilde{u} = 5.4$ m·s^{-1}, $\partial\tilde{p}/\partial n = 0$). The gradient-free boundary condition is set at outlet boundary ($\partial\tilde{u}_i/\partial n = 0$, $\partial\tilde{p}/\partial n = 0$). The no-slip condition is applied at the bottom surface ($\tilde{u}_i = 0$, $\partial\tilde{p}/\partial n = 0$). The settings of the boundary condition are summarized in Table 2.

Table 2. Settings of the boundary conditions.

Locations	Boundary Type	Expression
Outlet of the domain	Outflow	$\partial\tilde{u}_i/\partial n = 0, \partial\tilde{p}/\partial n = 0$
Lateral sides of the domain	Symmetry	$\partial\tilde{p}/\partial n = 0, \partial\tilde{u}/\partial n = 0, \partial\tilde{w}/\partial n = 0, \tilde{v} = 0$
Top of the domain	Symmetry	$\partial\tilde{p}/\partial n = 0, \partial\tilde{u}/\partial n = 0, \partial\tilde{v}/\partial n = 0,$
Inlet of the domain	Velocity inlet	$\tilde{v} = 5.4\ \mathrm{m\cdot s^{-1}}, \partial\tilde{u}/\partial n = 0, \partial\tilde{w}/\partial n = 0, \partial p/\partial n = 0$
Ground	Non-slip wall	$\partial\tilde{p}/\partial n = 0, \tilde{u}_i = 0$

2.2.4. Grid System

In terms of the grid size, in the vertical direction, the grid is stretched starting from a vertical grid spacing of 0.1 mm at the surface in both the fine-and coarse-grid domains in the smooth cases. When the ground is covered by vegetation canopy, 10 grids are adopted to divide the vertical space in the canopy. The resulted vertical grid resolution at the surface $z^+ < 1.0$ covers most of the domain except for the windward part of the hill where the maximum z^+ does not exceed 2. A horizontal grid size of 2.0 mm is used in the fine-grid region. In the rough-grid region, the horizontal grid shape is square, and a uniform grid size of 10 mm is applied. Downstream of the fine-grid domain and upstream of the roughness blocks, Δx is stretched at a ratio of 1.2. Total grid number is 2.2×10^7 for the smooth cases and 2.45×10^7 for the rough cases. The parameters determining the grid system are listed in Table 3.

Table 3. Parameters of the grid system.

Ground Condition	Δz_{min} in Fine Grid Domain (mm)	Δz_{min} in Coase Grid Domain (mm)	Horizontal Grid Resolution in Fine Grid Domain (mm)	Horizontal Grid Resolution in Coarse Grid Domain (mm)	Size of Fine Grid Domain L	Grid Number
Smooth	0.1	0.1	2	10	(12, 2, 3)	2.2×10^7
Rough	0.5	0.5	2	10	(12, 2, 3)	2.45×10^7

2.2.5. Solution Schemes

Finite volume method (FVM) is used in the present LES. The second-order central difference scheme is used for the convective and viscous terms, and the second-order implicit scheme is employed for the unsteady term [33]. Time-step size Δt is set as 0.0001 s, and in convective time units, $\Delta t^* = \Delta t U_\infty/L = 0.0058$. The Courant Friedrichs Lewy (CFL) number (Courant et al. 1928) is based on the time step size (Δt), velocity ($\tilde{u}_i$), and grid size (Δx_i), expressed as $C = \Delta t \Sigma \tilde{u}_i / \Delta x_i$. Here, the CFL number is limited to be less than 2 (i.e., $C_{max} = 2$) in the whole computational domain. The SIMPLE (semi-implicit pressure linked equations) algorithm, which was introduced by Ferziger & Peric [33], is applied to solve the discretized equations. The calculations are performed on 2PCs in parallel (Intel core i9-7980XE, 18 cores, 64G memory, Dell, Beijing, China). The simulation cost 2521 hours in total and the initial transient effects are found to disappear after 400 time units. Statistical convergence for the mean velocities is achieved when $|\langle\tilde{u}_{i-y}\rangle - \langle\tilde{u}_{i+y}\rangle/[(\langle\tilde{u}_{i-y}\rangle + \langle\tilde{u}_{i+y}\rangle)]/2| < 1\%$ in the near-wake of the 3D hill, which is over 350 time units, where $\langle\ \rangle$ means the time-averaging process, $-y$ and $+y$ mean the velocity at two points which are symmetrical position in lateral direction. For the fluctuations, u, v, and w, 350 time unit statistics still cannot show clear convergence, $10\% > |(u_{i-y} - u_{i+y})/[u_{i-y} + u_{i+y}]/2| > 5\%$, especially in the region $z < 1.0L$ for the streamwise component, which should be due to the much large turbulence in this region. However, the simulations of these 12 cases have cost us over 5 months based on the computational resources available in our group. Further increasing the statistical time can hardly be afforded by us. In the future, more detailed examination should be carried out using larger statistical time. Table 4 summarizes the numerical schemes in the present LES.

Table 4. Numerical schemes.

Time-discretization scheme	Second-order implicit scheme	Cs number	0.1
Space-discretization scheme	Finite-volume method Second-order central-difference scheme	SGS model	Smagorinsky-Lilly
Non-dimensional time step size: $\Delta t^* = \Delta t U_\infty / L$	0.0058	CFL number: $\Delta t \Sigma u_i / \Delta x_i$	<2
Turbulence model	LES	Software	ANSYS Fluent 14.0
Time for statistics	$400\Delta t^*$	Solution of the linearized equations	Preconditioned Conjugate Gradient + Algebraic Multigrid

2.3. Upcoming TBL

The resulted profiles of mean streamwise velocity, U, root mean square (r.m.s.) of streamwise velocity fluctuation, u, spanwise velocity fluctuation, v, and vertical velocity fluctuation, w, in the absence of the hills, are compared with those in the experiment by Ishihara et al. [28], as shown in Figure 3. U_{ref} determined as the mean streamwise velocity at the height of 0.16 m is applied to normalize the flow fields, which equals to 5.3 m·s^{-1} in the present LES and 5.2 m·s^{-1} in the wind-tunnel experiment. For the mean velocity profiles, there are some differences between the experimental data and those from the simulations. The difference is mainly below $z' = 0.1$ and $L = 10$ mm, where is very close to the ground. The introduction of the probe for measurement in the experiment will disturb the flow. That should be the reason of the clear differences between the experiment and the simulation. The difference between the simulation and the experiment above $0.1L$ is below 5% which is in the acceptable range of engineering applications. For the velocity fluctuations, the components in spanwise and the vertical directions from the simulations are comparable with those from the wind tunnel experiments. The maximum discrepancy is below 5%. However, for the streamwise component, relatively large discrepancies can be found in the region $0.6L < z < 1.0L$ and $0.1L < z < 0.4L$, and the maximum discrepancy reaches 10% at $z = 0.2L$. These discrepancies may result from the insufficient statistical time in the present simulations. In the wind tunnel experiments, the statistical time is10 s, which is over 2000 time units which is about 5 times as large as that in the simulation.

The boundary layer thickness in absence of the hills, δ, was about 0.36 m (both smooth and rough cases). The scale of the simulated boundary layer, λ, was about 1:1000, on the basis of the power spectra of the longitudinal velocity component. The wind speed outside the boundary layer was measured as $U_\infty = 5.8$ m·s^{-1}. As a result, the simulated boundary layer had a bulk Reynolds number:

$$Re_b = \frac{U_\infty \delta}{v} \tag{11}$$

which equals 1.4×10^5, where v was the kinematic viscosity of the air equaling 14.8×10^{-6} m^2·s^{-1}. The roughness height z_0 was 0.01 mm in the smooth cases and 0.3 mm in the rough cases. z_0 was determined by comparing the resulted mean streamwise velocity profile with the logarithmic law $U(z) = \frac{U_\tau}{\kappa} \ln \frac{z}{z_0}$, where $\kappa = 0.4$ was the Von-Kármán constant, $U_\tau = \sqrt{\tau/\rho} = 0.212$ m·s^{-1}, τ denotes the time averaged wall shear stress of the streamwise component. The main TBL parameters are summarized in Table 5.

Table 5. Parameters in the simulations for the smooth and rough ground conditions.

Ground Condition	Height of Grass h_r(mm)	Mean Wind Speed at the Contraction Exit U_{in}(m·s^{-1})	Boundary Layer Thickness δ(m)	Scale of the Boundary Layer λ	Wind Speed Outside the Boundary Layer U_∞(m·s^{-1})	Roughness Height z_0(mm)	Bulk Reynolds Number Re_b
Smooth	/	5.4	0.36	1:1000	5.8	0.01	1.4×10^5
Rough	5	5.4	0.36	1:1000	5.8	0.3	1.4×10^5

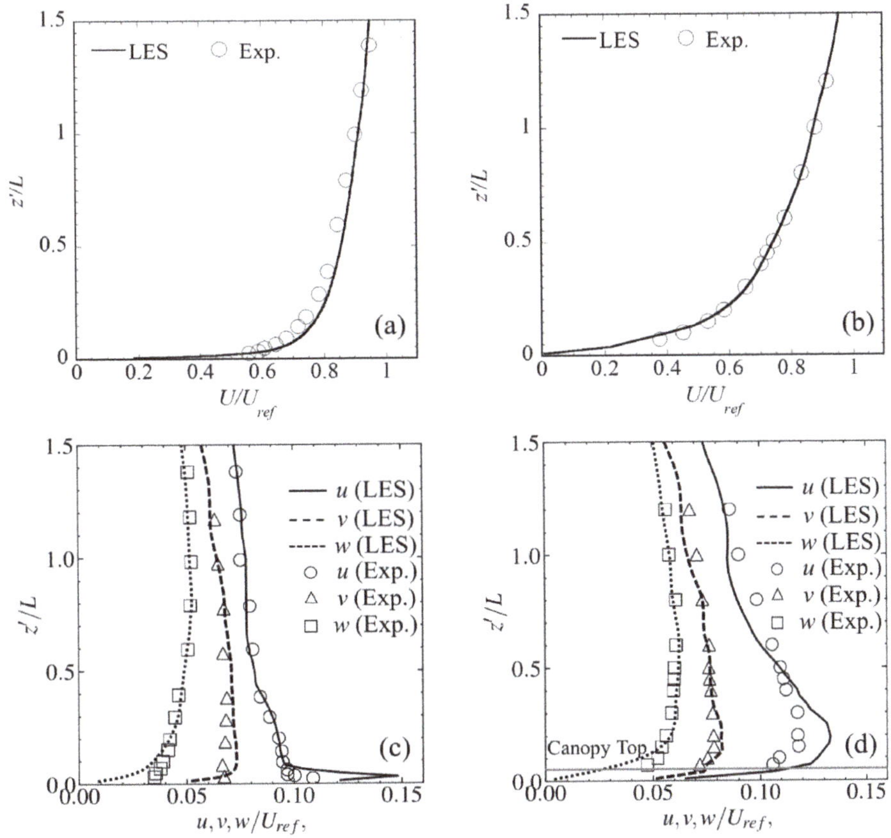

Figure 3. Comparison of the mean streamwise velocity over (**a**) smooth ground, and (**b**) rough ground, as well as the velocity fluctuations over (**c**) smooth ground, and (**d**) rough ground.

3. Numerical Results

In the following figures about the mean velocities profiles (Figures 4 and 5) and the fluctuations (Figures 6–8), the separation boundary defined as the connection of the reversion points on U profile is illustrated by solid red lines for the smooth hills and dashed red lines for the rough hills from LES. The corresponding pink lines are from the experiment by Ishihara et al. [28]. The shear layer center determined by the connection of the peak r.m.s. streamwise fluctuations, u, is illustrated by solid yellow lines for the smooth hills and dashed yellow lines for the rough hills from LES. The corresponding brown ones are from the experiment by Ishihara et al. [28]. The discussions about the results are mainly from five aspects, i.e., the roughness effects, the hill slope effects, the hill shape effects, the coupling between these effects, and the LES performance.

3.1. Mean Streamwise Velocity

As shown in Figure 4, the mean streamwise velocity, U, is found to decrease to nearly zero in the roughness region, but accelerate rapidly from the roughness top at the summit. It can be also identified that the vertical distance of the reversion point with reference to the local ground becomes higher as introducing the ground roughness, implying a larger recirculation bubble. This should be attributed to the fact that the ground roughness can produce more turbulence in the wake and in return prevent the flow from reattaching. However, this trend is not preserved for 80 mm high 2D hills, see Figure 4f, in which the recirculation bubble gets even smaller after introducing the ground roughness. It should result from a balance between two effects of the roughness, i.e., preventing the flow from reattaching, and decelerating the negative U in the bubble. The former effect may enlarge the bubble while the later one may shrink the bubble. For 80 mm high 2D hills, the deceleration effects of the roughness may be stronger than those preventing the flow from reattaching.

As the hill slope increases, the flow in the wake shows larger deceleration, while the acceleration of U in the vertical direction at the summit increases. In addition, at the lee-side, the recirculation bubble is significantly stretched and the speed-up of U above the recirculation bubble seems to extend further downstream for steeper hills, which is also an indication of a more stable recirculation bubble.

U seems insensitive to the change of the hill shape from 3D to 2D at the windward side and the summit. However, downstream the summit, the hill shape effect gets significant. Firstly, it is obvious that the blockage effects of 2D hills are more energetic than those of 3D hills, which is the result that the flow approaching the 3D hills can move around the lateral sides, whereas there is no such pass for the flow over 2D hills. Secondly, the near-ground acceleration of negative U is more obvious as the hill shape turns to 2D.

For the coupling between these effects, with increasing the slope, the roughness effects on U at the locations above $z = h$ becomes weaker. In addition, for the hills with low slope, the change of the hill shape from 3D to 2D does not show significant evolution of the flow fields (Figure 4a,d); however, as the hill slope increases, the hill shape effects turn evident (Figure 4c,f). Furthermore, for the hills with low and moderate slopes (Figure 4a,d, and Figure 4b,e), the roughness effects and the hill shape effects seem to be independent, whereas if the hill is steeply-sloped (Figure 4c,f), the roughness effects will be weakened as the hill alters from 3D to 2D.

For the performance of LES, importantly, it is obvious that the rougher the ground, the worse LES results will be. And the present LES shows larger discrepancies for 2D hills as compared with 3D hills. This should be due to the smaller eddies caused by the roughness or the stronger disturbance from 2D hills, which requires finer space and time resolutions to reproduce the flow structures. The discrepancies concentrate at the lee-side of the hills, while the separation points and the reattachment points are well predicted. The overall comparison between the numerical and the experimental results is fairly good.

3.2. Mean Vertical Velocity

Figure 5 shows that the introduction of ground roughness will weaken the acceleration of the mean vertical velocity, W, at the windward side. However, at the summit, the maximum W over the rough hills becomes even larger than that over the smooth hills, which is more obvious for 3D hills. And this maximum W occurs just at the roughness top. It is also clear that after introducing ground roughness, the recovery of W in the wake turns quicker, which should be the result of the stronger mixing effects from the roughness-generated turbulence. In addition, above the recirculation bubble, the negative W is weakened when the ground is rough.

As the hill is more steeply-sloped, the near-ground W acceleration at the windward side is obviously increased, mainly because the wind at the windward side almost flows following the shape of the hill. Importantly, the steeper the hill is, the sharper the near-ground W profiles at the windward side will be, which is the most obvious as the flow moves to the summit. In the wake region, the trend of stronger negative W with increasing the slope is evident due to the more energetic recirculation bubble. Importantly, at some locations, such as $x = 1.0L$, the downward W even almost equals the peak upward W at the summit for both 3D and 2D hills with low and moderate slopes (Figure 5a,b,d,e), which is however not observed for the steepest 2D hills (Figure 5).

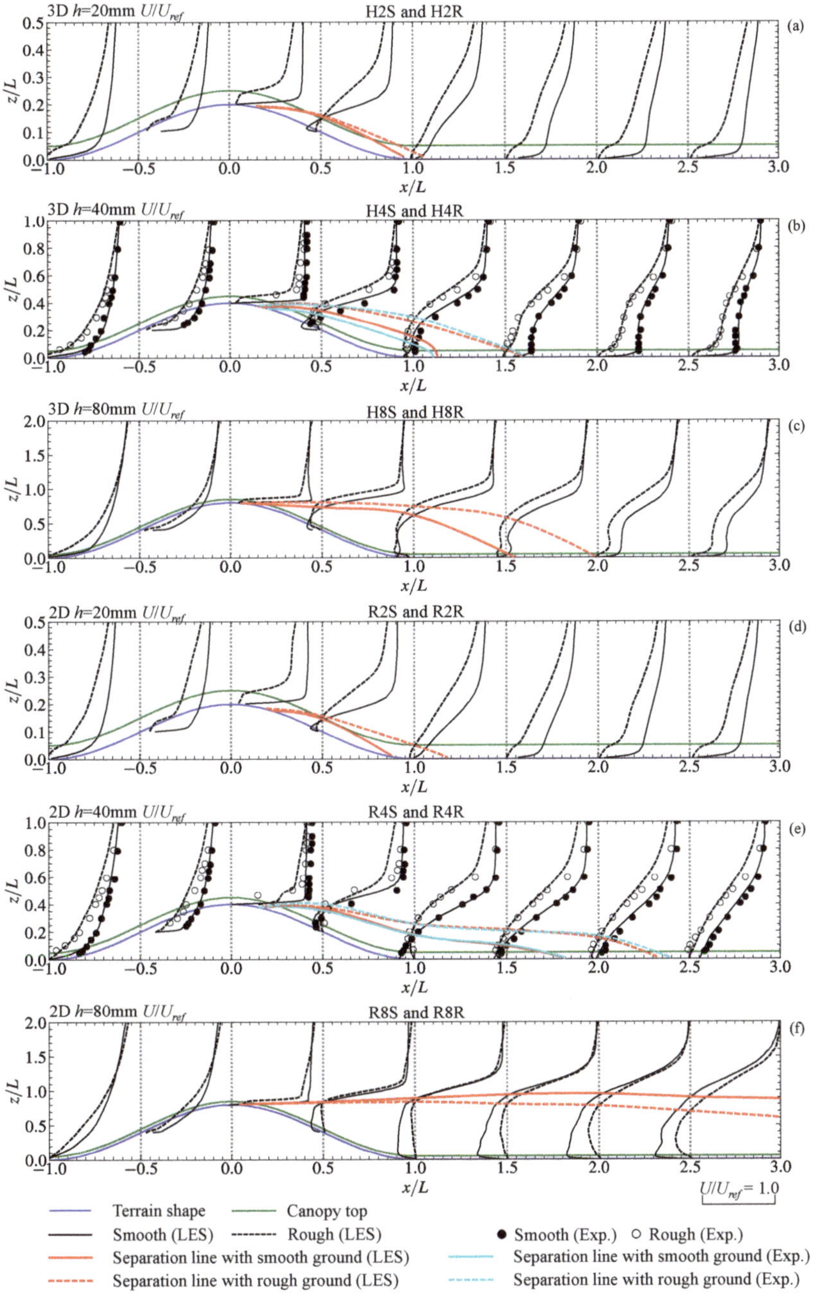

Figure 4. Vertical profiles of U, (**a**) 3D hills with h = 20 mm, (**b**) 3D hills with h = 40 mm, (**c**) 3D hills with h = 80 mm, (**d**) 2D hills with h = 20 mm, (**e**) 2D hills with h = 40 mm, and (**f**) 2D hills with h = 80 mm.

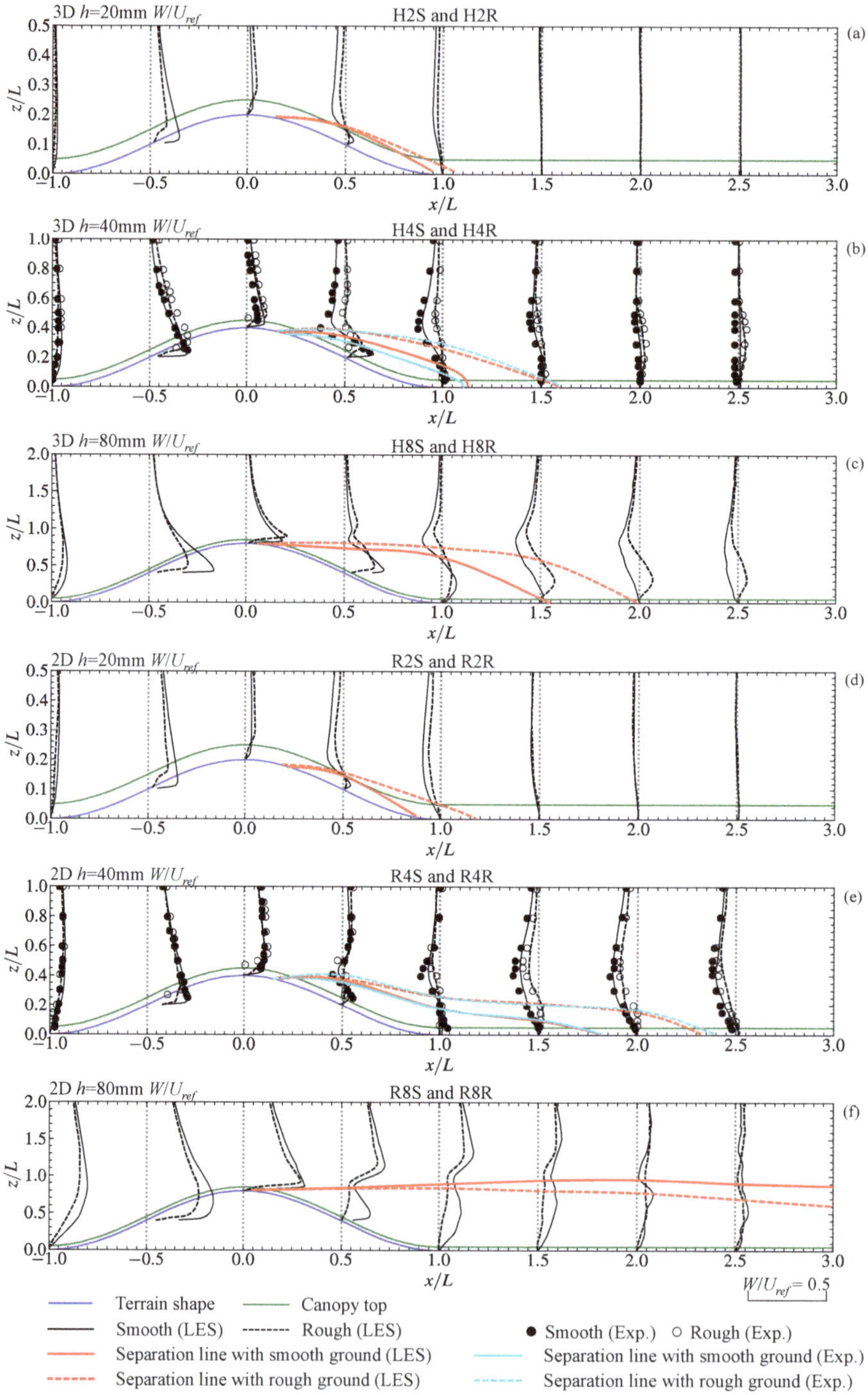

Figure 5. Vertical profiles of W, (**a**) 3D hills with h = 20 mm, (**b**) 3D hills with h = 40 mm, (**c**) 3D hills with h = 80 mm, (**d**) 2D hills with h = 20 mm, (**e**) 2D hills with h = 40 mm, and (**f**) 2D hills with h = 80 mm.

When the hill shape alters to 3D, W acceleration is found to extend to lower elevation at the windward side; however, W profiles become sharper. It is interesting that the downward W in the wake vanishes as the hill shape changes to 2D and the hill turns steeper (Figure 5f). This should result from the strong turbulence in the wake of the steep 2D hills.

As for the coupling between these effects, the distortion of W profiles owing to the ground roughness seems to be strengthened for the steeper hills, while it seems to be weakened as the hill changes from 3D to 2D. Importantly, at the summit of the hills, the acceleration of W gets quicker for 2D hills as the slope is increasing.

Similar to U, the discrepancies of W with the experimental data is concentrated in the near wake. However, different from U discrepancies, the large errors of W are found in the flow over the smooth hills. For 40 mm smooth 3D hill, the sharp near-wake deceleration of W cannot be predicted well and the negative W only reaches to about half of that measured in the experiment. Except this location, the predicted W shows satisfactory agreements with the experiment.

3.3. Fluctuations of Streamwise Velocity

It is evident that the strea·wise fluctuation, u, at $x = -L$ very close to the ground for the rough hills are smaller than the corresponding smooth hills, as seen from Figure 6, which should be the result of the limitation of the eddy development due to the drag effects in the roughness canopy. However, the flow can enter the roughness upper region and generate wavy structures near the roughness top, thus creating more turbulence above the roughness. And interestingly, in the wake region the elevation of the shear layer is higher for the rough hills, but at the shear layer, u over the smooth hills is larger.

As the slope increases, at the summit, u seems to be restrained, which is more obvious when the hill shape is 3D or the ground is rough. Furthermore, the peak u in the wake is found to be sharper when the hill is more steeply-sloped. It is also important that the wake depth grows quickly as the flow moves downstream; however, the increasing speed of the wake depth is slowed for the steeper hills. For the 80 mm smooth 3D hill (Figure 6c), the wake depth is even surprisingly decreased as the flow moves downstream.

After changing the hill shape to 2D, the location of the peak u is found to be lower than the corresponding 3D hills. However, this trend becomes not obvious for the steepest hills (Figure 6c,f). In addition, the peak u is obviously enhanced when the hill shape changes to 2D, which should attribute to the more stable recirculation bubble in the wake of 2D hills, providing stronger wind shears. This trend is even more obvious for the smooth hills.

Except the findings about the coupling of the ground roughness effects, hill slope effects, and the hill shape effects in the above presentations about u, we can also observe that in general the difference between the height of peak u of the smooth and rough hills gets smaller as the hill slope increases, and this trend becomes quicker as changing the hill shape from 3D to 2D, which may imply that in the real situation the forest may play a limited role in affecting the flow fields if the hill shape is about a 2D ridge and the hill is steep enough.

In LES, at the upstream footage of the hills, there is a large increase of u, implying the formation of a horseshoe vortex, but this feature is not observed in the experiments by Ishihara et al. [28]. This may be attributed to the difficulty setting the probes at this location or the fact that the existence of the probes weakens the horseshoe vortex. Interestingly, u at the hill crest shows a sharp peak in the present LES. However, this sharp peak is not measured in the experiments by Ishihara et al. [28], which should result from the limited observation points in the experiments or the disturbance of the flow fields very close the ground from the probes. Additionally, comparing with the experimental data, in the near-wake region the LES shows obvious overestimations for the 40 mm smooth 3D hill, and obvious underestimation for the 40 mm rough 2D hill, whereas these discrepancies decrease to the acceptable level as soon as the flow moves to the locations $x > 1L$.

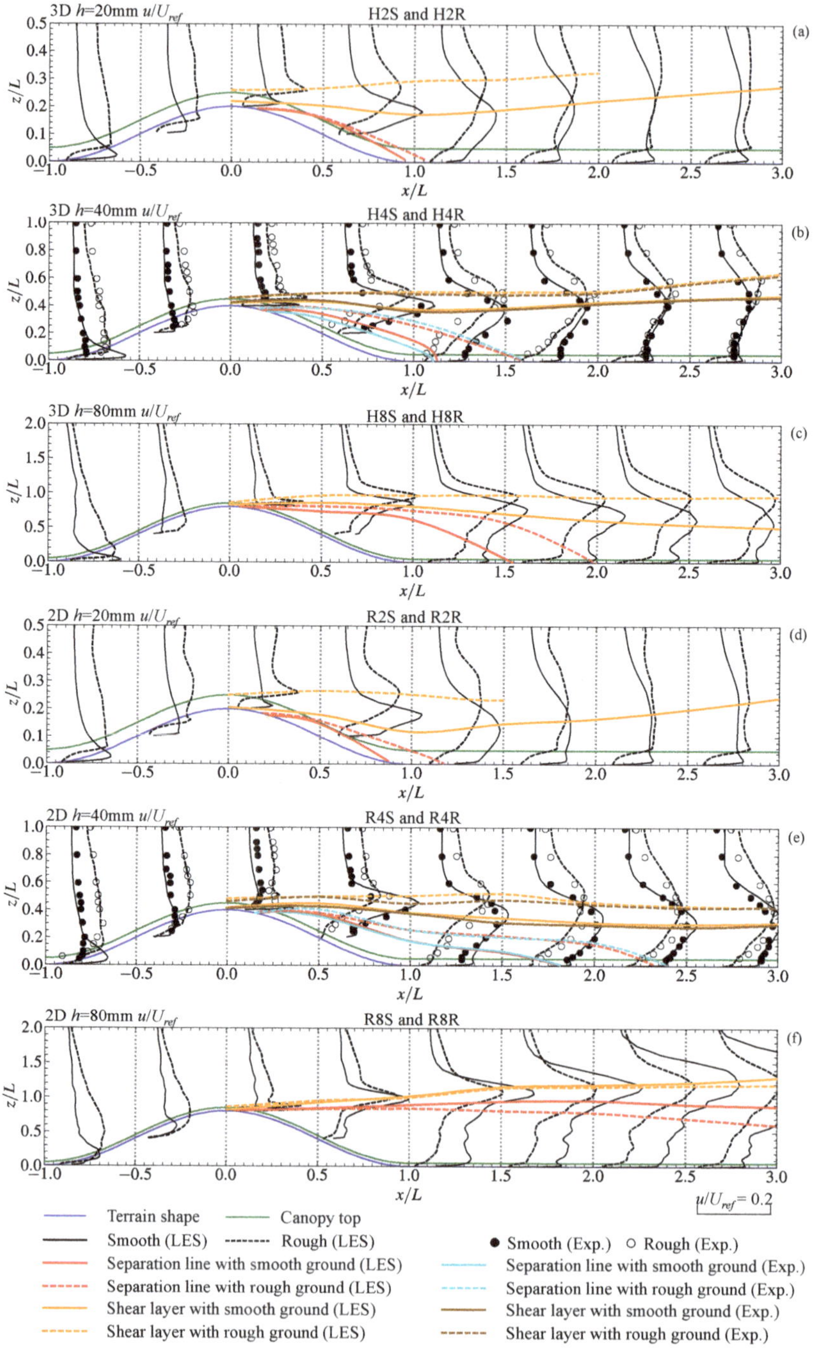

Figure 6. Vertical profiles of u, (**a**) 3D hills with h = 20 mm, (**b**) 3D hills with h = 40 mm, (**c**) 3D hills with h = 80 mm, (**d**) 2D hills with h = 20 mm, (**e**) 2D hills with h = 40 mm, and (**f**) 2D hills with h = 80 mm.

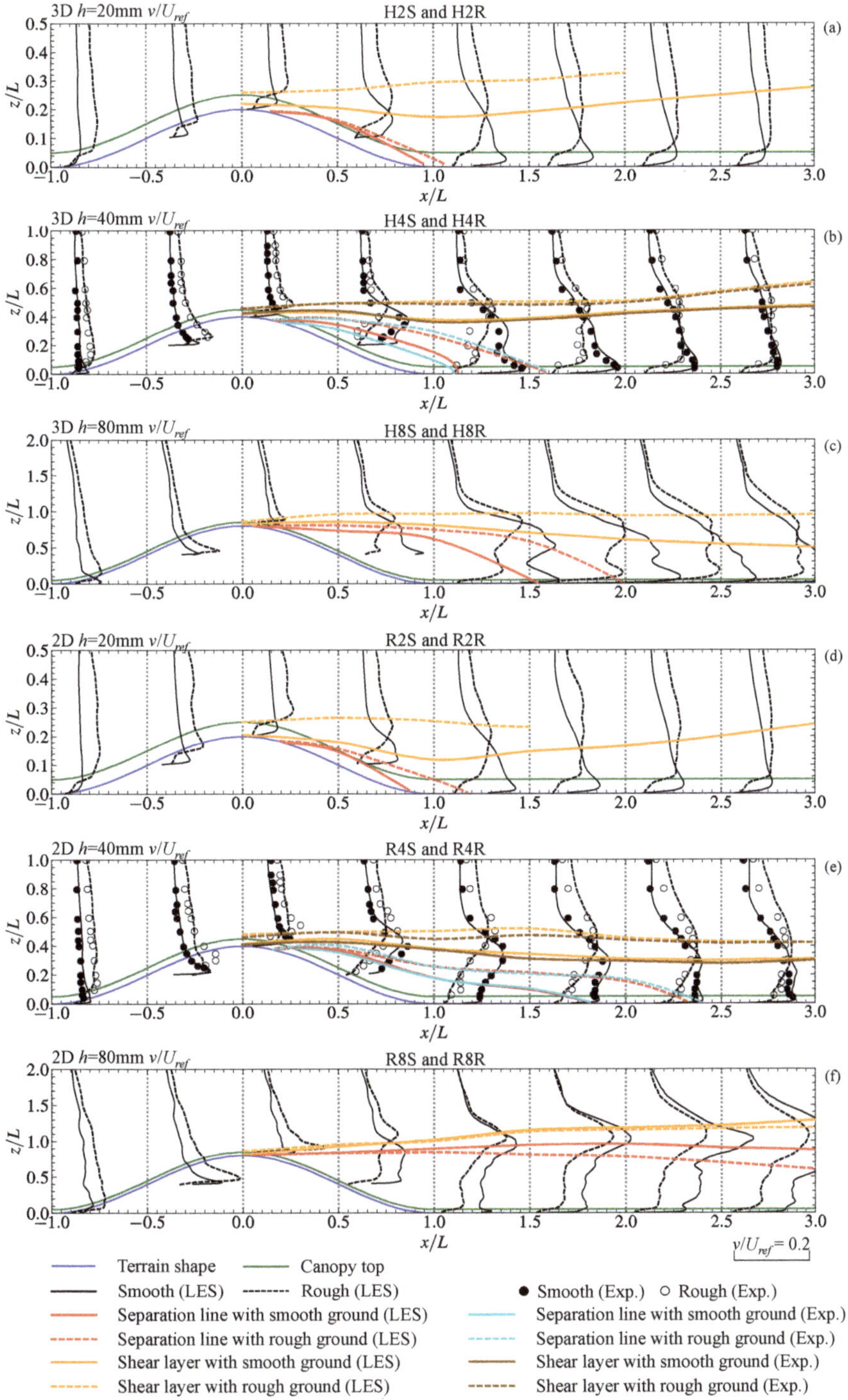

Figure 7. Vertical profiles of v, (**a**) 3D hills with h = 20 mm, (**b**) 3D hills with h = 40 mm, (**c**) 3D hills with h = 80 mm, (**d**) 2D hills with h = 20 mm, (**e**) 2D hills with h = 40 mm, and (**f**) 2D hills with h = 80 mm.

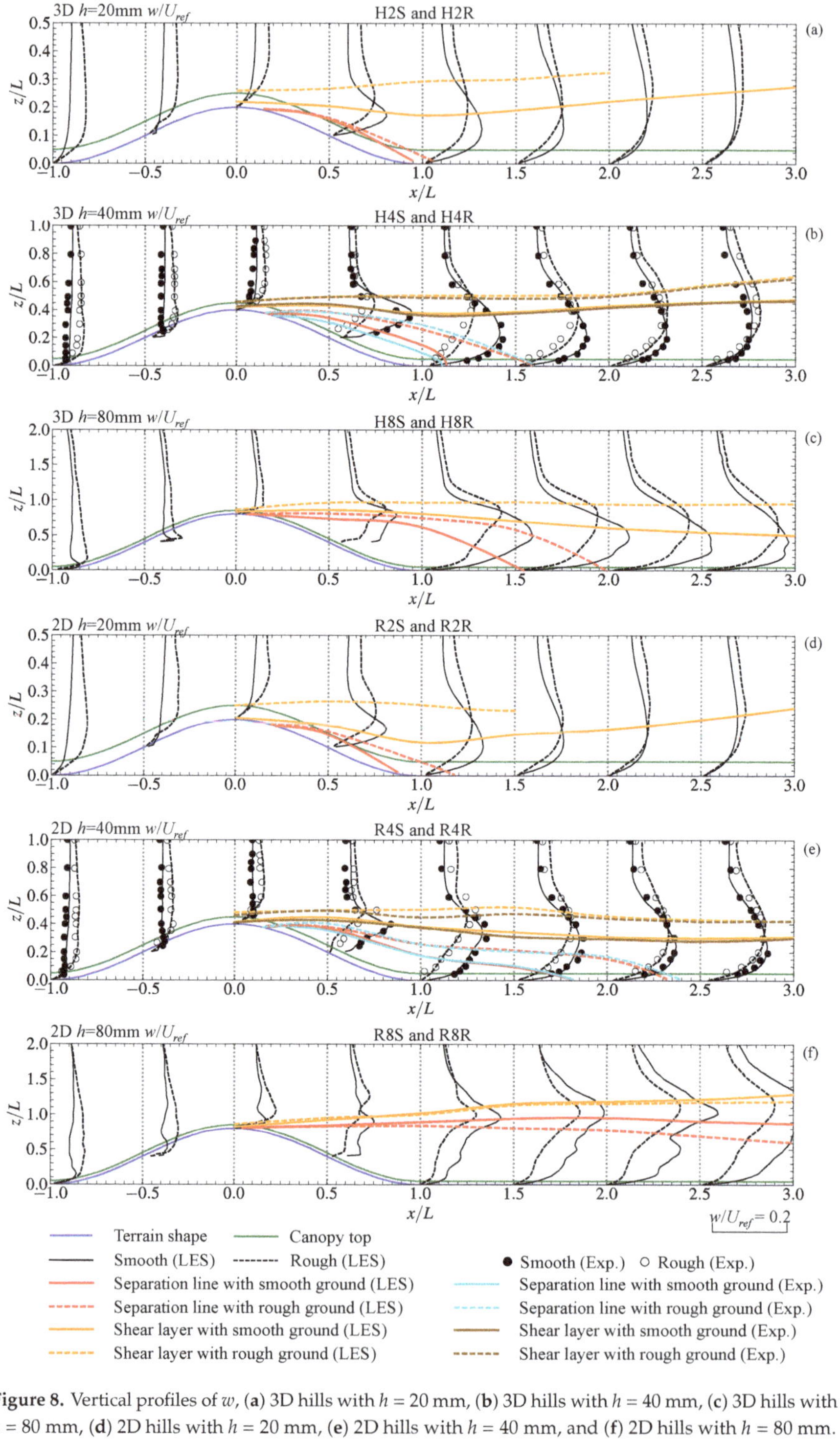

Figure 8. Vertical profiles of w, (**a**) 3D hills with h = 20 mm, (**b**) 3D hills with h = 40 mm, (**c**) 3D hills with h = 80 mm, (**d**) 2D hills with h = 20 mm, (**e**) 2D hills with h = 40 mm, and (**f**) 2D hills with h = 80 mm.

3.4. Fluctuations of Spanwise Velocity

For the spanwise fluctuation, v, shown in Figure 7, the similar trend as u can be observed. However, there are some unique features for v. Firstly, in the wakes, distinct v profiles can be identified for the 3D hills (Figure 7a–c). It is obvious that no matter how steep the 3D hill is, two peaks appear on the vertical v profiles with one locating near the ground and the other locating in the region close to the shear layer. The LES by Liu et al. has revealed the mechanism of the two peaks on v profiles [34], in which the secondary turbulence structures surrounding the major vortex core in the wake are believed to be the source. Two peaks of v still appear but get less obvious when the ground is rough. In addition, as increasing the hill slope, these two peaks become more obvious, indicating the more energetic secondary vortex.

Secondly, for 2D hills with h = 80 mm, it is surprising that v in the recirculation bubble is nearly a constant at each streamwise location, implying the full mixture from the hills and abruptly homogeneous turbulence, which is confirmed by plotting the instantaneous flow fields using Q-criteria in Figure 12. The performance of the present LES for predicting v is similar as u. Some errors can still be found in the near-wake region.

3.5. Fluctuations of Vertical Velocity

Figure 8 shows the distributions of vertical fluctuations, w, which are found to be very similar to those of u and v. Therefore, no detailed discussion about w is provided. In general, w is smaller than u and v, and there is no additional peak very close to the ground.

3.6. Fractional Speed-Up Ratio

Fractional speed-up ratio, ΔS, has been extensively applied to assess the impact of topography. ΔS is defined as:

$$\Delta S = \frac{U(x,\ z') - U_0(z')}{U_0(z')} \tag{12}$$

where $U(x,\ z')$ is the mean streamwise velocity at z', and $U_0(z')$ is the reference streamwise velocity at the same height in the absence of the topography. The horizontal distributions of ΔS at z' = 10 m and 50 m are illustrated in Figure 9. The selection of these two heights takes into account that 10 m is the reference height adopted in most of the wind-resistant guidelines, and 50 m is about the hub height of 1 MW or 1.5 MW wind turbines in operation in many hilly regions of China.

It can be clearly seen from Figure 9 that the introduction of ground roughness will obviously increase ΔS at $z' = 10$ m, where ΔS at the summit of rough hills can even reach to three times as large as that over the corresponding smooth hills. However, ΔS near the summit at $z' = 50$ m is not sensitive to the ground roughness condition. On the other hand, the deceleration of ΔS in the wake becomes stronger in general if the ground is rough even at high elevations, which should attribute to both the drag effects and the more energetic turbulence caused by the roughness.

With increasing the hill slope, ΔS at the summit increases obviously. ΔS ($z' = 10$ m) at the summit over rough hills reaches 0.4 for the low-slope hills, 0.7 for the moderate-slope hills, and 0.9 for the large-slope hills. However, it is worthwhile to point out that the area with positive ΔS turns smaller as the hill gets steeper. Furthermore, as increasing the hill slope, the deceleration area in the hill wakes will also expand.

It is also interesting that the hill shape shows unapparent effects on the maximum ΔS at the summit, the area of positive ΔS near the hill top, and the minimum ΔS in the near wake. However, obviously 2D hills will provide larger area with negative ΔS. The experimental data of ΔS are superimposed on Figure 9, where satisfactory agreement is achieved. The largest discrepancies are found in the wake of the 40 mm high 2D rough hill, which is about −0.3 in the LES while nearly −0.1 in the experiment.

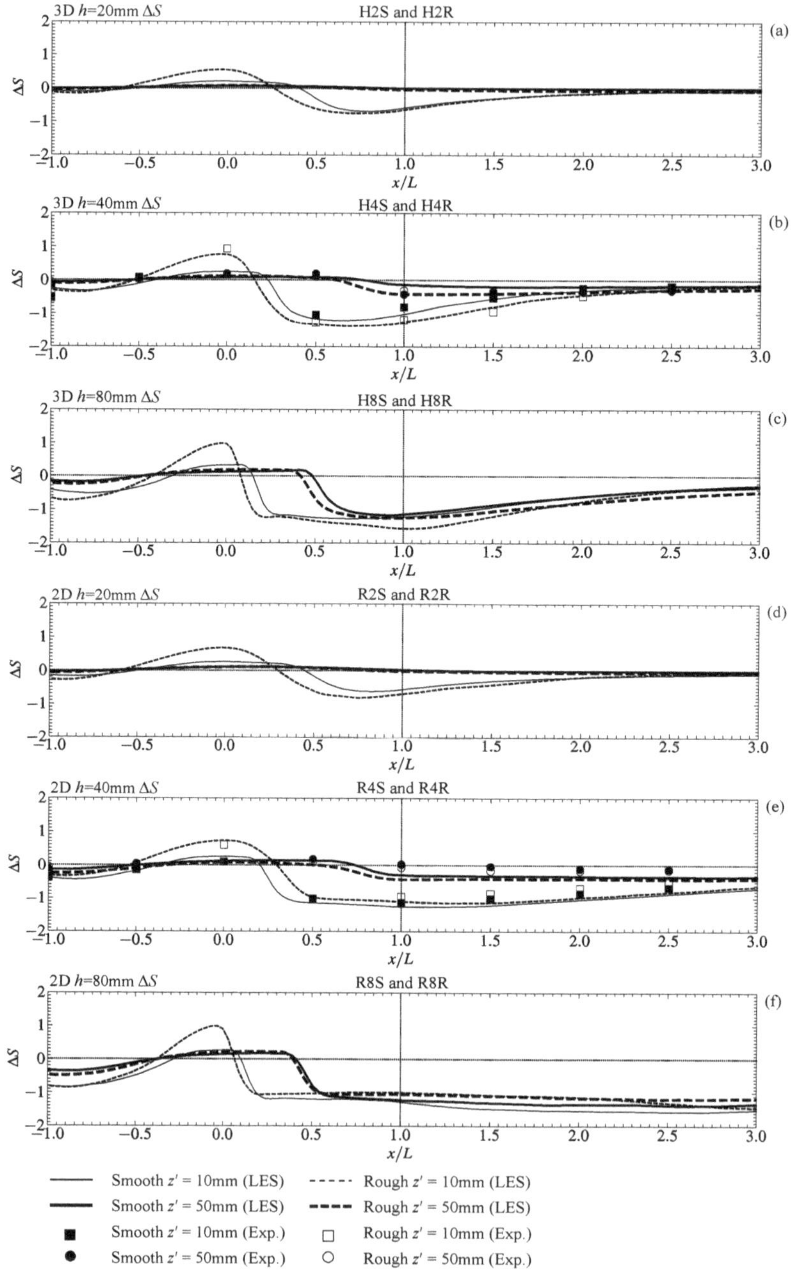

Figure 9. Fractional speed-up ratio ΔS at $z' = 10$ mm and 50 mm, (**a**) 3D hills with $h = 20$ mm, (**b**) 3D hills with $h = 40$ mm, (**c**) 3D hills with $h = 80$ mm, (**d**) 2D hills with $h = 20$ mm, (**e**) 2D hills with $h = 40$ mm, and (**f**) 2D hills with $h = 80$ mm.

3.7. Fluctuation Ratios

The fluctuation ratios are important parameters for determining the turbulent flow fields. In the present LES of the flow over flat ground, v/u and w/u are predicted as 0.81 and 0.49 respectively, close to the data suggested in IEC 61400-1 [35]. In case a wind turbine site is located within a complex terrain, IEC 61400-1 permits to increase the representative value by a factor C_{CT} defined as [35]:

$$C_{CT} = \frac{\sqrt{1 + (v/u)^2 + (v/u)^2}}{1.375} \tag{13}$$

C_{CT} is intended to account for the distortion of turbulence structure by complex terrains and to be estimated based on the site specific data. In the absence of site specific data, IEC 61400-1 recommends the use of a correction factor of 1.15 [35]. When C_{CT} = 1.15 and if the fluctuation ratios v/u and w/u are modified at the same rate, v/u and w/u become 1.0 and 0.7, respectively. The horizontal distributions of v/u in LES are shown in Figure 10. The symbols superimposed in Figure 10 are the data from wind-tunnel experiment by Ishihara et al. [28], and the dash-dotted lines indicate the data from IEC 61400-1 [35]. Same as the discussion about ΔS, the data on z' = 10 m and z' = 50 m are extracted.

Obviously, the introduction of ground roughness can hardly affect v/u, and the underestimation of v/u by IEC 61400-1 in the hill wake is quite obvious when the hill shape is 3D [35]. The results from LES can reach about 1.5 times as large as those in the guideline at z' = 10 m. Considering the fact that the suggested values in the guideline is based on the assumption of isotropic turbulence, the larger v/u in LES for 3D hills in the wake may indicate that the turbulence structure should be coherent instead of isotropic. From the snapshots of the instantaneous flow fields shown in Figure 12, the coherent turbulence structure over 3D hills is characterized mainly by a spanwise sway motion of the fluid, inducing a large spanwise fluctuation. However, this spanwise sway motion can hardly change the streamwise size of the eddies, therefore v/u becomes larger than that in the fluids characterized by isotropic eddies. We can also find that after changing the hill shape from 3D to 2D, the stronger blocking effects induce more energetic mixture, yielding nearly equal values of v/u as those in the guideline. However, it is important that at the summits, where most of the wind turbines are located, the data from the guideline are nearly consistent with the LES for both 2D and 3D hills.

As the hill slope increases, the underestimation from the guidelines seems to be enhanced, which is more obvious when the hill shape is 2D. For 80 mm 2D hills, the large underestimation from the guideline can be observed over almost the entire hill area except the locations near the summit. Furthermore, the difference between v/u over smooth hill and that over rough hill is enlarged as increasing the hill slope. At the high elevation, z' = 50 m, the predicted v/u is abruptly equal to that from IEC 61400-1 [35], which is also the indication of the different turbulence structures of the flow close the ground and that at high elevations.

The horizontal distributions of w/u are shown in Figure 11, which is found to be similar to the distributions of v/u. However, at the windward side, the large values found in v/u do not appear in w/u. And, different from v/u, w/u shows larger values in the far wake region for the steepest hills.

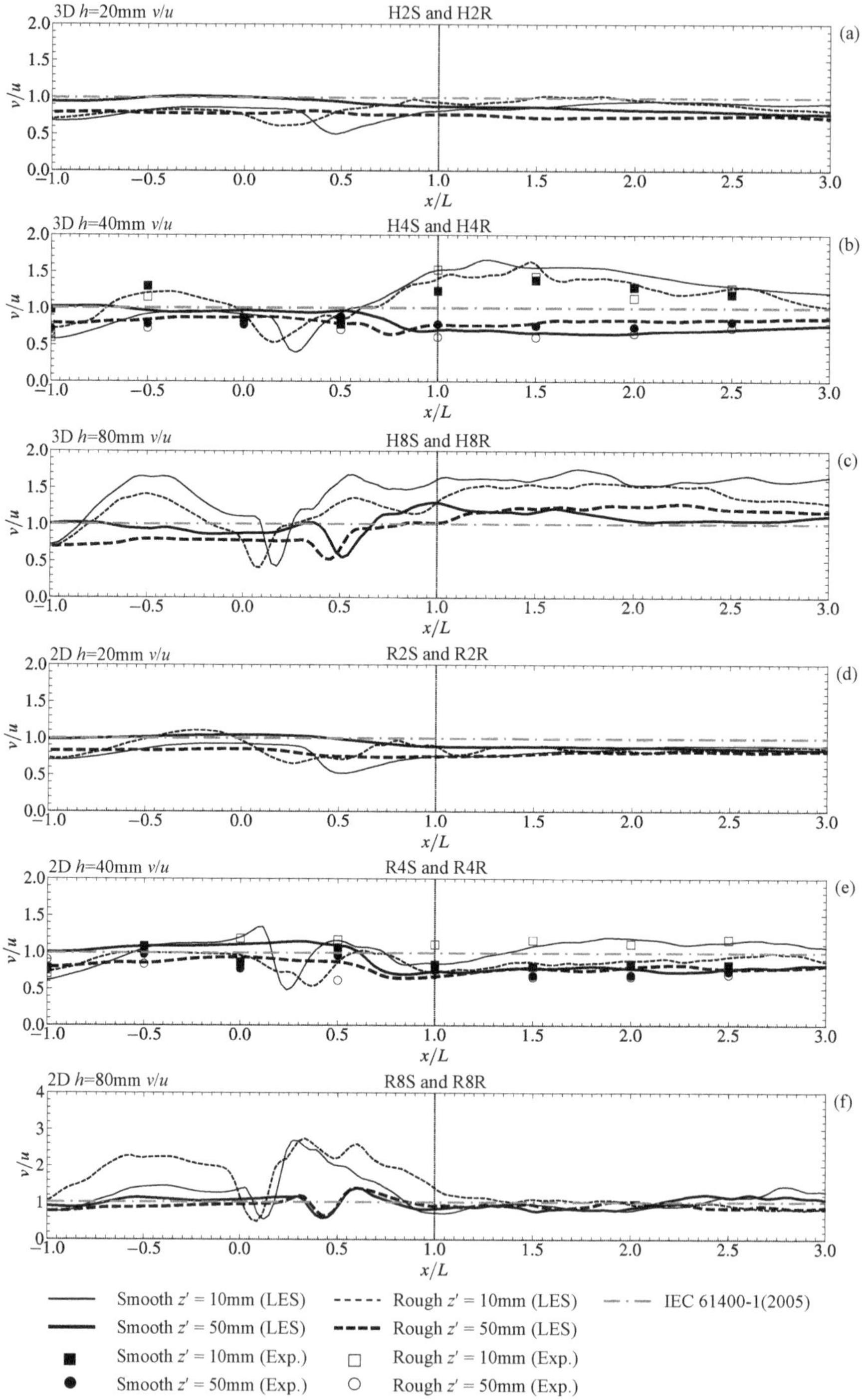

Figure 10. Fluctuation ratio v/u at $z' = 10$ mm and 50 mm, (**a**) 3D hills with $h = 20$ mm, (**b**) 3D hills with $h = 40$ mm, (**c**) 3D hills with $h = 80$ mm, (**d**) 2D hills with $h = 20$ mm, (**e**) 2D hills with $h = 40$ mm, and (**f**) 2D hills with $h = 80$ mm.

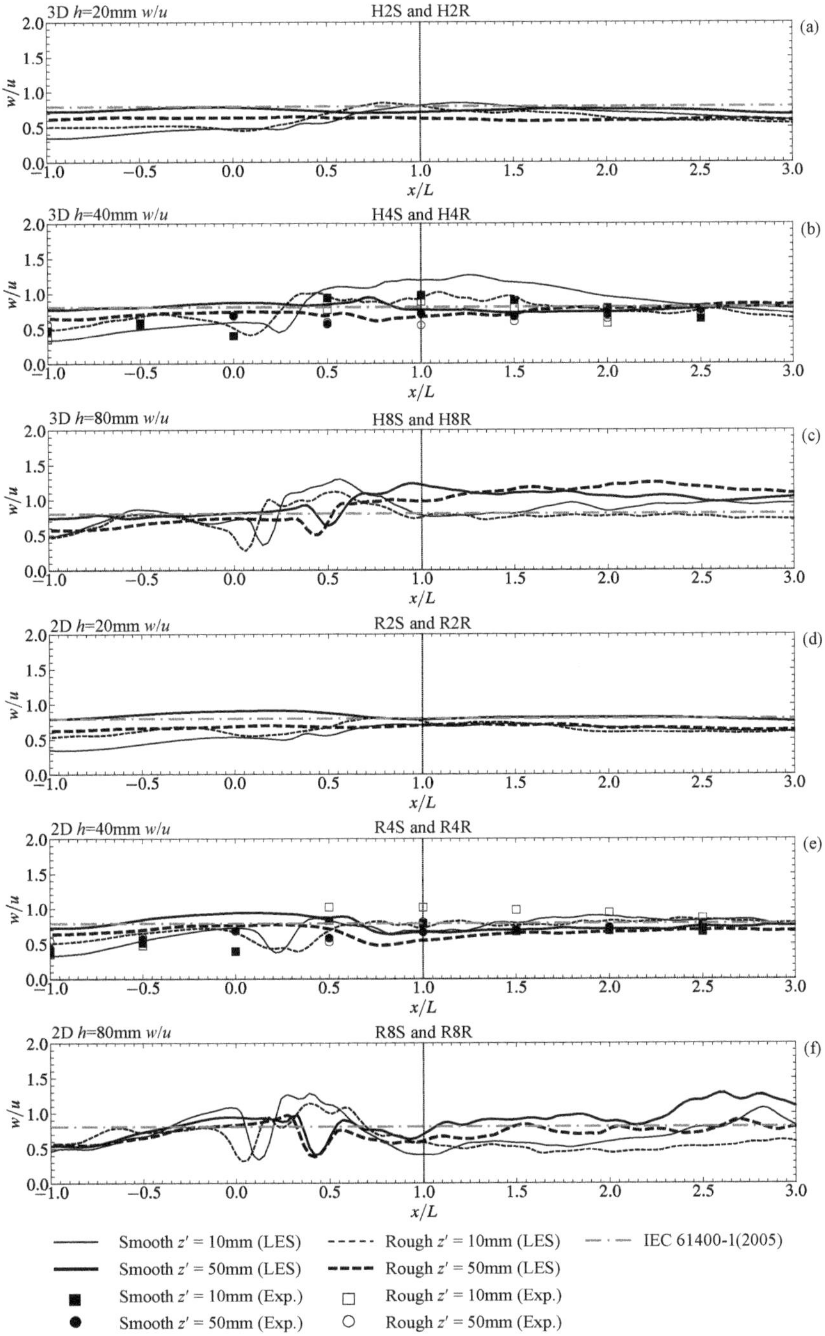

Figure 11. Fluctuation ratio w/u at $z' = 10$ mm and 50 mm, (**a**) 3D hills with $h = 20$ mm, (**b**) 3D hills with $h = 40$ mm, (**c**) 3D hills with $h = 80$ mm, (**d**) 2D hills with $h = 20$ mm, (**e**) 2D hills with $h = 40$ mm, and (**f**) 2D hills with $h = 80$ mm.

3.8. Instantaneous Flow Fields

The Q-criterion is further adopted to shed some lights on the instantaneous turbulent structures, which is defined as:

$$Q = \frac{1}{2}(S_{ij}S_{ij} - \Omega_{ij}\Omega_{ij}) \tag{14}$$

where S_{ij} and Ω_{ij} are the antisymmetric and symmetric components of the velocity-gradient tensor, respectively, quantifying the relative amplitude of the rotation rate as well as the strain rate of the flow, expressed as:

$$S_{ij} = \frac{1}{2}(\partial \tilde{u}_i / \partial x_j - \partial \tilde{u}_j / \partial x_i) \tag{15}$$

$$\Omega_{ij} = \frac{1}{2}(\partial \tilde{u}_i / \partial x_j + \partial \tilde{u}_j / \partial x_i) \tag{16}$$

The snapshots of instantaneous Q with values of 10000 (purple) and −10000 (green) are shown in Figure 12, where the thin yellow lines indicate the boundary of the hills, the yellow thick dashed lines show the major core of the coherent structure in the wakes, and the white dashed lines indicate the locations of the secondary vortex surrounding the major core.

For smooth 3D hills, as increasing the hill slope the major core of the coherent turbulence structure in the hill wake tends to show large spanwise sway motions, shown in Figure 12e. It is obvious that after introducing the ground roughness, the clear coherent turbulence structures are broken into small eddies. However, for the rough 3D hills, when the hill slope is large enough, the periodical vortex shedding will appear again, see Figure 12k.

Different evolution process of the turbulence structure in the hill wakes can be identified for 2D hills. When the hill is smooth and the hill slope is low, the major core of the wake vortex is nearly perpendicular to the ridge line. As the hill slope increases (Figure 12d), a kind of ejection-sweep structure of large scale occurs. Further increasing the hill slope, the ejection-sweep structure turns to be a wavy structure with the major core being parallel with the ridge line (Figure 12f). In addition, when the 2D hills are covered with roughness canopy, no clear coherent structure can be identified.

4. Conclusions

LES are adopted to study the flow fields over simplified topographies with different shapes, hill heights and rough conditions. The findings of this study are summarized below:

1) For the mean streamwise velocity, as the hill slope increases, the roughness effects get weaker; however, the hill shape effects become more evident. And if the hill is very steeply-sloped, the roughness effects will be further weakened as the hill changes from 3D to 2D.
2) For the mean vertical velocity, the distortion of the profiles owing to the ground roughness is strengthened for the steeper hills, while it seems to be weakened as the hill alters from 3D to 2D. Importantly, at the hill summits, the acceleration of W as increasing the slope becomes quicker for 2D hills.
3) For the fluctuations, at the summit, u seems to be restrained as increasing the slope, which is more obvious when the hill shape is 3D or the ground is rough. After changing the hill shape to 2D, the location of the peak u is lower than the corresponding 3D hills. Two peaks on v profiles still appear but become less obvious when the ground is rough. In addition, as increasing the hill slope, these two peaks turn more obvious.
4) For the fractional speed-up ratio, ΔS at the summit of rough hills can reach to three times as large as that over the corresponding smooth hills. With increasing the hill slope, ΔS at the summit increases obviously. However, the hill shape shows unapparent effects on the maximum ΔS at the summit. In addition, the area with positive ΔS becomes smaller as the hill gets steeper.
5) For the fluctuation ratios, the introduction of ground roughness can hardly affect v/u, and the underestimation of v/u by IEC 61400-1 in the hill wake is quite obvious when the hill shape is

3D [35]. After changing the hill shape from 3D to 2D, v/u becomes almost equal as those in the guideline.

6) For the turbulence structures, clear coherent turbulence structure can be identified for smooth 3D hills, and as increasing the hills slope, the major core of the coherent structure tends to show large spanwise sway motions. For 2D hills, when the ground is smooth and the slope is low, the major core of the wake vortex is nearly perpendicular to the ridge line, and as increasing the hill slope, a kind of ejection-sweep structure of large scale occurs. Further increasing the hill slope, a wavy structure with the major core being parallel with the ridge line appears. After introducing the ground roughness, the clear coherent turbulence structures are broken into small eddies.

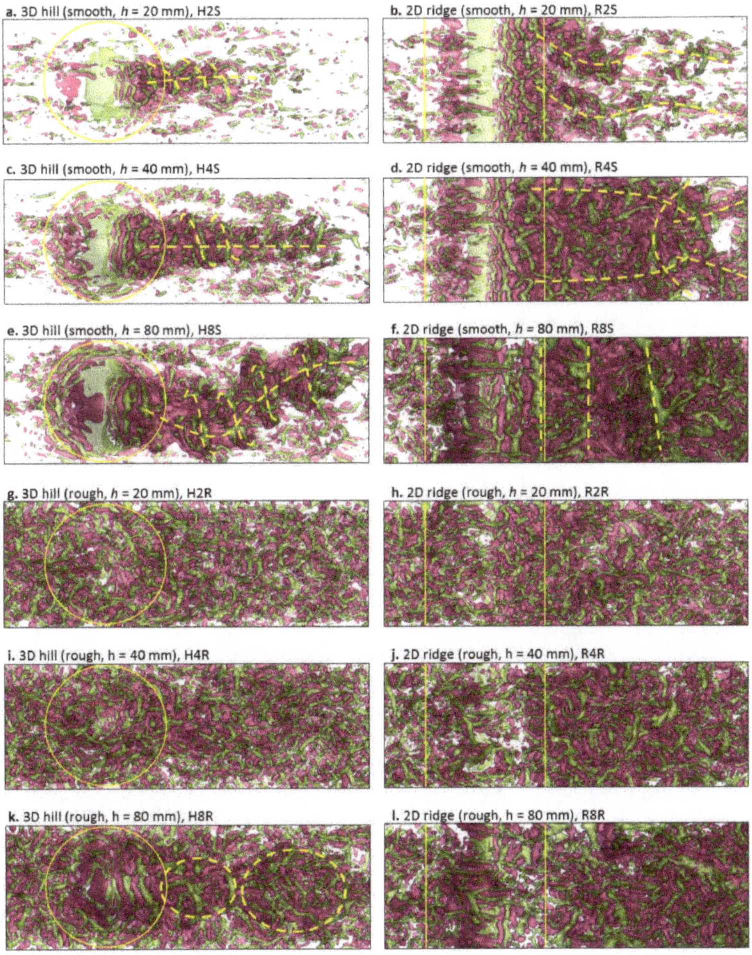

Figure 12. Instantaneous Q-criteria with values of 10000 (**purple**) and −10000 (**green**). The yellow solid lines indicate the location of the hills or ridges. The yellow dashed lines indicate the vortex cores in the wake of the topographies; (**a**) 3D hill (smooth, h = 20 mm), H2S, (**b**) 2D ridge (smooth, h = 20 mm), R2S, (**c**) 3D hill (smooth, h = 40 mm), H4S, (**d**) 2D ridge (smooth, h = 40 mm), R4S, (**e**) 3D hill (smooth, h = 80 mm), H8S, (**f**) 2D ridge (smooth, h = 80 mm), R8S, (**g**) 3D hill (rough, h = 20 mm), H2R, (**h**) 2D ridge (rough, h = 20 mm), R2R, (**i**) 3D hill (rough, h = 40 mm), H4R, (**j**) 2D ridge (rough, h = 40 mm), R4R, (**k**) 3D hill (rough, h = 80 mm), H8R, (**l**) 2D ridge (rough, h = 80 mm), R8R.

Author Contributions: Conceptualization, Z.L.; methodology, Z.L. and W.W.; software, Z.L. and W.W.; validation, Z.L. and W.W.; formal analysis, Z.L. and W.W.; investigation, Z.L.; resources, Z.L.; writing—original draft preparation, Z.L. and W.W.; writing—review and editing, Z.L. and Y.H.; supervision, Z.L.

Funding: This research was funded by the National Key Research and Development Plan of China (2016YFE0127900), the National Natural Science Foundations of China (51608220), and the Project of Innovation-driven Plan in Huazhong University of Science and Technology (2017KFYXJJ141).

Acknowledgments: The corresponding author sincerely thanks Huazhong University of Science and Technology for the support to this research.

Conflicts of Interest: The authors declare no conflict of interest.

References

1. Politis, E.; Prospathopoulos, J.; Cabezon, D.; Hansen, K.; Chaviaropoulos, P.; Barthelmie, R. Modeling wake effects in large wind farms in complex terrain: The problem, the methods and the issues. *Wind Energy* **2011**, *15*, 161–182. [CrossRef]
2. Carvalho, A.; Carvalho, A.; Gelpi, I.; Barreiro, M.; Borrego, C.; Miranda, A.; Pérez-Muñuzuri, V. Influence of topography and land use on pollutants dispersion in the Atlantic coast of Iberian Peninsula. *Atmos. Environ.* **2006**, *40*, 3969–3982. [CrossRef]
3. Davenport, A.G.; King, J.P.C. The influence of topography on the dynamic wind loading of long span bridges. *J. Wind Eng. Ind. Aerodyn.* **1990**, *36*, 1373–1382. [CrossRef]
4. Matusick, G.; Ruthrof, K.; Brouwers, N.; Hardy, G. Topography influences the distribution of autumn frost damage on trees in a Mediterranean-type Eucalyptus forest. *Trees* **2014**, *28*, 1449–1462. [CrossRef]
5. Loureiro, J.; Alho, A.; Freire, A. The numerical computation of near-wall turbulent flow over a steep hill. *J. Wind Eng. Ind. Aerodyn.* **2008**, *96*, 540–561. [CrossRef]
6. Britter, R.; Hunt, J.B.; Richards, K. Air flow over a two-dimensional hill: Studies of velocity speed-up, roughness effects and turbulence. *Q. J. R. Meteorol. Soc.* **1981**, *107*, 91–110. [CrossRef]
7. Pearse, J.; Lindley, D.; Stevenson, D. Wind flow over ridges in simulated atmospheric boundary layers. *Bound.-Layer Meteorol.* **1981**, *21*, 77–92. [CrossRef]
8. Brown, A.; Hobson, J.; Wood, N. Large-eddy simulation of neutral turbulent flow over rough sinusoidal ridges. *Bound.-Layer Meteorol.* **2001**, *98*, 411–441. [CrossRef]
9. Athanassiadou, M.; Castro, I. Neutral flow over a series of rough hills: A laboratory experiment. *Bound.-Layer Meteorol.* **2001**, *101*, 1–30. [CrossRef]
10. Finnigan, J.; Belcher, S. Flow over a hill covered with a plant canopy. *Q. J. R. Meteorol. Soc.* **2004**, *130*, 1–29. [CrossRef]
11. Ross, A.; Vosper, S. Neutral turbulent flow over forested hills. *Q. J. R. Meteorol. Soc.* **2005**, *131*, 1841–1862. [CrossRef]
12. Cao, S.; Tamura, T. Experimental study on roughness effects on turbulent boundary layer flow over a two-dimensional steep hill. *J. Wind Eng. Ind. Aerodyn.* **2006**, *94*, 1–19. [CrossRef]
13. Cao, S.; Tamura, T. Effects of roughness blocks on atmospheric boundary layer flow over a two-dimensional low hill with/without sudden roughness change. *J. Wind Eng. Ind. Aerodyn.* **2007**, *95*, 679–695. [CrossRef]
14. Ross, A. Large-eddy simulations of flow over forested ridges. *Bound.-Layer Meteorol.* **2008**, *128*, 59–76. [CrossRef]
15. Okaze, T.; Ono, A.; Mochida, A.; Kannuki, Y.; Watanabe, S. Evaluation of turbulent length scale within urban canopy layer based on LES data. *J. Wind Eng. Ind. Aerodyn.* **2015**, *144*, 79–83. [CrossRef]
16. Loureiro, J.; Freire, A. Note on a parametric relation for separating flow over a rough hill. *Bound.-Layer Meteorol.* **2009**, *131*, 309–318. [CrossRef]
17. Takahashi, T.; Ohtsu, T.; Yassin, M.; Kato, S.; Murakami, S. Turbulence characteristics of wind over a hill with a rough surface. *J. Wind Eng. Ind. Aerodyn.* **2002**, *90*, 1697–1706. [CrossRef]
18. Tamura, T.; Okuno, A.; Sugio, Y. Les analysis of turbulent boundary layer over 3d steep hill covered with vegetation. *J. Wind Eng. Ind. Aerodyn.* **2007**, *95*, 1463–1475. [CrossRef]
19. Cao, S.; Wang, T.; Ge, Y.; Tamura, Y. Numerical study on turbulent boundary layers over two-dimensional hills-effects of surface roughness and slope. *J. Wind Eng. Ind. Aerodyn.* **2012**, *104*, 342–349. [CrossRef]

20. Liu, Z.; Ishihara, T.; He, X.; Niu, H. LES study on the turbulent flow fields over complex terrain covered by vegetation canopy. *J. Wind Eng. Ind. Aerodyn.* **2016**, *155*, 60–73. [CrossRef]
21. Finardi, S.; Brusasca, G.; Morselli, M.; Trombetti, F.; Tampieri, F. Boundary-layer flow over analytical two-dimensional hills: A systematic comparison of different models with wind tunnel data. *Bound.-Layer Meteorol.* **1993**, *63*, 259–291. [CrossRef]
22. Hu, P.; Li, Y.; Han, Y.; Cai, S.; Xu, X. Numerical simulations of the mean wind speeds and turbulence intensities over simplified gorges using the SST k-ω turbulence model. *Eng. Appl. Comput. Fluid Mech.* **2016**, *10*, 361–374. [CrossRef]
23. Ma, Y.; Liu, H. Large-eddy simulations of atmospheric flows over complex terrain using the immersed-boundary method in the weather research and forecasting model. *Bound.-Layer Meteorol.* **2017**, *165*, 421–445. [CrossRef]
24. Ferreira, A.; Lopes, A.; Viegas, D.; Sousa, A. Experimental and numerical simulation of flow around two-dimensional hills. *J. Wind Eng. Ind. Aerodyn.* **1995**, *54*, 173–181. [CrossRef]
25. Griffiths, A.; Middleton, J. Simulations of separated flow over two-dimensional hills. *J. Wind Eng. Ind. Aerodyn.* **2010**, *98*, 155–160. [CrossRef]
26. Neff, D.; Meroney, R. Wind-tunnel modeling of hill and vegetation influence on wind power availability. *J. Wind Eng. Ind. Aerodyn.* **1998**, *74*, 335–343. [CrossRef]
27. Pirooz, A.; Flay, R. Comparison of speed-up over hills derived from wind-tunnel experiments, wind-loading standards, and numerical modelling. *Bound.-Layer Meteorol.* **2018**, *168*, 213–246. [CrossRef]
28. Ishihara, T.; Fujino, Y.; Hibi, K. A wind tunnel study of separated flow over a two-dimensional ridge and a circular hill. *J. Wind Eng.* **2001**, *89*, 573–576.
29. Lubitz, W.; White, B. Wind-tunnel and field investigation of the effect of local wind direction on speed-up over hills. *J. Wind Eng. Ind. Aerodyn.* **2007**, *95*, 639–661. [CrossRef]
30. Liu, Z.; Ishihara, T.; Tanaka, T.; He, X. LES study of turbulent flow fields over a smooth 3-D hill and a smooth 2-D ridge. *J. Wind Eng. Ind. Aerodyn.* **2016**, *153*, 1–12. [CrossRef]
31. Ishihara, T.; Qi, Y. Numerical study of turbulent flow fields over steep terrains by using a modified delayed detached eddy simulations. *Bound.-Layer Meteorol.* **2019**, *170*, 45–68. [CrossRef]
32. Ishihara, T.; Kazuki, H.; Susumu, O. A wind tunnel study of turbulent flow over a three-dimension steep hill. *J. Wind Eng. Ind. Aerodyn.* **1999**, *83*, 95–107. [CrossRef]
33. Ferziger, J.H.; Peric, M. Computational Methods for Fluid Dynamics. *Phys. Today* **1997**, *50*, 80–84. [CrossRef]
34. Liu, Z.; Cao, S.; Liu, H.; Ishihara, T. Large-Eddy simulations of the flow over an isolated three-dimensional hill. *Bound.-Layer Meteorol.* **2019**, *170*, 415–441. [CrossRef]
35. *IEC, Wind turbines—Part 1: design requirements 61400−1*, 3rd ed.; International Electrotechnical Commission: Geneva, Switzerland, 2005.

energies

MDPI

Article

Turbulent Flow Fields Over a 3D Hill Covered by Vegetation Canopy Through Large Eddy Simulations

Zhenqing Liu [1,*], Yiran Hu [1], Yichen Fan [1], Wei Wang [2] and Qingsong Zhou [1]

1 School of Civil Engineering & Mechanics, Huazhong University of Science and Technology, Wuhan 430074, China

2 Department of Architecture and Building Engineering, Tokyo Institute of Technology, Yokohama, Kanagawa 1528550, Japan

* Correspondence: liuzhenqing@hust.edu.cn

Received: 14 August 2019; Accepted: 16 September 2019; Published: 23 September 2019

Abstract: The flow fields over a simplified 3D hill covered by vegetation have been examined by many researchers. However, there is scarce research giving the three-dimensional characteristics of the flow fields over a rough 3D hill. In this study, large eddy simulations were performed to examine the coherent turbulence structures of the flow fields over a vegetation-covered 3D hill. The numerical simulations were validated by the comparison with the wind-tunnel experiments. Besides, the flow fields were systematically investigated, including the examinations of the mean velocities and root means square of the fluctuating velocities. The distributions of the parameters are shown in a three-dimensional way, i.e., plotting the parameters on a series of spanwise slices. Some noteworthy three-dimensional features were found, and the mechanisms were further revealed by assessing the turbulence kinetic energy budget and the spectrum energy. Subsequently, the instantaneous flow fields were illustrated, from which the coherent turbulence structures were clearly identified. Ejection-sweep motion was intensified just behind the hill crest, leading to a spanwise rotation. A group of vertical rotations were generated by the shedding of the vortex from the lateral sides of the hill.

Keywords: turbulent structure; computational fluid dynamics (CFD); large eddy simulations (LES); 3D hill; canopy; flow fields

1. Introduction

As we all know, turbulence has a very important research position in many fields, and the turbulent flow field on the topography in different fields is closely watched by many researchers, including the wind turbine sittings [1], the pollution diffusions [2], estimation of aerodynamic loadings on structures [3], identifications of the tree damage with high risk [4], as well as the forest fire propagation [5]. Given the complexity of the flow over real complex topographies, many studies have been conducted to clarify the turbulent boundary layer (TBL) flow over simplified hills. The relevant research is primarily about the smooth topographies. Very little research is about the flow over topographies with the canopy covered, which can be divided into three groups, i.e., the flat topography, the 2D ridge, and the 3D hill.

Dupont and Brunet examined the coherent structures in canopy edge flow over flat terrain by large eddy simulations (LES) [6]. In their research, the development of turbulent structures was revealed, and we can see from the schematic that many of them are strikingly similar to the development of coherent structures observed in mixed layers. Takahashi et al. investigated the turbulence characteristics of the flow over a 2D ridge with a rough surface in wind tunnel experiments [7] and the effects of roughness density were studied. It was found that the mean velocity profiles were not sensitive to the density of the roughness and the roughness served as strong windbreaks, weakening the wind velocity near the hill surface. Cao and Tamura experimentally studied the roughness effects on TBL flow over a 2D ridge [8], in which above the top of the rough hill the acceleration ratio was slightly larger than the smooth hill and because the separation bubble extends to the downstream of the hill, the reattachment length was also larger than the smooth hill. Subsequently, Cao and Tamura further examined the effects of ground roughness on the flow over a 2D ridge with or without sudden roughness change [9]. In their research, four situations were considered, i.e., (i) smooth hill in smooth flow, (ii) rough hill in rough flow, (iii) smooth hill in rough flow, and (iv) rough hill in smooth flow. It was concluded that the velocity deficit varied with roughness conditions on the hill surface or inflow area, thereby creating a completely different turbulence structure. Dupont et al. studied the turbulent flow over a 2D ridge with a vegetation canopy cover using LES [10], in which the coherent structures were identified. According to their research, the turbulence in the wake region is mainly caused by the superposition of the following three turbulent structures: (i) the large turbulent structure caused by the high shear layer, (ii) a turbulent structure caused by a back pressure gradient caused on the back of the ridge, and (iii) a turbulent structure due to the presence of the canopy. Ishihara et al. conducted a study of the flow over a 2D ridge and a 3D hill under smooth and rough ground by wind tunnel experiment [11]. Detailed information was provided, covering the inflow condition, the mean, as well as the fluctuating velocities. Afterwards, Ishihara and Hibi examined the flow over a rough 3D hill by Reynolds-averaged Navier–Stokes (RANS) models [12]. Standard k-ε model and Shin's nonlinear model were tested and the performance of Shin's nonlinear model was found to be much better than k-ε model. Tamura et al. adopted LES to reproduce the experimental data by Ishihara et al. [11,13], and the LES results were consistent with the experiments both for the smooth and rough hills. Most recently, Liu et al. numerically studied the flow over a blocks-covered flat ground [14], a vegetation-covered flat ground, a vegetation-covered 3-D hill, as well as a real forest-covered terrain. The method yielding the turbulent inflow, determining the grid spacing, and the settings of LES were validated.

However, most of the studies on the flow over a rough 3D hill only provided the flow information on the symmetry plane. There has been scarce research giving the 3D characteristics of the flow over a rough 3D hill. Therefore, systematic information of the 3D flow fields will be provided in the present study. The representative parameters, i.e., the mean velocities and root means square (r.m.s.) of the velocity fluctuations were examined. Furthermore, the three-dimensional features and the mechanisms were revealed by investigating the turbulence kinetic energy (TKE) budget and the spectrum energy.

2. Numerical Model

2.1. Governing Equations

Time-dependent Navier-Stokes (N-S) equations of large eddy simulations in Cartesian coordinates (x, y, z) are:

$$\frac{\partial \rho \tilde{u}_i}{\partial x_i} = 0, \tag{1}$$

$$\frac{\partial \rho \tilde{u}_i}{\partial t} + \frac{\partial \rho \tilde{u}_i \tilde{u}_j}{\partial x_j} = \frac{\partial}{\partial x_j}\left(\mu \frac{\partial \tilde{u}_i}{\partial x_j}\right) - \frac{\partial \tilde{p}}{\partial x_i} - \frac{\partial \tilde{\tau}_{ij}}{\partial x_j}, \tag{2}$$

where $\tilde{p}$ and $\tilde{u}_i$ represent the filtered pressure and velocities, respectively, ρ denotes the density, μ denotes the viscosity, and $\tilde{\tau}_{ij}$ is the sub-grid scale (SGS) stress. To close the equations for the filtered velocities, a model for $\tilde{\tau}_{ij}$ is required:

$$\tilde{\tau}_{ij} = -2\mu_t \tilde{S}_{ij} + \frac{1}{3}\tilde{\tau}_{kk}\delta_{ij}, \tag{3}$$

$$\tilde{S}_{ij} = \frac{1}{2}\left(\frac{\partial \tilde{u}_i}{\partial x_j} + \frac{\partial \tilde{u}_j}{\partial x_i}\right), \tag{4}$$

where $\tilde{S}_{ij}$ denotes the rate-of-strain tensor, μ_t is the SGS turbulent viscosity, and δ_{ij} is the Kronecker delta. The Smagorinsky–Lilly model is used to determine the SGS turbulent viscosity [15]:

$$\mu_t = \rho L_s^2 |\tilde{S}| = \rho L_s^2 \sqrt{2\tilde{S}_{ij}\tilde{S}_{ij}}, \tag{5}$$

$$L_s = \min\left(\kappa d,\ C_s V^{1/3}\right), \tag{6}$$

where L_s is the mixing length, κ is the von Kármán constant (0.42), d is the distance to the wall, and V is the cell volume. Here, C_s is set as 0.1 according to Iizuka and Kondo [16].

When the cells are in the viscous sublayer, the shear stresses are yielded from:

$$\frac{\tilde{u}}{\tilde{u}_\tau} = \frac{\rho \tilde{u}_\tau y}{\mu}. \tag{7}$$

Accurately modeling the inflow is essential for the success of simulating the flow in the 3D hill wake. In the present study, the upstream blocks arrangement in the simulations were identical to those in the experiments by Ishihara et al. [12]. Besides this, the distance between the roughness blocks to the location where the 3D hills were placed was also set identically to that in the wind tunnel experiment. The additional drag force is used to simulate the roughness elements:

$$\frac{\partial \rho \tilde{u}_i}{\partial t} + \frac{\partial \rho \tilde{u}_i \tilde{u}_j}{\partial x_j} = \frac{\partial}{\partial x_j}\left(\mu \frac{\partial \tilde{u}_i}{\partial x_j}\right) - \frac{\partial \tilde{p}}{\partial x_i} - \frac{\partial \tilde{\tau}_{ij}}{\partial x_j} + \tilde{f}_{u,i}, \tag{8}$$

where the drag force term $\tilde{f}_{u,i} = C_d A_f\, \rho\, \tilde{u}_i|\tilde{u}|$, A_f is the frontal area of the roughness blocks, $C_d = 100$ to model the large drag effects in the solid roughness blocks, consistent with the method by Lee and Sung [17]. The geometry of the computational domain and the grid system can be used from the previous work, ignoring the effects of different rough block layouts, which saves manpower and material resources and, according to the relevant experimental theory, only the function that determines geometry of the rough block needs to be modified.

To model the canopy, the drag force term, $\tilde{f}_{u,i}$, is substituted in the momentum equations in the same form as Equation (8). The drag force term is determined by $\tilde{f}_{u,i} = C_d\ A_f\ \rho\ \tilde{u}_i|\tilde{u}|$, where $C_d = 0.2$ denotes the drag force coefficient, $A_f = 0.6\ \mathrm{m}^{-1}$ is the leaf-area density, $|u|$ is the velocity magnitude. C_d and A_f are determined according to the study by Ishihara and Qi [18].

2.2. Configurations

Before providing the detailed configurations of the computational model, the wind tunnel experiments by Ishihara et al. should be briefly introduced, which is the case the present LES aimed to reproduce [11]. During the experiment, a neutral layered atmospheric boundary layer (ABL) was simulated by two 60 mm high cube arrays followed by 20 m and 10 mm high cubes, covered on the 1.2 m test section. The rough element was covered with an artificial turf with a height of 5 mm and the rest of the floor of the test building was 5.8 m long. At the contraction exit, the mean wind speed $U_{in} = 5.4\ \mathrm{m\ s}^{-1}$. The boundary layer thickness, δ, was nearly 0.36 m at the hill center, and the scale, λ, was about 1:1000 of the ABL, based on the power spectra of the longitudinal velocity component. The wind speed outside the boundary layer was measured as $U_\infty = 5.8\ \mathrm{m\ s}^{-1}$. As a result, the simulated boundary layer had a bulk Reynolds number, $Re_b = U_\infty\delta/\nu = 1.4 \times 10^5$, and the friction Reynolds number $Re_\tau = U_\tau z_0\ /\ \nu = 6.4$, where, $U_\tau = \sqrt{\tau\ /\ \rho} = 0.32\ \mathrm{m\ s}^{-1}$, τ is the shear stress in an arbitrary layer, $z_0 = 0.3$ mm refers to the roughness height determined by comparing the resulted mean streamwise velocity profile based on the logarithmic law: $U(z) = \frac{U_\tau}{\kappa} \ln \frac{z}{z_0}$.

The computational domain is shown in Figure 1. The origin point (0, 0, 0) was 3.4 m downstream from the roughness blocks. The 3D hill was placed at the origin point, exhibiting a shape: $z_s(x,\ y) = h\cos^2\pi(x^2 + y^2)^{1/2}/2L$ when $(x^2 + y^2)^{1/2} < L$, and $z_s(x,\ y) = 0$ when $(x^2 + y^2)^{1/2} > L$, where $h = 40$ mm and $L = 100$ mm. $z\prime = z - z_s(x, y)$, was employed to determine the height above the ground. A domain size in spanwise direction of approximately $L_y = 1.8\delta$ was adopted. $L_y = 1.8\delta$ was further confirmed to be enough, since the two-point correlations converged to zero in the half-width of

L_y. As for the necking zone, in the upstream region, a buffer zone (i.e., 2.0 m long, 5.6δ) was used to absorb upward-propagating wave disturbances. Vertical domain size was the same as that in the wind tunnel, L_z = 2.5δ. The outlet was set at 6.7δ downstream from the origin point (0, 0, 0). A grid nesting procedure was adopted. Two nested grid domains (coarse and fine) were adopted, as shown by the red dashed lines in Figure 1. The fine grid domain covered a range of (L_x', L_y', L_z') = (12L, 2L, 3L).

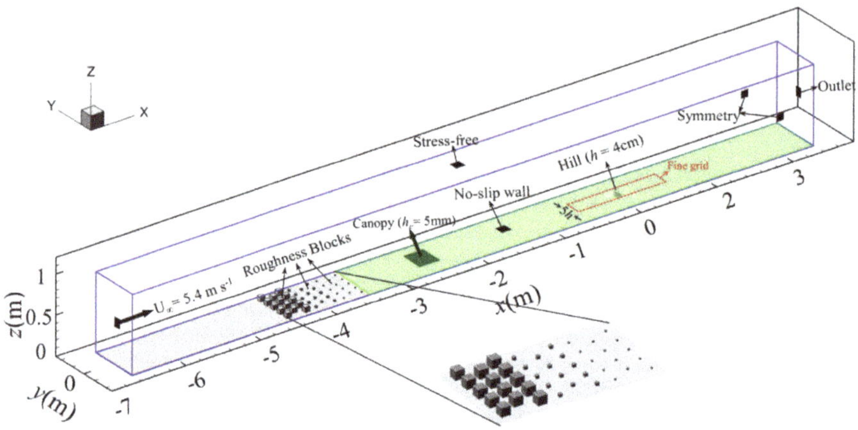

Figure 1. Configurations of the numerical model. The volume enclosed by solid blue lines is the computational domain and the area enclosed by the red dashed rectangular is the fine grid region, 5 h in width and 30 h in length. The original point (0, 0, 0) of the model is at the center of the fine grid, where the hill with a height of 40 mm is mounted. The solid dark squares with arrows indicate the boundary conditions. The green area indicates the regions with a vegetation canopy cover.

2.3. Mesh of the Model

For the grid size, in the vertical direction, the grid was stretched starting with a vertical grid spacing of 0.2 mm on the surface in both the fine and coarse grid domains. The resulting vertical grid resolution on the surface $z^+ = U_\tau \Delta z_{\min} / \nu < 1.0$ covered most of the domain, except for the windward part of the hill where the maximum z^+ was less than 2. The vertical grid should be also sufficiently fine enough to capture the turbulent flow fields induced by the large wind shear, which was located at the height of about h. The vertical size of the wind shear-induced eddies is proportional to the shear length-scale $L_s = U(z)/[dU(z)/dz]$, where U denotes the mean streamwise velocity. From a pre-modeling of the case with Δz = 10 mm in the shear layer region, L_s was found to be nearly 8 mm. In the present LES, Δz was finally set to be 4 mm at z' = 80 mm. In the region z' < 80 mm, the vertical grid was increased by a hyperbolic function. Exceedingly large aspect ratios ($\Delta x/\Delta z$, $\Delta y/\Delta z$) may cause numerical errors in the horizontal direction and some distortions of the eddies. The value of 10 was adopted in the present LES. Given $\Delta z_{\min}$ = 0.2 mm, Δx and Δy in the fine grid domain should be about 2 mm. Moreover, in the grid independence examinations, the models with Δxy = 5.65, 4.0, 2.8, and 2.0 mm were calculated. The meshes Δxy = 2.8 and 2.0 mm gave roughly the same results, suggesting the grid-independent results were achieved. From this result, we got a more convincing conclusion, that is, in the LES of the current research field, the aspect ratio can very well estimate the actual situation of the turbulent structure on the 3D hill. Thus, a grid size 2.0 mm in horizontal direction was applied in the fine grid region. Finally, the corresponding nondimensional grid resolutions of fine domain in x, y, and z directions were x^+ = [3.2, 23], y^+ = [3.2, 23], z^+ = [0.23, 1.9]. Downstream of the fine grid domain and upstream of the roughness blocks, Δx was stretched at a ratio of 1.2. Total grid number was 2.6×10^7. The vertical slice showing the grid distribution is provided in Figure 2. The mean skin friction coefficient was determined as:

$$C_f = \frac{\tau_w}{\frac{1}{2}\rho U_h^2} \tag{9}$$

where, τ_w is the mean shear stress on the wall, as shown in Figure 3, where it is clear that the maximum C_f is located at the front side of the hill reaching to about 0.003.

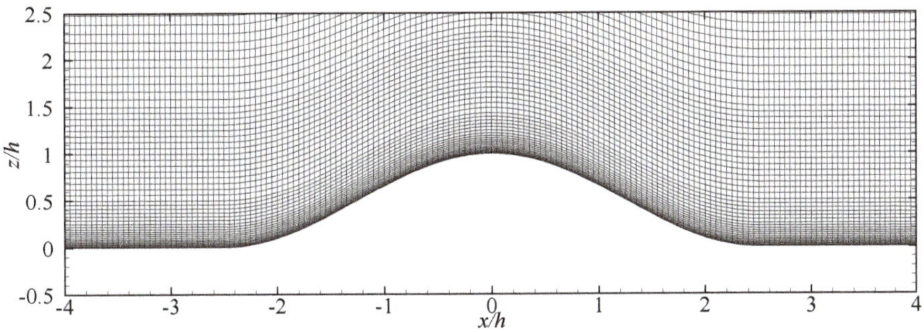

Figure 2. Mesh on the vertical slice crossing the center of the topographies.

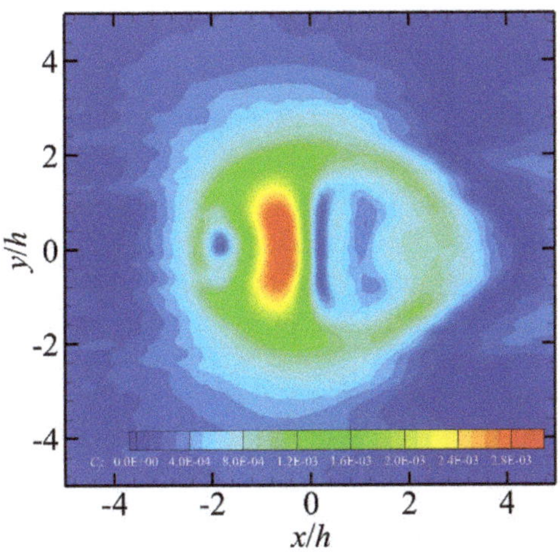

Figure 3. Distributions of mean skin friction coefficient around the hill.

2.4. Boundary Conditions

A stress-free condition was used at the top and the spanwise sides. Uniform wind flow with a 5.4 m s^{-1} was set at the inlet. The gradient-free boundary condition was set at the outlet. The no-slip condition was applied at the bottom. The stress-free condition at the top and the symmetry condition at the lateral sides were different than the no-slip condition in the wind tunnel. However, considering the very thin TBL at those smooth walls in the wind tunnel (at a scale of 10^{-3} m) and the fact that the induced turbulence at the walls are far from the region of interest, these boundary condition differences are supposed to have negligible effects on the LES solutions. By validating the numerical model, artificially changing the boundary conditions in wind tunnel to those adopted in the present LES was proven to be satisfactory.

2.5. Solution Schemes

Finite volume method (FVM) was applied in LES. For the convection and viscous terms in the LES method, the second-order center difference format was used for calculation, while for the viscous terms, the second-order implicit format was used for processing. Time step size Δt was set as 0.0001 s and in convective time units, $\Delta t^* = \Delta t U_h/h = 0.01$, where $U_h = 3.9$ m s^{-1} is the mean velocity at the point ($x = 0$, $y = 0$, $z = h$) in the absence of the 3D hill. The Courant–Friedrichs–Lewy (CFL) number is based on the time step size (Δt), velocity (u_i), and grid size (Δx_i), expressed as $C = \Delta t \Sigma u_i/\Delta x_i$. It is worth noting that the CFL number is limited to no more than 2 (i.e., $C_{max} = 0.9$) throughout the computational domain. For solving discrete equations, it is possible to use the SIMPLE (semi-implicit pressure linked equations) algorithm, which is very effective and popular. Due to the huge amount of computation, the entire calculation process needs to be performed on a high performance 3PCs in parallel. The simulation process takes 435 hours to complete, but after simulating nearly 400 time units, it was found that the initial transient effects disappeared. Statistical convergence was achieved when $\left|\left\langle u_{i-y}\right\rangle - \left\langle u_{i+y}\right\rangle / \left(\left\langle u_{i-y}\right\rangle + \left\langle u_{i+y}\right\rangle\right)/2\right| < 1\%$ in the near wake of the 3D hill, which was over 350 time units, where $\langle\,\rangle$ denotes the time-averaging process.

Quantifying the relative importance of the SGS viscosity υ_{sgs} is important for determining whether only the resolved fluctuations are sufficient for the statistics. The equivalent eddy-viscosity υ_e, evaluated as $\upsilon_e \left\|\frac{2}{3}k\delta_{ij} - \left\langle u_i' u_j'\right\rangle\right\| / \left\|2\left\langle S_{ij}\right\rangle\right\|$, where $k = \frac{1}{2}u_i' u_i'$, u_i' is the deviation from $\langle u_i\rangle$, and the norm $\|\,\|$ of a matrix ϕ_{ij} denotes $\left(\phi_{ij}\phi_{ij}\right)^{1/2}$, can be adopted to examine the relative importance between the resolved fluctuations and the modeled ones using the SGS model. In the present LES, $\upsilon_e/\upsilon_{sgs}$ in the 3D hill wake was found to be less than 5%, indicating that the effects of the SGS model on the present resolved LES results were negligible. As a result, only the resolved eddies in the present LES were considered for the statistics.

3. Numerical Results

3.1. Generated TBL

The profiles of incoming TBL, both in terms of $\langle u\rangle$ and r.m.s. of fluctuating velocities, σ_{u_i}, are critical for a precise reproduction of 3D hill wake flow, playing significant roles in the separation and reattachment, and in return affecting the coherent turbulence structures. Thus, before the detailed discussions of the turbulence properties in the wakes of the 3D hill, it is worthwhile to examine the performance of the adopted incoming TBL generation method. Firstly, it is noteworthy to get an insight of the instantaneous turbulent flow fields before the presentation of the flow fields statistics. Figure 4 displays a snapshot of two iso-surfaces of spanwise vorticity $\omega_y = \partial w/\partial x - \partial u/\partial z = -15$ s^{-1} and 15 s^{-1} of the TBL in the absence of 3D hills at 400 time units. As soon as the flow travels downstream to the roughness block region, strong coherent turbulence structures are induced. Further propagating downstream, the flow experiences an obvious boundary-layer growth, together with which, the streak structures are enhanced; on the contrary, the ejection-sweep motions are shredded.

After the flow reached an equilibrium state (400 time units), statistics of $\langle u\rangle$ and σ_{u_i} were collected over 400 time units. In addition, assuming statistics in spanwise direction is homogeneous, the profiles of $\langle u\rangle$ and σ_{u_i} are further smoothed by averaging them over all $y \in (-0.5L,\ 0.5L)$ locations at each $(0, z)$ position. The resulted profiles of $\langle u\rangle$ and σ_{u_i}, as well as the comparison with those in the experiment by Ishihara et al. [11], are shown in Figure 5, where the normalization, using h and U_{4h}, which is the same as the experiment, is performed. $U_{4h} = 5.3$ m s^{-1} in the present LES and 5.2 m s^{-1} in the wind tunnel experiment. The comparisons show good agreement.

3.2. Instanteneous Flow Fields

The Q-criterion was adopted to clarify the instantaneous turbulent structures, $Q = 1/2(S_{ij}S_{ij} - \Omega_{ij}\Omega_{ij})$, where $S_{ij} = 1/2(\partial \tilde{u}_i/\partial x_j - \partial \tilde{u}_j/\partial x_i)$ and $\Omega_{ij} = 1/2(\partial \tilde{u}_i/\partial x_j + \partial \tilde{u}_j/\partial x_i)$, respectively. The snapshots of iso-surfaces of the normalized Q equaling to 3.0 and −3.0 at $tU_h/h = 116$ and 118 are plotted in Figure 6.

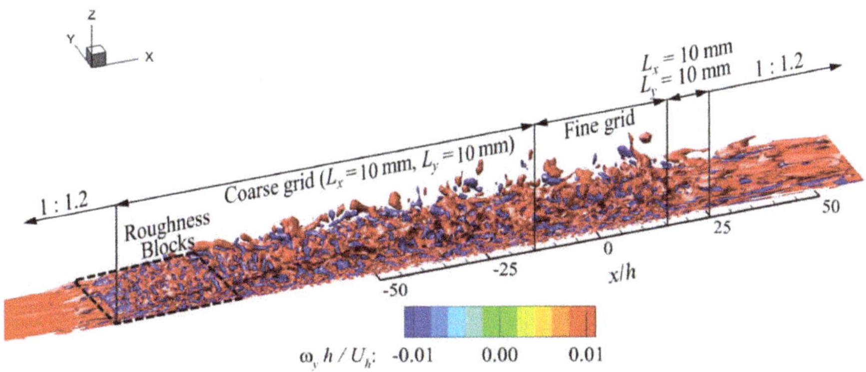

Figure 4. The instantaneous flow fields with the absence of the hill. The visualization by the normalized vorticity in spanwise direction, $\omega_y h/U_h$, is applied. Two iso-surfaces with $\omega_y h/U_h = -0.01$ and $\omega_y h/U_h = 0.01$ are plotted. The grid sizes and the grid spacing ratio are sketched.

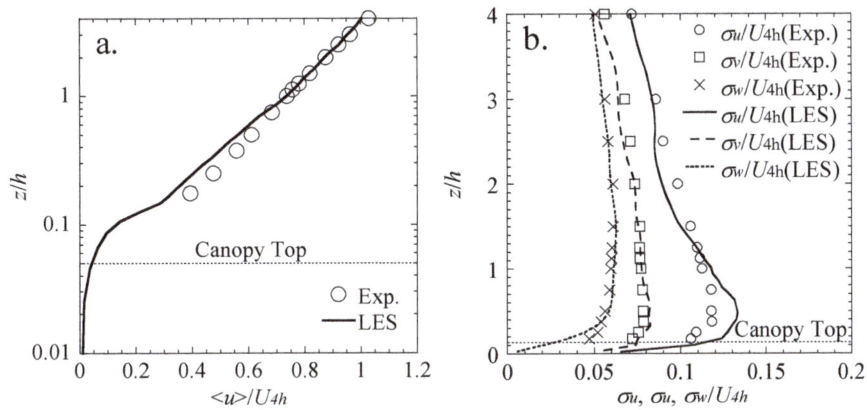

Figure 5. Vertical profiles of (**a**) normalized mean streamwise velocity, $\langle u\rangle/U_{4h}$, in which y axis is plotted in logarithmic form and (**b**) normalized fluctuations of streamwise component, σ_u/U_{4h}, spanwise component, σ_v/U_{4h}, and vertical component, σ_w/U_{4h} at $x = 0$ and $y = 0$ in the absence of the hills. The data from the wind tunnel experiment by Ishihara et al. [11] are superimposed to verify the accuracy of the generated upwind turbulent boundary layer (TBL) flow. The dotted line exhibits the canopy top.

The iso-surfaces were then colored by the normalized instantaneous streamwise velocities. The red dashed line represents the main vortex core downstream of the hill. Obviously, a pair of vortices sheds from the lateral sides of the hill. Also, the energy of the shedding at each lateral side of the hill seems to vary periodically, forming a Karman vortex street, as usually observed in the wake of bluff bodies. In addition to the shedding from the lateral side, some spanwise vortices shedding from the crest, indicated by the yellow dashed lines, can also be identified. These vortices are primarily due to the ejection of the flow from the hill top and the Kelvin–Helmholtz instability. The lateral shedding and the spanwise vortices at the upper boundary of the wake yield an area with quite small velocities and weak vortices, leading to the formation of a dead flow area. Reviewing the depictures of the averaged vorticities and connecting the visualization of the instantaneous flow fields, the mean coherent structures of the flow over the hill covered by vegetations can be sketched in Figure 6c, where V1 and V4 are the canopy-induced vortices, V2 is the vortex distorted by the topography, the vortex caused by the ejection-sweep motions is indicated by V3, and V4 are the vortices shedding from the lateral sides of the hill.

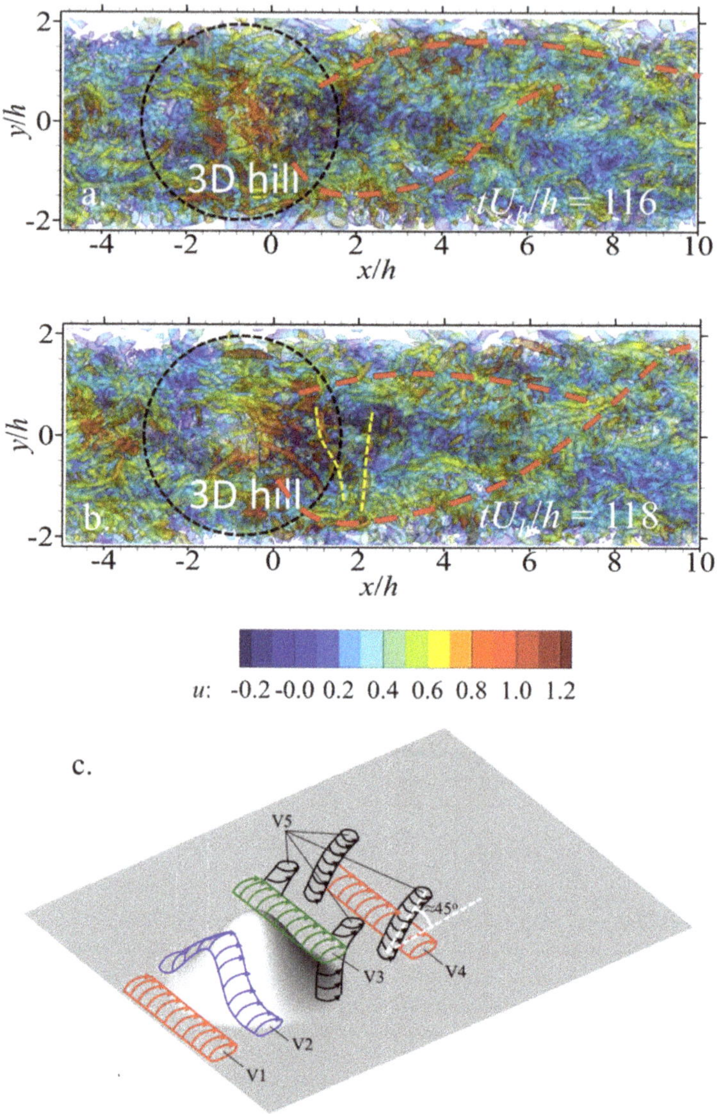

Figure 6. Instantaneous flow fields at (**a**) $tU_h/h = 116$ and (**b**) $tU_h/h = 118$ visualized by iso-surfaces of Q-criteria with normalized values 3.0 and −3.0. A 50% transparent effect was applied for the iso-surfaces of Q values to show the flow structures clearly. The iso-surfaces are further colored by the instantaneous streamwise velocity. The red dashed lines represent the center of the vortices shedding from the lateral sides of the hill and the yellow dashed lines show the vortices shedding from the hill crest. The centers of the vortices were determined by velocity gradient eigenmodes method. x and y axes are normalized by h. (**c**) Sketch of the flow structure. V1 and V4 are vortices induced by the canopy, V2 is the vortex distorted by the topography, V3 is the vortex shedding from the hill crest, V5 is the vortex shedding from the lateral sides of the hill.

3.3. LES Statistics

In this section, the numerically predicted mean velocities and fluctuations on $y = 0$ plane are firstly presented to validate the numerical model adopted in the present LES. Subsequently, in order to provide a general view of the 3D flow structures in the hill wake, a 3D view of the iso-surfaces of enstrophy are plotted, followed by the illustrations of the time-averaged velocities in streamwise direction, $\langle u \rangle$, vertical direction $\langle w \rangle$, the r.m.s. of the streamwise fluctuation, σ_u, spanwise fluctuation, σ_v, as well as vertical fluctuation, σ_w, at several spanwise slices.

The profiles of $\langle u_i \rangle$ and σ_{u_i} are plotted in Figure 7, where the normalization, using h and U_{4h}, the same as the experiment, was adopted. The flow accelerates both in streamwise and vertical directions on the way up to the hill top. $\langle w \rangle$ reaches the maximum at $x = -1h$, which is due to the largest hill slope here. Still at $x = -1h$, the fluctuations of the flow in the canopy are the most energetic, which suggests that the upcoming flow becomes very unstable at the upwind hill side. This unstable feature is probably because it is much easier for the flow to penetrate the canopy due to the blockage effects of the hill. As a result, the interaction between the flow and the canopy becomes much stronger.

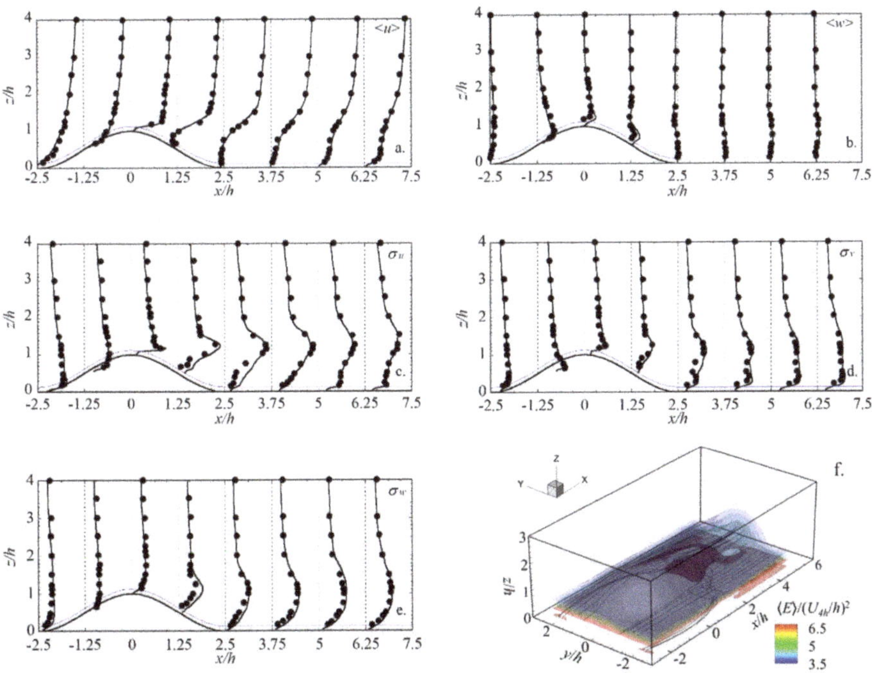

Figure 7. Vertical profiles of: (**a**) normalized mean streamwise velocity, $\langle u \rangle / U_{4h}$; (**b**) normalized mean vertical velocity, $\langle w \rangle / U_{4h}$; (**c**) normalized r.m.s of fluctuating streamwise velocity, σ_u / U_{4h}; (**d**) normalized r.m.s of fluctuating streamwise velocity, σ_v / U_{4h}; (**e**) normalized r.m.s of fluctuating vertical velocity, σ_w / U_{4h}, at x equals to $-2.5h$–$6.25h$ with a step size of 1.25 h. The unit in the x-axis represents the unit in the plotting of $\langle u \rangle / U_{4h}$ and $\langle w \rangle / U_{4h}$, and 0.2 in the plotting of σ_u / U_{4h}, σ_v / U_{4h} and σ_w / U_{4h}. The solid lines are from the present LES and the circles are from the wind tunnel experiment by Ishihara et al. [12]. The averaged three-dimensional flow structures are shown in Figure 7f by the normalized mean enstrophy with the mean streamlines superimposed. The blue solid line indicates the canopy top. x, y, and z axes have been normalized by h.

At the summit, $x = 0$, the speed-up of $\langle u \rangle$ is the largest. Interestingly, a sharp peak in the profile of σ_u occurs at the canopy top of the summit. This should result from the ejection of the flow as well as the rapid distortion of the streamwise eddies. However, $\langle u_i \rangle$ and σ_{u_i} quickly decrease to about 0 when the flow penetrates the canopy under the drag effects from the canopy, formulated by the drag force terms in the momentum–balance equations. When the flow propagates to the lee side of the hill, the mean velocities are distorted significantly. On one hand, inversion points on the profiles of $\langle u \rangle$ and $\langle w \rangle$ occur due to the presence of a recirculation bubble. On the other hand, the vertical gradient of $\langle u \rangle$ does not decrease monotonously as those upstream of the hill summit. In fact, $\partial\langle u \rangle/\partial z$ firstly decreases and then increases with the rise in the elevation, forming an inflection point, at which $\partial\langle u \rangle/\partial z$ reaches the maximum, suggesting large wind shear and turbulence productions. This can be confirmed by the near wake σ_u profiles, of which the maximum is just located at the inflection points of the $\langle u \rangle$ profiles.

Additionally, similar to the flow over an isolated smooth 3D hill observed by Liu et al. [19], two peaks on σ_v profiles appear, also for the rough 3D hill, with one located at the canopy top and the other in the shear layer region. Comparing the low-elevation peaks of σ_v downstream and upstream of the hill, the magnitudes of the peaks are roughly identical, which differs with that in the smooth 3D hill case, implying different sources of low-elevation peaks of σ_v. For the smooth case, Liu et al. concluded that the different rotation directions between the inner core and the outer core of the wake vortex are the sources [19]. For the rough one, however, due to the nearly same value between the near-ground peak σ_v upstream and downstream the hill, the production of turbulence from the canopy should be the source. This also indicates the different coherent structures in the wakes of the smooth and rough hills.

Different from the 2D ridge, the flow fields over an isolated 3D hill are indeed three dimensional. The three-dimensional flow fields over a smooth 3D hill have been revealed and the coherent structures have been clarified in the study by Liu et al. [19]. However, the research considering flow fields over an isolated 3D hill with vegetation covered from a three-dimensional view is lacking. Before the presentation of the three-dimensional distributions of the parameters, it is worthwhile to have a general view about the structures of the mean flow fields over the 3D rough hill. As shown in Figure 7f, three enstrophy iso-surfaces with normalized values of 3.5, 4.5, 5.5, and 6.5 are shown. To illustrate the flow structures clearly, transparent effects are applied, and the streamlines determined by the space–time averaged velocities are plotted as well.

Surprisingly, the most energetic rotations were not located on the central plane $y = 0$, but $0.5h$ away from it. From the later discussion about the averaged vorticities, the rotations in vertical and streamwise directions were found to be the major sources of the large enstrophy. That is the reason why the strongest enstrophy does not occur at $y = 0$ plane, where only the mean spanwise vorticity has values because of the symmetry of the flow. In the far wake region, the shapes of the enstrophy iso-surfaces seem to be the projections of the 3D hill, and obviously the effects of the hill remain strong even at $x = 5h$. Furthermore, from the depicture of the streamlines, it can be found that the near-ground flow converges to the very center of the wake as it moves downstream from the lateral sides, and then its direction is changed upstream and upward in the recirculation bubble. Furthermore, we can see that the flow at relatively high elevations can hardly penetrate the bubble. However, the convergence of the flow to the $y = 0$ plane is still obvious due to the negative pressure in the central wake.

Furthermore, it is necessary to mention that the time-averaged flow fields should be symmetrical if the sampling time is sufficient. However, the symmetry of the flow can hardly be achieved in the numerical simulations, even when the sampling lasts for 400 time units. Therefore, space averaging is further carried out to obtain the symmetry of the flow. For the symmetrical parameters, such as $\langle u \rangle$, $\langle w \rangle$, σ_{u_i}, etc., $\phi = [\phi\ (y) + \phi\ (-y)]/2$, and for the anti-symmetry parameters, such as $\langle v \rangle$, $\phi = [\phi\ (y) - \phi\ (-y)]/2$.

To illustrate the three-dimensional flow structures in a much clearer way, eight spanwise slices from $x = -2.5\ h$ to $x = -6.25\ h$ with a step size of 1.25 h were selected, which just coincide with the locations where the profiles in Figure 7 are shown. Each slice was centered at $y = 0\ h$ with a spanwise size of 5 h and vertical size of 4 h. The red dashed lines and the blue dashed lines superimposed on the slices indicate the locations where the flow recovers to the inactive boundary layer (IBL) flow and the locations where TKE shows the peak, respectively. Accordingly, the region enclosed by the red and blue dashed lines can be considered as the shear-layer region. In the present study, the wake depth (h_w) was quantitatively calculated as $|\langle u\rangle_{\text{down}}(h_w) - \langle u\rangle_{\text{up}}(h_w)| / \langle u\rangle_{\text{up}}(h_w) < 0.05$, where $\langle u\rangle_{\text{down}}$ is $\langle u\rangle$ downstream of the hill, and $\langle u\rangle_{\text{up}}$ is $\langle u\rangle$ upstream of the hill.

From the averaged parameters shown in Figure 8, some important features were not captured by the vertical profiles on $y = 0$ slice. Firstly, it is apparent that the area influenced by the topography not only increases in the vertical direction, but also expands in the spanwise direction with the increase in x, while this influence is weakened and almost dismisses when $x \approx 12h$ (not shown in the figures). Secondly, it is important that $\langle v\rangle$ is accelerated both at the upwind and lee sides of the hill. The magnitude of $\langle v\rangle$ can even reach 0.2 U_h. It can be further found that the streamwise distance to the hill center of the upstream peak $\langle v\rangle$, 1.25 h, differs from that downstream, 2.5h, which results from the appearance of the recirculation bubble. This recirculation bubble acts as an obstacle, and the flow majorly moves in the region outside of the bubble. The positive $\langle v\rangle$ in $-y$ part and the negative $\langle v\rangle$ in $+y$ part imply the presence of a pair of vortices pointing to positive and negative vertical directions, respectively. Finally, on the way up to the summit, $\langle w\rangle$ shows sole peak at $y = 0$, while two pairs of peaks $\langle w\rangle$ emerged downstream from the hill crest. One pair was located at the shear layer with negative value, and the other located in central wake with positive value, revealing the presence of the vortices pointing to the streamwise direction.

Similar distributions can be detected for σ_p, σ_u, and σ_v, as shown in Figure 9. The concentrations of σ_p, σ_u, and σ_v in the shear layer are obvious. However, the distributions for σ_w are distinct, whose peaks are centered just below the shear layer center. In addition, it is worth mentioning that σ_p and σ_{u_i} suddenly increase at the upwind footage of the hill, revealing the presence of a horseshoe vortex. This horseshoe vortex was also identified in the smooth 3D hill case in the study by Liu et al. [19], implying that the disturbance of the flow from the canopy cannot eliminate the formation of the horseshoe vortex due to the topography.

3.4. TKE Budget

Studying roughness–turbulence interactions is one of the motivations of the present study and the TKE budget provides information about the gain or loss of the TKE, as has been examined in the studies considering the effects of roughness or the particle bed on the turbulent flow fields [20–22]. In those studies, it was found that the roughness modulates the near-bed turbulence; produces streamwise structures, which undergo distortion and breaking; and reduce the large-scale anisotropy. A double averaging of the flow field reveals spatial inhomogeneities at the roughness scale and alternate paths of energy transportation in the turbulent kinetic energy (TKE) budget. The TKE budget is formulated as:

$$\overline{u}_j\frac{\partial k}{\partial x_j} = -\frac{1}{\rho}\frac{\partial u_i'\overline{p}}{\partial x_i} - \frac{1}{2}\frac{\partial u_j'u_j'u_i'}{\partial x_i} - u_i'u_j'\frac{\partial \overline{u}_i}{\partial x_j} - \nu\frac{\partial u_i'}{\partial x_j}\frac{\partial u_i'}{\partial x_j} \tag{10}$$

where the advection term A is $-\langle u_i\rangle\frac{\partial k}{\partial x_j}$, the pressure diffusion term, $-\frac{1}{\rho}\frac{\partial u_i'\langle p\rangle}{\partial x_i}$, is too small to be considered, the turbulence transportation term T is $-\frac{1}{2}\frac{\partial u_j'u_j'u_i'}{\partial x_i}$, the turbulence production term P is $-u_i'u_j'\frac{\partial \overline{u}_i}{\partial x_j}$, and the dissipation term D is $-\nu\frac{\partial u_i'}{\partial x_j}\frac{\partial u_i'}{\partial x_j}$.

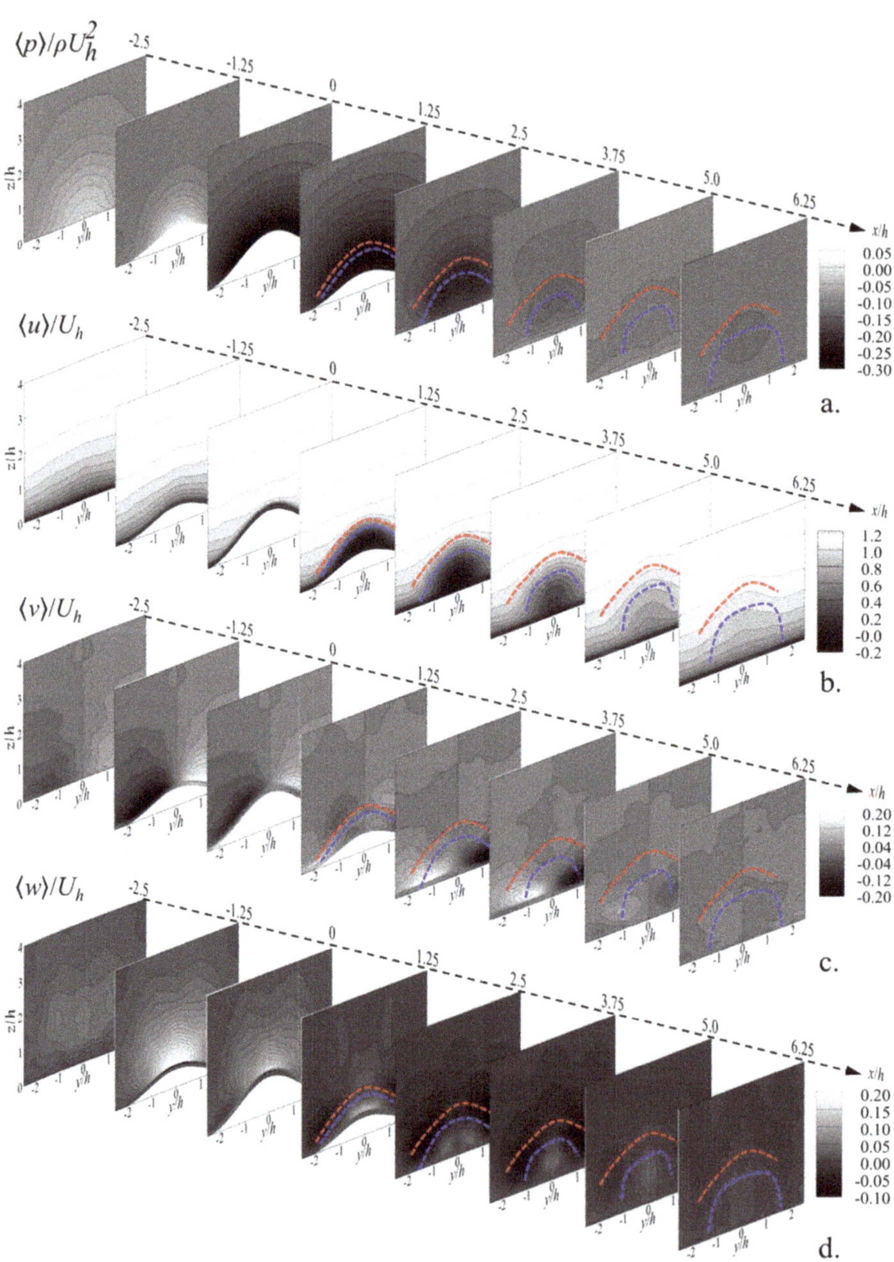

Figure 8. Distributions of normalized (**a**) averaged pressure, $\langle p \rangle$ / ρU_h^2; (**b**) averaged streamwise velocity, $\langle u \rangle / U_h$; (**c**) averaged spanwise velocity, $\langle v \rangle / U_h$; (**d**) averaged vertical velocity, $\langle w \rangle / U_h$, on the slices of $x = -2.5h$–$6.25h$ with a step size of $1.25h$. The width of the slices is $5h$ centering at the $y = 0$, and the height is $4h$. The red dashed lines represent the upper boundary of the wake and the blue dashed lines indicate the center of the shear layer. x, y, and z axes have been normalized by h.

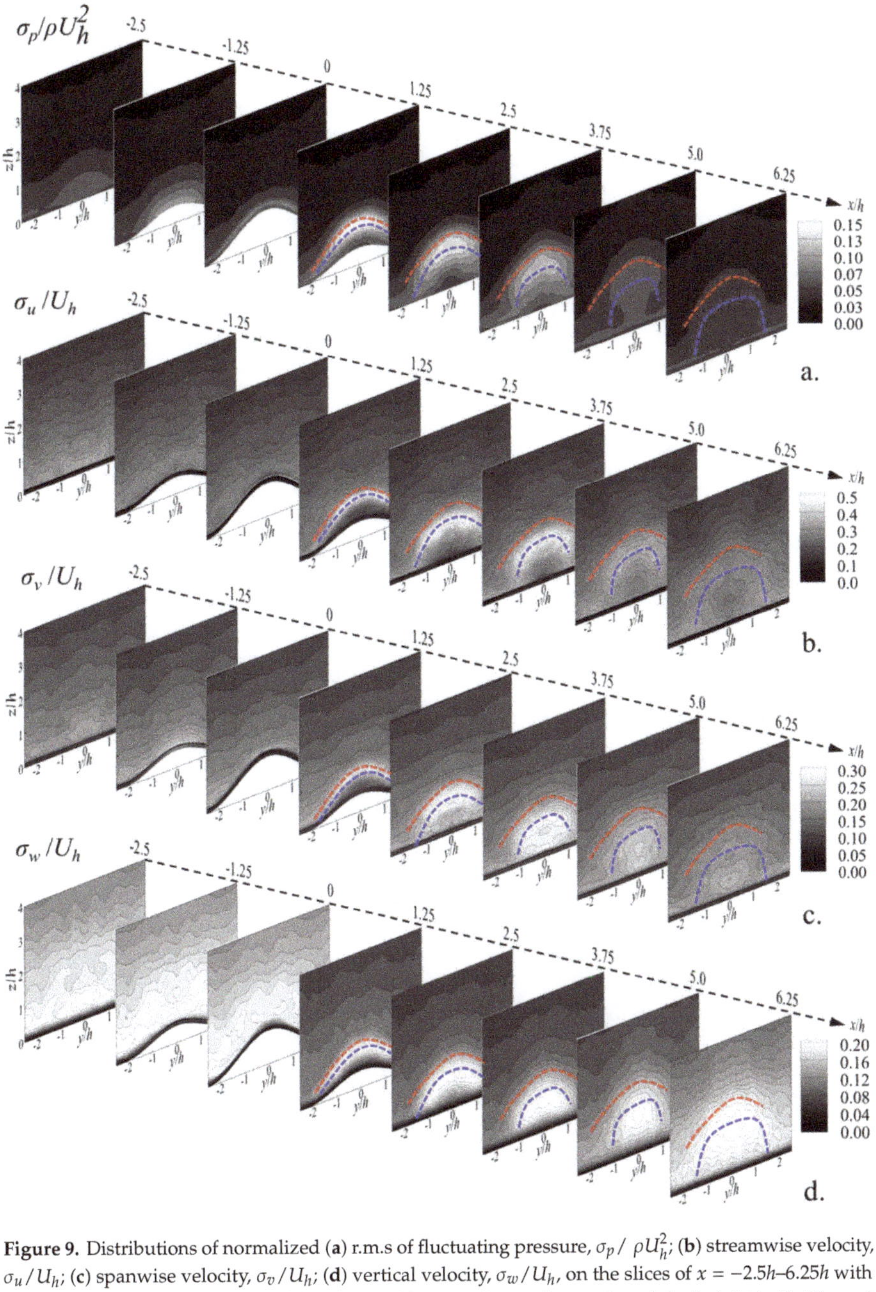

Figure 9. Distributions of normalized (**a**) r.m.s of fluctuating pressure, $\sigma_p/\ \rho U_h^2$; (**b**) streamwise velocity, σ_u/U_h; (**c**) spanwise velocity, σ_v/U_h; (**d**) vertical velocity, σ_w/U_h, on the slices of $x = -2.5h$–$6.25h$ with a step size of 1.25h. The width of the slices is 5h, centering at the $y = 0$, and the height is 4h. The red dashed lines indicate the upper boundary of the wake and the blue dashed lines indicate the center of the shear layer. x, y, and z axes have been normalized by h.

Their distributions on the slice of $y = 0$ are shown in Figure 10. The dominant term was found to be the turbulence production. A formula based on a turbulence generation term, this value is primarily determined by the shear of the average flow. Moreover, kinetic energy is generated in the opposite direction of the Reynolds stress, which not only eliminates the kinetic energy of the average flow, but also transfers the kinetic energy that is eliminated to the fluctuating velocity fields. Therefore, it is apparent that the maximum of the shear production P is at the entry of the flow into the hill wake, where a sudden flow ejection occurs and the shear layer begins to develop. Apart from the concentrations of the turbulence production in the shear layer region, a secondary concentration of P can be identified at the canopy top. The production P and dissipation D are directly responsible for the gain and loss of k in the transport equation, respectively. It can be seen that the dissipation is negative in the global process. This is because the turbulent dissipation acts as a sink for the TKE budget. It can act as a resistance to fluctuating stress and can also convert kinetic energy into internal energy. Interestingly, different from P, the concentration of D is observed at the locations just below the shear layer. Moreover, there are abruptly no dissipations above the shear layer, revealing the much more organized instantaneous turbulence structures here. The secondary dominant term is A. It is negative in the shear layer but positive in the wake core region, whose distribution is similar to the turbulence transportation term.

3.5. Spectrum

Spectral analysis provides information about how energetic the structures are, as a result perhaps shedding some light on the dynamics of the eddy motions. Figure 11 displays the one-dimensional pre-multiplied spectra of three velocity components nS_u, nS_v, and nS_w, further normalized by $\langle u'^2 \rangle$, $\langle v'^2 \rangle$, and $\langle w'^2 \rangle$, respectively. The horizontal axis is normalized by U_h/h, making it convenient to clarify the relative sizes of the eddies to those of the hills. The time signals of the flow at P1~P3 were recorded over a period of 400 time units. The maximum entropy method (MEM) was employed to obtain smoothed S_u. MEM is an extrapolated spectral density estimation technique based on segments of known autocorrelation functions with unknown lags [23–25]. In particular, it is necessary to mention that with a restricted number of poles, MEM will smooth the spectrum somewhat, which is important to identify the peaky regions in the plotting. However, if the number of poles is extremely small, the important information of the energetic motions may be smoothed, while if the number of poles is extremely large, round-off error can be a problem, leading to the introduction of additional peaks shifting with a phase of sine wave. Finally, 250 poles were selected to achieve a compromise between the smoothness of the spectra and the width of the frequency covered, which has also been selected in the previous study by Liu et al. [14].

In general, S_{u_i} in the present LES exhibits a −2/3 slope in the inertial subrange, covering at least one decade, which is in consistent with Kolmogorov's hypothesis. However, there is a steep decrease beyond this inertial range (nh/U_h > 1), suggesting that the present LES do not have the capabilities to resolve the small-scale eddies in a broad dissipation range due to the filtering effects from the SGS model. The spectra of streamwise, spanwise, and vertical velocities at P1 peak at different wavenumbers, which are about 0.1, 0.2, and 0.6, respectively. Whereas, moving the reference point to P2, a clear shift of peak S_u to high wavenumber region is identified, indicating the large-scale streamwise velocity breaks into smaller scales at the hill top. This shift can also be observed in the spectrum of the spanwise velocity. It is of interest that at the crest, the energies of the fluctuating velocities are condensed, which is revealed by steeper caves on the spectrum distributions, suggesting the generation of the energetic coherent structures. Significantly, the peak spectra of the three velocity components seem to occur at nearly the same wavenumber ($nh/U_h \approx 0.2$). Further moving the reference point to P3, the spectra curves do not vary much in comparison with those at P2. However, the curves are significantly smoothened at P3 and the peaks of the spectra are much easier to be identified. It is the signature that the coherent structures in the IBL flow are seldom transported into the shear layer. The similar spectra at P2 and P3 also suggest that the topography-generated coherent structures have significant effects on the velocities upstream.

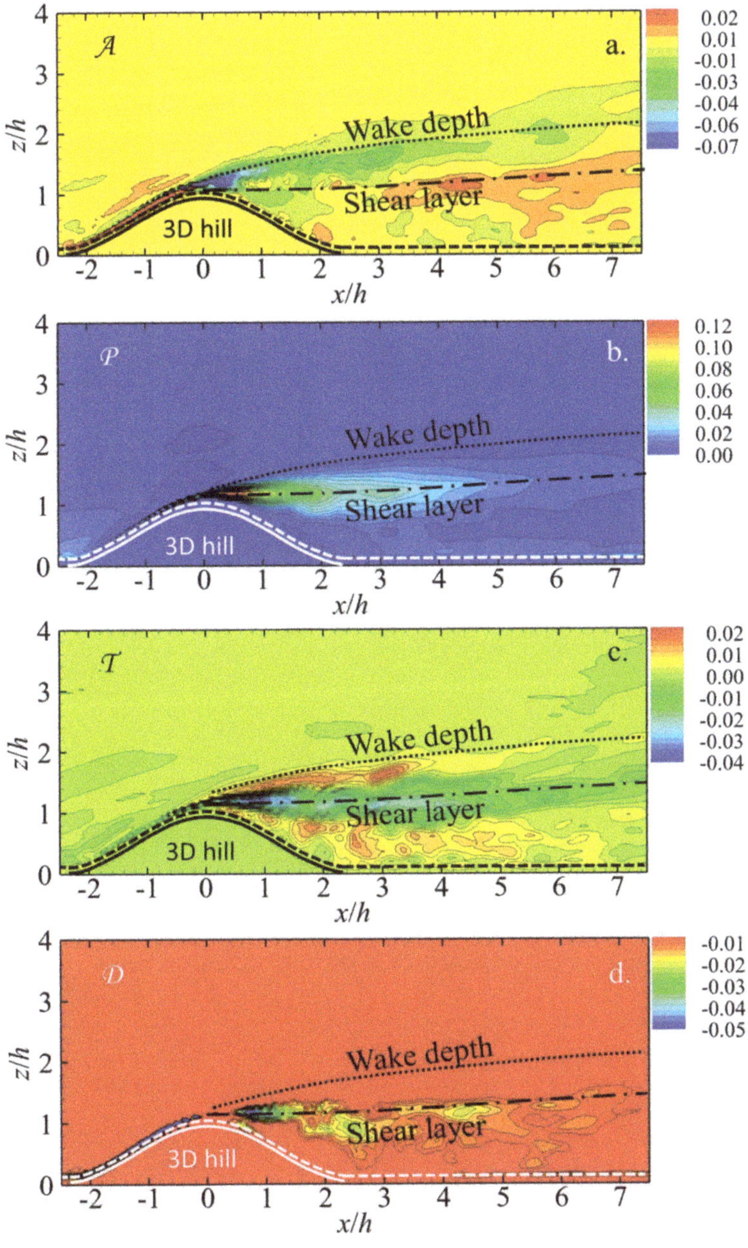

Figure 10. Distributions of normalized turbulence kinetic energy (TKE) budget on the slice of y = 0 with x-axis ranging from $-2.5h$ to $7.5h$ and z-axis ranging from $0h$ to $4h$, (**a**) advection, A; (**b**) turbulence production, P; (**c**) turbulence transportation, T, and (**d**) dissipation, D. The solid line indicates the hill shape, dashed line is the canopy top, dotted line shows the upper boundary of the wake, and the dash-dotted line implies the center of the shear layer.

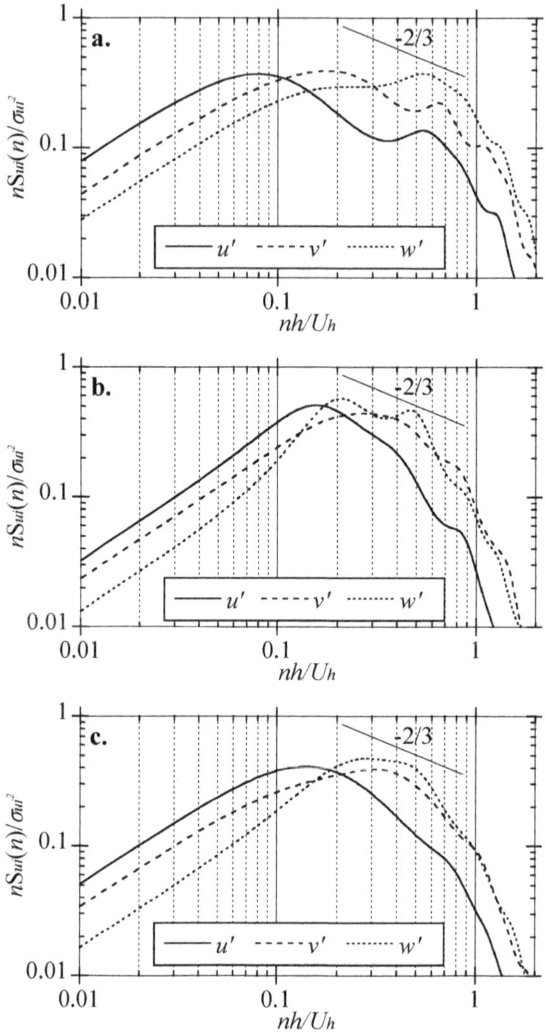

Figure 11. One-dimensional pre-multiplied spectra of three velocity components nS_u, nS_v, and nS_w at the monitoring points: (**a**) P1; (**b**) P2; and (**c**) P3. The pre-multiplied spectra nS_u, nS_v, and nS_w are further normalized by $\langle u'^2 \rangle$, $\langle v'^2 \rangle$, and $\langle w'^2 \rangle$, respectively, and the frequency n is normalized by U_h/h. The inclined lines with a slope of 2/3 are drawn to show the inertial subrange of the spectra.

4. Conclusions

In the present study, the 3D mean and turbulent flows over a hill with vegetation cover were simulated and analyzed. The major conclusions are summarized as follows.

- The production of turbulence from the canopy is the source of the secondary peak spanwise fluctuations in the hill wake. Moreover, two pairs of peak vertical velocity downstream of the hill crest were identified from the three-dimensional view of the mean velocities. One pair was located at the shear layer, and the other located in the central wake.
- The energies of the fluctuating velocities are condensed, represented by steeper caves on the spectrum distributions, suggesting the generation of the energetic coherent structures. Additionally,

the similar spectra at P2 and P3 are the indications that the topography-generated coherent structures have significant effects on the velocities upstream.

- A pair of the lateral shedding are the reason for the double peaks on the distributions of σ_u, Sk_u, Ku_u, and R_{uu}. The turbulence structures in the hill consist of the lateral shedding, the shedding from the hill crest, as well as the canopy-induced rotations. The lateral shedding and the shedding from the hill crest yield an area with quite small velocities and weak vortices, leading to the formation of a dead flow area.

Author Contributions: Conceptualization, Z.L. and Y.H.; methodology, Z.L.; software, Y.H.; validation, W.W., Z.L. and Y.F.; formal analysis, Y.F.; investigation, Q.Z.; resources, Z.L.; data curation, Q.Z.; writing—original draft preparation, Z.L.; writing—review and editing, Y.H.; visualization, Y.F.; supervision, Z.L.; project administration, Z.L.; funding acquisition, Z.L.

Funding: This research was funded by National Key Research and Development Plan of China, grant number 2016YFE0127900, National Natural Science Foundations of China, grant number 51608220, and the Project of Innovation-driven Plan in Huazhong University of Science and Technology, grant number 2017KFYXJJ141.

Conflicts of Interest: The authors declare no conflict of interest.

References

1. Politis, E.S.; Prospathopoulos, J.; Cabezon, D.; Hansen, K.S.; Chaviaropoulos, P.K.; Barthelmie, R.J. Modeling wake effects in large wind farms in complex terrain: The problem, the methods and the issues. *Wind Energy* **2011**, *15*, 161–182. [CrossRef]
2. Carvalho, A.C.; Carvalho, A.; Gelpi, I.; Barreiro, M.; Borrego, C.; Miranda, A.I.; Pérez-Muñuzuri, V. Influence of topography and land use on pollutants dispersion in the Atlantic coast of Iberian Peninsula. *Atmos. Environ.* **2006**, *40*, 3969–3982. [CrossRef]
3. Davenport, A.G.; King, J.P.C. The influence of topography on the dynamic wind loading of long span bridges. *J. Wind Eng. Ind. Aerod.* **1990**, *36*, 1373–1382. [CrossRef]
4. Matusick, G.; Ruthrof, K.X.; Brouwers, N.C.; Hardy, G.S.J. Topography influences the distribution of autumn frost damage on trees in a Mediterranean-type Eucalyptus forest. *Trees* **2014**, *28*, 1449–1462. [CrossRef]
5. Lopes, A.M.G.; Cruz, M.G.; Viegas, D.X. FireStation—An integrated software system for the numerical simulation of fire spread on complex topography. *Environ. Modell. Softw.* **2002**, *17*, 269–285. [CrossRef]
6. Dupont, S.; Brunet, Y. Edge Flow and Canopy Structure: A Large-Eddy Simulation Study. *Bound. Layer Meteorol.* **2008**, *126*, 51–71. [CrossRef]
7. Takahashi, T.; Ohtsu, T.; Yassin, M.F.; Kato, S.; Murakami, S. Turbulence characteristics of wind over a hill with a rough surface. *J. Wind Eng. Ind. Aerod.* **2002**, *90*, 1697–1706. [CrossRef]
8. Cao, S.; Tamura, T. Experimental study on roughness effects on turbulent boundary layer flow over a two-dimensional steep hill. *J. Wind Eng. Ind. Aerod.* **2006**, *94*, 1–19. [CrossRef]
9. Cao, S.; Tamura, T. Effects of roughness blocks on atmospheric boundary layer flow over a two-dimensional low hill with/without sudden roughness change. *J. Wind Eng. Ind. Aerod.* **2007**, *95*, 679–695. [CrossRef]
10. Dupont, S.; Brunet, Y.; Finnigan, J.J. Large-eddy simulation of turbulent flow over a forested hill: Validation and coherent structure identification. *Q. J. R. Meteorol. Soc.* **2008**, *134*, 1911–1929. [CrossRef]
11. Ishihara, T.; Fujino, Y.; Hibi, K. A wind tunnel study of separated flow over a two-dimensional ridge and a circular hill. *J. Wind Eng.* **2001**, *89*, 573–576.
12. Ishihara, T.; Hibi, K. Numerical study of turbulent wake flow behind a three-dimensional steep hill. *Wind Struct.* **2002**, *5*, 317–328. [CrossRef]
13. Tamura, T.; Cao, S.; Okuno, A. LES Study of Turbulent Boundary Layer Over a Smooth and a Rough 2D Hill Model. *Flow Turbul. Combust.* **2007**, *79*, 405–432. [CrossRef]
14. Liu, Z.; Ishihara, T.; Tanaka, T.; He, X.H. LES study of turbulent flow fields over a smooth 3-D hill and a smooth 2-D ridge. *J. Wind Eng. Ind. Aerod.* **2016**, *153*, 1–12. [CrossRef]
15. Ferziger, J.H.; Peric, M. *Computational Method for Fluid Dynamics*; Springer: Berlin, Germany, 2002.
16. Iizuka, S.; Kondo, H. Large-eddy simulations of turbulent flow over complex terrain using modified static eddy viscosity models. *Atmos. Environ.* **2006**, *40*, 925–935. [CrossRef]
17. Lee, S.H.; Sung, H.J. Direct numerical simulation of the turbulent boundary layer over a rod-roughened wall. *J. Fluid Mech.* **2007**, *584*, 125–146. [CrossRef]

18. Ishihara, T.; Qi, Y.H. Numerical study of turbulent flow fields over steep terrains by using a modified delayed detached eddy simulation. *Bound. Layer Meteorol.* **2018**, *170*, 1–24. [CrossRef]
19. Liu, Z.; Cao, S.; Liu, H.; Ishihara, T. Large-eddy simulations of mean and turbulent flow and coherent structures over an isolated 3D hill. *Bound. Layer Meteorol.* **2018**, *170*, 1–27.
20. Ghodke, D.C.; Apte, V.S. DNS study of particle-bed–turbulence interactions in an oscillatory wall-bounded flow. *J. Fluid Mech.* **2018**, *792*, 232–251. [CrossRef]
21. Ghodke, D.C.; Apte, V.S. Roughness effects on the second-order turbulence statistics in oscillatory flows. *Comput. Fluids.* **2018**, *162*, 160–170. [CrossRef]
22. Ghodke, D.C.; Apte, V.S. Spatio–temporal analysis of hydrodynamic forces on the particle bed in an oscillatory flow environment. *J. Fluid Mech.* **2018**, *841*, 167–202. [CrossRef]
23. Childers, D. *Modern Spectrum Analysis*; Chapter II; IEEE Press: New York, NY, USA, 1978.
24. Kay, S.; Marple, S. Spectrum Analysis: A Modern Perspective. *Proc. IEEE* **1981**, *69*, 1380–1419. [CrossRef]
25. Press, W.; Teukolsky, S.; Vetterling, W.; Flanney, B. *Numerical Recipes 3rd Edition: The Art of Scientific Computing*; Cambridge University Press: Cambridge, UK, 2007; pp. 681–685.

energies

MDPI

Article

Radiation Heat Transfer in a Complex Geometry Containing Anisotropically-Scattering Mie Particles

Ali Ettaleb [1], Mohamed Ammar Abbassi [1], Habib Farhat [2], Kamel Guedri [3], Ahmed Omri [1], Mohamed Naceur Borjini [4], Marjan Goodarzi [5,*] and M. M. Sarafraz [6]

1 Unité de Recherche «Matériaux, Energie et Energies Renouvelables» (MEER), Faculté des Sciences de Gafsa, B.P.19-2112 Zarroug-Gafsa, Tunisie; alietaleb@gmail.com (A.E.); MedAmmar.Abbassi@enim.rnu.tn (M.A.A.); omri.ahmed233@gmail.com (A.O.)

2 Laboratoire des Etudes des Milieux Ionisés et Réactifs Avenue Ibn-Eljazzar, 5019 Monastir, Tunisie; habibfarhat@ipeim.rnu.tn

3 Mechanical Engineering Department, College of Engineering and Islamic Architecture, Umm Al-Qura University, Makkah 21955, Saudi Arabia; kamel.guedri@enim.rnu.tn

4 Unité de Métrologie et des Systèmes Energétiques, Ecole Nationale d'Ingénieurs de Monastir, Université de Monastir, 5000 Monastir, Tunisie; naceur.borjini@issat.rnu.tn

5 Sustainable Management of Natural Resources and Environment Research Group, Faculty of Environment and Labour Safety, Ton Duc Thang University, Ho Chi Minh City, Vietnam

6 School of Mechanical Engineering, University of Adelaide, Adelaide SA 5005, South Australia, Australia; mohammadmohsen.sarafraz@adelaide.edu.au

* Correspondence: marjan.goodarzi@tdtu.edu.vn; Tel.: (+1)-502-432-0339

Received: 7 September 2019; Accepted: 16 October 2019; Published: 19 October 2019

Abstract: This study aims to numerically investigate the radiation heat transfer in a complex, 3-D biomass pyrolysis reactor which is consisted of two pyrolysis chambers and a heat recuperator. The medium assumes to be gray, absorbs, emits, and Mie-anisotropically scatters the radiation energy. The finite volume method (FVM) is applied to solve the radiation transfer equation (RTE) using the step scheme. To treat the complex geometry, the blocked-off-region procedure is employed. Mie equations (ME) are applied to evaluate the scattering phase function and analyze the angular distribution of the anisotropically scattered radiation by particles. In this study, three different states are considered to test the anisotropic scattering impacts on the temperature and radiation heat flux distribution. These states are as: (i) Isotropic scattering, (ii) forward and backward scattering and (iii) scattering with solid particles of different coals and fly ash. The outcomes demonstrate that the radiation heat flux enhances by an increment of the albedo and absorption coefficients for the coals and fly ash, unlike the isotropic case and the forward and backward scattering functions. Moreover, the particle size parameter does not have an important influence on the radiation heat flux, when the medium is thin optical. Its effect is more noticeable for higher extinction coefficients.

Keywords: radiation; blocked-off-region procedure; heat recuperation; anisotropic scattering; mie particles

1. Introduction

For a vast range of engineering applications, radiation is a substantial method of heat transfer. Especially, in high temperature equipment like furnaces, boilers, gas turbine combustors, and nuclear reactors, where the combustion generating luminous flames includes combustion gases and other particles. Where the scattering is mostly anisotropic, the particles emit, absorb and scatter radiant energy. Therefore, the necessity for analysis of radiation heat transfer leads to an increase demand for developing well-designed radiation models, applicable to arbitrary shaped multi-dimensional geometries and capable of treating anisotropic characteristics in participating media. In recent decades, many researchers have tried to calculate the radiation transfer equation (RTE) in multidimensional complicated geometries. Considering computation costs and precision, three most adopted methods could be recalled the discrete transfer, the discrete ordinates and the finite-volume methods. The first description of the discrete transfer method (DTM) is presented by Lockwood and Shah [1] and applied later to complex geometries by the cell-blocking process according to Cartesian coordinates [2], and nonorthogonal grid systems [3]. Chai et al. [4] used the DOM with the blocked-off-region procedure. Fiveland and Jesse [5] have accomplished a formulation of the discrete ordinates method (DOM) using finite element associated with curvilinear grids. The finite volume method (FVM) was adapted with various procedures that treated the problem of irregular geometries. The FVM was developed with nonorthogonal coordinate systems [6], the blocked-off-region [7–9], and the spatial-multiblock [10] procedures. Coelho et al. [11] modelled the radiation heat transfer in enclosures including blocks of narrow thicknesses by the above-mentioned methods. Guedri et al. [12] investigated the impacts of baffles on radiation heat transfer in the 2-D and 3-D complicated geometries. The authors examined two different schemes: The STEP and CLAM schemes. Furthermore, they treated the effect of change of the absorption and albedo coefficients on the temperature profiles as well as net radiation heat flux distributions in a 3-D biomass pyrolysis reactor. The similar study is done by Abbassi et al. [13] in a 2-D complex geometry. They examined the baffles shadow and soot volume fraction impacts on the temperature profiles and radiation heat flux. In all previous works, the problem of anisotropic scattering is processed in a simple way. Mengüç and Viskanta [14] analyzed the radiation exchanges in a 3-D rectangular enclosure housing radiatively participating mixture of gases and anisotropic scattering particles applying the first and third-order spherical harmonics approximation. The delta-Eddington model is employed to define the scattering phase function. Kim and Lee [15,16] studied the impact of the anisotropic scattering in a 2-D rectangular enclosure using the S-N discrete ordinates scheme. The scattering phase function is expanded in a series of Legendre polynomials. Results indicated that the phase function anisotropy has a vital importance in the radiation heat transfer while the non-symmetric boundary conditions are considered in the problem. Farmer and Howell [17] used the Monte Carlo approach to anticipate the radiation heat transfer within general inhomogeneous media that represents both highly spectral and anisotropic scattering behaviors. Guo and Maruyama [18] investigated the scaled isotropic scattering radiation in a 3-D inhomogeneous medium by the radiation element method. In a 2-D rectangular enclosure, Trivic et al. [19–21] coupled the FVM to calculate the radiation transfer equation employing Mie equations for evaluating the scattering phase function. This model can be applied to any given particle parameters regardless the previous designed analytical expressions for the scattering function. It is also adopted for the radiation heat transfer in a 3-D geometry with grey and non-grey anisotropically scattering media. In this work, the authors used two different numerical schemes to solve the radiation transfer equation: (i) The finite volume method with a highly refined angular discretization and (ii) the combination of zone method and the Monte Carlo statistical simulation method, and for the non-gray gases, they used the Smith's weighted sum of gray gases model (WSGGM) for a hypothetical gas represented by 5 gray gases. Outcomes from the two methods are found in an acceptable agreement. Hunter and Guo [22] presented a technique for the normalization of the phase function to meet the scattered energy conservation constraint and thus, minimize the numerical errors generated by the discretization of the integral numerical term in the RTE. This technique is simpler and applied to the 3-D FVM to enhance radiation calculation precision and efficiency in anisotropically scattering media. The outcomes matched well to those achieved by the Monte Carlo and discrete-ordinates methods for a cubic enclosure confining a highly-anisotropic scattering mediumproblem. The authors achieved an acceptable compromise between the FVM results by applying the normalization technique, and the two different methods by little effect on computational efficiency. Guedri et al. [23] formulated and applied the FTn Finite Volume Method (FTn FVM) for transient radiation in a 3-D absorbing, emitting, and anisotropically scattering medium. The outcomes illustrated that FTn FVM decreases mostly the ray impacts and its accuracy is higher than the standard FVM. Moreover, FTn FVM has shorter convergence time compared to the standard FVM for all cases using both STEP and CLAM schemes.

The aim of this study is to understand the anisotropic scattering phenomenon impact on the radiation heat transfer in a 3-D complex geometry. The enclosure is a pilot plant for biomass pyrolysis reactor consisted of two pyrolysis chambers and a heat recuperator. The FVM is used in relation to the blocked-off-region process and Mie equations to evaluate the scattering phase function (SPF). The heat recuperator includes a gray, absorbing–emitting, and anisotropically scattering medium. Solid particles of many different coals and fly ash are formed during the pyrolysis process. The main advantage of the proposed procedure is that it can be readily applied to radiating particles of any types within the Mie scattering approximation limits, given the geometrical and optical particle parameters like the complicated index of refraction, wavelength of the incident radiation and particle mean diameter.

2. Mathematical Formulations

2.1. Radiation Transfer Equation

In this study, an absorbing, emitting and anisotropically scattering gray medium, comprising of a mixture of gas and particles inside an enclosure is considered. For such a situation, the integro-differential radiation transfer equation (RTE) is given by [6]:

$$\frac{dI(\vec{r},\vec{\Omega})}{ds} = -\beta(\vec{r})I(\vec{r},\vec{\Omega}) + S(\vec{r},\vec{\Omega}) \tag{1}$$

The RTE resulting from an energy balance performed for a volume element, means that radiation intensity along a given path $\vec{\Omega}$ gaines radiation energy by absorption and in-scattering but decreases by emission and out-scattering of radiation energy.

$\beta(\vec{r})$ and $S(\vec{r},\vec{\Omega})$ are respectively the local extinction coefficient and source function given by

$$\beta(\vec{r}) = \kappa(\vec{r}) + \sigma(\vec{r}) \tag{2a}$$

$$S(\vec{r},\vec{\Omega}) = \kappa(\vec{r})I_b(\vec{r}) + \frac{\sigma(\vec{r})}{4\pi}\int_{4\pi} I(\vec{r},\vec{\Omega})\Phi(\vec{\Omega},\vec{\Omega}')d\Omega' \tag{2b}$$

The radiation intensity $I(\vec{r},\vec{\Omega})$ depends on the Cartesian coordinates (x, y, z) of the position vector $\vec{r}$, and the polar and azimuthal angles (θ, ϕ), that characterise the radiant intensity direction, defined by the unit vector

$$\vec{\Omega} = \sin\theta\cos\phi\vec{e}_x + \sin\theta\sin\phi\vec{e}_y + \cos\theta\vec{e}_z \tag{3}$$

The FVM is used to discretize the RTE. This method comprises of dividing the computational area into $N_x \times N_y \times N_z$ control volumes and the spherical space into $N_\theta \times N_\phi$ control solid angles. The initial solid angle is analytically computed by

$$\Delta\Omega^l = \int_{\theta^{l-}}^{\theta^{l+}}\int_{\phi^{l-}}^{\phi^{l+}} \sin\theta d\theta d\phi \tag{4}$$

The following finite volume formula may be achieved by considering the constant magnitude of the radiation intensity and varying its orientation through the control volume and control angle.

$$\sum_m I_m^l \Delta A_m N_{c,i}^l = (-\beta I^l + S^l)\Delta V_P \tag{5}$$

where

$$N^l_{c,i} = \frac{1}{\Delta\Omega^l}\int_{\Delta\Omega^l}(\vec{\Omega}^l \cdot \vec{n}_i)d\Omega^l \tag{6}$$

Equation (7) is achieved using the STEP scheme which is obtained by Chai et al. [24] in 3D discretization.

$$a^l_P I^l_P = a^l_W I^l_W + a^l_E I^l_E + a^l_S I^l_S + a^l_N I^l_N + a^l_F I^l_F + a^l_R I^l_R + b^l_P \tag{7}$$

where

$$a^l_I = \Delta A_m \| - N^l_{c,m}, 0\|,\ \mathrm{I} = \mathrm{W, E, S, N, F\ and\ R}$$

$$a^l_P = \sum_m \Delta A_m \| N^l_{c,m}\|, 0 + \beta \Delta V_P$$

$$b^l_P = S^l \Delta V_P$$

$$S^l = \kappa I_b + \frac{\sigma}{4\pi}\sum_{l'=1}^{L} I^{l'}_P \overline{\Phi}^{ll'} \Delta\Omega$$

The quantity $\overline{\Phi}^{l'l}$ is defined as the average scattering phase function (ASPF) from control angle l' (incident angle) to control angle l (scattering angle). It is expressed as follows

$$\overline{\Phi}^{l'l} = \frac{\int_{\Delta\Omega^l}\int_{\Delta\Omega^{l'}} \Phi(\vec{\Omega}^l, \vec{\Omega}^{l'}) d\Omega^l d\Omega^{l'}}{\Delta\Omega^l \Delta\Omega^{l'}} \tag{8a}$$

$\overline{\Phi}^{l'l}$ can be estimated by the previous equation discretization [25]

$$\overline{\Phi}^{l'l} = \frac{\sum_{l_s=1}^{L_s}\sum_{l'_s=1}^{L'_s} \Phi^{l'_s l_s} \Delta\Omega^{l'_s} \Delta\Omega^{l_s}}{\Delta\Omega^l \Delta\Omega^{l'}} \tag{8b}$$

where $\Delta\Omega^{l'_s}$ and $\Delta\Omega^{l_s}$ are the sub-control angles.

The radiant intensities are calculated as follows

$$I^l_P = \frac{a^l_W I^l_W + a^l_E I^l_E + a^l_S I^l_S + a^l_N I^l_N + a^l_F I^l_F + a^l_R I^l_R + b^l_P}{a^l_P} \tag{9}$$

When the below convergence criterion is satisfied, the iterative procedure is terminated for all cases:

$$\frac{\left| I^l_P - I^{l_0}_P \right|}{I^l_P} \leq 10^{-5} \tag{10}$$

where l_0 and l stand for the previous and current iterations, respectively.

2.2. Boundary Conditions

The radiation boundary conditions are defined by the below equation for a gray, diffusely reflecting, opaque, and emitting surface:

$$I(\vec{r}, \vec{\Omega}) = \varepsilon(\vec{r}) I_b(\vec{r}) + \frac{\rho(\vec{r})}{\pi}\int_{\vec{n}.\vec{\Omega}'<0} I(\vec{r}, \vec{\Omega}) \left| \vec{\Omega}'.\vec{n} \right| d\Omega' \tag{11a}$$

The following equation represents the discretized form of Equation (11).

$$I^l = \varepsilon_w I_b + \frac{\rho_w}{\pi}\sum_{l'} I^{l'} N^{l'}_w \Delta\Omega^{l'} \tag{11b}$$

2.3. Scattering Phase Function

The phase function describes the angular distribution of the scattered radiation which is denoted by Φ. This function illustrates the possibility that an incident ray beam centered around the direction $\vec{\Omega}'$, is scattered towards the propagation direction $\vec{\Omega}$.

In the spherical particle size compared with the incident beam wavelength ($x_p \approx 1$), the Mie theory applies and thus, the anisotropic scattering incident is described by the phase function described by [26]

$$\Phi(\Theta) = 2\frac{i_1 + i_2}{x_p^2 Q_{sca}} \tag{12}$$

Θ represents the scattering angle between the incomingand the outgoing directions respectively $\vec{\Omega}$ and $\vec{\Omega}'$, evaluated using the following expression

$$\cos\Theta = \mu\mu' + (1-\mu^2)^{1/2}(1-\mu'^2)\cos(\varphi' - \varphi) \tag{13}$$

(i_1, i_2) are the non-dimensional polarized intensities; Q_{sca} is the efficiency factor for scattering and x_p is the particle size parameter. In this study, the value of x_p is dependent on wavelength and the particle radius, it is used to calculate the typical wavelength of λ = 3.1415 µm as suggested by Modest [27]. Indeed, the definition of Mie theory equations for homogeneous spherical particles, and all the appropriate variables linked to Equation (12) are represented in detail by [19]. The iterative calculations will be stopped as indicated by Deirmendjian et al. [26], when $n = 1.2x_p + 9$.

3. Blocked-Off-Region Procedure

The blocked-off-region process is consisted of appending artificial areas to the real physical domain to gain a simple formation where the RTE solution is very simple to create. This process has been expanded for conduction and convection heat transfer, and after that, developed to radiation transfer by Chai et al. [6,7] and Borjini et al. [8].

An extra source term is added to discern real areas from artificial ones as follows:

$$S_{bloc} = S_C + S_P I_P^l \tag{14}$$

The extra source term is chosen for a specified black boundary as follows: $(S_C, S_P) = (0,0)$ and $(MI_b, -M)$ for the real areas and the fictitious ones, respectively. Curved and inclined shapes can be dealt with applying this process by a grid mesh refinement.

4. Validation

To validate our implemented numerical code prior to its application for case study problems (i.e., 3-D complex geometry), the results of Trivic and Amon [20] are used. The authors solved the RTE by combining the zone (ZM) and Monte Carlo (MC) solution methods where the SPF is described by Mie Equations (ME). The utilized software in this research is coded using the FORTRAN computing language.

A cubic enclosure with a length of 1.0 is considered. The calculation area is discretized into (15 × 15 × 15) uniform control volumes, and (16 × 20) control solid angle. The rear wall is black and hot with $E_{br} = 1$. All the other walls are assumed to be black and cold. The enclosure confines a pure scattering medium with $\beta = 1$ m^{-1}. In this section, the quantities are non-dimensional. It is noted that X= x/L, Y= y/L and Z= z/L are the non-dimensional coordinates and L is the cube side.

Four various states are investigated: One isotropic with the designation ISO and three anisotropic states with the scattering phase functions denoted with F1, B1, and carbon [19].

All the particle information which is used to evaluate the scattering phase functions F1 and B1 are taken from [28] and presented in Table 1. Letters F and B denote the forward and backward-directed scattering phase functions, respectively.

Table 1. Data for evaluation of the scattering phase functions by Mie equations.

Scattering Phase Function	F1	B1
Particle size parameter x_p	5	1
Real part of complex refractive index n	1.33	very large, taken 10^8
Imaginary part of complex refractive index k	0	0

The variation of the considered scattering phase functions versus the scattering angle are demonstrated in Figure 1a,b. It is shown that the scattered radiation by fly ash particles for small particle size parameter presents a similar tends as B1 phase function. As x_p increases, the scattered radiation tends to be directed towards the forward direction. F1 phase function seems to be adequate to represent the fly ash scattering behavior when $x_p = 5$.

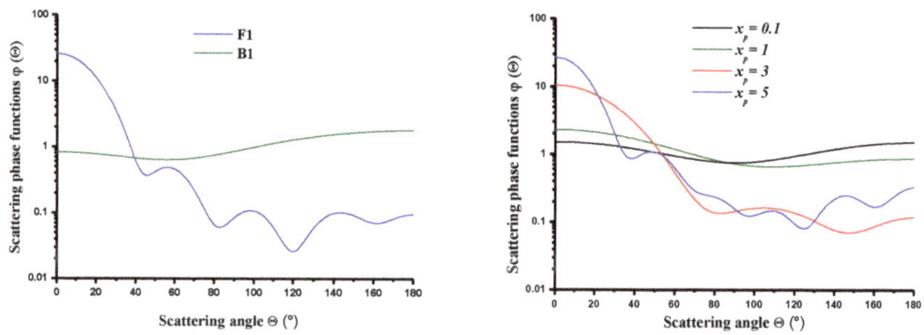

(**a**) (**b**)

Figure 1. Scattering phase functions: (**a**) Function F1 and B1, (**b**) fly ash for different size parameters.

Figure 2a–c presents the net radiation heat flux density in **y** direction $q^*_{ys} = q_{ys}/E$ at the middle of bottom wall, along the z coordinate. The (FVM + ME) predictions corresponding to the isotropic case (Figure 2a) and carbon (Figure 2d), clearly follows the (ZM + MC) results. Figure 2b depicts that the F1 function predictions for function F1 of the heat flux with (FVM + ME) follow those obtained with the aid of (ZM+MC). The maximum discrepancy observed does not exceed 26%. For the scattering function B1, the achieved result is underestimated compared with benchmark.

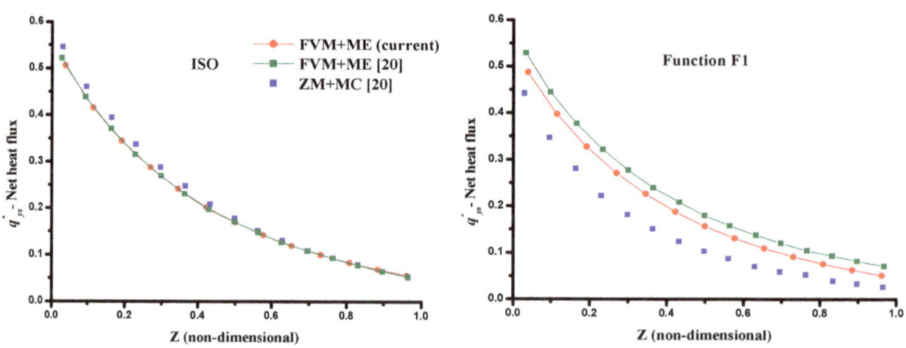

(**a**) (**b**)

Figure 2. *Cont.*

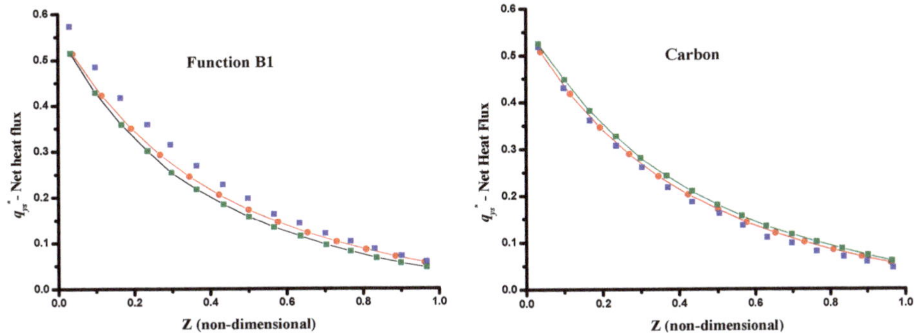

(c) (d)

Figure 2. Dimensionless radiation heat flux density in y-direction at the middle of south wall (X= 0.5, Y= 0.0).

The z-component of the dimensionless net radiation heat flux density $q^*_{zf} = q_{zf}/E$, at the middle of the cold front wall, along x-coordinate is shown in Figure 3a–c. First, the isotropic case, shown in Figure 3a presents an overestimation around 10.8% for the (FVM+ME) results as compared with the (ZM + MC) solutions. In Figure 3b, where the case of function F1 is shown, our results present an overestimation around 39% with benchmark (ZM + MC). Figure 3c shows the (FVM+ME) results corresponding to B1 phase function which agree well with the reference predictions. Figure 3d presents the (FVM+ME) predictions for carbon, and shows an overestimation ranging between 18.2% and 21.3%.

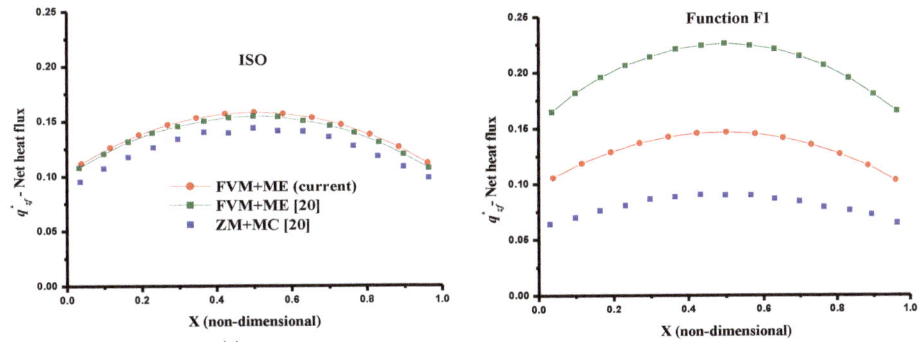

(a) (b)

Figure 3. *Cont.*

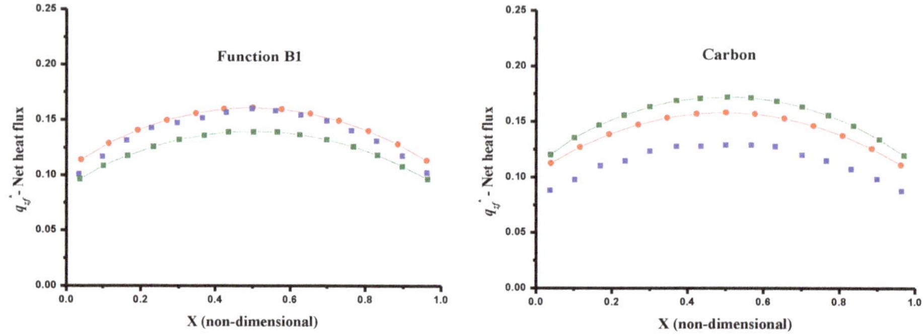

(c) (d)

Figure 3. Dimensionless radiation heat flux density in z-direction at the middle of front cold wall (Y= 0.5, Z= 1.0).

It should be noted that the results found by Trivic and Amon [20], using the FVM for the resolution of the RTE and ME for evaluating the scattering phase function, are also shown on the previous figures.

5. A 3D Complex Heat Recuperator of Biomass Pyrolysis Fumes

The biomass pyrolysis pilot plant is composed of two metallic rooms for biomass pyrolysis and a heat recuperator, as described in detail by Abbassi et al. [12,13,28] (Figure 4a,b).The heat recuperator consists of a parallelepiped metallic construction with dimensions (1.8, 1.2, 1.2) m, including two cylindrical chambers and 17 baffles in direct contact with gases emitted from the combustion of biomass pyrolysis fumes.

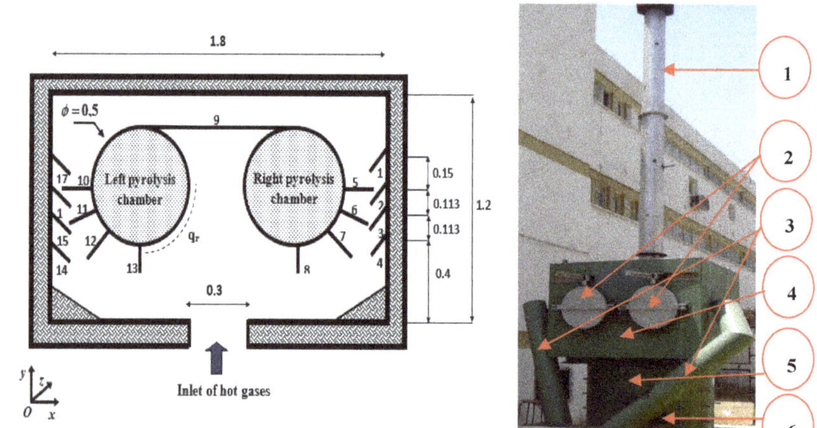

(a) (b)

Figure 4. (**a**) Geometrical characteristics of the heat recuperator projected in z-plane, (**b**) Pilot plan. 1. Chimney. 2. Pyrolysis chambers. 3. Pyrolysis gases channels. 4. Heat exchanger. 5. Pyrolysis gases combustor. 6. Pyrolysis gases and air combustor inlet.

6. Result and Discussion

6.1. Grid Independence

The heat recuperator contains an absorbing-emitting and anisotropically scattering medium. The temperature of all the walls was evaluated equal to 300 K with an emissivity equal to $\epsilon = 0.8$, except the wall at $y = 0$ and for $0.75 < x < 1.05$ m, where the temperature was assumed 1000 K with a blackbody surface. The baffles are black and were taken at a prescribed temperature of 300 K. The two pyrolysis chambers are considered as black and at a uniform temperature 500 K. This choice is made based on experimental measurements.

In this section, it is aimed to examine the effect of varying the spatial and angular meshes for the isotropic case. The step scheme is employed for all calculations. First, a spatial grid refinement is made as $(N_x \times N_y \times N_z) = (46 \times 44 \times 11)$, $(62 \times 58 \times 11)$, $(92 \times 86 \times 11)$, and $(184 \times 172 \times 11)$, for a fixed angular grid $(N_\theta \times N_\phi) = (4 \times 20)$. The choice of the grid mesh $(92 \times 86 \times 11)$, offers a satisfactory compromise between accuracy and computation time. Also, the number of control solid angles is varied as $(N_\theta \times N_\phi) = (4 \times 12)$, (4×20), (4×24), and (4×20) for the retained spatial grid.It is noted that it does not have a great effect on the CPU time, and consequently, a grid of the type $(N_\theta \times N_\phi) = (4 \times 20)$ seems to be appropriate.

In the following, it is set out to investigate the impact of the scattering anisotropy of particles, the absorption and scattering albedo coefficients on the gas temperature and net radiation flux distributions. The solution method of the RTE for the pilot plant of Figure 4a,b is (FVM+ME). To achieve this, the forward and backward scattering phase functions (F1, B1) are considered. Also, the predictions are analyzed for real engineering fuels and fly ash. The particles are consisting of carbon, anthracite, bituminous and lignite.

6.2. Absorption Coefficient and Scattering Albedo Effects

In this section, a comparison between isotropic and two anisotropic scattering functions (F1, carbon) is made. Figure 5a–g shows the temperature contours at plane $z = 0.6$ m, for six cases. It should be mentioned that the size parameter for carbon is $x_p = 1$. First, for an absorption coefficient $\kappa = 1$ m^{-1} (states 1,3, and 5), when the albedo coefficient is relatively weak (i.e., $\omega = 0.1$), it is observed that the temperatures distributions have the same pattern for the two-phase functions: Isotropic case as well as carbon. The impact of scattering anisotropy is negligible, as it refers to an optically thin body with little scattering. Furthermore, it is seen that higher temperatures are distributed near the inlet of the hot-gases region, and at the vicinity of the baffles while the temperature distribution tends to be uniform in the recuperator core region. When the albedo increases, it is seen that the temperature near the lateral baffles remains nearly constant (i.e., 700 K). In the rest of the recuperator, the difference between the temperature contours of the two phase functions is clear. This difference is pronounced with a larger albedo coefficient. For carbon, it can be noted that the isothermal line 700 K (in bold) tends towards the central region. Also, the temperature increases near the horizontal baffle and the cold wall. However, for the isotropic case, contours behavior is completely the opposite. This may be due to the fact that radiation is equally scattered while carbon scattering phase function seems to be a backward-directed function. For an absorption coefficient $\kappa = 5$ m^{-1} (cases 2, 4 and 6), on the one hand, it can be seen a remarkable deformation of the isothermal lines close to the inlet of the hot gases, towards the right pyrolysis room for carbon case especially when $\omega = 0.9$. In addition, there is an increase of the temperature namely in the centre of the recuperator and at the horizontal baffle. On the other hand, for the isotropic case, there is a decline of the temperature in the physical system mainly for a large value of the albedo coefficient, and it is shown a stagnation zone with the highest albedo coefficient, where the temperature about 410 K. In fact, in these regions, the thermal inertia is important, and radiation thermal energy tends to be stored. This phenomenon is perceived when the absorption and the albedo coefficients are high enough.

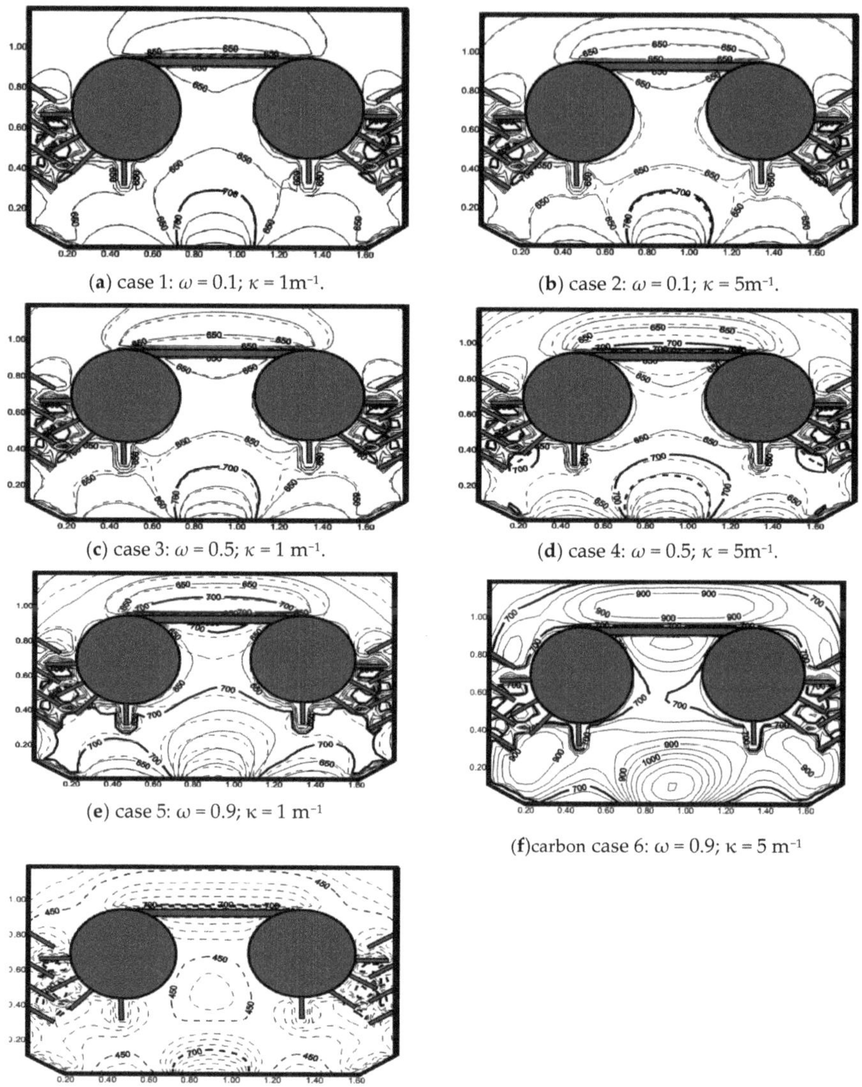

(a) case 1: ω = 0.1; κ = 1m^{-1}.

(b) case 2: ω = 0.1; κ = 5m^{-1}.

(c) case 3: ω = 0.5; κ = 1 m^{-1}.

(d) case 4: ω = 0.5; κ = 5m^{-1}.

(e) case 5: ω = 0.9; κ = 1 m^{-1}

(f)carbon case 6: ω = 0.9; κ = 5 m^{-1}

(g) isotropic case 6: ω = 0.9; κ = 5 m^{-1}

Figure 5. Effects of the absorption and albedo coefficients on the medium temperature pattern: Comparison between the isotropic case and carbon. Dashed lines: Isotropic, continuous lines: carbon. (**a**) case 1: $\omega = 0.1$; $\kappa = 1\ \text{m}^{-1}$, (**b**) case 2: $\omega = 0.1$; $\kappa = 5\ \text{m}^{-1}$, (**c**) case 3: $\omega = 0.5$; $\kappa = 1\ \text{m}^{-1}$, (**d**) case 4: $\omega = 0.5$; $\kappa = 5\ \text{m}^{-1}$, (**e**) case 5: $\omega = 0.9$; $\kappa = 1\ \text{m}^{-1}$, (**f**) carbon case 6: $\omega = 0.9$; $\kappa = 5\ \text{m}^{-1}$, (**g**) isotropic case 6: $\omega = 0.9$; $\kappa = 5\ \text{m}^{-1}$.

Figure 6a–g depicts the forecasted net radiation heat flux density contours. For all cases of carbon, it is shown that the radiant energy is directly from the hot regions to the cold ones, and the radiation heat flux is enhanced in all the regions of the recuperator and reachesa maximum value equal 61.47 kW·m^{-2}. For the isotropic case, there is a remarkable fall of the radiation heat flux mainly at the inlet of hot gases with a maximum value of 33.82 kW·m^{-2}. Moreover, the appearance of stagnation zones is noticed for two states.

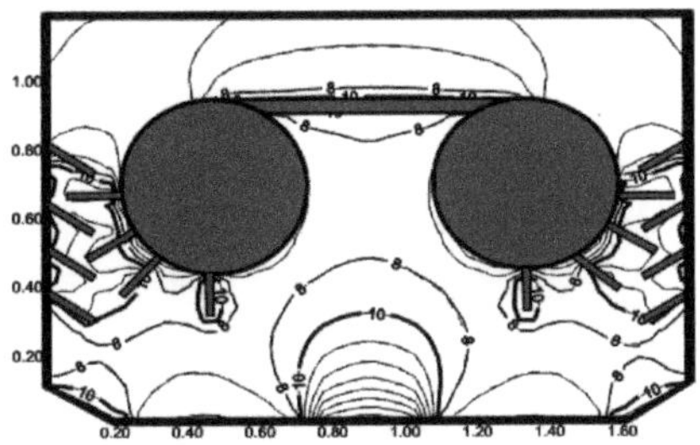

(a) case 1: $\omega = 0.1$; $\kappa = 1\ \text{m}^{-1}$.

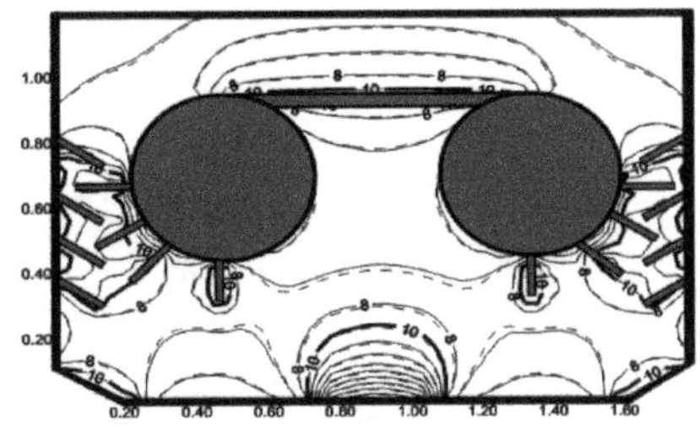

(b) case 2: $\omega = 0.1$; $\kappa = 5\text{m}^{-1}$.

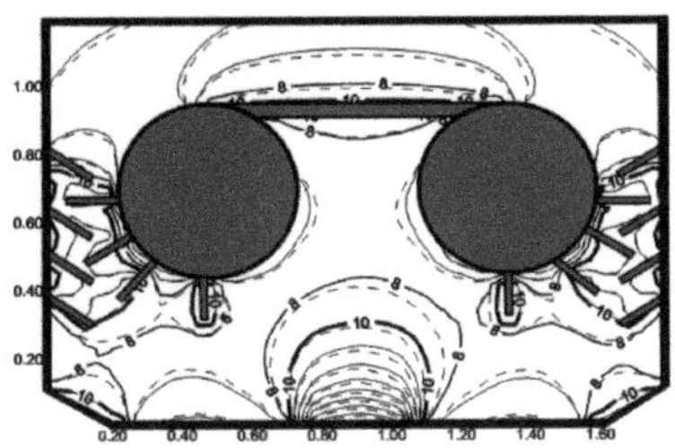

(c) case 3: $\omega = 0.5$; $\kappa = 1\text{m}^{-1}$.

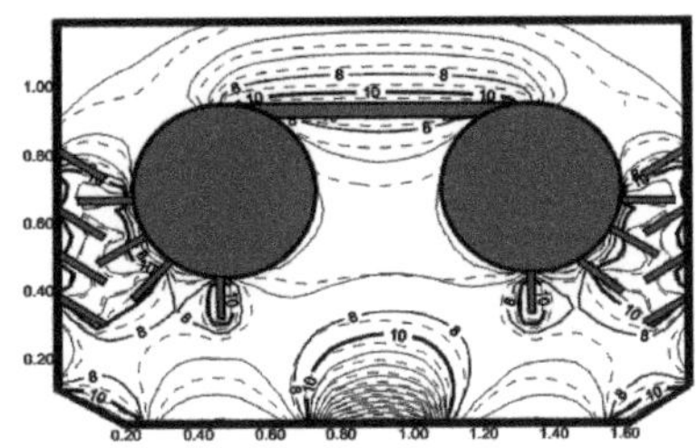

(d) case 4: $\omega = 0.5$; $\kappa = 5\ \text{m}^{-1}$.

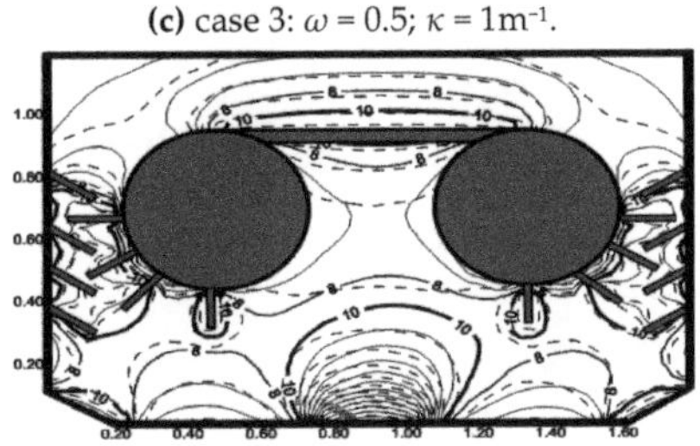

(e) case 5: $\omega = 0.9$; $\kappa = 1\ \text{m}^{-1}$.

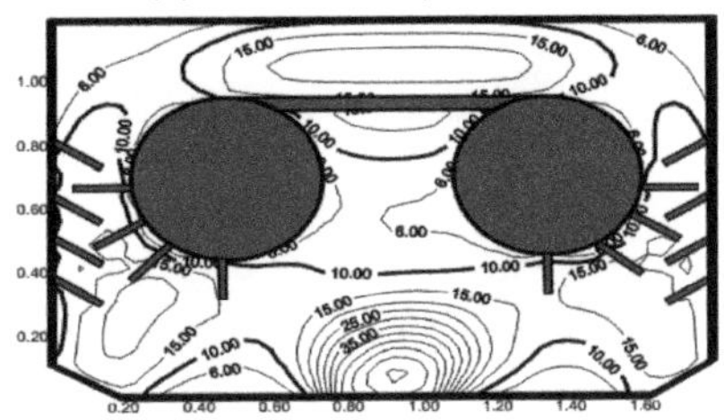

(f) carbon case6: $\omega = 0.9$; $\kappa = 5\ \text{m}^{-1}$

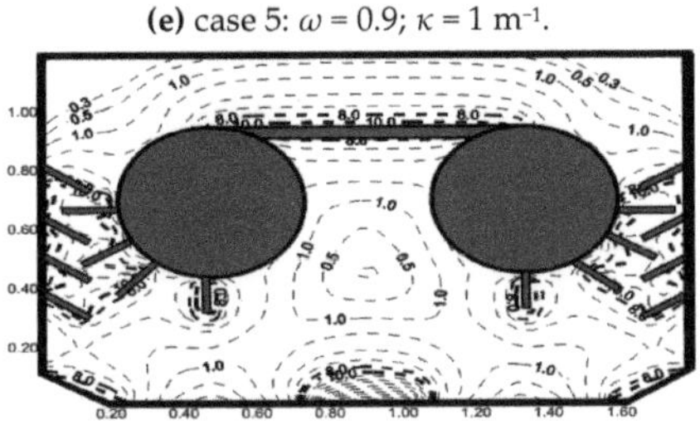

(g) isotropic case 6: $\omega= 0.9$; $\kappa = 5\ \text{m}^{-1}$

Figure 6. Effects of the absorption and albedo coefficients on the net radiation heat flux density distribution. Dashed lines: Isotropic, continuous lines: carbon. (**a**) case 1: $\omega = 0.1$; $\kappa = 1\ \text{m}^{-1}$, (**b**) case 2: $\omega = 0.1$; $\kappa = 5\ \text{m}^{-1}$, (**c**) case 3: $\omega = 0.5$; $\kappa = 1\ \text{m}^{-1}$, (**d**) case 4: $\omega = 0.5$; $\kappa = 5\ \text{m}^{-1}$, (**e**) case 5: $\omega = 0.9$; $\kappa = 1\ \text{m}^{-1}$, (**f**) carbon case 6: $\omega = 0.9$; $\kappa = 5\ \text{m}^{-1}$, (**g**) isotropic case 6: $\omega = 0.9$; $\kappa = 5\ \text{m}^{-1.}$

Figure 7a–g gives the isothermal lines for the function F1 and the isotropic case. It can be noticed that for the lower amount of the albedo coefficient, the anisotropy effect is imperceptible. As the albedo coefficient increases, a shift of the isotherms is observed corresponding to F1 function as compared to those of the isotropic case and the medium cools down. Also, a deformation of the isotherms is noted, which are close to the inlet of hot gases, toward the right pyrolysis chamber. This that the propagation of radiation is privileged to the right direction this behavioris more important when the absorption coefficient is greater.

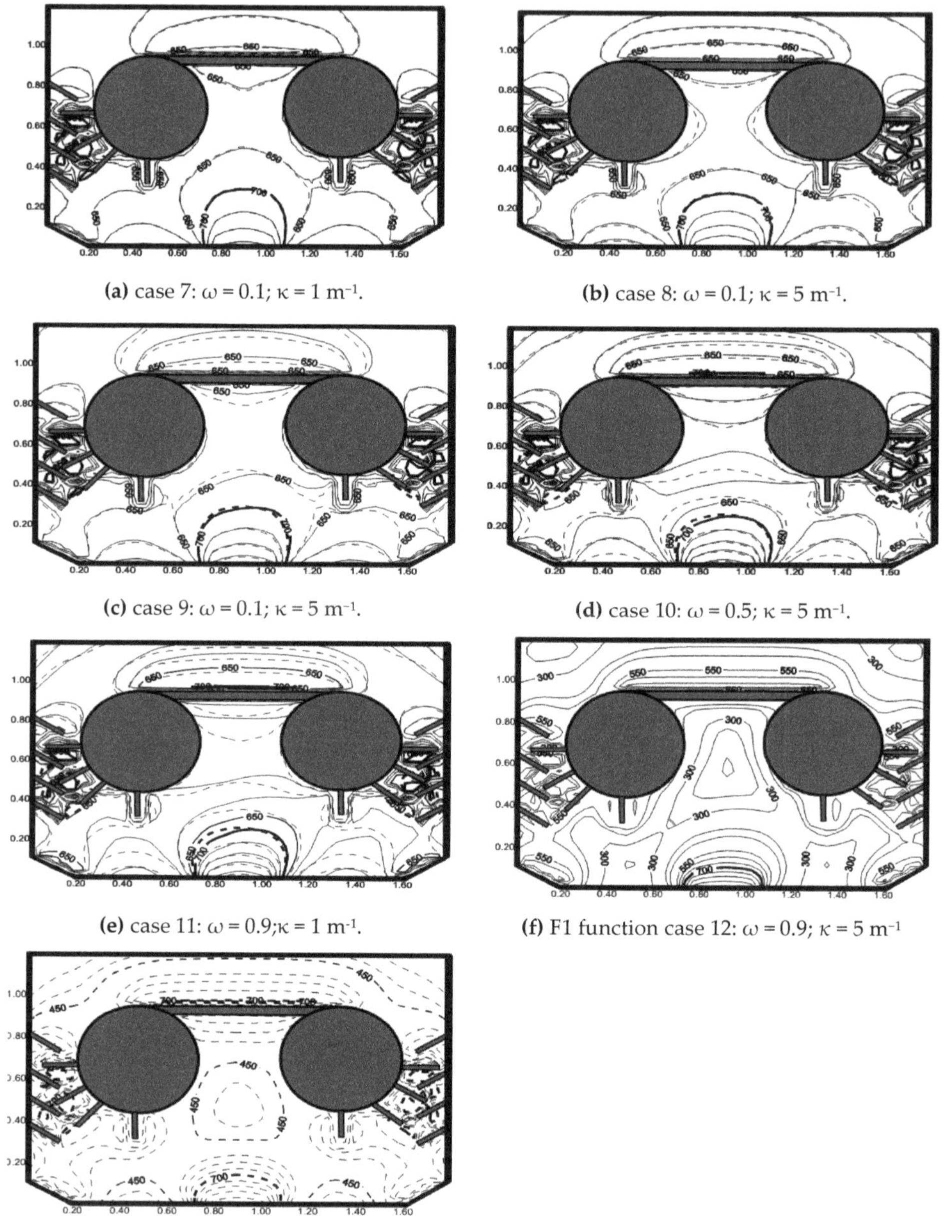

(a) case 7: ω = 0.1; κ = 1 m-1. **(b)** case 8: ω = 0.1; κ = 5 m-1.

(c) case 9: ω = 0.1; κ = 5 m-1. **(d)** case 10: ω = 0.5; κ = 5 m-1.

(e) case 11: ω = 0.9;κ = 1 m-1. **(f)** F1 function case 12: ω = 0.9; κ = 5 m-1

(g) isotropic case 12: ω = 0.9; κ = 5 m-1

Figure 7. Effects of the absorption and albedo coefficients on the medium temperature pattern: Comparison between the isotropic case and F1 function. Dashed lines: Isotropic, continuous lines: carbon. (**a**) case 7: $\omega = 0.1$; $\kappa = 1\ m^{-1}$, (**b**) case 8: $\omega = 0.1$; $\kappa = 5\ m^{-1}$, (**c**) case 9: $\omega = 0.1$; $\kappa = 5\ m^{-1}$, (**d**) case 10: $\omega = 0.5$; $\kappa = 5\ m^{-1}$, (**e**) case 11: $\omega = 0.9$; $\kappa = 1\ m^{-1}$, (**f**) F1 function case 12: $\omega = 0.9$; $\kappa = 5\ m^{-1}$, (**g**) isotropic case 12: $\omega = 0.9$; $\kappa = 5\ m^{-1}$.

Figure 8a–g shows the net radiation heat flux density contours for F1 and isotropic function. It can be seen that the radiation energy has conducted to the hot area when the absorption and albedo coefficients increase, especially for the anisotropic case. Moreover, a stagnation zone with an unrealistic net radiation heat flux value of –0.25 kW·m^{-2} is seen, indeed, an artificial environment is studied. Furthermore, it should be pointed that the average temperature and net radiation heat flux gradually decrease as the albedo coefficient enhances for the isotropic state and function F1, while they increase for the carbon case due to the backward scattering behavior of carbon particles. Tables 2 and 3 summarize the aforementioned results and also, include results for function B1.

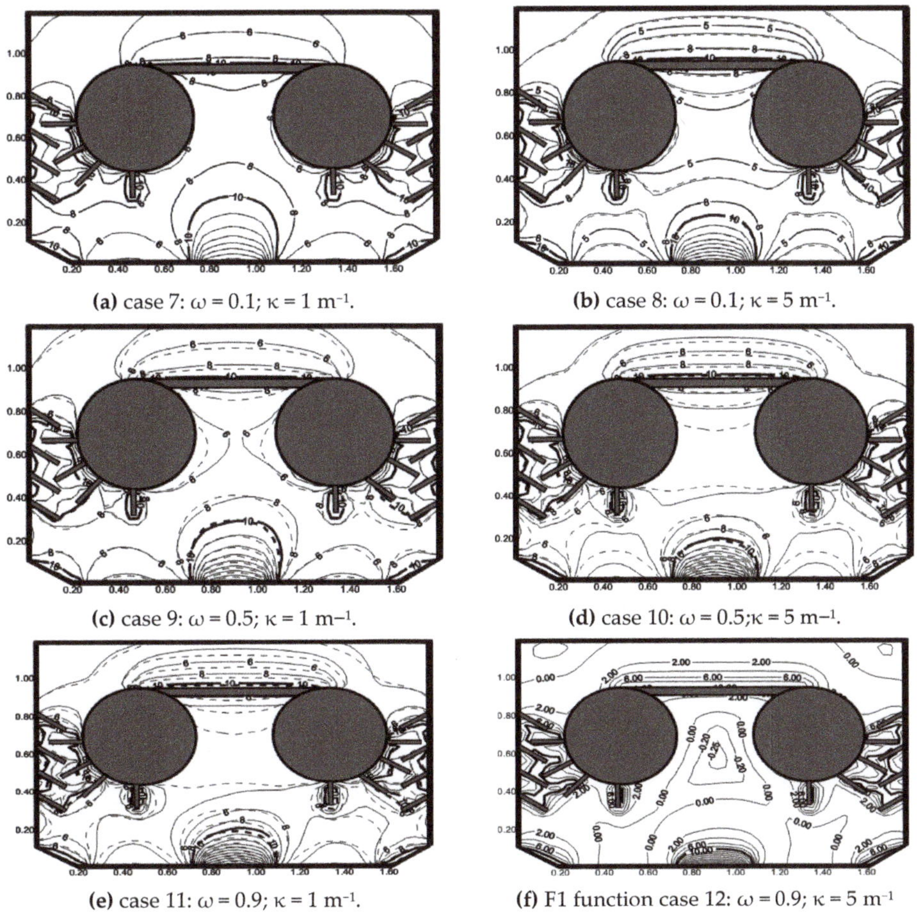

(a) case 7: $\omega = 0.1$; $\kappa = 1$ m^{-1}.

(b) case 8: $\omega = 0.1$; $\kappa = 5$ m^{-1}.

(c) case 9: $\omega = 0.5$; $\kappa = 1$ m^{-1}.

(d) case 10: $\omega = 0.5$; $\kappa = 5$ m^{-1}.

(e) case 11: $\omega = 0.9$; $\kappa = 1$ m^{-1}.

(f) F1 function case 12: $\omega = 0.9$; $\kappa = 5$ m^{-1}

Figure 8. *Cont.*

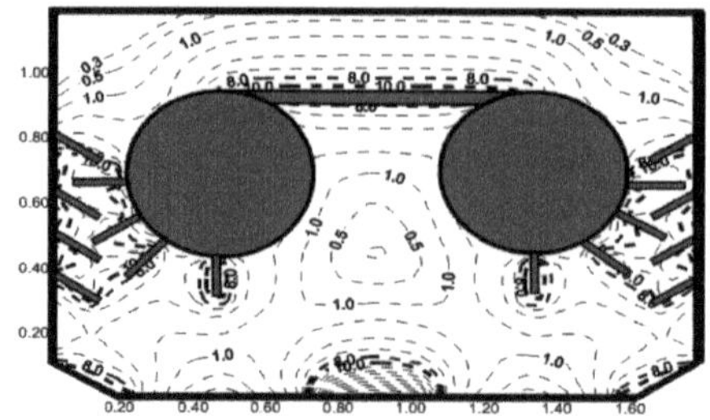

(g) isotropic case 12: $\omega = 0.9$; $\kappa = 5\text{m}^{-1}$

Figure 8. Effects of the absorption and albedo coefficients on the net radiation heat flux density distribution. Dashed lines: Isotropic, continuous lines: carbon. (**a**) case 7: $\omega = 0.1$; $\kappa = 1\ \text{m}^{-1}$, (**b**) case 8: $\omega = 0.1$; $\kappa = 5\ \text{m}^{-1}$, (**c**) case 9: $\omega = 0.5$; $\kappa = 1\ \text{m}^{-1}$, (**d**) case 10: $\omega = 0.5$; $\kappa = 5\ \text{m}^{-1}$, (**e**) case 11: $\omega = 0.9$; $\kappa = 1\ \text{m}^{-1}$, (**f**) F1 function case 12: $\omega = 0.9$; $\kappa = 5\ \text{m}^{-1}$, (**g**) isotropic case 12: $\omega = 0.9$; $\kappa = 5\ \text{m}^{-1}$.

Table 2. Average temperature and net radiation heat flux density value for different albedo values ($\kappa = 1\ \text{m}^{-1}$).

Albedo		$\omega = 0.1$	$\omega = 0.5$	$\omega = 0.7$	$\omega = 0.9$
ISO	T_{Av} (K)	585.10	585.01	584.29	577.00
	q_{Av} (kW·m^{-2})	6.60	6.43	6.21	5.47
B1	T_{Av} (K)	585.12	585.16	584.37	575.19
	q_{Av} (kW·m^{-2})	6.60	6.46	6.25	5.51
F1	T_{Av} (K)	585.25	585.39	583.12	558.34
	q_{Av} (kW·m^{-2})	6.60	6.45	6.19	5.07
Carbon	T_{Av} (K)	585.50	588.68	593.27	619.60
	q_{Av} (kW·m^{-2})	6.62	6.61	6.62	7.12

Table 3. Average temperature and net radiation heat flux density value for different albedo values ($\kappa = 5\ \text{m}^{-1}$).

Albedo		$\omega = 0.1$	$\omega = 0.5$	$\omega = 0.7$	$\omega = 0.9$
ISO	T_{Av} (K)	582.29	577.00	567.71	523.15
	q_{Av} (kW·m^{-2})	5.90	5.47	5.03	4.00
B1	T_{Av} (K)	582.26	576.04	564.54	510.53
	q_{Av} (kW·m^{-2})	5.91	5.5	5.03	3.84
F1	T_{Av} (K)	581.54	565.83	538.07	438.71
	q_{Av} (kW·m^{-2})	5.88	5.21	4.44	2.84
Carbon	T_{Av} (K)	584.47	598.97	623.96	724.18
	q_{Av} (kW·m^{-2})	5.99	6.25	6.92	11.75

It is observed that for $\kappa = 5\ \text{m}^{-1}$, the anisotropic character has a little effect on the average temperature and net radiation heat flux with a weak albedo coefficient (i.e., $\omega = 0.1$). When varying the albedo from $\omega = 0.1$ to $\omega = 0.9$, the average net radiation heat flux diminishes of 35.02% for the backward function, and 51.70% for the forward function. Actually, the size parameter of F1 function is greater than that of the B1 function, which explains the increase of the scattered energy. Furthermore, one can state that the medium temperature and flux are more sensitive to scattering rather than absorption phenomena. However, it can be seen carbon particles have a reverse effect as compared to the previous phase functions. This behavior is mainly due to the imaginary part of the complex index that is responsible of absorption of radiation. More radiation heat is transferred to the recuperator medium implies less boundary heat losses.

The graphs presented in Figure 9a–c describe the radiation heat flux density q_r on the walls of the left pyrolysis room (Figure 4a) for z = 0.60 m, where the same conditions as previously are considered. It is presented that the wall radiation heat flux density enhances, when approaching the horizontal baffle due to the latter blocks the radiation energy between the two pyrolysis rooms. Moreover, for the lower value of the ω, the anisotropic scattering character is negligible. When increasing ω from 0.1 to 0.9, the carbon phase function enhances the radiation heat emit from the hot surface, whereas the forward and backward scattering phase functions try to decrease it. It is noticed an overprediction of 42.50% for the carbon and an underprediction equal to 20.34% for function F1 as compared to the isotropic case. Yet, the impact of the backward-scattering phase function is found relatively less than the forward-scattering function, regarding the isotropic scattering state. This is foreseeable as forward scattering accentuates the radiation heat exchange between surface and volume zones.

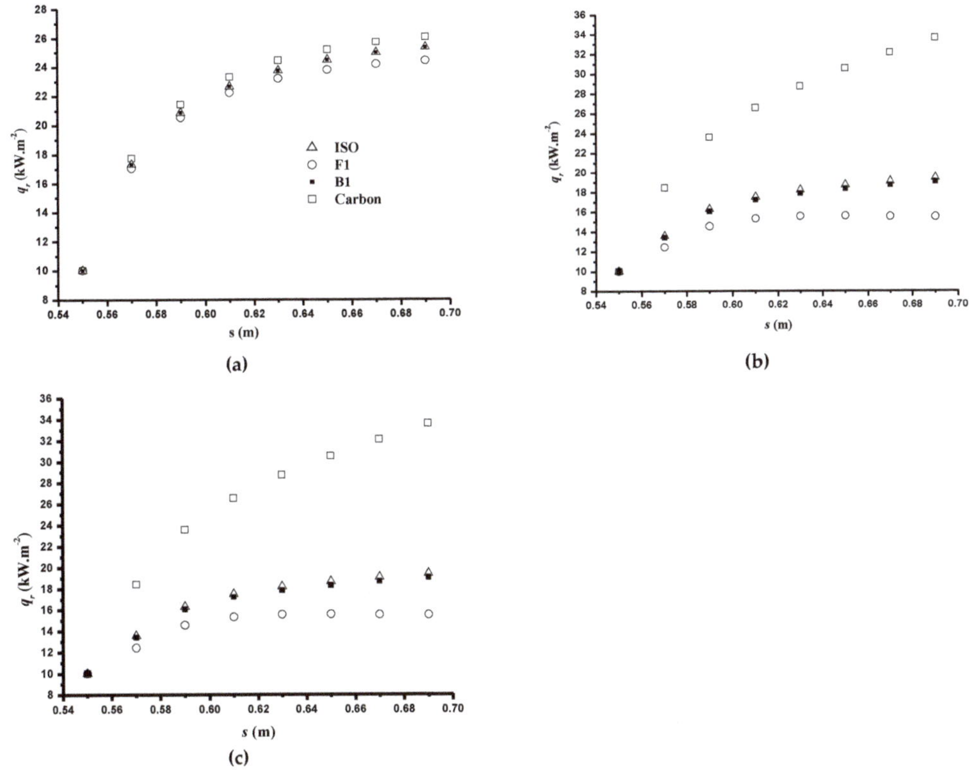

Figure 9. Distributions of the net radiation heat flux density profils on the walls of the left pyrolysis room (z = 0.6 m, κ = 1 m^{-1}): (**a**) ω = 0.1, (**b**) ω = 0.5, (**c**) ω = 0.9.

6.3. Particles Type Effect

The effect of the fly ash and other coals like anthracite, bituminous and lignite on the net radiation flux density contours at the front wall z = 0 mis analyzed. Table 4 are listed the corresponding data for evaluating of scattering phase functions related to the near infrared region [28].

Table 4. Complex refractive index for different coals and ashes in-near infrared region.

Coal and Ash	Carbon	Anthracite	Bituminous	Lignite	Ash
Particle size parameter x_p	1	1	1	1	1
Real part of complex refractive index n	2.20	2.05	1.85	1.70	1.50
Imaginary part of complex refractive index k	1.120	0.540	0.220	0.066	0.020

Numerical results in Figure 10a–d are calculatedfor an absorption coefficient $\kappa = 5\ \mathrm{m}^{-1}$ and a particle size parameter of unity. The scattering phase function curves of all the particles could be obtained from [19]. It can be noted that the anthracite is the best in terms of enhancing the radiation heat emitted from the hot wall while allows the maximum radiation heat to be transferred to biomass inside the two pyrolysis rooms. Moreover, it is stated that the medium anisotropy effect is the weakest for the fly ash due to its small imaginary part complex index of refraction, thus resulting in a more similar distribution of heat flux in the medium.

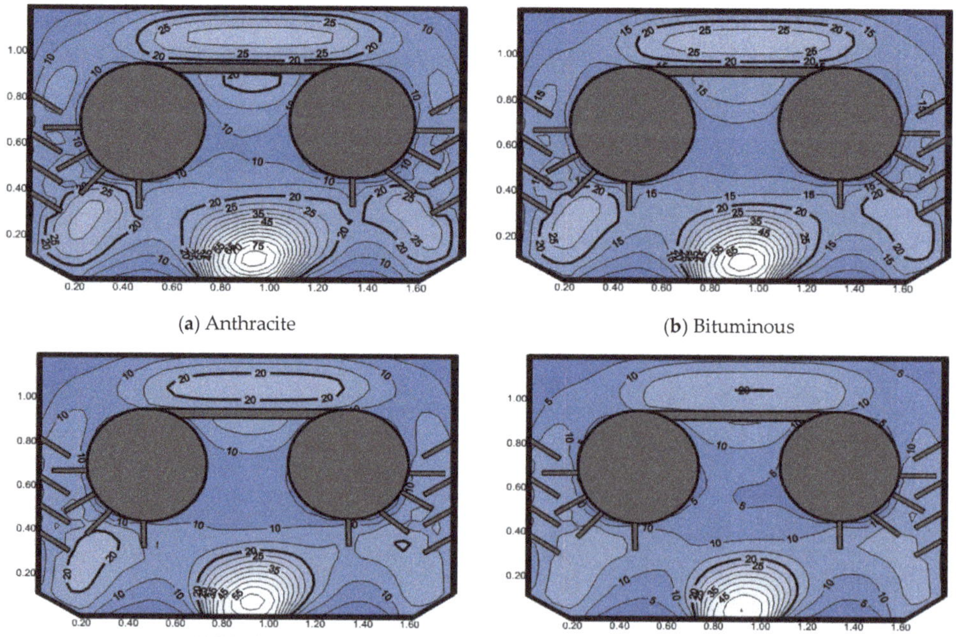

Figure 10. Isothermal profilesfor different particles materials at the plane $z = 0.0$ m ($x_p = 1$, $\omega = 0.9$, $\kappa = 5\ \mathrm{m}^{-1}$). (**a**) Anthracite, (**b**) Bituminous, (**c**) Lignite, (**d**) Fly Ash.

Table 5 depicts the average temperature and net radiation heat flux for the four coal types and fly ash. It can be noted that the anthracite has the highest impact. The average temperature and net radiation heat flux increase is equal to 25.18% and 62.70% for the albedo coefficients of $\omega = 0.1$ and $\omega = 0.9$, respectively. All the considered coals and fly ash have similar trends as the carbon, because they almost lead to the same angular distribution of the scattered radiation.

Table 5. The average temperature and net radiation heat flux density value for different albedo values (κ = 1 m^{-1}).

Albedo		$\omega = 0.1$	$\omega = 0.5$	$\omega = 0.7$	$\omega = 0.9$
Anthracite	T_{Av} (K)	584.84	603.29	636.35	783.78
	q_{Av} (kW·m^{-2})	6.00	6.41	7.43	16.09
Bituminous	T_{Av} (K)	584.70	601.67	631.83	760.44
	q_{Av} (kW·m^{-2})	6.00	6.35	7.23	14.25
Lignite	T_{Av} (K)	584.51	599.4	625.18	729.51
	q_{Av} (kW·m^{-2})	5.99	6.27	6.97	12.09
Fly ash	T_{Av} (K)	584.33	597.41	619.45	703.79
	q_{Av} (kW·m^{-2})	5.98	6.20	6.75	10.53

Figure 11a,b illustrates the predictions of the radiation flux divergence along the centerline (0.9, 0.6, z) m for different particle materials ($x_p = 1$), an albedo coefficient $\omega = 0.9$ and two absorption coefficient values. It can be noted in Figure 11a that all the curves have the same shape and they are high in the center region. The radiation source term profiles follow the same trend as in [19]. Moreover, the difference between the highest value (for anthracite) and the lowest value (for fly ash) varies between 15.07% and 16.82% for $\kappa = 1$ m^{-1} and around 49% for the relatively optically thick medium ($\kappa = 5$ m^{-1}). Further radiation energy is transferred through the medium whereas the optical thickness is enhanced. The highest amount of the radiation flux divergence corresponds to the anthracite which is multiplied by a factor of 8, when increasing the absorption coefficient from $\kappa = 1$ to $\kappa = 5$ m^{-1}. This is might be due to the large value of the imaginary part of anthracite's complex index of refraction.

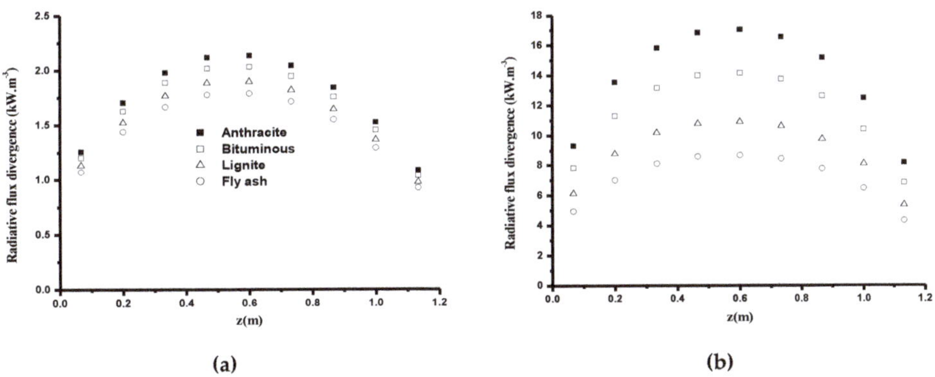

Figure 11. Profiles of the radiation flux divergence along the centerline for different particle materials ($x_p = 1$, $\omega = 0.9$): (**a**) $\kappa = 1$ m^{-1}, (**b**) $\kappa = 5$ m^{-1}.

6.4. Particle Size Parameter Effect

Figure 12a,b demonstrates the predictions of the radiation heat flux for fly ash in *y*-direction, having various particle size parameters: (a) q_{yz} at the middle of the south hot surface along (0.9, 0.0, z) m and (b) q_{yx} at the middle of the north surface along (x, 1.2, 0.6) m with a purely scattering medium (ω = 1). The profiles of the involved phase functions are presented in Figure 2. One can easily notice that when the particle size parameter increases, the forward scattering behavior becomes more significant. According to Figure 12a, the largest radiation heat flux density, produced by the highest value of the particle size parameter. The radiation heat flux difference between the smallest and the largest particle size parameter is equal to 6.64%. While, in Figure 12b this difference is the largest at the location x = 0.9 m, (i.e., 6.67%). Approaching the ends, where the side baffles are located, the difference becomes more and more weak. The mean temperature calculated at the plane z = 0 varies between 585.56 K and 588.99 K, and the mean radiation heat flux varies between 6.64 kW·m^{-2} and 6.81 kW·m^{-2}. It is concluded that by increasing the size parameter, the evolution of the temperature and net radiation heat flux density transfer is imperceptible.

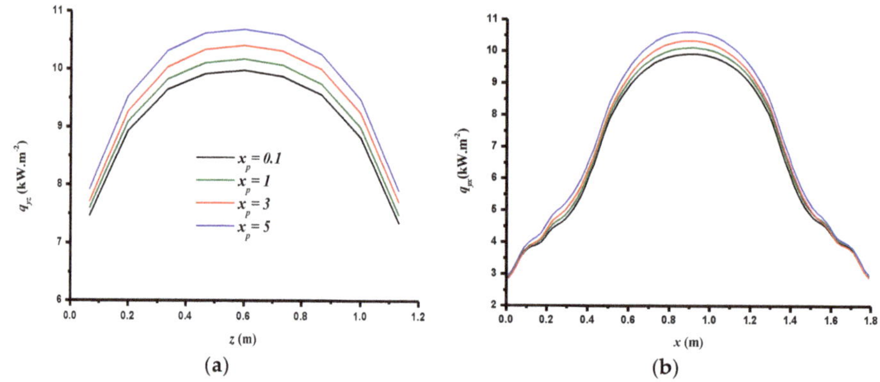

Figure 12. Profiles of the net predicted y-net radiation flux density for fly ash with different particle size parameters (ω = 0.9, β = 1 m^{-1}): (**a**) (0.9, 0.0, z) m and (**b**) (x, 1.8, 0.6) m.

Figure 13 shows the impact of the fly ash particle size parameter on the radiation heat flux density q_{yx}, at the middle of the north wall, for different optical thicknesses. The anisotropic scattering effect is not considerable for $\beta = 1$, because it is compatible with a thin optical state without much scattering occurring. The effect of the particle size parameter is more pronounced for higher extinction coefficients, (i.e., $\beta = 10$). Moreover, when β increase, the radiation heat flux difference between the smallest and the largest particle size parameter become more important, mainly at the location x = 0.9 m where it reaches 13.84%.

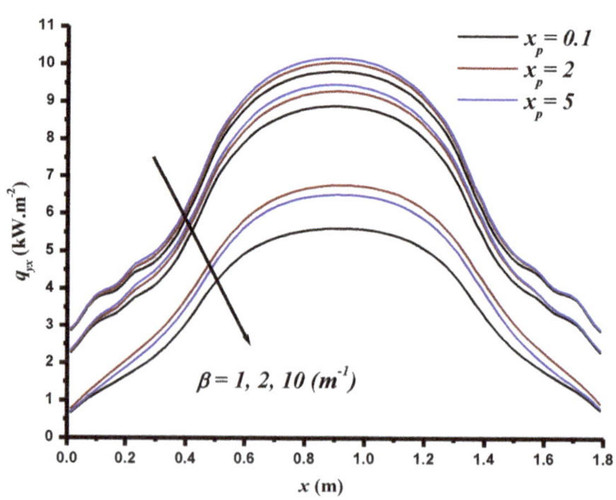

Figure 13. Evolution of the y-net radiation heat flux density component along (x, 1.8, 0.6) m vs the fly ash particle size parameter (ω = 0.5).

7. Conclusions

In present research, the radiation heat transfer in a biomass pyrolysis pilot plant with an absorbing and anisotropic scattering medium is studied. The calculation procedure of the radiation transfer equation (RTE) is numerically done by coupling the Mie theory and the finite volume method (FVM) with the STEP scheme, using the blocked-off-region process. The numerical code was satisfactorily validated by comparison of the predictions with available data for a present work. The goal of present investigation is to study the anisotropic scattering phenomenon impact on the radiation heat transfer in the biomass pyrolysis pilot plant, and especially the two pyrolysis rooms. It is concluded that the anisotropic character of particles is negligible for small values of the albedo regardless the value of the absorption coefficient. When the albedo boosts, the net radiation heat flux diminishes for the forward and backward phase functions, and is underpredicted as compared to the isotropic case. This decrease is more important when the absorption coefficient changes from $\kappa = 1$ to $\kappa = 5\ \text{m}^{-1}$. The predictions for solid particles of different coals and fly ash which are very similar to the actual combustion processes are produced. It is obtained that the net radiation heat flux enhances by the increment of the albedo coefficient. This tendency is similar for all the coals and fly ash, because these particles have almost the same angular distribution of the scattered radiation for low or moderate absorption coefficient values. In addition, it is found that the radiation energy transferred to biomass inside the two pyrolysis rooms is enhanced when using the anthracite's particles. The impact of the particle size parameter on the radiation heat flux for fly ash was analyzed. It is concluded that the net radiation heat flux enhances by increment of the particle size parameter. Furthermore, for high extinction coefficient values, a remarkable decrement in the net radiation heat flux is obtained, and the radiation heat flux difference between the smallest and the largest particle size parameter become more significant.

The numerical code developed and validated in the present investigation can further be extended to a 3-D configuration for a more pragmatic modeling of the heat transfer in the biomass pyrolysis pilot plant. Another possible perspective of the present work is to use the present code for the study of radiating nanoparticles in the flow of nanofluids by coupling the work like [29–33] and the present code.

Author Contributions: Conceptualization, M.A.A., M.M.S., and M.G.; methodology, A.E. and H.F.; software, K.G. and A.O.; validation, A.E., H.F. and M.N.B.; formal analysis, A.E.; investigation, H.F; resources, M.A.A.; data curation, K.G.; writing—original draft preparation, M.A.A. and A.E.; writing—review and editing, M.M.S. and M.G.; visualization, M.N.B.; supervision, M.M.A.; project administration, M.G.

Funding: This research received no external funding.

Conflicts of Interest: The authors declare no conflict of interest.

Nomenclature

a	coefficient of the discretized equations
b	source term of the discretized equations
D	direction cosines
E	radiation emissive power
I	radiant intensity
k	imaginarypart of the complex index
M	a large number (M = 10^{20})
m	complex index of refraction, face of control volume
N_θ	number of angular discretization in the polar angle
N_ϕ	number of angular discretization in the azimuthal angle
n	real part of the complex index of refraction or index in infinite series
P_n	Legendre Polynomial of order n
q	radiation heat flux
Q	efficiency factor

$\vec{r}$	position vector
r	particle radius
S	source function
S_{bloc}, S_C, S_P	additional source terms for blocked-off region procedure
s	distance in the direction $\vec{\Omega}$ of radiant intensity; curvilinear coordinate.
T	temperature
x_p	particle size parameter
x, y, z	Cartesian coordinates
Greek Symbols	
β	extinction coefficient
ΔA	area of a control volume face
ΔV	control volume
$\Delta\Omega$	control solid angle
ε	emissivity
κ	absorption coefficient
λ	wavelength of incident radiation
μ	direction cosine in the x-direction
ρ	reflectivity
θ	polar angle
Θ	scattering angle
σ	scattering coefficient
ϕ	azimuthal angle
Φ	phase function
$\overline{\Phi}^{l'l}$	average scattering phase function
ω	scattering albedo coefficient
$\vec{\Omega}$	angular direction
Subscript	
l, l′	angular directions
*	dimensionless
Av	average
b	blackbody
ext, sca, abs	extinction, scattering and absorption, respectively
e, w, n, s, r, f	east, west, north, south, rear and front neighbours of control volume P
E, W, N, S, B, T	nodes around the nodal point P
w	wall
Abbreviations	
RTE	radiation transfer equation
SPF	scattering phase function
WSGGM	weighted sum of gray gases model
ZM	zone method

References

1. Lockwood, F.C.; Shah, N.G. A new radiation solution method for incorporation in general combustion prediction procedures. *Symp. Int. Combust.* **1981**, *18*, 1405–1414. [CrossRef]
2. Malalasekera, W.M.G.; Lockwood, F.C. Computer Simulation of the King's Cross Fire: Effect of Radiative Heat Transfer on Fire. *Proc. Inst. Mech. Eng. Part C* **1991**, *205*, 201–208. [CrossRef]
3. Malalasekera, W.M.G.; James, E.H. Radiative Heat Transfer Calculations in Three Dimensional Complex Geometries. *J. Heat Transf.* **1996**, *118*, 225–228. [CrossRef]
4. Chai, J.C.; Lee, H.S.; Patankar, S.V. Treatment of Irregular Geometries Using a Cartesian-Coordinates-Based Discrete-Ordinates Method, Radiative Heat Transfer. *Theory Appl.* **1993**, *244*, 49–54.
5. Fivel, W.A.; Jesse, J.P. Finite Element Formulation of the Discrete-Ordinates Method for Multidimensional Geometries, Radiative Heat Transfer. *Curr. Res.* **1994**, *276*, 49–57.

6. Chai, J.C.; Parthasarathy, G.; Lee, H.S.; Patankar, S.V. Finite Volume Radiative Heat Transfer Procedure for Irregular Geometries. *J. Heat Transf.* **1995**, *9*, 410–415. [CrossRef]
7. Chai, J.; Lee, H.S.; Patankar, S.V. Treatment of irregular geometries using a Cartesian coordinates finite volume radiative heat transfer procedure. *Numer. Heat Transf. Part B* **1994**, *26*, 225–235. [CrossRef]
8. Borjini, M.N.; Farhat, H.; Radhouani, M.S. Analysis of radiative heat transfer in a partitioned idealized furnace. *Numer. Heat Transf. Part A* **2003**, *35*, 467–495. [CrossRef]
9. Guedri, K.; Borjini, M.N.; Farhat, H. Modelization of Combined Radiative and Conductive Heat Transfer in Three-Dimensional Complex Enclosures. *Int. J. Numer. Methods Heat Fluid Flow* **2005**, *15*, 257–276. [CrossRef]
10. Chai, J.C.; Moder, J.P. Spatial-Multiblock Procedure for Radiation Heat Transfer. *Numer. Heat Transf. B* **1997**, *31*, 277–293. [CrossRef]
11. Coelho, P.J.; GonÇalves, J.M.; Carvalho, M.G.; Trivic, D.N. Modeling of Radiative Heat Transfer in Enclosures with Obstacles. *Int. J. Heat Mass Transf.* **1998**, *41*, 745–756. [CrossRef]
12. Guedri, K.; Abbassi, M.A.; Borjini, M.N.; Halouani, K. Application of the finite-volume method to study the effects of baffles on radiative heat transfer in complex enclosures. *Numer. Heat Transf. Part A* **2009**, *55*, 1–27. [CrossRef]
13. Abbassi, M.A.; Guedri, K.; Borjini, M.N.; Halouani, K.; Zeghmati, B. Modeling of radiative heat transfer in 2D complex heat recuperator of biomass pyrolysis furnace: A study of baffles shadow and soot volume fraction effects. *Int. J. Math. Comput. Nat. Phys. Eng.* **2014**, *8*, 450–460.
14. Mengüç, M.P.; Viscanta, R. Radiative transfer in three-dimensional rectangular enclosures containing inhomogeneous, anisotropically scattering media. *J. Quant. Spectrosc. Radiat. Transf.* **1995**, *33*, 533–549. [CrossRef]
15. Kim, T.K.; Lee, H.S. Effect of anisotropic scattering on radiative heat transfer in two-dimensional rectangular enclosures. *Int. J. Heat Mass Transf.* **1988**, *31*, 1711–1721. [CrossRef]
16. Kim, T.K.; Lee, H.S. Radiative heat transfer in two dimensional anisotropic scattering media with collimated incidence. *J. Quant. Spectrosc. Radiat. Transf.* **1989**, *42*, 225–238. [CrossRef]
17. Farmer, J.T.; Howell, J.R. Monte Carlo Prediction of Radiative Heat Transfer in Inhomogeneous, Anisotropic, Nongray Media. *AIAA J. Thermophys. Heat Transf.* **1994**, *8*, 133–139. [CrossRef]
18. Guo, Z.; Maruyama, S. Scaling anisotropic scattering in radiative transfer in three-dimensional nonhomogeneous media. *Int. Commun. Heat Mass Transf.* **1999**, *26*, 997–1007. [CrossRef]
19. Trivic, D.N.; O'Brien, T.J.; Amon, C.H. Modeling the radiation of anisotropically scattering media by coupling Mie theory with finite volume method. *Int. J. Heat Mass Transf.* **2004**, *47*, 5765–5780. [CrossRef]
20. Trivic, D.N.; Amon, C.H. Modeling the 3-D radiation of anisotropically scattering media by two different methods. *Int. J. Heat Mass Transf.* **2008**, *51*, 2711–2732. [CrossRef]
21. Trivic, D.N. 3-D radiation modeling of nongray gases–particles mixture by two different numerical methods. *Int. J. Heat Mass Transf.* **2014**, *70*, 298–312. [CrossRef]
22. Hunter, B.; Guo, Z. Improved treatment of anisotropic scattering in radiation transfer analysis using the finite volume method. *Heat Transf. Eng.* **2015**, *37*, 341–350. [CrossRef]
23. Guedri, K.; Al-Ghamdi, A.S.; Bouzid, A.; Abbassi, M.A.; Ghulman, H.A. Evaluation of the FTn finite volume method for transient radiative transfer in anisotropically scattering medium. *Numer. Heat Transf. Part A* **2015**, *68*, 1137–1154. [CrossRef]
24. Chai, J.C.; Lee, H.S.; Patankar, S.V. Finite volume method for radiation heat transfer. *J. Heat Transf.* **1994**, *8*, 419–425. [CrossRef]
25. Chai, J.C.; Patankar, S.V. Finite-volume method for radiation heat transfer. In *Advances in Numerical Heat Transfer 2*; Minkowycz, W.K., Sparrow, E.H., Eds.; Taylor & Francis: Boca Raton, FL, USA, 2000.
26. Deirmendjian, D.; Clasen, R.; Viezee, V. Mie scattering with complex index of refraction. *J. Opt. Soc. Am.* **1961**, *51*, 620–633. [CrossRef]
27. Abbassi, M.A.; Grioui, N.; Halouani, K.; Zoulalian, A.; Zeghmati, B. A practical approach for modelling and control of biomass pyrolysis pilot plant with heat recovery from combustion of pyrolysis products. *Fuel Process. Technol.* **2009**, *90*, 1278–1285. [CrossRef]
28. Modest, M.F. *Radiative Heat Transfer*; Academic Press: New York, NY, USA, 2013.
29. Togun, H.; Safaei, M.R.; Sadri, R.; Kazi, S.N.; Badarudin, A.; Hooman, K.; Sadeghinezhad, E. Numerical simulation of laminar to turbulent nanofluid flow and heat transfer over a backward-facing step. *Appl. Math. Comput.* **2014**, *239*, 153–170. [CrossRef]

30. Hosseini, M.; Safaei, M.R.; Estellé, P.; Jafarnia, S.H. Heat transfer of water-based carbon nanotube nanofluids in the shell and tube cooling heat exchangers of the gasoline product of the residue fluid catalytic cracking unit. *J. Therm. Anal. Calorim.* **2019**, 1–12. [CrossRef]
31. Togun, H.; Ahmadi, G.; Abdulrazzaq, T.; Shkarah, A.J.; Kazi, S.N.; Badarudin, A.; Safaei, M.R. Thermal performance of nanofluid in ducts with double forward-facing steps. *J. Taiwan Inst. Chem. Eng.* **2015**, *47*, 28–42. [CrossRef]
32. Safaei, M.R.; Togun, H.; Vafai, K.; Kazi, S.N.; Badarudin, A. Investigation of heat transfer enhancement in a forward-facing contracting channel using FMWCNT nanofluids. *Numer. Heat Transf. Part A Appl.* **2014**, *66*, 1321–1340. [CrossRef]
33. Gholamalizadeh, E.; Pahlevanzadeh, F.; Ghani, K.; Karimipour, A.; Nguyen, T.K.; Safaei, M.R. Simulation of water/FMWCNT nanofluid forced convection in a microchannel filled with porous material under slip velocity and temperature jump boundary conditions. *Int. J. Numer. Methods Heat Fluid Flow* **2019**. [CrossRef]

energies

MDPI

Article

Energy Dissipation in Stepped Spillways with Different Horizontal Face Angles

Yongqin Peng [1], Xujin Zhang [1], Hao Yuan [1,*], Xia Li [1], Chunhang Xie [1], Shuqing Yang [1] and Zhaoliang Bai [2]

1 Southwest Research Institute for Water Transport Engineering, Chongqing Jiaotong University, Chongqing 400074, China; pyqin_xks@163.com (Y.P.); xjzhang.cqjtu@gmail.com (X.Z.); lxhhu@cqjtu.edu.cn (X.L.); xiechunhang@126.com (C.X.); ysq20052707@163.com (S.Y.)

2 State Key Laboratory of Hydraulics and Mountain River Engineering, Sichuan University, Chengdu 610065, China; baizhaoliang@hotmail.com

* Correspondence: 18996152721@163.com

Received: 23 September 2019; Accepted: 19 November 2019; Published: 23 November 2019

Abstract: Energy dissipation is one of the most important factors in choosing stepped spillways. However, very few studies have investigated energy dissipation with different horizontal face angles. In this paper, the realizable k-ε turbulent model was used to study the flow field, energy dissipation rates and turbulent kinetic energy and its dissipation rate for different stepped spillways with five horizontal face angles in the skimming flow regions. Results showed that the field and direction of the flow were changed by the horizontal face angles of the stepped spillway, which produced some unique characteristics and thus caused better energy dissipation. The fluctuation of free water surface will be larger with increasing horizontal face angles and the energy dissipation rate decreases with an increasing unit discharge and increases for the enlargement of the horizontal face angles. This conclusion could provide a reference for the relevant research of V shaped stepped spillways.

Keywords: numerical simulation; horizontal face angle; energy dissipation rates; stepped spillway

1. Introduction

Stepped spillways, an energy dissipation structure widely used in hydraulic engineering, have better energy dissipation rates than smooth spillways [1]. The huge energy of the water flowing through the spillway downstream is likely to cause serious erosion downstream, so it is of great significance to dissipate energy in the spillway stage. The energy dissipation rate is an important factor in choosing an energy dissipation structure. Therefore, the energy dissipation in stepped spillways has been a research focus.

To date, most of the studies on the energy dissipation rate of traditional stepped spillways have mainly focused on the size [2], number, and arrangement of the step [3], flow state, channel slope, unit discharge, and so on. Chanson [4] believed that the flow pattern of a stepped spillway can be either falling or slip stream, and the two streams have different effects on different lengths of spillways. Abbasi et al. [5] numerically studied the influence of the number of steps, step height, and unit discharge on the energy dissipation rate of a step spillway using the standard k-ε turbulent model, and found that the energy dissipation rate decreased with an increase in the number of steps and unit discharge and increased as the step height and length increased. Rassaei et al. [6] and Tabari et al. [7] used the k-ε turbulent model and derived the same rules for different sizes of step spillways. Wu [8], who studied the stepped spillway with four channel slope, found that the energy dissipation rate could reach 60% with the slope range of 1:2 to 1:3 and significantly reduce when the channel slope goes beyond 1:2. Using the RNG k-ε turbulent model combined with the VOF method, Shahheydari et al. [9] found that the slope ratio also has a significant impact on the energy dissipation rate of the step spillway. Attarian et al. [10] found that the energy dissipation rate was affected not only by the height of the step, but also by the aeration amount with the realizable k-ε turbulent model.

The rolling, collision, and turbulent shearing between the water in the step spillway can improve the energy dissipation rate [11]. Guenther et al. [12] studied the characteristics of aeration concentration, vortex characteristics, and energy dissipation rate with four different types of stepped spillways by model tests. Mero et al. [13] suggested that the energy dissipation rate of both the upturned stepped body and the stepped horizontal plane were about twice that of the conventional body using model experiment. Barani et al. [14] found that the shape of the step has a significant influence on the energy dissipation rate of the spillway by using three different types of texts: Conventional, upswing, and cantilever step. The shape of the cantilever step also had a significant impact on the flow pattern and energy dissipation efficiency of the spillway [15]. Therefore, it is scientifically beneficial to modify the shape of the step spillway to increase its collision and roll to achieve energy dissipation.

A V-shaped stepped spillway can effectively increase the energy dissipation rate by enhancing the collision and rotation of the water flow. The energy dissipation rate of the V-shaped stepped spillway expressed an obvious advantage compared to a traditional stepped spillway with the same height and length of the step [16]. However, detailed studies on energy dissipation via the shape of the step in V-shaped step spillways are rare. The horizontal face angle is a significant factor in a V-shaped stepped spillway, which can clearly influence the energy dissipation rate. In this paper, we investigated the streamlines and energy dissipation rates of stepped spillways with different horizontal face angles of −30°, −15°, 0°, 15°, and 30°. Then, the relation of the energy dissipation rates with different horizontal face angles was obtained. These results can be used in choosing a stepped spillway with better energy dissipation rates and provide a reference for the design of a V-shaped stepped spillway.

2. Numerical Model

2.1. Layout of the Numerical Model

The numerical simulations were performed with the Fluent software and the layout of the numerical model is shown in Figure 1, which was composed of a pressed slope section, a smooth section, a transitional section, a stepped section, and a tail water section. The width of the stepped spillways was B = 40 cm, the inlet height was h = 12 cm, and the outlet height of the pressed slope section was 8 cm. In the stepped section, the step sizes of different stepped spillways were the same (6 cm in height and 12 cm in length). The only difference was the horizontal face angles. Here, five horizontal face angles (θ = −30°, −15°, 0°, 15°, and 30°) were studied. The tail water section was directly connected to the stepped section. There were 56 steps (all surfaces of steps were either horizontal or vertical) in each stepped spillway. These steps were named #1 to #56. The stepped spillway with θ = 0° was named a traditional stepped spillway.

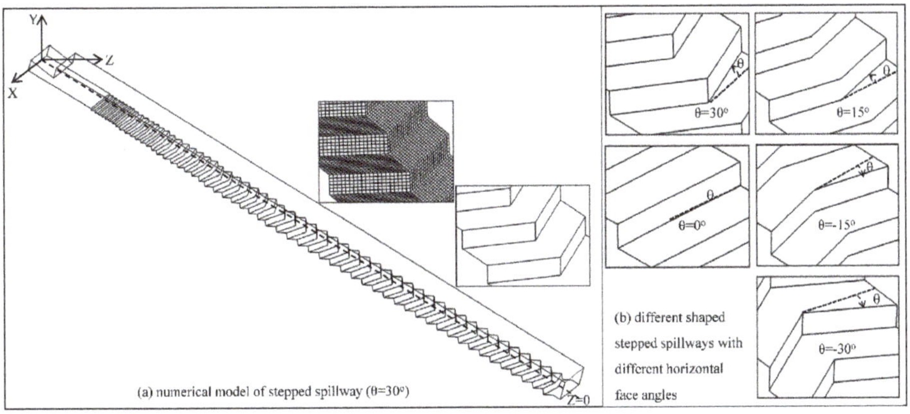

Figure 1. Layout of the numerical model.

2.2. Governing Equations

The realizable k-ε turbulent model was presented by Shih et al. [17] and is useful for simulating stepped flow [18–20]. The air–water interface was tracked by the volume of fluid (VOF) method. The equations for turbulent kinetic energy (k) and its dissipation rate (ε) are as follows:

k equation:

$$\frac{\partial(\rho k)}{\partial t} + \frac{\partial(\rho u_j k)}{\partial x_j} = \frac{\partial}{\partial x_j}[(\mu + \frac{\mu_t}{\sigma_k})\frac{\partial k}{\partial x_j}] + G_k + G_b - \rho\varepsilon - Y_M + S_k \tag{1}$$

ε equation:

$$\frac{\partial(\rho \varepsilon)}{\partial t} + \frac{\partial(\rho u_j \varepsilon)}{\partial x_j} = \frac{\partial}{\partial x_j}[(\mu + \frac{\mu_t}{\sigma_\varepsilon})\frac{\partial \varepsilon}{\partial x_j}] + \rho C_1 S_\varepsilon - \rho C_2 \frac{\varepsilon^2}{k + \sqrt{\nu\varepsilon}} + C_{1\varepsilon}\frac{\varepsilon}{k}C_{3\varepsilon}G_b + S_\varepsilon \tag{2}$$

The corresponding supplementary equation for different parameters in Equations (1) and (2) are as follows:

$$C_1 = \max[0.43, \frac{\eta}{\eta + 5}] \tag{3}$$

$$\eta = \frac{sk}{\varepsilon} \tag{4}$$

$$S = \sqrt{2S_{ij}S_{ij}} \tag{5}$$

$$S_{ij} = 0.5(\frac{\partial u_j}{\partial x_i} + \frac{\partial u_i}{\partial x_j}) \tag{6}$$

$$\rho = \alpha_w \rho_w + (1 - \alpha_w)\rho_a \tag{7}$$

$$\mu = \alpha_w \mu_w + (1 - \alpha_w)\mu_a \tag{8}$$

where α_w is the volume fraction of water; ρ_w and ρ_a are the densities of water and air, respectively; μ_w and μ_a are the viscosities of water and air, respectively; and $C_2 = 1.9$, $\sigma_k = 1.0$, and $\sigma_k = 1.2$ are the empirical constants.

2.3. Boundary Conditions and Solver

The water inlet was set as the velocity-inlet condition and its velocity was dependent on the unit discharges (q). The outlet boundary was treated as a pressure-outlet condition, where atmospheric pressure was assumed. For the wall boundary, a no-slip velocity condition was used, and the standard wall function was used to deal with the near-wall regions. The air boundary was set as the pressure-inlet condition, where atmospheric pressure was assumed. Pressure-based calculation and transient time were adopted in the solver and the SIMPLE algorithm was employed for the coupling of pressure and velocity. The computational domain was discretized using a structured grid, which is shown in Figure 1.

2.4. Validation Model

The grid convergence index (GCI) [21] with numbers of approximately 0.25 million, 0.60 million, and 1.95 million was were used to check the sensitivity of the grid in the numerical model. This method was run with Equations (9) and (10), and more details can be found in Celik [21].

$$GCI = \frac{1.25\left|\frac{\phi_1-\phi_2}{\phi_1}\right|}{(h_2/h_1)^P - 1} \tag{9}$$

$$P = \frac{1}{\ln(h_2/h_1)}\left|\ln\left|(\phi_3-\phi_2)/(\phi_2-\phi_1)\right| + \ln\left(\frac{(h_2/h_1)^P - \mathrm{sgn}((\phi_3-\phi_2)/(\phi_2-\phi_1))}{(h_3/h_2)^P - \mathrm{sgn}((\phi_3-\phi_2)/(\phi_2-\phi_1))}\right)\right| \tag{10}$$

where, ϕ_i is the solution for the ith grid, i is the selected numerical value, and h_i is the grid size. In this paper, three representative grids with 0.25 million, 0.60 million, and 1.95 million were used to calculate the GCI.

The calculation model of the horizontal face angle (θ = 30°) was chosen and the effect of the grid sizes on the uncertainty of the computational velocity distribution is shown in Figure 2, where the horizontal axis is the dimensionless velocity and the horizontal axis represents the dimensionless width of the step in the position of Z/B = 0.25. As shown in Figure 2, the maximum uncertainties in the velocity were approximately 7.3%. Thus, 0.60 million was chosen in this paper.

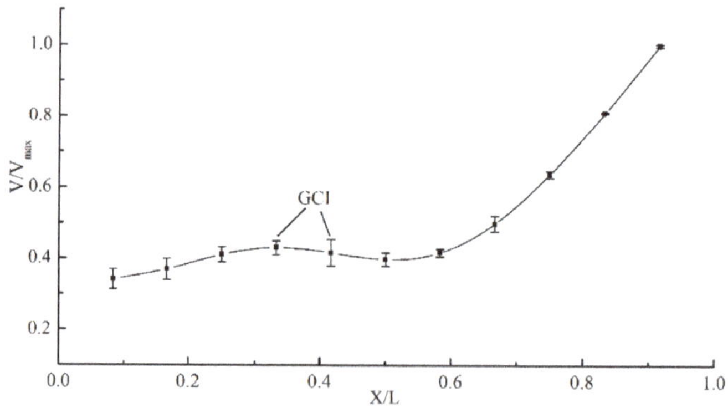

Figure 2. Grid convergence (GCI) values for different grid densities with velocity.

In order to test the accuracy of the numerical values, physical model experiments were made in the State Key Laboratory of Hydraulic and Mountain River Engineering, Sichuan University, Chengdu. This model was composed of an upper water tank, a stepped spillway (including transitional section and a stepped section), a tail water section, a measuring weir, and a reservoir. The size of the stepped spillway and the unit discharges were identical to those of the numerical model. Two horizontal face angles (θ = 15°, 30°) were chosen in the physical model experiments and the validation model is shown in Figure 3. Figure 3a–c is the physical model, the layout of model, and the layout of the pressure measurement point, respectively.

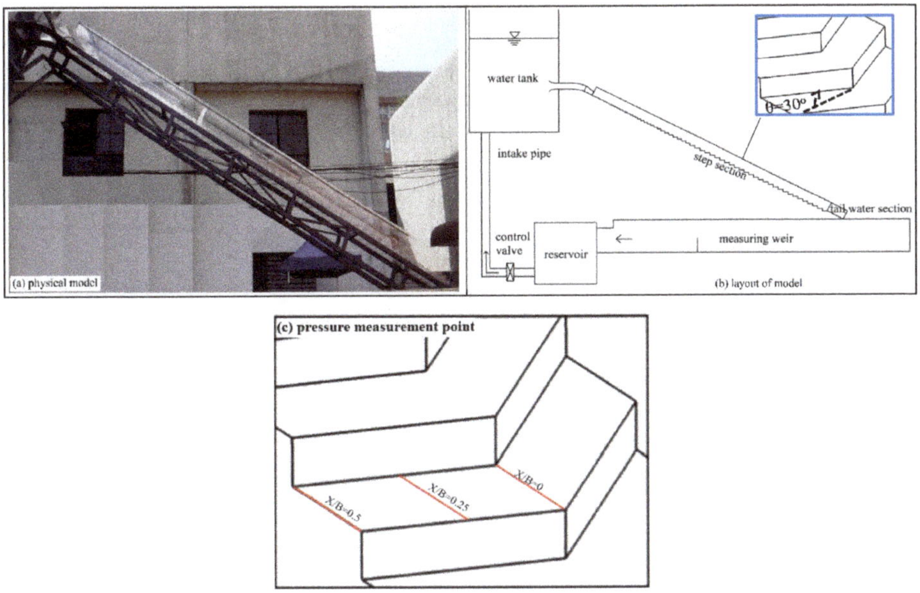

Figure 3. Layout of the validation model.

Tables 1 and 2 show a comparison of the pressures between the numerical and physical values on the horizontal surface and vertical surface, respectively. Here, X represents the distance of the pressure detecting points on the horizontal surface from the step's inner edge, L represents the length of the step, Y represents the distance of the pressure detecting points on the vertical surface from the step's lower edge, and H represents the height of the step. Table 3 features a comparison of the energy dissipation rates between numerical and physical values with various unit discharges. As seen in Tables 1–3, the maximum error of the pressure on the horizontal surface was 7.94%, the maximum error of the pressure on the vertical surface was 7.41%, and the maximum error of the energy dissipation rates was 6.7%. Although there were some errors, the accuracies were sufficient.

Table 1. Comparison of the pressures between the numerical and physical values on the horizontal surface (q = 0.489 m^2/s, θ = 30°).

X/L	X/B = 0			X/B = 0.25			X/B = 0.5		
	Physical	Numerical	Error (%)	Physical	Numerical	Error (%)	Physical	Numerical	Error (%)
0.08	−7.02	−7.58	7.94	8.02	8.58	6.95	13.39	13.34	−0.37
0.17	−9.48	−9.93	4.76	3.80	4.04	6.23	12.70	13.12	3.34
0.25	−12.72	−13.18	3.56	1.40	1.46	4.15	12.80	13.11	2.43
0.33	−15.48	−14.45	−6.65	3.42	3.22	−5.76	14.05	13.30	−5.35
0.42	−11.04	−11.83	7.19	8.21	8.46	3.14	14.24	13.67	−4.01
0.50	−2.20	−2.37	7.45	15.08	15.23	1.03	14.99	14.09	−6.01
0.58	11.67	12.42	6.41	22.48	21.64	−3.72	15.17	14.35	−5.43
0.67	31.65	32.87	3.85	27.98	26.28	−6.08	14.44	14.22	−1.53
0.75	51.92	52.66	1.41	30.63	28.53	−6.86	14.57	13.65	−6.30
0.83	61.08	63.15	3.39	31.24	28.86	−7.62	13.40	12.64	−5.68
0.92	60.57	57.25	−5.47	27.32	25.72	−5.88	11.85	11.19	−5.59

Table 2. Comparison of the pressures between the numerical and physical values on the vertical surface (q = 0.489 m^2/s, θ = 30°).

Y/H	X/B = 0			X/B = 0.25			X/B = 0.5		
	Physical	Numerical	Error (%)	Physical	Numerical	Error (%)	Physical	Numerical	Error (%)
0.83	−23.94	−25.70	7.35	−3.31	−3.53	6.86	8.16	7.65	−6.21
0.67	−11.58	−12.33	6.54	−1.49	−1.55	4.35	8.68	8.49	−2.14
0.50	−10.48	−11.25	7.41	−2.84	−2.64	−7.09	8.90	9.03	1.39
0.33	−9.12	−9.47	3.83	1.49	1.60	6.95	10.71	10.39	−3.01
0.17	−7.61	−7.26	−4.48	8.95	9.56	6.82	11.92	12.37	3.70

Table 3. Comparison of the energy dissipation rates between the numerical and physical values with various unit discharges.

Case	θ = 30°		
q(m^2/s)	Physical Value	Numerical Value	Error (%)
0.313	75.21	79.56	5.78
0.425	73.21	75.39	2.98
0.489	70.14	74.84	6.70
0.552	69.24	73.56	6.24
0.600	67.76	70.24	3.66

3. Results and Discussion

3.1. Streamlines

The direction of velocity, the basic characteristic in a V-shaped stepped spillway, varies greatly, and an analysis of streamlines helps us to understand the effects of energy dissipation of the stepped spillway in-depth. Figure 4 shows the streamlines of differently shaped steps for a one-step number (#43, for example). It can be seen that (1) when θ = 0°, the streamlines were parallel to the axial plane, so the free water surface did not change along the cross-section; (2) when θ > 0°, the streamlines were not parallel to the axial plane and extended from the sidewalls to the axial plane, so the water flow collided near the axial plane. However, the body of the stepped spillway with θ = 30° became larger than that in the stepped spillway with θ = 15°, so the collision of the water flow near the axial plane was more intense; and (3) when θ < 0°, the streamlines were also not parallel to the axial plane and extended from the axial plane to the sidewalls, so the water flow from the axial plane collided with the sidewalls. As a result, the free water surface would be higher near the sidewalls. However, the body of the stepped spillway with θ = −30° became larger than that in the stepped spillway with θ = −15°, so the collision of the water flow near the sidewalls was more intense.

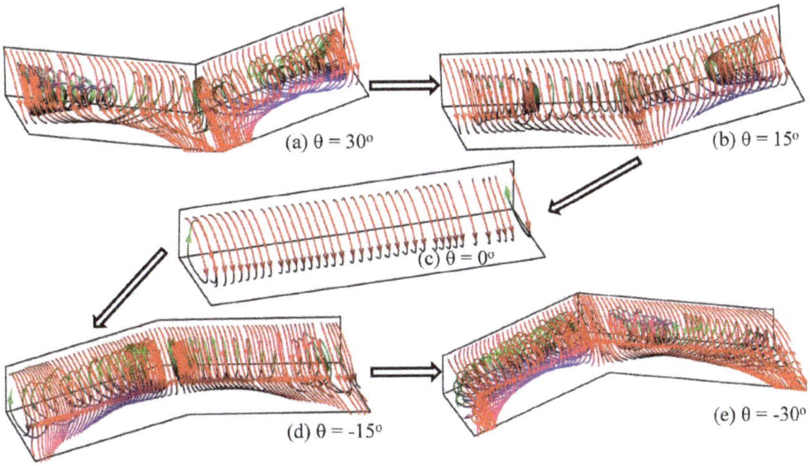

Figure 4. Streamlines of different shaped steps (#43).

3.2. Energy Dissipation Rate

As a result of the conservation of upstream and downstream energy, the ratio of the energy loss to the upstream energy is defined as the energy dissipation rate:

$$\eta = \frac{\Delta E}{E_1} \times 100\% = \frac{E_1 - E_2}{E_1} \times 100\% \tag{11}$$

where E_1, E_2 are the total energy in the beginning and end of the stepped section, respectively; $E_1 = \Delta h + v_1^2/(2g)$, $E_2 = v_2^2/(2g)$; where Δh is the difference in the height between the two sections, and v_1 and v_2 are the average velocities in the two sections.

Figure 5 shows the energy dissipation rates change with various unit discharges and horizontal face angles. It can be seen that (1) the energy dissipation rates in all shaped stepped spillways decreased with an increase of the unit discharges but the decrement rate was larger in the traditional stepped spillway; and (2) at a given unit discharge, as the angle increased, the energy dissipation rate initially decreased and then increased. Thus, the energy dissipation rate was the lowest in the traditional stepped spillway.

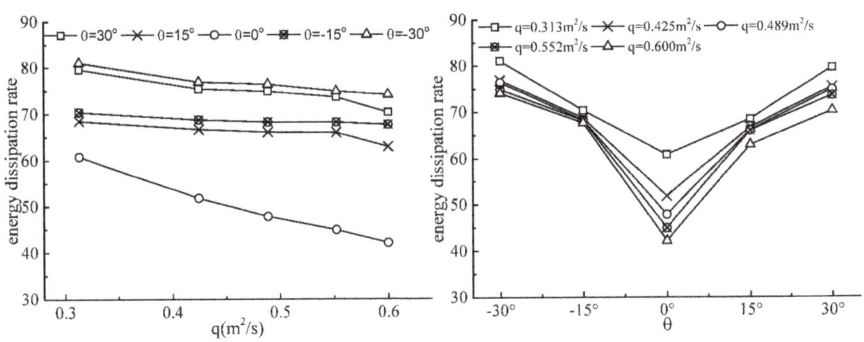

Figure 5. Energy dissipation rate changes with various unit discharges (**left**) and horizontal face angles (**right**).

3.3. Turbulence Kinetic Energy (k) and Its Dissipation Rate (ε)

Figure 6 shows the distribution of turbulence kinetic energy and its dissipation rate on the stepped surface of the V-shaped and inverted V-type step spillways, respectively. In this paper, the turbulent kinetic energy and dissipation rate from the 1-mm position to the wall were selected for analysis. It can be seen that obvious changes, which were shown on the step surface of the V-shaped and inverted V-shaped steps, appeared along the width of the spillway. The maximum of turbulence kinetic energy and its dissipation rate for the V-shaped step were found in the stepped horizontal plane in the vicinity of the side wall. However, those for the inverted V-shaped step were presented in a stepped horizontal plane near the central axis, from which a distinct three-dimensional feature is shown.

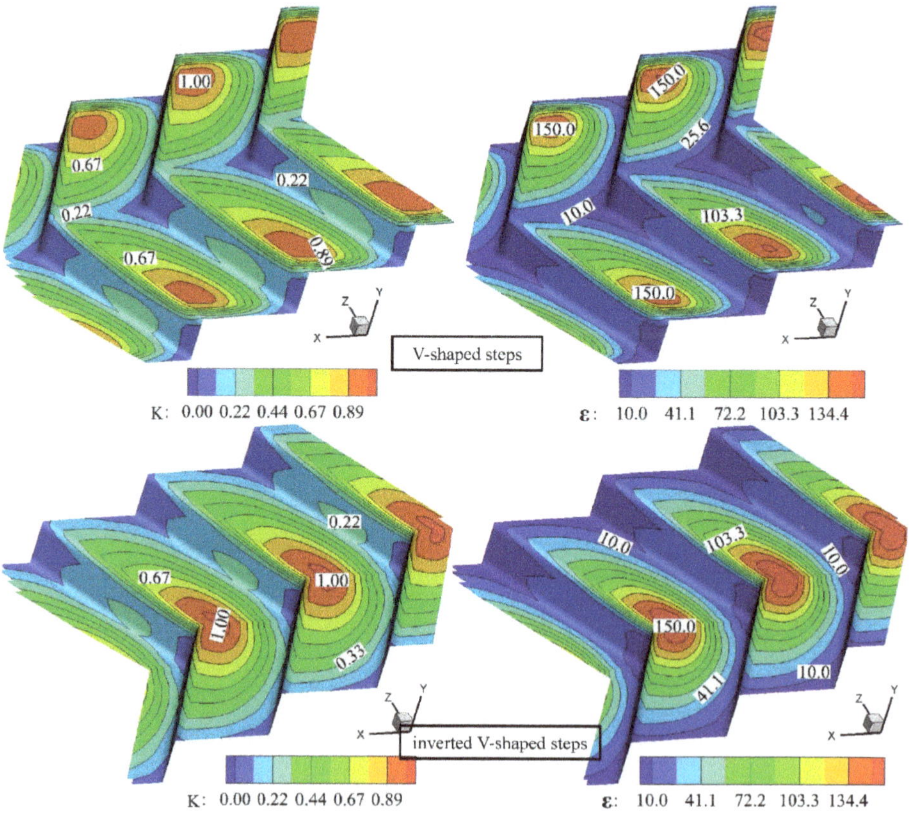

Figure 6. The distribution of turbulence kinetic energy (**left**) and its dissipation rate (**right**) on the stepped surface of V-shaped and inverted V-type step spillways (q = 0.425 m^2/s).

For the V-shaped step, the maximal turbulence kinetic energy appears on the stepped horizontal plane near the side wall because the water flow, falling near the side wall and moving to the central axis surface by swirling, intensified the turbulence on the step surface. At the same time, the energy loss caused by this collision was also greater than the others, so its dissipation rate also reached its maximum in the vicinity of the side wall. However, for the inverted V-shaped step, the maximum of the turbulence kinetic energy and its dissipation rate were shown in the stepped horizontal plane near the central axis, where the water fell into the central axis from the sidewall. The distribution of the turbulence kinetic energy and its dissipation rate showed a certain symmetry due to the symmetry of the structure for the V-shaped step and inverted V-shaped step.

Figure 7 shows the changes of the ratio of turbulence kinetic energy and its dissipation rate with different unit discharges and horizontal face angles. In this figure, k_{max} and ε_{max} indicate the maximum turbulence kinetic energy and its maximum dissipation rate in the corresponding calculation conditions, respectively; $\varepsilon_{max}/k_{max}$ indicates the ratio of its dissipation rate and turbulence kinetic energy, which can reflect the changes in energy dissipation rates because when the unit discharge is larger and has a more complex structure caused by horizontal face angles, the turbulence kinetic energy and the its dissipation rate are larger for the increasing fluctuating velocity.

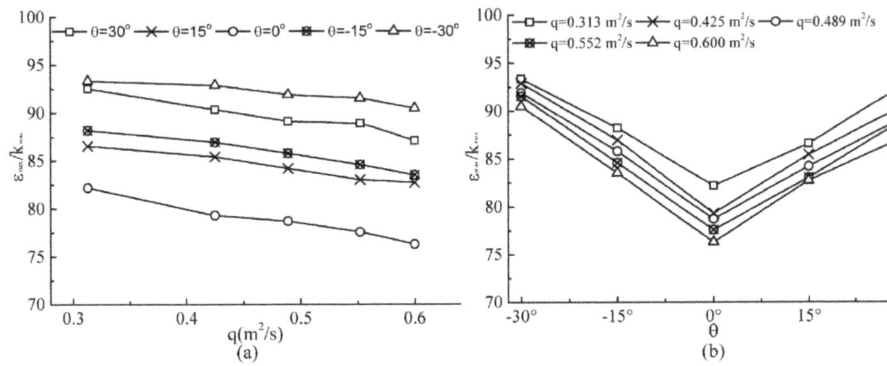

Figure 7. The changes of the ratio of its dissipation rate and turbulence kinetic energy with different unit discharges (**a**) and horizontal face angles (**b**).

As is shown in Figure 7a, in all shaped stepped spillways, $\varepsilon_{max}/k_{max}$ decreased with an increase of the unit discharge. This suggests that the increase of the dissipation rate was less than that of the turbulence kinetic energy due to their increase in unit discharges. Therefore, the energy dissipation rates decreased with the increasing of the unit discharge.

As shown in Figure 7b, a given unit discharge with increasing angles, the $\varepsilon_{max}/k_{max}$, initially decreased and then increased. The minimum $\varepsilon_{max}/k_{max}$ was observed at $\theta = 0°$, which is slightly smaller in a stepped spillway with $\theta > 0°$ than that in a stepped spillway with $\theta < 0°$ for equal absolute values of the angles. Thus, with increasing horizontal face angles, the energy dissipation rates initially decreased and then increased and were slightly smaller in a stepped spillway with $\theta > 0°$ than that in a stepped spillway with $\theta < 0°$ for equal absolute values of the angles, as shown in Figure 5.

4. Conclusions

This paper investigated the effects of varying horizontal face angles on energy dissipation rates by comparing the flow field, energy dissipation rates, turbulence kinetic energy, and turbulence dissipation rate with different horizontal face angles. The conclusions obtained are as follows:

(1) The fluctuation of a free water surface will be larger with the larger absolute values of the angles in the V shaped stepped spillway. The fluctuations will be higher in the vicinity of the axial plane or sidewalls for $\theta > 0°$ or $\theta < 0°$.

(2) The energy dissipation rate increases with the absolute values of the horizontal face angles and decreases as the unit discharge increases. The energy dissipation rate of the traditional stepped spillway is the minimum in all kinds of stepped spillways.

(3) The flow field and the flow direction can be changed by the horizontal face angles of the stepped spillway, which produces some unique characteristics, such as unique vortex structures, which can cause better energy dissipation. These results will be useful in choosing a better stepped spillway for energy dissipation.

Author Contributions: Conceptualization, Y.P. and H.Y.; Data curation, X.Z., X.L., C.X., S.Y. and Z.B.; Formal analysis, H.Y.; Methodology, H.Y. and C.X.; Project administration, Y.P. and X.Z.; Resources, X.L.; Supervision, X.Z. and Z.B.; Writing—original draft, Y.P. and H.Y.; Writing—review and editing, Y.P. and H.Y.

Funding: This project was supported by the National Natural Science Foundation of China (Grant No. 51709293).

Conflicts of Interest: The authors declare no conflict of interest.

Nomenclature

B	model width
g	acceleration of gravity
GCI	Grid convergence index
H	height of the step
i	number of Grid in GCI
k	turbulence kinetic energy
L	length of the step
P	pressure
q	unit discharge
t	time
v	velocity
X	distance of the pressure detecting point on horizontal surface from the step's inner edge
Y	distance of the pressure detecting points on vertical surface from the step's lower edge
C_2	empirical constant
C_u	0.09
$C_{1\varepsilon}$	1.44
$C_{2\varepsilon}$	1.92
E_1	total energy in the first step of a stepped section
E_2	the total energy at a section below the stepped section
G_b	turbulent energy caused by average velocity gradient
G_k	turbulent energy caused by lift force
h_i	grid size in GCI
k_{max}	maximum turbulence kinetic energy
S_K	user-defined source item
S_ε	user-defined source item
ui	mean velocity component in the *i*th direction
v_1	average velocities in the first step of a stepped section
v_2	average velocities in a section below the stepped section
Y_M	contribution of compressible turbulent fluctuation expansion to overall dissipation rate
ε	turbulence dissipation rate
μ	dynamic viscosity
θ	horizontal face angle
η	energy dissipation rate
ρ	mean density
ρ_a	densities of air
ρ_w	densities of water
α_w	volume fraction of water
ε_{max}	maximum turbulence dissipation rates
μ_a	viscosities of air
μ_w	viscosities of water
σ_k	empirical constant,1.0
σ_ε	empirical constant
ϕ_i	solution about the *i*th grid
Δh	difference in height between two sections
ΔE	difference in energy between two sections

References

1. Anwar, N.; Dermawan, V. Physical hydraulic model investigation of flow and energy dissipation on smooth and stepped spillway with steep slope. *Int. J. Acad. Res.* **2011**, *3*, 727.
2. Felder, S.; Chanson, H. Energy dissipation down a stepped spillway with non-uniform step heights. *J. Hydraul. Eng.* **2011**, *137*, 1543–1548. [CrossRef]
3. Morovati, K.; Eghbalzadeh, A.; Soori, S. Numerical study of energy dissipation of polled stepped spillways. *Civ. Eng. J.* **2016**, *2*, 208–220. [CrossRef]
4. Chanson, H. Comparison of energy dissipation between nape and skimming flow regimes on stepped chutes. *J. Hydraul. Res.* **1994**, *32*, 213–218. [CrossRef]
5. Abbasi, S.; Kamanbedast, A.A. Investigation of Effect of Changes in Dimension and Hydraulic of Stepped Spillways for Maximization Energy Dissipation. *World Appl. Sci. J.* **2012**, *18*, 261–267.
6. Rassaei, M.; Rahbar, S. Numerical flow model stepped spillways in order to maximize energy dissipation using FLUENT software. *IOSR J. Eng.* **2014**, *4*, 17–25. [CrossRef]
7. Tabari, M.M.R.; Tavakoli, S. Effects of stepped spillway geometry on flow pattern and energy dissipation. *Arab. J. Sci. Eng.* **2016**, *41*, 1215–1224. [CrossRef]
8. Wu, P.; Wang, B.; Chen, Y.L.; Zhou, Q. Energy Dissipation of Skimming Flow over Stepped Chutes with Different Slopes. *J. Sichuan Univ.* **2012**, *44*, 24–29. (In Chinese)
9. Shahheydari, H.; Nodoshan, E.J.; Barati, R.; Moghadam, M.A. Discharge coefficient and energy dissipation over stepped spillway under skimming flow regime. *KSCE J. Civ. Eng.* **2015**, *19*, 1174–1182. [CrossRef]
10. Attarian, A.; Hosseini, K.; Abdi, H.; Hosseini, M. The effect of the step height on energy dissipation in stepped spillways using numerical simulation. *Arab. J. Sci. Eng.* **2014**, *39*, 2587–2594. [CrossRef]
11. Zhang, Z.C.; Zeng, D.Y.; Liu, Y.F. Expermiental Research on the Presented Water-Surface Curve of Skmiming Flow on Stepped Spillways and Energy Disspation. *Chin. J. Appl. Mech.* **2005**, *22*, 30–36. (In Chinese)
12. Guenther, P.; Felder, S.; Chanson, H. Flow aeration, cavity processes and energy dissipation on flat and pooled stepped spillways for embankments. *Environ. Fluid Mech.* **2013**, *13*, 503–525. [CrossRef]
13. Mero, S.; Mitchell, S. Investigation of energy dissipation and flow regime over various forms of stepped spillways. *Water Environ. J.* **2016**, *31*, 127–137. [CrossRef]
14. Barani, G.A.; Rahnama, M.B.; Sohrabipoor, N. Investigation of flow energy dissipation over different stepped spillways. *J. Phys. Chem. C* **2005**, *117*, 7123–7137.
15. Jahad, U.A.; Al-Ameri, R.; Das, S. Energy dissipation and geometry effects over stepped spillways. *Int. J. Civ. Eng. Technol.* **2016**, *7*, 188–198.
16. Tian, Z.; Xu, W.L.; Yu, T.; Zhou, H.H.; Xiao, F.R. Characteristics of Energy Dissipation for V shaped Stepped Spillway. *J. Sichuan Univ.* **2010**, *42*, 21–25. (In Chinese)
17. Shih, T.H.; Liou, W.W.; Shabbir, A.; Yang, Z.; Zhu, J. A new k-ε eddy viscosity model for high Reynolds number turbulent flows. *Comput. Fluids* **1995**, *24*, 227–238. [CrossRef]
18. Qian, Z.D.; Hu, X.Q.; Huai, W.X.; António, A. Numerical simulation and analysis of water flow over stepped spillways. *Sci. China Technol. Sci.* **2009**, *52*, 1958–1965. (In Chinese) [CrossRef]
19. Duangrudee, K.; Chaiyuth, C.; Pierre, Y.J. Numerical simulation of flow velocity profiles along a stepped spillway. *Inst. Mech. Eng.* **2014**, *227*, 327–335.
20. Bai, Z.L.; Zhang, J.M. Comparison of different turbulence models for numerical simulation of pressure distribution in v-shaped stepped spillway. *Math. Probl. Eng.* **2017**, *2017*, 1–9. [CrossRef]
21. Celik, I.B.; Ghia, U.; Roache, P.J.; Freitas, C.J. Procedure for estimation and reporting of uncertainty due to discretization in CFD applications. *J. Fluids Eng.* **2008**, *130*, 078001.

energies

MDPI

Article

Optimization Design of the Impeller Based on Orthogonal Test in an Ultra-Low Specific Speed Magnetic Drive Pump

Fei Zhao [1,2], Fanyu Kong [1,*], Yisong Zhou [1], Bin Xia [2] and Yuxing Bai [3]

1 National Research Center of Pumps, Jiangsu University, Zhenjiang 212013, China; zhaof@wxit.edu.cn (F.Z.); m13215555321@163.com (Y.Z.)
2 School of Mechanical Technology, Wuxi Institute of Technology, Wuxi 214121, China; xia_bin2007@126.com
3 School of Automotive & Rail Transit, Nanjing Institute of Technology, Nanjing 211167, China; yxbai1012@163.com
* Correspondence: kongfy2918@sohu.com

Received: 19 October 2019; Accepted: 10 December 2019; Published: 13 December 2019

Abstract: To improve the hydraulic performance in an ultra-low specific speed magnetic drive pump, optimized design of impeller based on orthogonal test was carried out. Blades number Z, bias angle in peripheral direction of splitter blades θ_s, inlet diameter of splitter blades D_{si}, and deflection angle of splitter blades α were selected as the main factors in orthogonal test. The credibility of the numerical simulation was verified by prototype experiments. Two optimized impellers were designed through the analysis of orthogonal test data. The internal flow field, pressure fluctuation, and radial force were analyzed and compared between optimized impellers and original impeller. The results reveal that impeller 7 ($Z = 5$, $\theta_s = 0.4\theta$, $D_{si} = 0.75D_2$, $\alpha = 0°$) could increase the head and efficiency, compared to the original impeller, by 2.68% and 4.82%, respectively. Impeller 10 ($Z = 5$, $\theta_s = 0.4\theta$, $D_{si} = 0.55D_2$, $\alpha = 0°$) reduced the head by 0.33% and increased the efficiency by 8.24%. At design flow rate condition, the internal flow of impeller 10 was the most stable. Peak-to-peak values of pressure fluctuation at the volute tongues of impeller 7 and impeller 10 were smaller than those of the original impeller at different flow rate conditions (0.6 Q_d, 1.0 Q_d and 1.5 Q_d). Radial force distribution of impeller 10 was the most uniform, and the radial force variance of impeller 10 was the smallest.

Keywords: ultra-low specific speed magnetic drive pump; orthogonal test; splitter blades; optimized design; pressure fluctuation; radial force

1. Introduction

With the characteristics of no leakage and compact structure, magnetic drive pumps are widely used in aerospace, chemical, pharmaceutical, and military industries [1–4]. The efficiency of magnetic drive pumps is usually lower than that of ordinary mechanical seal pumps. Therefore, improving its efficiency is of great significance for saving energy and reducing environmental pollution.

In recent years, research has been conducted on the influence of splitter blades on performance and internal flow of pumps. ХлопеНКоВ.п.р[5] improved head and efficiency by welding short blade structures on the impeller. The research results of Kergourlay G et al. [6] found that the internal velocity and pressure distribution of the centrifugal pump impeller with splitter blades are more uniform, which increases the pump head by 10% to 15%. Gölcü M. [7–9] analyzed the influence of splitter blades on the performance of deep well centrifugal pumps. The results show that the use of splitter blades can effectively reduce energy consumption.

Shigemitsu T. et al. [10,11] improved the design of a small centrifugal pump in the form of adding splitter blades. It was found that the flow near the impeller outlet became uniform after the splitter blades were used, while the efficiency of the volute increased and the vortex loss decreased.

Yuan [12–14] discussed the principle and results of adding splitter blades based on the real flow in centrifugal pumps, the analysis results indicated that the splitter blades that deviated toward the suction surface of ordinary blades improved pump head and efficiency, adding splitter blades in the impeller moved the high efficiency point of the pump to the direction of large flow rate, and adding splitter blades in a screw-type centrifugal pump could decrease the radial forces on the impeller, but also decrease oscillations in volute as well. The influence of blade number was not considered in the analysis.

Zhang [15,16] investigated the influences of different positions of splitter blades on the performance of a centrifugal pump, two different splitter blade schemes were proposed: One located in the middle of the channel and the other having a deviation angle to the suction side of the long blade. The results showed that adding splitter blades can improve the tangential component distribution of absolute velocity at impeller outlet, and the streamline has a better consistency with the blades shape when splitter blades deviated to the long blade, which decreases energy loss.

Gu [17] and Cui [18] studied the effects of splitter blades on the performance and internal flow of low specific speed centrifugal pumps. Gu [17] found that both the vorticity and energy loss were enlarged around the volute tongue significantly after the blades passed by the cutwater, and the splitter blades produced more energy dissipation and unsteadiness than main blades. Cui [18] found that, at the design flow rate condition, both the efficiency and the head of the impeller with eight long blades were higher than those of the impeller with four long blades plus four splitter blades. While Cui [18] did not analyze the influence of deflection angle of splitter blades and bias angle in peripheral direction of splitter blades in the impeller optimization

Guo [19], Yuan [20] analyzed the influence of splitter blades on the internal pressure fluctuation and flow field of high-speed centrifugal pumps. Guo [19] clearly elucidated the anti-cavitation performance and the mechanism of bubble evolution of the high-speed centrifugal pump with splitter blades. Yuan [20] found that splitter blades obviously harmonized the turbulence kinetic energy distribution, and the pressure fluctuations distributed more evenly in the impeller, which enhanced the steady flow in the high-speed pump.

Jia et al. [21] carried out experiments to analyze the effect of the splitter blades on performance characteristics of a Francis turbine, and the results obtained were compared with those with normal blades. The results revealed that splitter blades increased the efficiency by approximately 2%, and they reduced the pressure fluctuation in the vaneless area under high-head operating conditions.

Although scholars around the world have carried out extensive and in-depth research on the splitter blades of ordinary centrifugal pumps, there is little research on the splitter blades of ultra-low specific speed magnetic driven centrifugal pumps. Based on the impeller of the prototype pump and the method of orthogonal test, nine different impellers are proposed in this paper. Blades number Z, bias angle in peripheral direction of splitter blades θ_s, inlet diameter of splitter blades D_{si}, and deflection angle of splitter blades α are selected as the main factors in the orthogonal test. Through the combination of numerical simulation and experiment, the influence of splitter blades on the performance of the ultra-low specific speed magnetic driven centrifugal pump is studied. The optimized impellers and the original impeller are analyzed and compared from internal flow field, pressure fluctuation, and radial force.

2. Materials and Methods

2.1. Orthogonal Test Design

The basic parameters of the magnetic drive pump for special engineering are as follows: Design flow rate Q_d is 8 $m^3 \cdot h^{-1}$, rated head H_d is 70 m, rotational speed n is 2900 $r \cdot min^{-1}$, specific speed n_s is

21, transmission medium is ethylene glycol aqueous solution, and temperature is normal. The main hydraulic dimensions of the impeller are as follows: Impeller outer diameter D_2 is 230 mm, outlet width b_2 = 5.2 mm, inlet diameter D_j is 40 mm, blade inlet placement angle is 24°, and outlet placement angle is 34°. Four long blades and four short blades are evenly distributed in the impeller. The inlet diameter of the splitter blades is 150 mm. The impeller and the internal magnetic rotor are integrated.

As the heart of the magnetic drive pump, the impeller is used to increase the pressure energy of the liquid. Different design parameters of splitter blades affect the performance of pump. According to previous research results, it is considered that the geometric parameters of splitter blades that affect the performance of magnetic drive pump are: The bias angle in peripheral direction of splitter blades θ_s, the inlet diameter of splitter blades D_{si}, the deflection angle of splitter blades α, and the blades number Z.

The increase of the number of blades has a great influence on the energy conversion of magnetic drive pump. The increase of the number of blades increases the finite blade correction coefficient, increases the head, and plays a decisive role in the performance change. Therefore, the number of blades Z was chosen as the first factor of orthogonal test.

The bias angle in the peripheral direction of splitter blades will affect the velocity distribution in the impeller passage, which will affect the performance of the magnetic drive pump. Therefore, the bias angle in the peripheral direction of splitter blades θ_s was chosen as the second factor of orthogonal test

The inlet diameter of splitter blades is related to the action length of splitter blades. Theoretically, the longer the splitter blades, the larger the head, so the inlet diameter D_{si} of splitter blades was selected as the third factor of the orthogonal test. The deflection angle of splitter blades α affects the velocity distribution of the impeller outlet, which has a significant influence on the performance of the magnetic drive pump. The deflection angle of splitter blades was selected as the fourth factor of the orthogonal test. Four selected factors of the orthogonal test are presented in Figure 1.

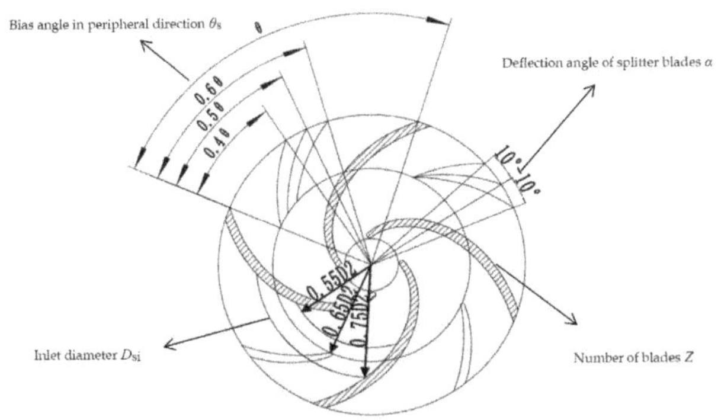

Figure 1. Schematic diagram of the four factors of the orthogonal test.

Table 1 lists three levels of the four factors in the orthogonal test. The factors of Z, θ_s, D_{si}, and α are expressed by code A, B, C, and D, respectively. Values of each factor are set within a certain range of the original impeller design value.

Table 1. Level table of the orthogonal test factors.

Levels	A	B	C	D
	Z	θ_s	D_{si}	α (°)
1	3	0.4θ	$0.55D_2$	−10
2	4	0.5θ	$0.65D_2$	0
3	5	0.6θ	$0.75D_2$	10

According to the selected four factors and three levels, the L_9 (3^4) orthogonal design schemes are established, as shown in Table 2.

Table 2. Orthogonal test schemes.

NO.	A	B	C	D	Parameter			
					Z	θ_s	D_{si}	α (°)
1	A_1	B_1	C_1	D_1	3	0.4θ	$0.55\,D_2$	−10
2	A_1	B_2	C_2	D_2	3	0.5θ	$0.65\,D_2$	0
3	A_1	B_3	C_3	D_3	3	0.6θ	$0.75\,D_2$	10
4	A_2	B_1	C_2	D_3	4	0.4θ	$0.65\,D_2$	10
5	A_2	B_2	C_3	D_1	4	0.5θ	$0.75\,D_2$	−10
6	A_2	B_3	C_1	D_2	4	0.6θ	$0.55\,D_2$	0
7	A_3	B_1	C_3	D_2	5	0.4θ	$0.75\,D_2$	0
8	A_3	B_2	C_1	D_3	5	0.5θ	$0.55\,D_2$	10
9	A_3	B_3	C_2	D_1	5	0.6θ	$0.65\,D_2$	−10

According to orthogonal test schemes, the three-dimensional geometric models of nine impellers were constructed with UG NX9.0. Figure 2 shows water bodies of nine impellers.

Figure 2. Water bodies of impellers corresponding to orthogonal test schemes.

2.2. Calculation Model

The compact magnetic drive pump features a direct-connected construction. The impeller and the inner magnet are integral and are rotating parts. The pump shaft and the isolation sleeve are integrated and are stationary parts. The structure ensures that the magnetic drive pump has a small size, light weight, no leakage, and reliable operation. The configuration of the magnetic drive pump is shown in Figure 3.

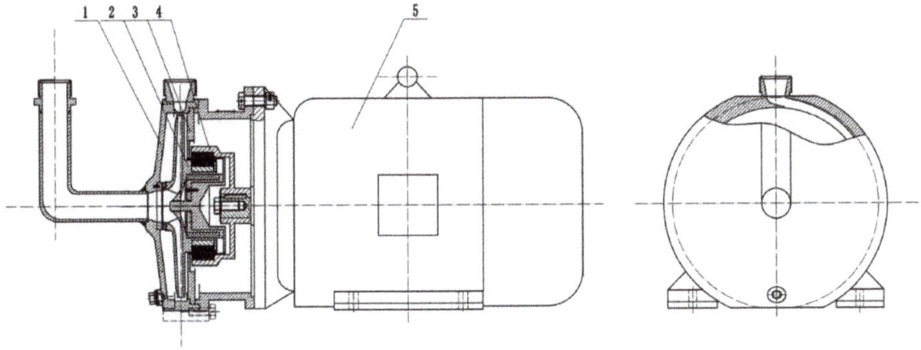

Figure 3. Configuration of the magnetic drive pump. 1—pump shell; 2—impeller; 3—isolation sleeve; 4—external magnetic rotor; 5—motor.

The computational domain of the whole flow field mainly includes the inlet pipe fluid, the impeller internal fluid, the pump internal fluid, the cooling circulating fluid in the pump, and the outlet pipe fluid. As shown in Figure 4, the inlet and outlet sections were extended appropriately in order to obtain stable inlet and outlet flow.

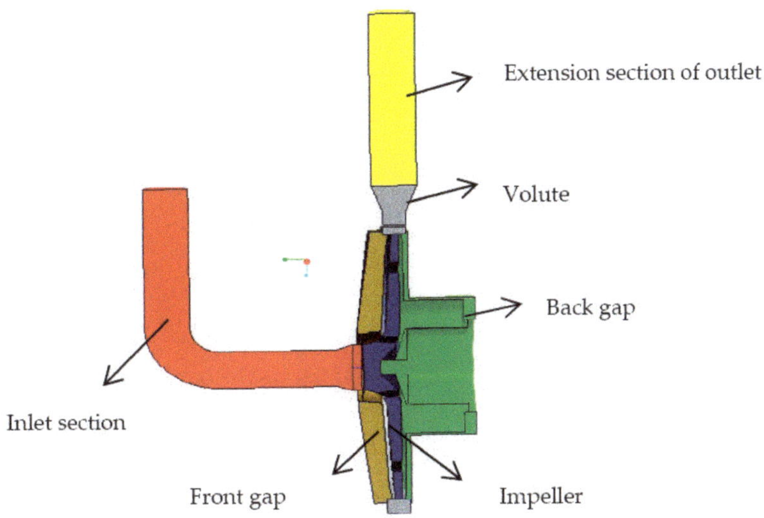

Figure 4. Calculation domain of magnetic drive pump flow field.

2.3. Mesh Generation and Independence Verification

Hexahedral structured meshes were generated in the flow field by ANSYS-ICEM 14.5 (ANSYS, Inc., Canonsburg, PA, USA). Meshes in the impeller blade wall were refined. The meshes of magnetic drive pump are shown in Figure 5. In order to reduce the computational load and ensure the accuracy of numerical calculation results, mesh independence analysis was carried out. Four groups of meshes with different numbers were divided, and ANSYS-CFX 14.5 (ANSYS, Inc., Canonsburg, PA, USA) was used to simulate the performance at design flow rate condition. The head and efficiency were selected as evaluation indicators. Mesh independence analysis is shown in Table 3. When the total number of grids is greater than 1,254,650, the head and efficiency of the magnetic drive pump change little, so 1,254,650 is chosen as the number of grids for numerical simulation.

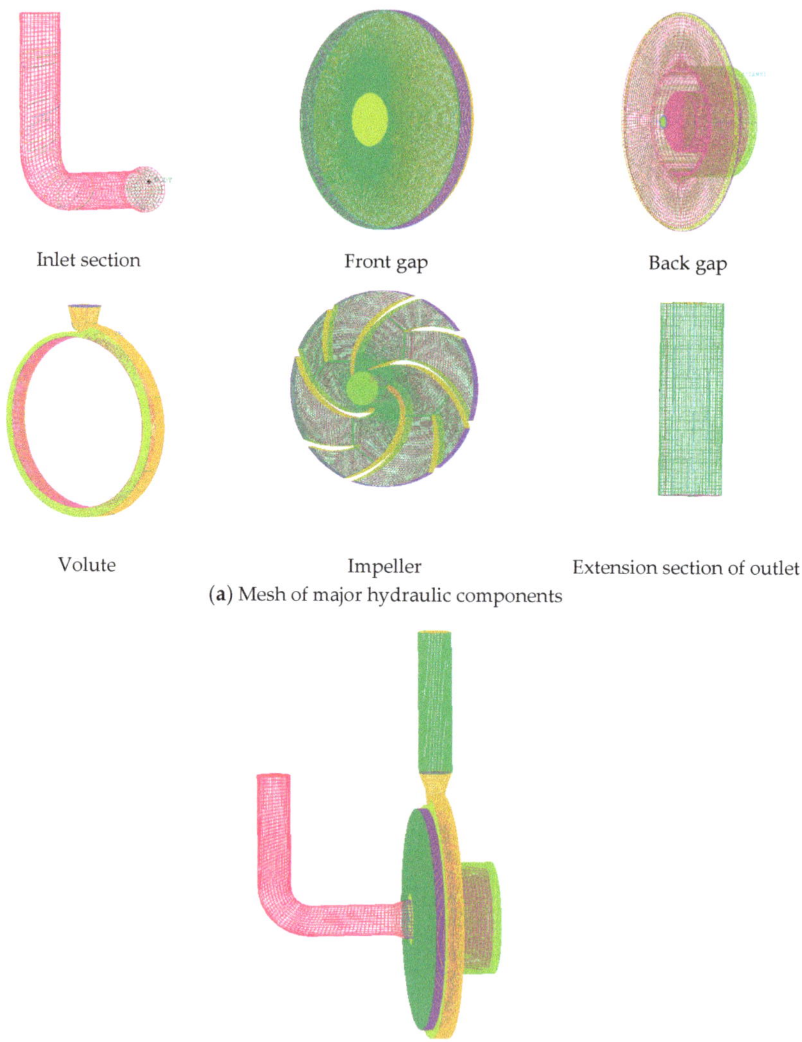

(a) Mesh of major hydraulic components

(b) Mesh assembly

Figure 5. Field mesh of the magnetic drive pump.

Table 3. Mesh independence analysis.

Number of Grids	Head (m)	Efficiency
575,681	77.84	0.3865
960,887	78.48	0.3929
1,254,650	78.94	0.4067
1,638,560	78.98	0.4065

2.4. Turbulence Model

In actual engineering, all flow problems should satisfy the laws of conservation of mass, momentum, and energy. The fluid flow studied in this paper is a three-dimensional incompressible turbulent flow. The turbulence model selected for numerical simulation is the SST (Shear Stress Transport) $k-\omega$ model, which is a mixture of model $k-\varepsilon$ and model $k-\omega$. It not only has the reliability of the model $k-\omega$ in calculating the viscous flow in the near-wall region, but also has the accuracy of the model $k-\varepsilon$ in calculating the free flow of the far-field head [22]. The basic governing equations of the SST $k-\omega$ model are as follows [23,24]:

$$\frac{\partial(\rho k)}{\partial t}+\frac{\partial}{\partial x_i}(\rho u_i k)=\frac{\partial}{\partial x_i}\left[\left(\mu+\frac{\mu_i}{\sigma_k}\right)\frac{\partial k}{\partial x_i}\right]+P_k-\beta'\rho k\omega \tag{1}$$

$$\frac{\partial(\rho\omega)}{\partial t}+\frac{\partial}{\partial x_i}(\rho u_i\omega)=\frac{\partial}{\partial x_i}\left[\left(\mu+\frac{\mu_i}{\sigma_k}\right)\frac{\partial \omega}{\partial x_i}\right]+D_m+\alpha\frac{\omega}{k}P_k-\beta\rho\omega^2 \tag{2}$$

where $\beta' = 0.09$, $\alpha = 5/9$, $\beta = 0.075$ and $\sigma_k = 2$, ρ stands for density (kg·m^{-3}), P_k represents turbulent productivity [25,26].

2.5. Steady Computation Settings

The assembly mesh of computational domain was imported in ANSYS-CFX and calculation type was defined as a constant calculation. The magnetic drive pump impeller was set to the rotation domain, and the other calculation fields were set to the stationary calculation domain.

The import boundary definition was set on the inlet face of the inlet section, and the boundary condition type was set to the static pressure inlet. The pressure was set to a standard atmospheric pressure and the turbulence intensity was chosen to be 5%, corresponding to the moderate turbulence. The exit boundary was defined on the exit face of the outlet extension and the boundary condition type was set to the mass flow outlet.

The interface between the impeller rotation domain and other calculation domains was set to the dynamic–static calculation domain interface, the type was set to frozen rotor type, and the other interfaces were set to the static–static calculation domain interface.

The impeller speed of the magnetic drive pump was set to 2900 r/min. According to the actual temperature of the medium delivered in the pump, the simulated medium was water and the water temperature was set to 25 °C. The wall condition was set to a non-slip solid wall condition, and the wall roughness was set to 15 μm (equivalent value of surface roughness in pump cavity). The time item for solving the parameter was set to the automatic time step and the residual convergence value was set to 10^{-5}.

2.6. Test Bench for Magnetic Drive Pump

The prototype pump experiment was carried out at the special pump test bench in the National Pump Engineering Technology Research Center of Jiangsu University. Schematic diagram of the magnetic drive pump test bench structure is shown in Figure 6, and Figure 7 is the scene of magnetic drive pump experiment.

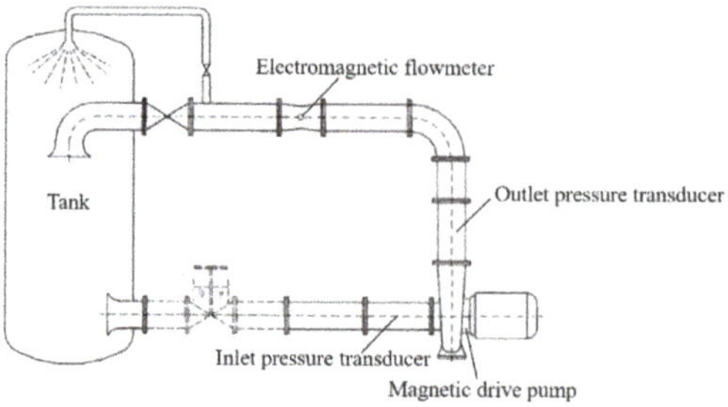

Figure 6. Magnetic drive pump test bench structure.

Figure 7. Drive pump test bench.

OPTIFLUX 2100 electromagnetic flowmeter was used in the magnetic drive pump test bench to realize the instantaneous measurement of flow. Its measurement range is 0 to 30 $m^3 \cdot h^{-1}$, the measurement accuracy is less than 0.5%, and the uncertainty of the flow measurement is ±0.2%. The WIKA S10 pressure transmitter is used to measure the instantaneous pressure at the inlet and outlet of the pump. The range of the pressure transducer at the inlet and outlet of the pump is −1 to 1 MPa and 0 to 1.6 MPa, respectively. Its measurement accuracy is less than 0.25%, and the uncertainty of the pressure measurement is ±0.1%.

The output signals of flow and pressure sensors are 4–20 mA current signals, which are collected by high-speed data acquisition card and stored in PC for data processing.

3. Results and Discussion

3.1. Experimental Verification

To verify the reliability of the numerical calculation, the numerical calculation results at different flow rate conditions were compared with the experimental data of the magnetic drive pump prototype. The external characteristic curves drawn by the experimental data at different flow rate conditions were compared with the ANSYS-CFX numerical simulation results, as shown in Figure 8.

From the flow-head curve, it can be seen that the simulation calculation agrees well with the experiment near the design flow rate condition, and there are slight deviations between the numerical calculation and the experimental results under other flow rate conditions. The reason for the deviation may be that the stability of the flow field is poor and the boundary layer is separated under non-rated conditions. In this condition, if numerical simulation still uses the steady flow model, the error would occur. The simulation efficiency agrees well with the experiment efficiency near the design flow rate condition. In the flow rate range of 0.6 and 1.4 Q_d, the maximum relative error of efficiency is 4.4% (at 0.6 Q_d). In conclusion, the general trends of the simulation and experiment results are consistent, and the numerical simulation results of the whole flow field in the magnetic drive pump are true and credible.

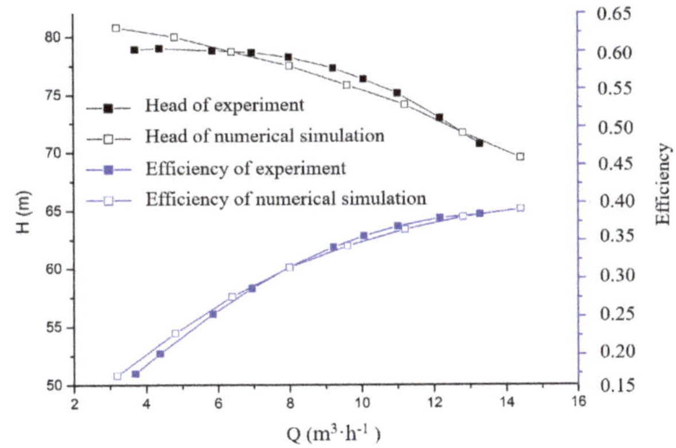

Figure 8. Characteristic curve of numerical simulation and prototype experiment.

3.2. Internal Flow Analysis of Impeller

The internal flow condition of the impeller reflects the advantages and disadvantages of the hydraulic design. The nine impeller schemes were numerically simulated, respectively, and the internal flow cloud diagram of the impellers at design flow rate condition are shown in Figure 9.

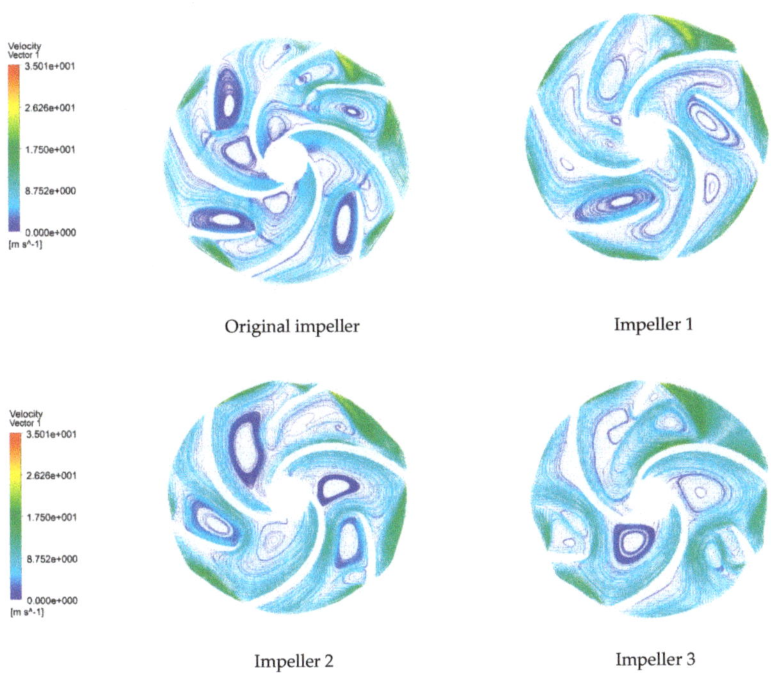

Figure 9. *Cont.*

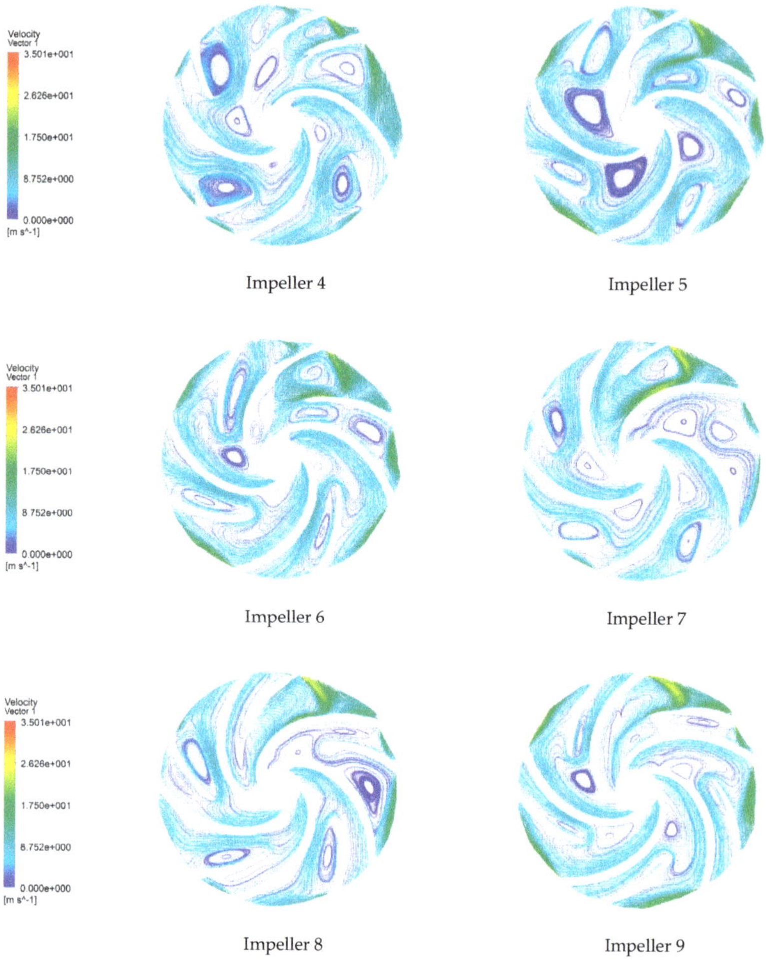

Figure 9. Flow cloud diagram of the nine impellers at design flow rate condition.

From Figure 9, it can be seen that, due to the low specific speed of the compact magnetic drive pump and the narrow inner passage of the impeller, there are vortices in the impeller of all schemes. Schemes 1–3 have larger low-speed vortices and larger high-speed flow areas at the outlet of impellers, resulting in worse internal flow conditions. Schemes 4–6 are similar to the internal flow condition of the original design. Vortexes appeared on the back of the splitter blade and inside the working face of the main blade, and the flow is not uniform. The internal flow of the impellers in Schemes 7–9 is more uniform than that of other schemes, and there are less internal vortices. The flow condition in the flow passage is better improved with the increase of the blades, and the energy loss is reduced. The flow condition of Scheme 7 is optimal in all schemes, with the least impeller exit wake region and vortex.

3.3. *Range and Variance Analysis of Performance Indexes*

The efficiency and head of each scheme impeller obtained by numerical simulation are shown in Table 4, in which scheme 0 corresponds to the original design impeller.

Table 4. Efficiency and head of different scheme.

Scheme Number	Head (m)	Efficiency
0	77.98	0.4212
1	71.23	0.4368
2	72.78	0.4215
3	74.41	0.3743
4	76.57	0.4129
5	77.03	0.4156
6	77.75	0.4076
7	80.07	0.4415
8	77.67	0.4429
9	75.62	0.4446

The range analysis of head and efficiency was conducted based on the orthogonal test results in Table 5. In Table 5, H represents the performance index of head (m), η represents the performance index of efficiency in percentage, K represents the sum of the test results at the corresponding level, k represents the average value of the test results at the corresponding level, and R represents the range value of k.

Table 5. Range analysis of head and efficiency.

		Factor						Factor			
		A	B	C	D			A	B	C	D
Index		Z	θ_s	D_{si}	α	Index		Z	θ_s	D_{si}	α
H	K_1	218.42	227.87	226.65	223.88	η	K_1	123.26	129.12	128.73	129.7
	K_2	231.35	227.48	224.97	230.6		K_2	123.61	128	127.9	127.06
	K_3	233.36	227.78	231.51	228.65		K_3	132.9	122.65	123.14	123.01
	k_1	72.81	75.96	75.55	74.63		k_1	41.09	43.04	42.91	42.23
	k_2	77.12	75.83	74.99	76.87		k_2	41.20	42.67	42.63	42.35
	k_3	77.78	75.93	77.17	76.22		k_3	44.3	40.88	41.05	41.00
	R_H	4.97	0.13	2.18	2.24		R_η	3.21	2.16	1.86	1.35

According to the principle of the orthogonal test, the influence of the factors on the result is directly proportional to the magnitude of the range. Therefore, the sequence of factors affecting the head of the magnetic drive pump is ADCB, namely, blades number, deflection angle, inlet diameter of splitter blades, and bias angle in the peripheral direction. The sequence of factors affecting the efficiency of the magnetic drive pump is ABCD, namely, blades number, bias angle in the peripheral direction, inlet diameter of splitter blades, and deflection angle. The effects of various factors on head and efficiency are shown in Figures 10 and 11, respectively.

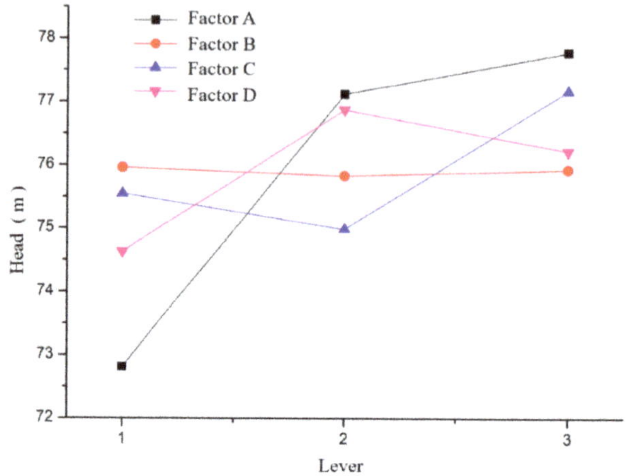

Figure 10. Curve.

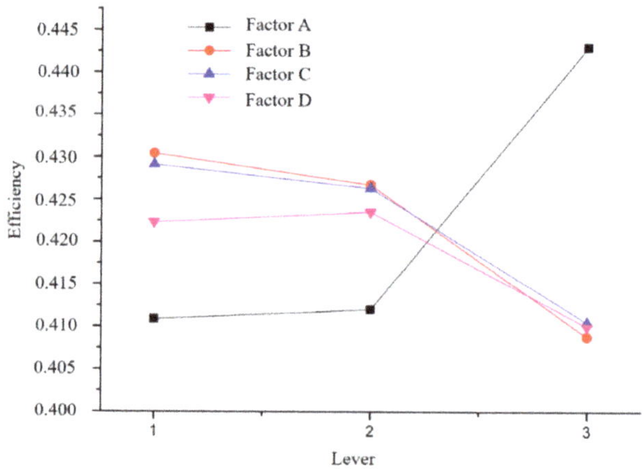

Figure 11. Curve.

The influence of each factor level on the performance of the magnetic drive pump is as follows:

(1) For factor A, the head increases with the increase of the number of blades. The head increases sharply from A_1 to A_2, while the head increases slowly from A_2 to A_3. The head reaches the maximum at A_3. The efficiency increases with the increase of the number of blades. Efficiency increases slowly from A_1 to A_2, while the efficiency curve increases sharply from A_2 to A_3. The efficiency at A_3 is the highest. Considering the influence of factor level on head and efficiency, A_3 (the number of blades Z = 5) is the best choice.

(2) For factor B, the change of bias angle in the peripheral direction has no obvious effect on head. The efficiency decreases gradually with the increase of bias angle in the peripheral direction. The efficiency reaches the highest when the bias angle is $0.4\theta(B_1)$. Therefore, it is reasonable to select the bias angle of 0.4θ.

(3) For factor C, the head is the highest at C_3 (the inlet diameter of splitter blades $D_{si} = 0.75D_2$), while the efficiency is the lowest at C_3. The head at C_1 (the inlet diameter of splitter blades $D_{si} = 0.55D_2$) is slightly lower than that at C_3, while the efficiency at C_1 is the highest. If more consideration

is given to head, C_1 is chosen. If the head at C_1 and C_3 has met the design requirements, C_3 is more energy-saving.

(4) For factor D, the variation trend of the head curve and efficiency curve is basically the same. When the deflection angle of splitter blades is 0° (D_2), the head and efficiency both reaches the maximum; therefore, the best deflection angle was 0°.

The Taguchi method and ANOVA (Analysis of Variance) were applied to find the proportion of the influence of each orthogonal factor on the performance indexes. The Taguchi method is an optimal design method based on the orthogonal test, which was proposed by Dr. Taguchi of Japan. The Taguchi method can optimize the design of multiple objectives. Through the establishment of the orthogonal test table and the ANOVA of the orthogonal test results, the best combination of design parameters can be obtained with the least number of tests [27,28]. Analysis of variance of factors is conducted based on orthogonal test results (Tables 4 and 5), and the analysis results are given in Table 6.

Table 6. Variance of each factor and percentage of influence on performance indexes.

Factors	Head		Efficiency	
	Variance Value	Influence Ratio (%)	Variance Value	Influence Ratio (%)
A (Z)	4.857	73.602	2.214	51.972
B (θ_s)	0.003	0.046	0.890	20.892
C (D_{si})	0.854	12.941	0.670	15.728
D (α)	0.885	13.411	0.486	11.408
Sum	6.599	100	4.26	100

As can be seen in Table 6, factor A (73.602%) has the decisive influence on head. Factor D (13.411%) and factor C (12.941%) influence the head greatly, while factor B (0.046%) has the least influence on the head.

The results of variance analysis show that the influence proportion of each factor on efficiency are: Factor A (51.972%), factor B (20.892%), factor C (15.728%), and factor D (11.408%), respectively.

Factor A has a greater influence on both head and efficiency than other factors. For factor C, the influence ratio on efficiency (15.728%) is higher than the influence on head (12.941%).

Based on the Taguchi method, combined with the data in Tables 5 and 6, two sets of optimized impeller combinations are obtained. The combination of $A_3B_1C_3D_2$ has the best head, while the combination of $A_3B_1C_1D_2$ has the best efficiency. Table 7 lists the parameters of the two optimized impellers.

Table 7. Parameters of the two optimized impellers.

Combination	Parameter			
	Z	θ_s	D_{si}	α (°)
$A_3B_1C_3D_2$	5	0.4 θ	0.75 D_2	0
$A_3B_1C_1D_2$	5	0.4 θ	0.55 D_2	0

The above analysis shows that the better design schemes of split blades is $A_3B_1C_3D_2$ and $A_3B_1C_1D_2$. $A_3B_1C_3D_2$ is Scheme 7 of the orthogonal test. The combination scheme of $A_3B_1C_1D_2$ did not appeared in the test schemes, and is newly named as Scheme 10. Modeling and numerical simulation of Scheme 10 was conducted. Figure 12 is the water body model of the impeller of Scheme 10, and Figure 13 is its inner flow cloud diagram. It could be seen that the inner vortex of the impeller of Scheme 10 is less and the streamline is more stable than that of Scheme 7 (shown in Figure 9).

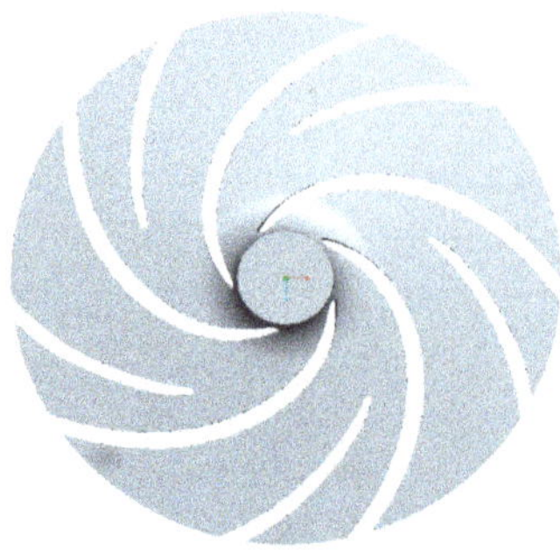

Figure 12. Model of impeller 10.

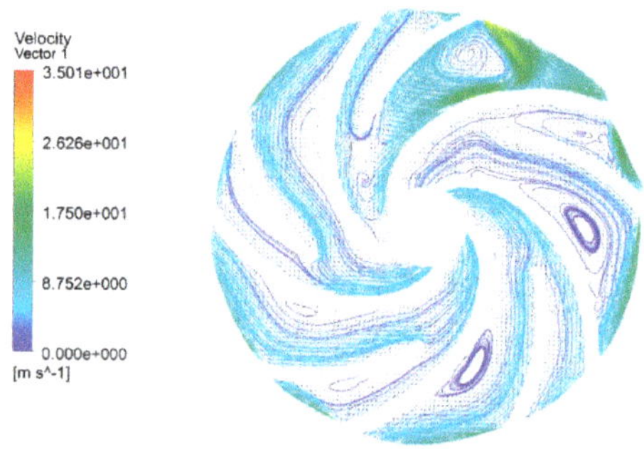

Figure 13. Flow cloud diagram of impeller 10.

Comparisons of impeller performance between Scheme 7 ($Z = 5$, $\theta_s = 0.4\theta$, $D_{si} = 0.75D_2$, $\alpha = 0°$) and Scheme 10 ($Z = 5$, $\theta_s = 0.4\theta$, $D_{si} = 0.55D_2$, $\alpha = 0°$) is shown in Figure 14. The head of Scheme 10 is 77.12 m, which is lower than that of Scheme 7 (80.07 m). The efficiency of Scheme 10 is 45.59%, which is higher than that of Scheme 7 (44.15%). Scheme 7 increases the head and efficiency by 2.68% and 4.82%, respectively. Scheme 10 reduces the head by 0.33% and increases the efficiency by 8.24%. The data shows that the decrease of inlet diameter of splitter blades improves the flow condition inside the impeller, reduces the energy loss, and improves the efficiency.

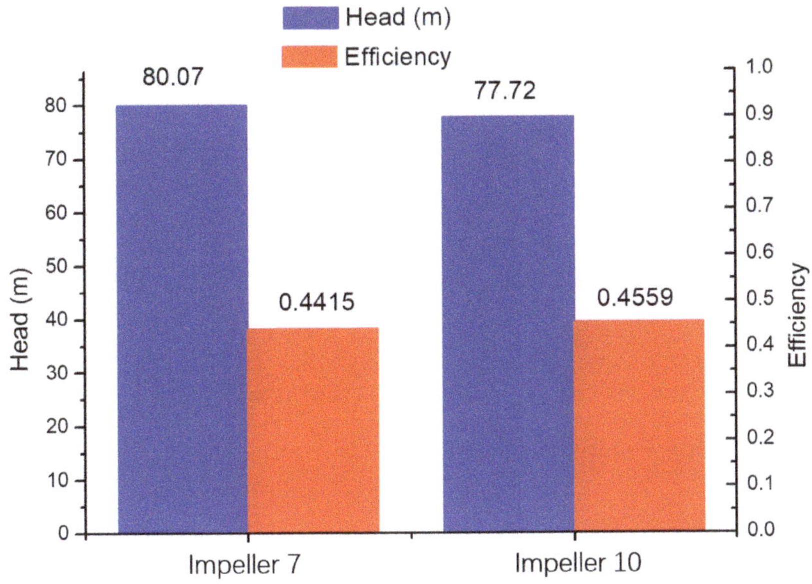

Figure 14. Performance comparison of impeller 7 and impeller 10.

3.4. Analysis of Pressure Fluctuation

Pressure fluctuation is the main factor affecting the stable operation of the pump. The transient numerical calculation of the whole flow field of the magnetic drive pump was carried out. A pressure fluctuation monitoring point was set at the volute tongue. The location of the monitoring point is shown in Figure 15. The steady calculation results were taken as the initial condition of transient calculation. Because the rated speed of the impeller is 2900 r/min, considering the economy of calculation, the final time step was 1.7241×10^{-4} s, namely, the impeller rotated 3° per time step.

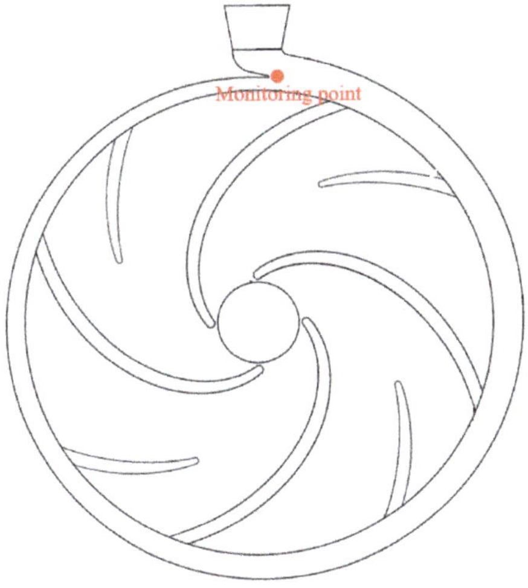

Figure 15. Location of monitoring point.

The pressure fluctuation of the original impeller, impeller 7, and impeller 10 at different flow rate conditions (0.6 Q_d, 1.0 Q_d and 1.5 Q_d) in one period is presented in Figure 16. It can be seen from Figure 16 that the instantaneous low pressure appears in the original impeller, and the pressure fluctuation of the original impeller is the most disordered. The peak-to-peak values of pressure fluctuation of impellers 7 and 10 are smaller than those of the original impeller.

At small flow rate condition (0.6 Q_d), the pressure fluctuation amplitudes of the three impellers are close. The pressure fluctuation of impellers 7 and 10 are more stable than those of the original impeller at design and large flow rate conditions, and the average pressure values of impellers 7 and 10 are higher than those of the original impeller. The pressures of impellers 7 and 10 are higher than those of the original impeller, which bring higher heads to the magnetic drive pump.

Comparing the pressure fluctuation of impellers 7 and 10 further, the peak-to-peak value of impeller 10 is slightly smaller than that of impeller 7 at small flow rate condition (0.6 Q_d), and the pressure fluctuation performance of impeller 10 is more stable. At design flow rate condition (1.0 Q_d), the average pressure amplitude of impeller 10 is slightly smaller than that of impeller 7, which is consistent with the conclusion of the higher head of impeller 7 mentioned above. At large flow rate condition (1.5 Q_d), the pressure fluctuation of impeller 7 is the most consistent with that of impeller 10, and the fluctuation amplitude, maximum peak value, and minimum peak value are close to each other.

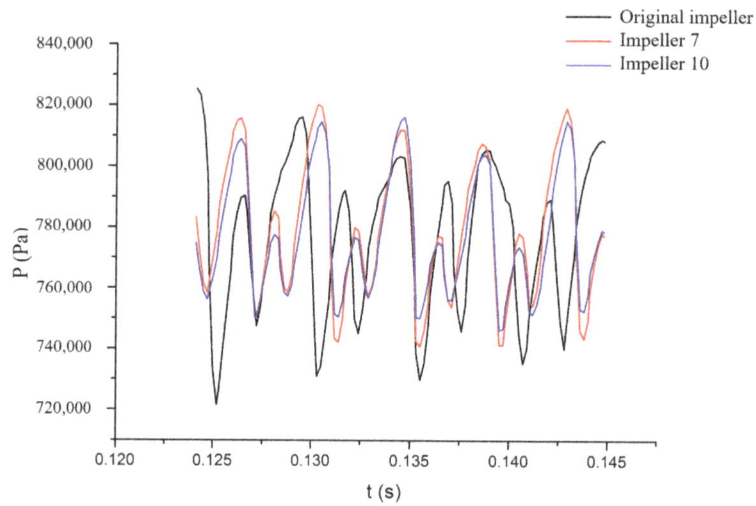

(**a**) 0.6 Q_d

Figure 16. *Cont.*

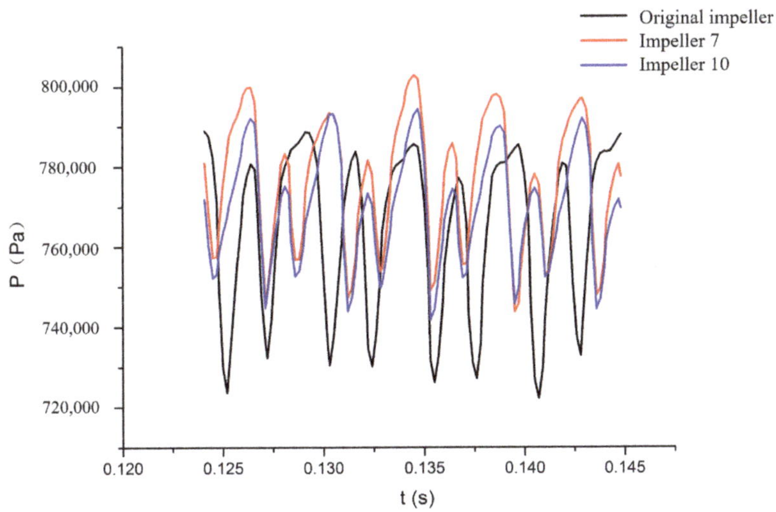

(b) 1.0 Q_d

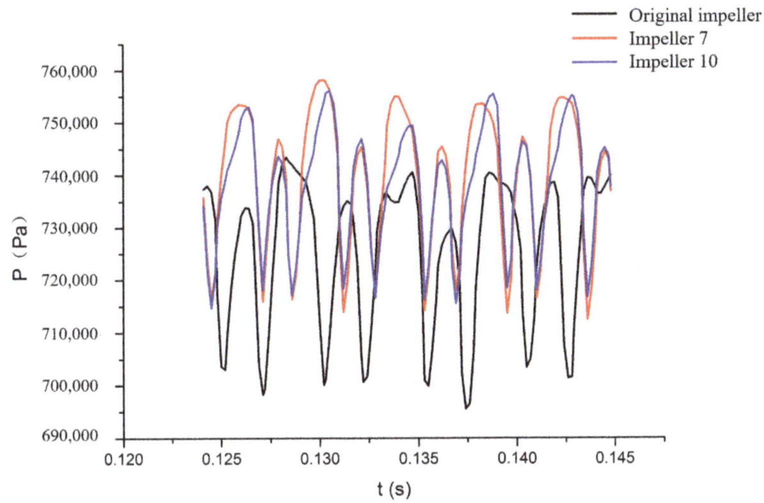

(c) 1.5 Q_d

Figure 16. Comparison of pressure fluctuation of different impellers.

3.5. Analysis of Radial Force

The radial forces of the original impeller, impeller 7, and impeller 10 at different flow rate conditions are shown in Figure 17. It can be seen from the figure that the radial force amplitudes of the three impellers are similar at different flow rate conditions, and the radial force gradually decreases as the flow rate increases. By comparison, the radial force circumferential distribution of impeller 10 is the most uniform at each flow rate condition.

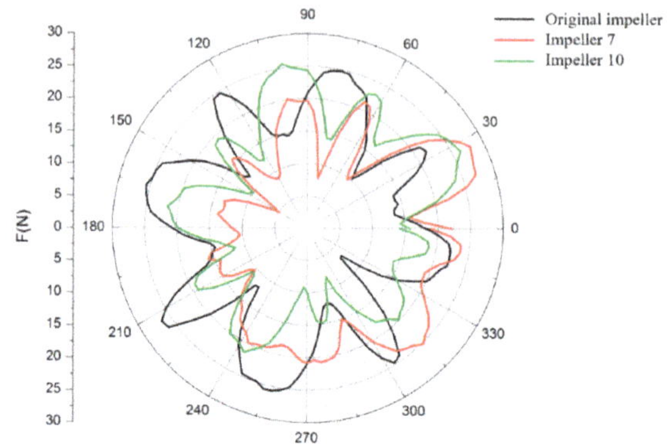

(a) 0.6 Q_d

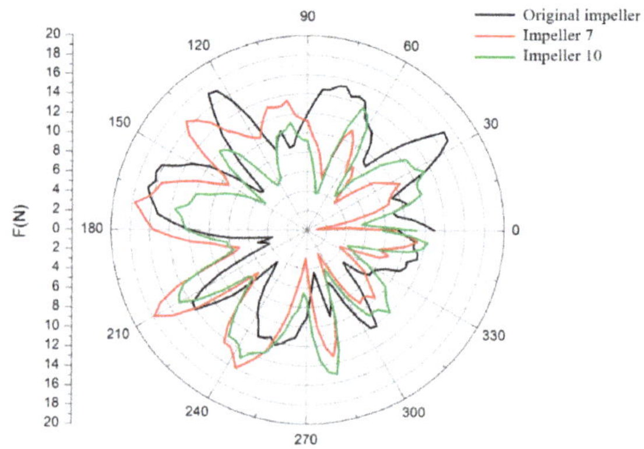

(b) 1.0 Q_d

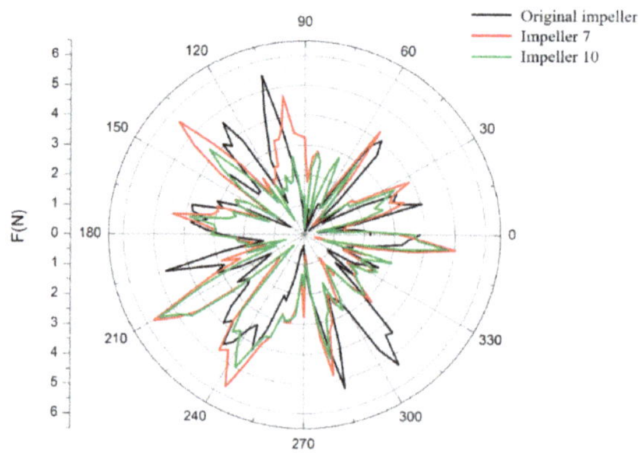

(c) 1.5 Q_d

Figure 17. Comparison of radial force of different impellers.

The variance analysis of the radial forces of different impellers was carried out. Figure 18 shows the variance histograms of the three impellers at different flow rate conditions. It could be seen from the figure that impeller 10 has the smallest variance and impeller 7 has the largest variance at different flow rate conditions, so the radial force distribution of impeller 10 is the most stable among the three impellers.

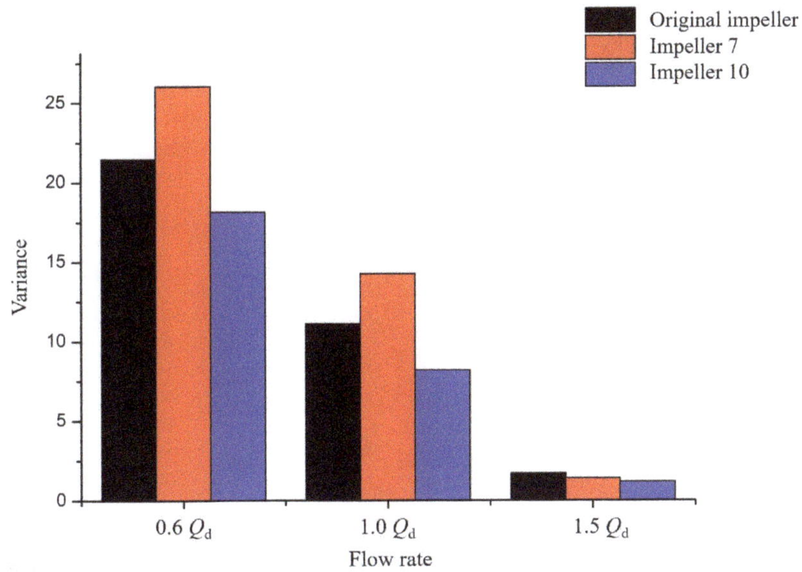

Figure 18. Diagram of radial force for different impellers.

4. Conclusions and Future Work

With the characteristics of electromagnetic transmission, no leakage, and compact structure, magnetic drive pumps are widely used in aerospace, chemical, pharmaceutical, and military industries. The efficiency of magnetic drive pumps is usually lower than that of ordinary mechanical seal pumps. To improve the performance of an ultra-low specific speed magnetic drive pump, combining numerical simulation with experiment, the optimized design of the impeller was carried out. The blades number Z, the bias angle in the peripheral direction of splitter blades θ_s, the inlet diameter of splitter blades D_{si}, and the deflection angle of splitter blades α were taken as the main factors in the orthogonal test, and two sets of better impeller design parameters were found. The optimized impellers and the original impeller were compared in internal flow, pressure fluctuation, and radial force. The accuracy of the numerical simulation was verified by prototype experiments. In the flow rate range of 0.6 Q_d and 1.4 Q_d, the maximum relative errors of head and efficiency were 2.1% and 4.4%, respectively. The main conclusions were drawn as follows:

(1) The factors affecting the head of magnetic drive pump in descending order were: Blades number, deflection angle, inlet diameter of splitter blades, and bias angle in the peripheral direction. The factors affecting efficiency in descending order were: Blades number, bias angle in the peripheral direction, inlet diameter of splitter blades, and deflection angle.

(2) Impeller 7 ($Z = 5$, $\theta_s = 0.4\theta$, $D_{si} = 0.75D_2$, $\alpha = 0°$) increased the head and efficiency by 2.68% and 4.82%, respectively. Impeller 10 ($Z = 5$, $\theta_s = 0.4\theta$, $D_{si} = 0.55D_2$, $\alpha = 0°$) reduced the head by 0.33% and increases the efficiency by 8.24%.

(3) At small flow rate condition (0.6 Q_d), the pressure fluctuation amplitudes of the three impellers were close. At design flow rate condition (1.0 Q_d) and large flow rate condition (1.5 Q_d), the pressure fluctuations of impeller 7 and impeller 10 were more stable than that of the original impeller, and the average pressure of each new impeller was higher than that of the original impeller, respectively.

(4) At flow rate conditions of 0.6, 1.0, and 1.5 Q_d, impeller 10 had the smallest radial force variance, and its radial force distribution was more stable than those of the original impeller and impeller 7.

This paper did not analyze the disk energy loss and axial force distribution of the optimized impellers. In the future, numerical simulation and experiments will be combined to further analyze the performance of optimized impellers. In addition, how to reduce the noise and vibration of the magnetic driven centrifugal pump is the next research direction.

Author Contributions: F.Z. and F.K. presented the optimal scheme and designed the experiments. F.Z. and Y.Z. made the numerical simulation and performed the experiment. F.Z., Y.B. and B.X. analyzed the data. F.Z. wrote the paper.

Funding: This research was funded by: the Priority Academic Program Development (PAPD) of Jiangsu Higher Education Institutions; the Industry-University-Research Cooperation Project of Jiangsu Province; the Excellent scientific and technological innovation team of Jiangsu Higher Education Institutions (Intelligent Manufacturing Equipment Design and Engineering Application); the National Natural Science Foundation of China, grant number 11602097; the Top-Notch Academic Programs Project of Jiangsu Higher Education Institutions, grant number PPZY2015A086; the Natural Science Foundation of the Jiangsu Higher Education Institutions of China, grant number 19KJD470006; and the Scientific Research Project of Wuxi Institute of Technology, grant number ZK201804.

Conflicts of Interest: The authors declare no conflicts of interest.

References

1. Brodersen, S. Magnetic-drive pumps: Technology and new applications. *Hydrocarb. Process.* **2001**, *80*, 45–48.
2. Thompson, K. What the future holds for magnetic drive pumps. *World Pumps.* **2007**, *495*, 28–29. [CrossRef]
3. Kong, F.Y.; Shen, X.K.; Wang, W.T.; Zhou, S.Q. Performance Study Based on Inner Flow Field Numerical Simulation of Magnetic Drive Pumps with Different Rotate Speeds. *Chin. J. Mech. Eng.* **2012**, *25*, 137–143. [CrossRef]
4. Borovsky, J.E.; Horne, R.B.; Meredith, N.P. The contribution of compressional magnetic drive pumping to the energization of the Earth's outer electron radiation belt during high-speed-stream-driven storms. *J. Geophys. Res. Sp. Phys.* **2017**, *122*, 12072–12089.
5. ХлопеНКоВ.п.р.; Zhuang, C.R. Translated. On the issues of the optimization of the structure of centrifugal pumps. *Drain. Irrig. Mach.* **1988**, *6*, 13–17.
6. Kergourlay, G.; Younsi, M.; Bakir, F.; Rey, R. Influence of splitter blades on the flow field of a centrifugal pump: Test-analysis comparison. *Int. J. Rotat. Mach.* **2007**, *13*, 85024. [CrossRef]
7. Gölcü, M.; Pancar, Y.; Sekmen, Y. Energy saving in a deep well pump with splitter blade. *Energy Convers. Manage.* **2006**, *47*, 638–651. [CrossRef]
8. Gölcü, M. Artificial neural network based modeling of performance characteristics of deep well pumps with splitter blade. *Energy Convers. Manag.* **2006**, *47*, 3333–3343. [CrossRef]
9. Gölcü, M.; Usta, N.; Pancar, Y. Effects of Splitter Blades on Deep Well Pump Performance. *J. Energy Res. Technol.* **2007**, *129*, 169–176. [CrossRef]
10. Shigemitsu, T.; Fukutomi, J.; Nasada, R.; Kaji, K. The effect of blade outlet angle on performance and internal flow condition of mini turbo-pump. *J. Therm. Sci.* **2011**, *20*, 32–38. [CrossRef]
11. Shigemitsu, T.; Fukutomi, J.; Wada, T.; Shinohara, H. Performance analysis of mini centrifugal pump with splitter blades. *J. Therm. Sci.* **2013**, *22*, 573–579. [CrossRef]
12. Yuan, S.Q.; Zhang, Y.Z. Tests on centrifugal pump with deviated splitting vanes. *Trans. Chin. Soc. Agric. Mach.* **1995**, *1995*, 79–83.
13. Yuan, S.Q.; Zhang, J.F.; Yuan, J.P.; Fu, D.Y. Orthogonal testal study effect of main geometry factors of splitter blades on pump performance. *J. Drain. Irrig. Mach. Eng.* **2008**, *2008*, 1–5.
14. Yuan, S.Q.; Zhou, J.J.; Yuan, J.P.; Zhang, J.F.; Xu, Y.P. Numerical Simulation on Radial Hydraulic Forces for Screw-type Centrifugal Pump with Splitter Blade. *Trans. Chin. Soc. Agric. Mach.* **2012**, *43*, 37–42.
15. Zhang, J.F. Numerical Forecast and Research on the Design Method for Centrifugal Pumps with Splitter Blades. Ph.D. Thesis, Jiangsu University, Zhenjiang, China, 2007.
16. Zhang, J.F.; Li, G.D.; Mao, J.Y.; Yuan, S.Q.; Qu, Y.F.; Jia, J. Effects of the outlet position of splitter blade on the flow characteristics in low-specific-speed centrifugal pump. *Adv. Mech. Eng.* **2018**, *10*, 1–12. [CrossRef]
17. Gu, Y.D.; Yuan, S.Q.; Pei, J. Effects of the impeller–volute tongue interaction on the internal flow in a low-specific-speed centrifugal pump with splitter blades. *Proc. Inst. Mech. Eng. Part A J. Power Energy.* **2018**, *232*, 170–180. [CrossRef]

18. Cui, B.L.; Chen, F.; Ge, M.Y. Internal flow characteristics and performance test of low specific-speed centrifugal pumps. *J. Drain. Irrig. Mach. Eng.* **2016**, *34*, 375–380.
19. Guo, X.M.; Zhu, L.; Zhu, Z.C. Numerical and experimental investigations on the cavitation characteristics of a high-speed centrifugal pump with a splitter-blade inducer. *J. Mech. Sci. Technol.* **2015**, *29*, 259–267. [CrossRef]
20. Yuan, Y.; Yuan, S.Q. Analyzing the effects of splitter blade on the performance characteristics for a high-speed centrifugal pump. *Adv. Mech. Eng.* **2017**, *9*, 1–11. [CrossRef]
21. Jia, Y.; Wei, X.; Wang, Q.; Cui, J.; Li, F. Experimental Study of the Effect of Splitter Blades on the Performance Characteristics of Francis Turbines. *Energies* **2019**, *12*, 1676. [CrossRef]
22. Wang, F.J. *Principle and Application of CFD Software for Computational Fluid Dynamics Analysis*; Tsinghua University Press: Beijing, China, 2004; pp. 110–130.
23. Menter, F.R. Zonal two equation k-turbulence models for aerodynamic flows. *Am. Inst. Aeronaut. Astronaut. Pap.* **1993**, *7*, 2906.
24. Menter, F.R. Ten years of experience with the SST turbulence model. *Turbul. Heat Mass Transf.* **2003**, *4*, 625–632.
25. Xie, L.H.; Zhao, X.Y. *ANSYS CFX Fluid Analysis and Simulation*; Publishing House of Electronics Industry: Beijing, China, 2013; pp. 12–19.
26. Dai, C.; Kong, F.Y.; Dong, L. Pressure fluctuation and its influencing factors in circulating water pump. *J. Cent. South Univ.* **2013**, *20*, 149–155. [CrossRef]
27. Wang, A.M.; Wen, J. Application for Optimal Designing of Interior Permanent Magnet Synchronous Machines by Using Taguchi Method. *J. North Chin. Electr. Power Univ.* **2016**, *43*, 39–44.
28. Ahmadzadeh, M.; Mortazavi, S.; Saniei, M. Applying the Taguchi Method for Investigating the Phase-Locked Loop Dynamics Affected by Hybrid Storage System Parameters. *Energies* **2018**, *11*, 226. [CrossRef]

energies

MDPI

Article

Investigation of Dilution Effect on CH_4/Air Premixed Turbulent Flame Using OH and CH_2O Planar Laser-Induced Fluorescence

Li Yang [1,2,*], Wubin Weng [3], Yanqun Zhu [4], Yong He [4], Zhihua Wang [4] and Zhongshan Li [3]

1 Key Laboratory for Technology in Rural Water Management of Zhejiang Province, Zhejiang University of Water Resources and Electric Power, Hangzhou 310018, China
2 The College of Electrical Engineering, Zhejiang University of Water Resources and Electric Power, Hangzhou 310018, China
3 Division of Combustion Physics, Lund University, P.O. Box 118, S-22100 Lund, Sweden; wubin.weng@forbrf.lth.se (W.W.); zhongshan.li@forbrf.lth.se (Z.L.)
4 State Key Laboratory of Clean Energy Utilization, Zhejiang University, Hangzhou 310012, China; yqzhu@zju.edu.cn (Y.Z.); heyong@zju.edu.cn (Y.H.); wangzh@zju.edu.cn (Z.W.)
* Correspondence: yanglizju06@zju.edu.cn

Received: 22 November 2019; Accepted: 7 January 2020; Published: 9 January 2020

Abstract: Diluting the combustion mixtures is one of the advanced approaches to reduce the NO_x emission of methane/air premixed turbulent flame, especially with high diluents to create a distributed reaction zone and mild combustion, which can lower the temperature of reaction zone and reduce the formation of NO_x. The effect of N_2/CO_2 dilution on the combustion characteristics of methane/air premixed turbulent flame with different dilution ratio and different exit Reynolds number was conducted by OH-PLIF and CH_2O-PLIF. Results show that the increase of dilution ratio can sharply reduce the concentration of OH and CH_2O, and postpone the burning of fuel. Compared with the ultra-lean combustion, the dilution weakens the combustion more obviously. For different dilution gases, the concentration of OH in the combustion zone varies greatly, while the concentration of CH_2O in the unburned zone is less affected by different dilution gas. The CO_2 dilution has a more significant effect on OH concentration than N_2 with the given dilution ratio, but a similar effect on the concentration of CH_2O in the preheat zone of flame. However, dilution does not have much influence on the flame structure with the given turbulent intensity.

Keywords: dilution; turbulent flame; premixed; OH; CH_2O; planar laser-induced fluorescence

1. Introduction

Energy and environmental issues have become significant topics in recent decades. Many efforts have been done to reduce the emission of pollutions caused by the utilization of fossil fuel, such as NO_x, SO_x, and soot. In NO_x reduction, one of those advanced approaches is diluting the combustion mixtures [1–4] especially with high temperature diluents to create a distributed reaction zone and mild combustion [2,5,6] which can lower the temperature of the reaction zone and reduce the formation of NO_x. The dilution can be achieved with exhaust gas recirculation (EGR), steam, or low calorific value (LCV) fuels utilization, which normally contain a massive amount of incombustible gas, such as N_2, CO_2, and H_2O. These gases make the chemical reaction rate drop significantly and combustion process out of the normal flame-let regime more easily [4–11]. Recently, many studies about the effect of dilution on the characteristics of combustion have been performed, such as turbulent burning velocity, flame stability, flame structure, and pollution emission [2,12–16].

Among these researches, laser diagnostic techniques were adopted widely, and much essential information of flames has been obtained, e.g., the structure of turbulent flames [1–3,17,18], OH radical laser measurement [19–22], and the intermediate product CH_2O [20,23–33]. Dally et al. [34] and Medwell et al. [2] used planar laser induced fluorescence (PLIF) and Rayleigh scattering (RS) measured the flame structure of turbulent non-premixed jet flames. Kobayshi et al. [3] did some research on the methane/air flames with the dilution of 10% CO_2 with OH-PLIF. Wang et al. [35] calculated the local radius of curvature, fractal inner cutoff scale, and local flame angle using OH-PLIF images of methane/air flames with 10% CO_2/H_2O dilution. In addition, Han et al. [11] did some work of the dilution effect of N_2 and CO_2 on the flames in a swirl-stabilized combustor.

The dilution ratio of the above-mentioned studies was all less than 20%, and the concentration of OH was studied without the important intermediate product CH_2O. Therefore, in the current work, in order to get some knowledge of the effect of high dilution ratio on the premixed turbulent flames, dilution ratio was raised up to 50% to gain a high Kariovitaz number (Ka) with high Reynolds number. Ka number is the ratio of the rotational time of the turbulent minimum vortex (Komogorov vortex) to the chemical reaction time of the flame. The greater the Ka, the faster the turbulent mixing rate compared to the chemical reaction rate of the flame. The flame with a high rate of dilution was also compared with the flame with extreme low equivalence ratio ($\varphi = 0.4$), which can obtain some fundamental understanding about the interaction between turbulent transfer and chemical reaction in distributed reaction regimes including mild combustion [1,5,6]. In addition, the spatial distribution of OH and CH_2O of several jet flames were imaged by PLIF to have a look at the detail structure of the reaction zone from the downstream position 5d (d: Diameter of jet tube) to 37d. OH results were mainly used for the flame front structure obtaining and thought as the temperature indicator, and CH_2O results were employed as the marker of the low temperature zone of flames [2].

2. Experiment Setup

The experiment setup employed in the present work is mainly comprised of a burner and a laser diagnostic system, as shown in Figure 1. A water cooled McKenna burner with a centre jet tube was utilized. The jet tube was 1 mm in diameter and its exit were 4 mm above the surface of the McKenna burner. With this tube, the jet flame of the mixture presented in Table 1 was generated. It was surrounded by a co-flow comprised of hot flue gas which was generated from a laminar premixed flame of methane/air ($\varphi = 0.9$) locating about 2 mm below the jet tube exit. The gas supply was controlled by several mass flow controllers (Alicat) to obtain designed mixtures with the given Reynolds number. Air was emulated with 21% O_2 and 79% N_2 and D_{diluent}(dilution ratio) was given by,

$$D_{\text{diluent}} = V_{\text{diluent}}/V_{\text{sum}} \tag{1}$$

where V_{diluent} is the volume of diluent (N_2 or CO_2) and V_{sum} is the total volume of the gas mixture.

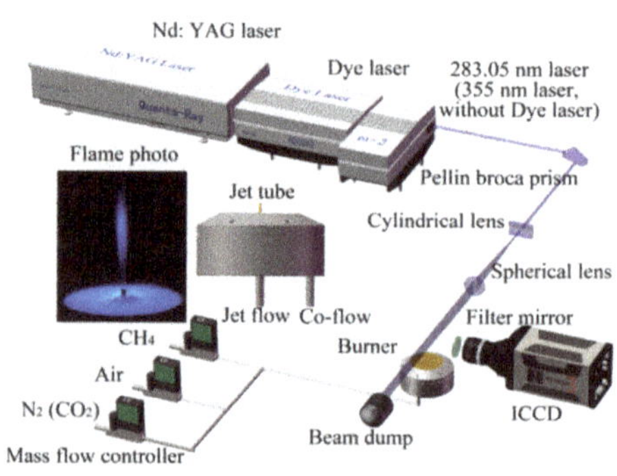

Figure 1. Schematic of experiment setup.

Table 1. Flame operation conditions.

Experiment Cases	Premixed Mixture Composition				Reynolds Number (*Re*)
	CH_4%	Air%	N_2%	CO_2%	
Flame 1 (φ = 0.4/DN_2 = 0%)	4.0	96.0	0	0	6000
Flame 2 (φ = 0.9/DN_2 = 0%)	8.6	91.4	0	0	6000
Flame 3 (φ = 0.9/DN_2 = 30%)	6.0	64.0	30.0	0	6000
Flame 4 (φ = 0.9/DN_2 = 50%)	4.3	45.7	50.0	0	6000
Flame 5 (φ = 0.9/DCO_2 = 50%)	4.3	45.7	0	50	6000
Flame 6 (φ = 0.9/DN_2 = 50%)	4.3	45.7	50.0	0	3000
Flame 7 (φ = 0.9/DCO_2 = 50%)	4.3	45.7	0	50	3000

The PLIF system was adopted for OH and CH_2O measurement. A frequency-double Quanta-Ray Nd:YAG laser (PRO-250-10H, Spectra physics, Santa Clara, CA, USA) was used to pump a dye laser (ND6000, Continuum, Boston, MA, USA) to generate 283.049 nm UV laser(with pulse energy of 15 mJ) with a frequency doubler. This UV laser was made to a vertical laser sheet with a length of 32 mm passing through the center of flame vertically to excite Q_1 (8) line of OH with the transition of $A^2\Sigma \leftarrow X^2\Pi(1,0)$. The corresponding fluorescence was captured by an intensified charge coupled device (ICCD) camera (PI MAX 3, Princeton Instruments, Trenton, NJ, USA) with a 1024 × 1024 pixel array. A UV lens (105 mm focal length, Nikon, Tokyo, Japan) and two combined Schott filters (WG305 and UG11) were used to obtain the signal around 308 nm. For the measurement of CH_2O in the flame, the frequency-triple Quanta-Ray Nd:YAG laser was used to generate a 355 nm laser with pulse energy around 150 mJ, which was also formed to a 32 mm high vertical laser sheet passing through the flame. Its fluorescence was collected by an ICCD camera and a Nikon objective (f/4.5, 50 mm) through a Schott long-pass filter (GG395).

3. Results and Discussions

3.1. Emission Spectra Analysis

The photographs and emission spectra of the flames under different conditions were shown in Figure 2. Compared with Flame 2 (φ = 0.9, D_{diluent} = 0), there is almost no visible light in Flame 1 (φ = 0.4, D_{diluent} = 0) due to the property of flameless combustion [1]. The flames with large amount of dilution, such as Flame 4 (φ = 0.9, N_2D_{diluent} = 50%) and Flame 5 (φ = 0.9, CO_2D_{diluent} = 0), also had much less visible light, showing very close to the characteristic of Flame 1.

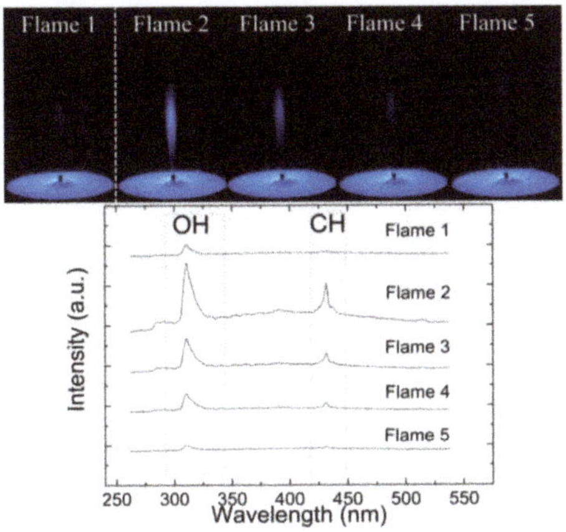

Figure 2. Photos and emission spectra of flames.

The emission spectra of these flames from 260 to 540 nm were captured by a spectrometer (Ocean optics 2000) focusing on the downstream position 10d with a spherical lens. As shown in Figure 2, it mainly contains the radiation from OH radical (310 nm) and CH radical (431 nm). Flame 1 and Flame 5 only have some light emission of OH radical, which is consistent with the observation by Duwig [1]. Comparing the flames with different dilution ratios, significant difference in the strength of emission spectra was observed. With a large amount of dilution, both the signals from OH and CH were dropped significantly. The OH signal of Flames 4 and 5 have the similar strength to Flame 1, but their CH signal still exists. From Table 1, it can be seen that their content of CH_4 is very close to that of Flame 1, but theirO_2 fraction is much lower, which affects the chemical path in the reaction and makes diluted combustion different from ultra lean combustion.

3.2. Image Process

Figure 3a presented a typical OH-PLIF (Planar Laser Induced Fluorescence) instantaneous image of Flame 4. Those instantaneous images show the structures of turbulent flames. The color bar represents the signal intensity of radical, the bigger the color number, the stronger the signal intensity. According to the intensity of OH fluorescence signal, the wrinkled flame interface can be well recognized, which were processed with the adaptive threshold method: First, background signal was subtracted and the effect of uneven distribution of laser energy was removed from the original image using MATLAB. Second, the spot was assigned to 1 when the intensity of the OH fluorescence signal was more than 0, or the spot was assigned to 0. Therefore, a black and white picture can be obtained and the burned region was white while the unburned region was black. Consequently, the flame front contours were obtained by extracting the boundary of black and white, as shown in Figure 3b. In the present experiment, 500 transient images were collected for each case. The overlaying value of 500 single shots was presented in Figure 3c which can be used in statistical analysis with averaging and root mean square (RMS) values to analyze the flame front distribution.

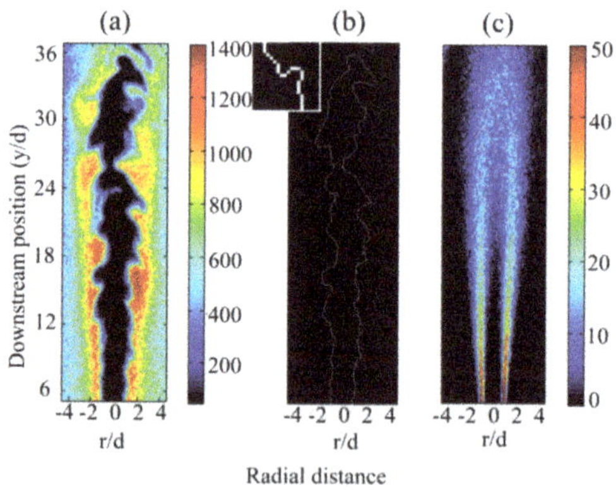

Figure 3. (**a**) Instantaneous image of OH-PLIF; (**b**) flame front of the instantaneous image; (**c**) overlaying of 500 flame front boundaries.

Additionally, the wrinkle ratio (W) of these flames was also calculated out with those flame front boundaries. Wrinkle ratio is a significant parameter to demonstrate the scale of turbulent flame front fractality. The bigger the wrinkle ratio, the stronger the turbulent intensity, the more flame front fractality. The calculated method was given by,

$$W = L/h \tag{2}$$

in the formula, h is the given flame height at vertical distance, which is determined as 1 mm in the current experiment. L is the flame front length within a given flame height of h at a given downstream position. The distance between two neighbouring pixels with corner connection was counted as $\sqrt{2}$ times of that of the pixels with side connection, as shown in Figure 3b.

3.3. Analysis of PLIF Images

Figure 4 presents the typical instantaneous PLIF images of CH_2O (a) and OH (b) in Flame 1–4, Flame 2–4 were used as a dilution effect analysis and Flame 1 was adopted as a comparison case. Reynolds number (*Re*) was kept around 6000 based on exit velocity, tube diameter, and the viscosity of corresponding gas mixture.

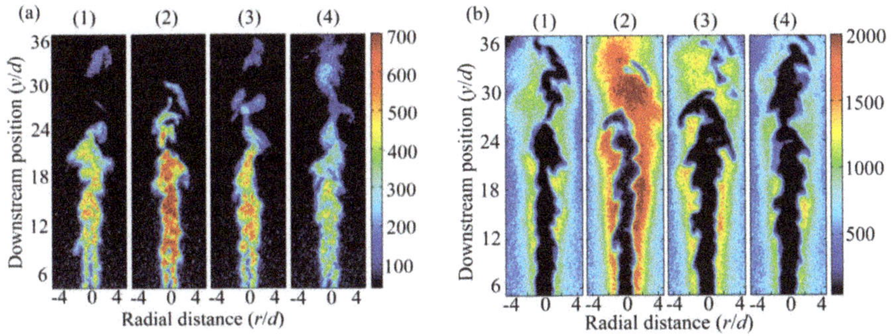

(a) **(b)**

Figure 4. (**a**) CH_2O-PLIF instantaneous images and (**b**) OH-PLIF instantaneous images of (1) Flame 1; (2) Flame 2; (3) Flame 3; (4) Flame 4.

From these figures, it is clear that the CH_2O layers are surrounded by OH layers. CH_2O was generated in the middle of the flame during the preheating stage which was earlier than OH. In the flame front zone, OH was generated and the distribution of OH shows the conjunction between burnt and unburned regions, thought as an indicator of flame region [36]. In all cases, OH signal shows relatively stronger as close to the flame front, especially in wrinkle pockets and flame tip zone. The unburned region and the thickness of OH layer expanded from the flame bottom to the tip. The parts of flames below ~10dareshown to be smooth and above ~10dthey are wrinkled to form many eddies due to turbulent effect. When the flame was diluted, the strength of OH signal dropped and unburned zone expanded significantly. The similar effect was observed in ultra lean flame, Flame 1. In addition, CH_2O layers became more thickened from the tube exit and started to merge together at the downstream position around 10d. At the tip of the flames, CH_2O started to be consumed out. The strength of CH_2O signal was also weakened notably by dilution.

The averaging and RMS results of these flames were obtained with 500 single shots of instantaneous OH ad CH_2O images to the statistical analysis, which can be used to analyze the flame front distribution. Root mean square (RMS) values can be used to indicate the degree of dispersion of the measurement sample, in this case it refers to the OH and CH_2O radical signal, which was real-time and online measured. Therefore, the root mean square (RMS) values indicated the position of the flame front with maximum possibility. The radial distribution at the downstream position of 18d is presented in Figure 5. At this axial position, CH_2O already merged together making the peak of the mean value appear in the central position and the peak of RMS value show at the side of flames corresponding to the CH_2O consuming zone, since at this region, the distribution of CH_2O concentration changes a lot with time due to strong fluctuation by turbulence. The peak of mean value of OH locates around the radial distance of 2d, where there is almost no CH_2O left. The peak of RMS value of OH is around 1d, close to that of CH_2O, where OH was mainly generated.

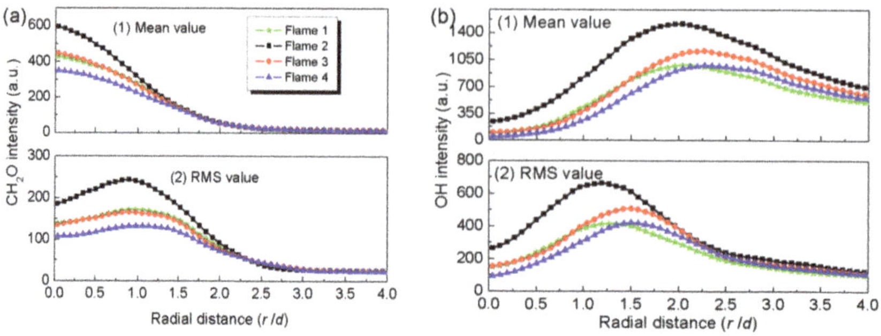

Figure 5. The averaged and root mean square (RMS) value of (**a**) CH_2O and (**b**) OH at the downstream position of 18d.

The mean and RMS value of CH_2O and OH decline with dilution and the peak of OH shifts about 0.5d to the burnt region side, which indicates the expanding of the unburned region. The distribution curve of CH_2O of Flame 1 almost overlaps that of Flame 3, even though their composition is notably different as shown in Table 1. For these two flames, the OH distribution curve also overlaps each other in the flame center. However, Flame 3 has a bigger peak value of OH than Flame 1 caused by its higher methane fraction and larger calorific value inducing higher flame temperature. With the same reason, the peak value of OH of Flames 1 and 4 are very close to each other. However, comparing to Flame 1, the combustion of Flame 4 delayed significantly according to the distribution of OH. The main difference is oxygen fraction in premix. It is found that, even though all the flames are under-lean condition and the oxygen in the premixed mixture are abundant for fixed methane fraction, oxygen fraction still has some influence on the combustion. The main effect happened in the preheat reaction zone, making the reaction stronger and more CH_2O to be generated with the same fraction of methane. Since that, the combustion occurs earlier as observed above. However, it has almost no effect on the burnt region, because the content of methane dominants the calorific value of the mixture and affects the heat release and the temperature of flames. It is worth paying attention to the averaged and RMS value of OH that may be affected by the OH signal of co-flow comprised of hot flue gas, which was generated from a laminar premixed flame of methane/air (φ = 0.9) located about 2 mm below the jet tube exit.

Figure 6 shows the statistics characteristic of the flame front of Flame 1 to 4. The results were obtained from those OH-PLIF instantaneous images with the process shown in Figure 3. In Figure 6a, the plotted flame front distribution value was obtained at downstream position 18d. The front distribution value represents the OH radical signal intensity of the flame front statistics images, and the unit "a.u." is the unit of signal intensity. Figure 6(b1) shows the maximum value of the front distribution at different given vertical position. It indicated the position of flame front with the maximum possibility. As the premix was diluted, the flame expanded and the combustion delayed obviously. Hence, the mean volume of the flame region increased and the mean fuel consumption rate decreased with less methane fraction in mixtures as concluded by [3]. The flame front of Flame 1 was much closer to the jet center than that of Flames 3 and 4, even though Flame 1 has same CH_2O signal as Flame 3 and the same peak of OH signal as Flame 4. In Figure 6(b2) the wrinkle ratio was obtained along the axial direction, which was used to characterize the front fractality structure of turbulent flame. The change in the wrinkle ratio is small until the tip of the flame. The big drop around the tip was caused by the flame merging and reaction intensity decline. With dilution, the wrinkle ratio has a slight drop in the part lower than 22d.

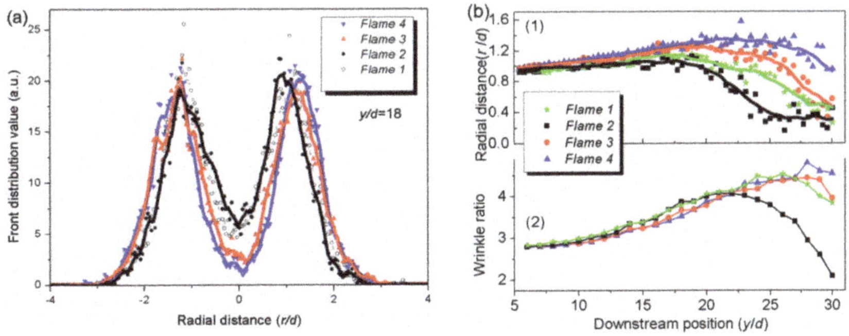

Figure 6. (**a**) Flame front distribution, (**b1**) flame front position and (**b2**) wrinkle ratio of Flame 1, Flame 2, Flame 3, and Flame 4.

Typical instantaneous images of (a) CH_2O-PLIFand (b) OH-PLIF of the flames with 50% N_2 or 50% CO_2 dilution are shown in Figure 7. Figure 7(a1,a3,b1,b3) show the flames (Flames 6 and 4) diluted by N_2 and the other figures show the flames (Flames 7 and 5) diluted by CO_2. Additionally, different Reynolds number (3000 and 6000) was adopted for the turbulent effect study.

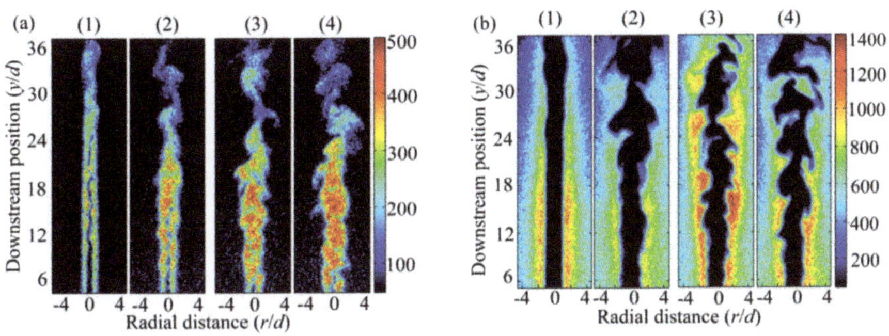

Figure 7. (**a**) CH_2O-PLIF instantaneous images and (**b**) OH-PLIF instantaneous images of (1) Flame 6; (2) Flame 7; (3) Flame 4; (4) Flame 5.

Comparing with the flames diluted by N_2, the ones diluted by CO_2 have an OH signal with much less strength. As turbulent intensity enhanced at higher Reynolds number, the flame front was more folded. At the Reynolds number of 6000, the structure is very similar for N_2 and CO_2 cases. When the Reynolds number was changed to 3000, there is no big change in CO_2 dilution flames. However, the folded structure in the N_2 dilution flames almost disappeared. This phenomenon can also be observed based on the structure of CH_2O distributions. The reason was thought to be the drop of the chemical reaction rate and fuel consumption rate caused by CO_2 addition, similar to the CO_2 effect on the laminar burning velocity [37,38].

Figure 8 presents the distribution of the mean and RMS values of OH and CH_2O along the radial direction at the downstream position of 18d of the flames with different dilution and Reynolds number. The structure of Flame 6 (DN_2 = 50%, *Re* = 3000) is significantly different from other flames, as described above with instantaneous images, which was kept in laminar flame style and its distribution of OH and CH_2O is much close to those as one dimension laminar flames [37,38]. For the flames with different Reynolds number, the mean value of their signal can be increased due to turbulent intensity enhancement, and can also be weakened due to their reaction zone expanding. The RMS value was enhanced significantly with Reynolds number indicating stronger fluctuation of flames. The strength and the thickness of CH_2O increased at higher Reynolds number due to the enhancement of mixing between burnt and unburned gas which benefited the reaction in the preheat zone. Comparing the flames with different dilution content at Reynolds number of 3000, the RMS value of CH_2O and OH of Flame 7 (DCO_2 = 50%, Re = 3000) is much larger than Flame 6 due to the folded flame front of Flame 7 as observed above. The mean value of CH_2O in Flame 7 is larger than that of Flame 6, but that of OH is smaller. The wrinkled front enhanced the preheat reaction, but the larger heat capacity of CO_2 makes the flame temperature lower and the reaction rate slower, which is consistent with the result of Roy [39]. At the Reynolds number of 6000, there is no difference in the mean value of CH_2O between Flames 4 and 5, but the OH value was weakened and the combustion was delayed in Flame 5, indicating less heat release and weakened reaction in flame zone.

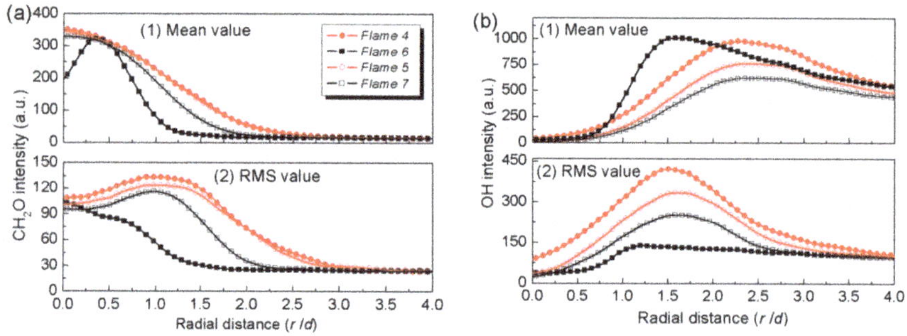

Figure 8. (**a**) Mean and RMS value of CH_2O and (**b**) Mean and RMS value of OH at downstream position of 18d.

In Figure 9a, the distribution of the flame front of Flames 4, 5, and 7 overlapped except Flame 6 having a peak at the position of 1d. Combined with Figure 9b, Flame 6 had less front fluctuation as mentioned before. It indicated that the dilution difference has almost no influence on the front fluctuation under turbulent condition. When the downstream position was above 20d, the flame expanded significantly with CO_2 dilution as shown in Figure 9(b1) and the wrinkle ratio dropped in the entire flame shown in Figure 9(b2), which indicates that the combustion was delayed and flame front structure was smoothed as dilution changed from N_2 to CO_2.

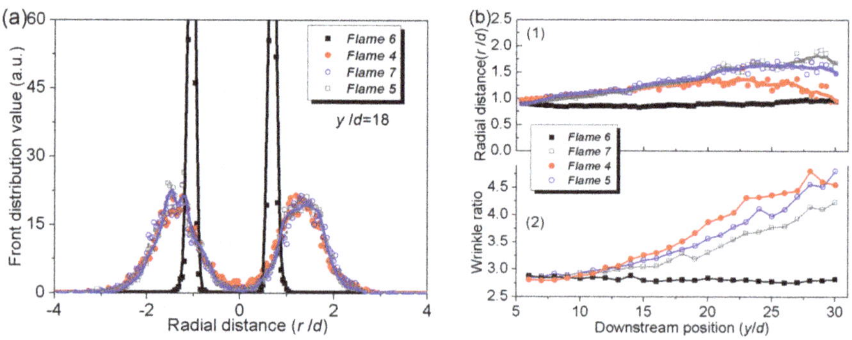

Figure 9. (**a**) Flame front distribution, (**b1**) flame front position (**b2**) wrinkle ratio of Flame 4, Flame 5, Flame 6, and Flame 7.

4. Conclusions

In current work, the effect of N_2/CO_2 dilution on the combustion characteristics of methane/air premixed turbulent jet flame with different dilution ratio and different exit Reynolds number were conducted using a water cooled McKenna burner with a centre jet tube. In order to obtain the knowledge about the effect of high dilution on the premixed turbulent flame, 50% dilution gas was used in the experiment. The distribution of free radical OH and combustion intermediate product CH_2O in several turbulent jet flames were measured by OH-PLIF and CH_2O-PLIF. The OH results were mainly used to obtain the structure of flame front and the distribution of CH_2O was mainly used to study the low temperature zone of flames. Results show that the increasing of dilution ratio can sharply reduce the concentration of OH and CH_2O, and postpone the burning of fuel. Compared with the ultra-lean combustion, the dilution weakens the combustion more obviously. For different dilution gases, the concentration of OH in the combustion zone varies greatly, while the concentration of CH_2O in the unburned zone is less affected by different dilution gas. The CO_2 dilution has a more significant effect on OH concentration than N_2 with the given dilution ratio, but a similar effect on the concentration of CH_2O in the preheat zone of the flame. However, dilution does not have much influence on the flame structure with the given turbulent intensity.

Author Contributions: Data curation, W.W.; Funding acquisition, Y.Z.; Methodology, Y.H.; Project administration, Y.Z.; Supervision, Z.W. and Z.L.; Writing-original draft, W.W.; Writing-review and editing, L.Y. and Y.H. All authors have read and agreed to the published version of the manuscript.

Funding: This research was funded by the National Natural Science Foundation of China [51621005] and the Program of Introducing Talents of Discipline to Universities [B08026], and the APC was funded by Key Laboratory for Technology in Rural Water Management of Zhejiang Province, Zhejiang University of Water Resources and Electric Power.

Acknowledgments: The authors would like to thank editor and the reviewers who provided many valuable comments to improve the paper. Li Yang would also like to thank the help of Wubin Weng and the supervisor of Zhihua Wang.

Conflicts of Interest: The authors declare no conflict of interest.

References

1. Duwig, C.; Li, B.; Li, Z.S.; Aldén, M. High resolution imaging of flameless and distributed turbulent combustion. *Combust. Flame* **2012**, *159*, 306–316. [CrossRef]
2. Medwell, P.R.; Kalt, P.A.; Dally, B.B. Simultaneous imaging of OH, formaldehyde, and temperature of turbulent nonpremixed jet flames in a heated and diluted coflow. *Combust. Flame* **2007**, *148*, 48–61. [CrossRef]
3. Kobayashi, H.; Hagiwara, H.; Kaneko, H.; Ogami, Y. Effects of CO_2 dilution on turbulent premixed flames at high pressure and high temperature. *Proc. Combust. Inst.* **2007**, *31*, 1451–1458. [CrossRef]
4. Yoo, H.; Park, B.Y.; Cho, H.; Park, J. Performance Optimization of a Diesel Engine with a Two-Stage Turbocharging System and Dual-Loop EGR Using Multi-Objective Pareto Optimization Based on Diesel Cycle Simulation. *Energies* **2019**, *12*, 4223. [CrossRef]
5. Pramanik, S.; Ravikrishna, R.V. Effect of N2 dilution and preheat temperature on combustion dynamics of syngas in a reverse-flow combustor. *Exp. Therm. Fluid Sci.* **2020**, *110*, 109926. [CrossRef]
6. Karimi, H.; Mardani, A. Investigation of fuel dilution in ethanol spray MILD combustion. *Appl. Therm. Eng.* **2019**, *159*, 113898.
7. Yang, Z.; Yu, X.; Peng, J.; Wang, L.; Dong, Z.; Li, X.; Sun, S.; Meng, S.; Xu, H. Effects of N_2, CO_2 and H_2O dilutions on temperature and concentration fields of OH in methane Bunsen flames by using PLIF thermometry and bi-directional PLIF. *Exp. Therm. Fluid Sci.* **2017**, *81*, 209–222. [CrossRef]
8. Abbasi-Atibeh, E.; Bergthorson, J.M. The effects of differential diffusion in counter-flow premixed flames with dilution and hydrogen enrichment. *Combust. Flame* **2019**, *209*, 337–352. [CrossRef]
9. Wang, D.; Ji, C.; Wang, S.; Meng, H.; Yang, J. Chemical effects of CO_2 dilution on CH_4 and H_2spherical flame. *Energy* **2019**, *185*, 316–326. [CrossRef]
10. Song, Z.; Zhang, X.; Hou, X.; Hu, S. Relationship of the combustion characteristics of natural gas-hydrogen/carbon dioxide mixtures with the ion current and pressure parameters. *J. Energy Inst.* **2019**, *92*, 1014–1022. [CrossRef]
11. Han, Y.; Cai, G.; Wang, H.; Bruno, R.; Abdelkrim, B. Flow characterization and dilution effects of N_2 and CO_2 on premixed CH_4/air flames in a swirl-stabilized combustor. *Chin. Phys. B* **2014**, *3*, 386–399.

12. Lee, M.C.; Seo, S.B.; Yoon, J.; Kim, M.; Yoon, Y. Experimental study on the effect of N_2, CO_2, and steam dilution on the combustion performance of H_2 and CO synthetic gas in an industrial gas turbine. *Fuel* **2012**, *102*, 431–438. [CrossRef]
13. Jourdaine, P.; Mirat, C.; Caudal, J.; Lo, A.; Schuller, T. A comparison between the stabilization of premixed swirling CO_2-diluted methane oxy-flames and methane/air flames. *Fuel* **2016**, *201*, 431–438.
14. Xie, Y.; Wang, X.; Wang, J.; Huang, Z. Explosion behavior predictions of syngas/air mixtures with dilutions at elevated pressures: Explosion and intrinsic flame instability parameters. *Fuel* **2019**, *255*, 115724. [CrossRef]
15. Sahu, A.; Wang, C.; Jiang, C.; Xu, H.; Ma, X.; Xu, C.; Bao, X. Effect of CO_2 and N_2 dilution on laminar premixed MTHF/air flames: Experiments and kinetic studies. *Fuel* **2019**, *255*, 115659. [CrossRef]
16. Zheng, J.; Zhang, Z.; Huang, Z.; Hu, E.; Tang, C.; Wang, J. Numerical study on combustion of diluted methanol-air premixed mixtures. *Chin. Sci. Bull.* **2010**, *55*, 882–889. [CrossRef]
17. Wang, Z.H.; Li, B.; Ehn, A.; Sun, Z.W.; Li, Z.S.; Bood, J.; Aldén, M.; Cen, K.F. Investigation of flue-gas treatment with O_3 injection using NO and NO_2 planar laser-induced fluorescence. *Fuel* **2010**, *89*, 2346–2352. [CrossRef]
18. Luque, J.; Jeffries, J.B.; Smith, G.P.; Crosley, D.R. Quasi-simultaneous detection of CH_2O and CH by cavity ring-down absorption and laser-induced fluorescence in a methane/air low-pressure flame. *Appl. Phys. B* **2001**, *73*, 731–738. [CrossRef]
19. Elbaz, A.M.; Roberts, W.L. Experimental study of the inverse diffusion flame using high repetition rate OH/acetone PLIF and PIV. *Fuel* **2016**, *165*, 447–461. [CrossRef]
20. Johchi, A.; Naka, Y.; Shimura, M.; Tanahashi, M.; Miyauchi, T. Investigation on rapid consumption of fine scale unburned mixture islands in turbulent flame via 10 kHz simultaneous CH-OH PLIF and SPIV. *Proc. Combust. Inst.* **2015**, *3*, 3663–3671. [CrossRef]
21. Coriton, B.; Steinberg, A.M.; Frank, J.H. High-speed tomographic PIV and OH PLIF measurements in turbulent reactive flows. *Exp. Fluids* **2014**, *55*, 1743. [CrossRef]
22. Wellander, R.; Richter, M.; Aldén, M. Time-resolved (kHz) 3D imaging of OH PLIF in a flame. *Exp. Fluids* **2014**, *55*, 1764. [CrossRef]
23. Li, H.; Li, B.; Gao, Q.; Zhang, D.; Li, X.; Li, Z. Plane laser-Induced fluorescence for flame surface density calculation of OH/CH_2O:A comparative study. *J. Combust. Sci. Technol.* **2018**, *24*, 523–527.
24. Zhu, J.; Zhao, G.; Long, T.; Sun, M.; Li, Q.; Liang, J. Simutaneous OH and CH_2O PLIF imaging of flame structures. *J. Exp. Fluid Mech.* **2016**, *30*, 55–60.
25. Yan, H.; Zhang, S.; Xu, X.; Li, F.; Lin, X. Flame structure and dynamics characters investigation by OH and CH_2O planar laser-induced fluorescence in the swirl combustor. *J. Aerosp. Power* **2019**, *34*, 894–907.
26. Li, X.; Li, H.; He, K.; Lin, H.; Cheng, L. Analysis of characteristic spectrum of CN/CH/CHO/CH_2O/NCN in induction period of methane explosion. *J. Chin. Coal Soc.* **2014**, *39*, 2042–2046.
27. Wang, Z.; Stamatoglou, P.; Zhou, B.; Aldén, M.; Bai, X.S.; Richter, M. Mattias Richtera Investigation of OH and CH_2O distributions at ultra-high repetition rates by planar laser induced fluorescence imaging in highly turbulent jet flames. *Fuel* **2018**, *234*, 1528–1540. [CrossRef]
28. Alzueta, M.U.; Glarborg, P. Formation and destruction of CH_2O in the exhaust system of a gas engine. *Environ. Sci. Technol.* **2003**, *37*, 4512–4516. [CrossRef] [PubMed]
29. Wang, S.; Davidson, D.F.; Hanson, R.K. High-temperature laser absorption diagnostics for CH_2O and CH_3CHO and their application to shock tube kinetic studies. *Combust. Flame* **2013**, *160*, 1930–1938. [CrossRef]
30. Schroeder, J.R.; Crawford, J.H.; Fried, A.; Walega, J.; Weinheimer, A.; Wisthaler, A.; Müller, M.; Mikoviny, T.; Chen, G.; Shook, M.; et al. New insights into the column CH_2O/NO_2 ratio as an indicator of near-surface ozone sensitivity. *J. Geophys. Res. Atmos.* **2017**, *122*, 8885–8907. [CrossRef]
31. Osborne, J.R.; Ramji, S.A.; Carter, C.D.; Peltier, S.; Hammack, S.; Lee, T.; Steinberg, A.M. Simultaneous 10 kHz TPIV, OH PLIF, and CH_2O PLIF measurements of turbulent flame structure and dynamics. *Exp. Fluids* **2016**, *57*, 65. [CrossRef]
32. Hammack, S.D.; Carter, C.D.; Skiba, A.W.; Fugger, C.A.; Felver, J.J.; Miller, J.D.; Gord, J.R.; Lee, T. 20kHz CH_2O and OH PLIF with stereo PIV. *Opt. Lett.* **2018**, *5*, 1115–1118. [CrossRef] [PubMed]
33. Ma, X.; Wang, Z.; Jiang, C.; Jiang, Y.; Xu, H.; Wang, J. An optical study of in-cylinder CH_2O and OH chemiluminescence in flame-induced reaction front propagation using high speed imaging. *Fuel* **2014**, *134*, 603–610. [CrossRef]

34. Dally, B.B.; Karpetis, A.N.; Barlow, R.S. Structure of turbulent non-premixed jet flames in a diluted hot coflow. *Proc. Combust. Inst.* **2002**, *29*, 1147–1154. [CrossRef]
35. Wang, J.; Matsuno, F.; Okuyama, M.; Ogami, Y.; Kobayashi, H.; Huang, Z. Flame front characteristics of turbulent premixed flames diluted with CO_2 and H_2O at high pressure and high temperature. *Proc. Combust. Inst.* **2013**, *34*, 1429–1436. [CrossRef]
36. Yang, L.; Wang, Z.; Zhu, Y.; Li, Z.; Zhou, J.; Huang, Z.; Cen, K. Premixed jet flame characteristics of syngas using OH planar laser induced fluorescence. *Chin. Sci. Bull.* **2011**, *56*, 2862–2868. [CrossRef]
37. Weng, W.B.; Wang, Z.H.; He, Y.; Whiddon, R.; Zhou, Y.J.; Li, Z.S.; Cen, K.F. Effect of N_2/CO_2 dilution on laminar burning velocity of H_2-CO-O_2 oxy-fuel premixed flame. *Int. J. Hydrog. Energy* **2015**, *40*, 1203–1211. [CrossRef]
38. Wang, Z.H.; Weng, W.B.; He, Y.; Li, Z.S.; Cen, K.F. Effect of H_2/CO ratio and N_2/CO_2 dilution rate on laminar burning velocity of syngas investigated by direct measurement and simulation. *Fuel* **2015**, *141*, 285–292. [CrossRef]
39. Roy, R.; Gupta, A.K. Flame structure and emission signature in distributed combustion. *Fuel* **2020**, *262*, 116460. [CrossRef]

energies

MDPI

Article

An Experimental Study on the Jet Pressure Performance of Organ–Helmholtz (O-H), Self-Excited Oscillating Nozzles

Xiaochuan Wang [1,2], Yueqin Li [1,2], Yi Hu [1,2,*], Xiaolong Ding [1,2,3], Meijing Xiang [1,2] and Deng Li [1,2]

1 Hubei Key Laboratory of Waterjet Theory and New Technology, Wuhan University, Wuhan 430072, China; xcw001@whu.edu.cn (X.W.); 2019202080047@whu.edu.cn (Y.L.); xlding@whu.edu.cn (X.D.); xxh830@whu.edu.cn (M.X.); 2008lee@whu.edu.cn (D.L.)

2 School of Power and Mechanical Engineering, Wuhan University, Wuhan 430072, China

3 Chinese Ship Research and Design Center, Wuhan 430064, China

* Correspondence: huxiaoyi@whu.edu.cn

Received: 19 November 2019; Accepted: 3 January 2020; Published: 12 January 2020

Abstract: To make a better application of the self-excited oscillation jet, a compound organ–Helmholtz oscillation nozzle (OH nozzle) was designed to generate better pulse effects in the present study. The effects of geometric parameters on pressure characteristics of self-excited oscillation jets were investigated experimentally. The geometric parameters of OH nozzles were determined based on the design principle of the organ-pipe and Helmholtz nozzles. Various types of OH nozzles were tested to obtain the relationship between oscillation pressure and geometric parameters. Experimental results showed that some structural parameters are sensitive to the pump pressure, while others are not. The optimum geometric optimum parameters were obtained based on experimental results. The peak pressure and pressure pulsation amplitude are closely related to the target distance. The peak pressure decreases sharply with target distance at first and decreases with a further increase of the target distance. The peak pressure becomes stable beyond a certain range of target distance. With the increase of the target distance, the pressure pulsation amplitude increases at first and then decreases. The optimum target distance was determined based on experimental results to achieve better pulsation performance and has a value of 20 mm in the present study.

Keywords: self-excited oscillation jet; organ–Helmholtz nozzle; pulse waterjet; pressure pulsation amplitude

1. Introduction

Waterjet technology has been applied in a wide range of industries, including cutting [1], cleaning [2], surface treatment [3], coal and gas exploration [4,5], etc. Hu et al. [1] used an ultra-high-pressure water jet for cutting rubber, and it was found that the water jet was capable of recycling rubber materials without damaging the internal organizational structures of materials. Kohan et al. [2] investigated a jetting system to move the self-elevating mobile jack-up units, which are employed for offshore exploration and development purposes. Soyama et al. [3] used the waterjet to increase peening intensity, and the improvement of the fatigue strength of stainless steel was observed. Lin et al. [4] proposed the cross-borehole hydraulic slotting technique, and the minimum period of gas drainage was reduced in this way.

Waterjets are categorized into pulsed waterjets [6], abrasive waterjets [7], cavitating waterjets [8], rotating waterjets [9], and so on. Among them, pulsed waterjets, which can improve cutting efficiency because of the water-hammer effect, have attracted the interest of researchers [10]. Mechanical methods, including wobbling, rotating, and reciprocating, are the ordinary ways of producing pulsed waterjets. Though these methods have been successfully applied, the equipment for mechanical methods is complex and the reliability is not satisfactory [11–13]. In order to get more powerful waterjets with simpler devices, self-excited oscillation waterjets are becoming the hotspot in the field of waterjets. Different from mechanical methods, self-excited methods produce effective pulsed waterjets by specially designed nozzles, which turn continuous jets into discontinuous pulse jets without any moving parts [14,15]. Ding et al. [16] utilized a self-excited jet in surface peening and achieved better performance. Recently, Huang et al. [17] conducted experiments of the self-excited oscillating jets, and observed both a higher peak pressure and a higher erosion rate, indicating its high applicability for oil-gas exploration and development. Li utilized self-excited jets in petroleum engineering, and improvements in penetration rates were observed.

The nozzles are the key devices for generating self-excited oscillation jets, and their geometric parameters are closely related to the jet's performance. At present, organ-pipe nozzles and Helmholtz nozzles are the most common and efficient nozzles used to produce self-excited oscillation waterjets. The structural diagram and working principle of the organ-pipe nozzle are shown in Figure 1. The resonance chamber of the organ-pipe nozzle mainly consists of two co-axial cylindrical cavities. Fluid enters from the upper chamber and flows out through the lower chamber. Johnson et al. [18,19] found that organ-pipe nozzles could improve rock cutting capability and efficiency by nearly six times. It was demonstrated that self-excited oscillations would occur in the nozzle if the excitation frequency was near the fundamental frequency of the nozzle. Chahine et al. [20] found that the efficiency of organ-pipe nozzles was more than two orders of magnitude higher than that of conventional nozzles. Besides, he concluded that bubbles resulting from highly resonant organ-pipe nozzles were the main reason for improving the erosion capability [21]. Li et al. [22,23] studied the effects of area discontinuity at nozzle inlet on the oscillation characteristics of organ-pipe nozzles. And the experiment results showed that area enlargement and contraction have a positive effect on the pressure fluctuation characteristics of the jets.

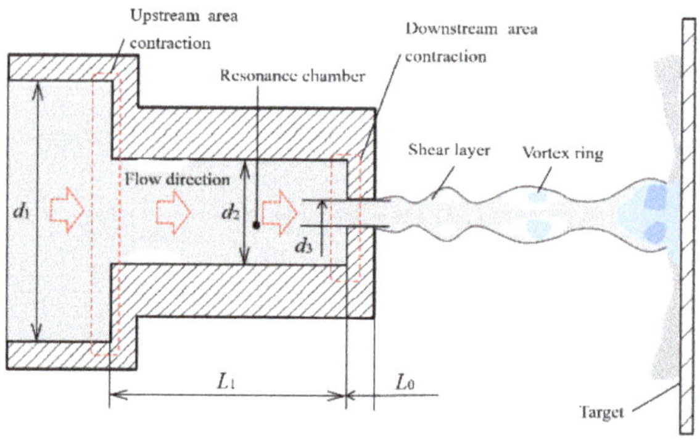

Figure 1. Structure diagram of an organ nozzle.

Helmholtz nozzles are also used to generate self-excited oscillation jets. The basic structure and working principle of the Helmholtz nozzle are shown in Figure 2. The resonance chamber is the structure where jets oscillation is generated and develops, and the cavity's shape is cylindrical. Compared to organ-pipe nozzles, Helmholtz nozzles have better jet oscillation performance, especially in the air, but they do not function well under high pump pressure [24]. Wilson et al. [25] first established a model of Helmholtz nozzles. He concluded that jet instability would form a large number of vortex rings, and the impact of the vortex rings was the main factor to generate pulsation in the nozzle. Crow et al. [26] found that the edge shear layer of jets showed an obvious trend of a structured vortex ring when the external excitation frequency matched with the natural frequency of the air jet. Rockwell et al. [27] studied the self-excited oscillation mechanism from four aspects: feedback, frequency, amplitude, and resonance. In the experiment introduced by Morel et al. [28], pressure fluctuations can exceed 5.6 times the value of jet dynamic pressure. This is owing to the resonance between the oscillating chamber of the Helmholtz nozzle and unstable jets, which amplifies the pressure pulsation. In recent years, experimental measures have been developed. Geveci et al. [29] used a three-dimensional particle video graph to measure the change rule of flow longitude in the pipeline cavitation system. They pointed out that the oscillation characteristics of moving fluid passing through the cavity in the pipeline were related to the instability of the fluid boundary layer and the fixed vibration characteristics of the cavity. This further confirms the important role of the nozzle oscillating chamber structure for the stroke pulse jet. Kolsek et al. [30] studied the influence of different structure parameters on the characteristics of the self-excited oscillation pulse waterjet, and found vortex structure and its pattern of change over time in the Helmholtz oscillation nozzles. Based on the thickness of the approach boundary layer, Ma et al. [31] proposed a model for predicting pressure fluctuations. Hu et al. [32] proposed a new methodology of generating pulsed air-water jets by entraining and mixing air into the cavity of the Helmholtz oscillator, and found that there was an optimum cavity length corresponding to the jet dynamic pressure.

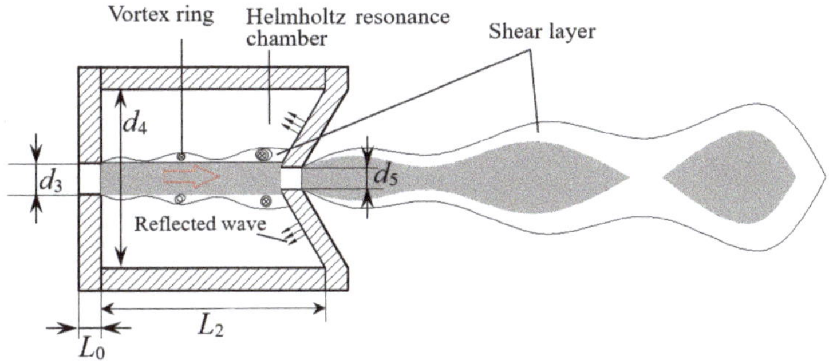

Figure 2. Structure diagram of a Helmholtz nozzle.

From the middle of the last century, a considerable amount of research has been carried out on self-excited oscillation waterjets, especially on organ-pipe nozzles and Helmholtz nozzles. However, it is hard to significantly improve certain nozzle pulsed performances based on the key parameter designing. At an international conference in the United States in 2007, Hlavac proposed a nozzle composed of an organ pipe and a Helmholtz chamber [33,34]. Double-chamber strengthened nozzles are studied in this article. It is an effective idea to use two chambers to generate self-excited oscillation pulse waterjets, as reasonable cooperation of both resonance chambers stimulates fluid resonance more obviously. To the best of our knowledge, there is, so far, little literature on this kind of composite nozzle, and the flow field structure and pressure pulsation rule inside the nozzle are rarely involved. In this study, the OH composite nozzle (organ–Helmholtz composite nozzles) experiments were carried out; the axial peak pressure and pressure oscillation pulsation amplitude of jet flow were taken as the parameters to evaluate the jet performance. This paper serves as a supplement to the previous research on waterjets and the beginning of the study into organ–Helmholtz composite nozzles.

2. Structural Principle of Self-Excited Oscillation Nozzle

The relationships between the structural parameters of different types of nozzles and their natural frequencies were analyzed based on the theory of fluid network model.

2.1. Structural Principle of Organ-Pipe Nozzles

The natural frequency is determined by geometry and material of the resonance chamber,

$$f = K_N \frac{c}{L_1}, \tag{1}$$

where c is the wave velocity, which is related to material of nozzle. L_1 is the length of oscillation cavity. K_N is the coefficient, which is related to the diameter of organ-pipe oscillation cavity.

$$K_N = F\left\{N, \left(\frac{d_1}{d_2}\right)^2, \left(\frac{d_2}{d_3}\right)^2\right\} = \left\{ \begin{array}{ll} \frac{2N-1}{4}, & if\left[\left(\frac{d_1}{d_2}\right)^2 \gg 1\ and \left(\frac{d_2}{d_3}\right)^2 \gg 1\right] \\ \frac{N}{2}, & if\left[\left(\frac{d_1}{d_2}\right)^2 \gg 1\ and \left(\frac{d_2}{d_3}\right)^2 \approx 1\right] \end{array} \right. . \tag{2}$$

According to the principle of hydroacoustics, the resonant standing wave is close to that of the fluid critical self-excited structure, which is determined by the critical Strouhal number of the nozzle. Their relationship is as follows:

$$S_t^* = f^* \cdot \frac{d_3}{v}, \tag{3}$$

where S_t^* is the critical Strouhal number, f^* is the critical oscillation frequency, d_3 is diameter of nozzle outlet, and v is the jet velocity. The jet velocity v can be expressed by the Mach number,

$$Ma = \frac{v}{c}. \tag{4}$$

From the above two equations, we can obtain

$$f^* = S_t^* \cdot \frac{Ma}{d_3} c. \tag{5}$$

When the resonant standing wave frequency is equal to or close to the nozzle natural frequency, a self-oscillating jet is formed. According to previous research [23,31], when the critical Strouhal number is a multiple of 0.3, a strong oscillation phenomenon can occur in the cavity. Combining Equations (3) and (5), the relationship between the length of the oscillating chamber and the diameter of the outlet can be obtained [18,35].

$$\frac{L_1}{d_3} = \frac{K_N}{MaS_t^*}. \tag{6}$$

In the self-excited oscillation case, S_t^* can be expressed with the use of S_L^*. According to the theory of Morel [29], S_L^* is given as follows.

$$S_t^* = \frac{S_L^* d_4}{L_2}; \tag{7}$$

$$S_L^* = (N - 0.25 - L_1 f)\frac{u_c}{u}. \tag{8}$$

The oscillation frequency is calculated through:

$$f = \frac{c}{2\pi}\sqrt{\frac{\pi d_3}{4V}}. \tag{9}$$

Substituting the length L_1 of the cavity and the diameter d_2 of the cavity into the Equation (8):

$$f = \frac{c}{2\pi d_2}\sqrt{\frac{d_3}{L_1}}. \tag{10}$$

The Strouhal number can be substituted into the above equation.

$$\frac{d_2}{d_3} = \frac{1}{2\pi S_t Ma}\sqrt{\frac{d_3}{L_1}}. \tag{11}$$

According to the experiment, when the Strouhal number is 0.6, the cavity length to diameter ratio is 0.8, so the oscillation effect is obvious.

2.2. Structural Principle of Helmholtz Nozzles

When S_t is a critical value, the jet ejected from the nozzle becomes a discontinuous vortex circulation. The natural frequency relationship of the Helmholtz resonator is calculated as follows:

$$f = \frac{c}{2\pi d_4}\sqrt{\frac{d_3}{L_2}}. \tag{12}$$

Equation (11) can be expressed using S_t and Ma as follows:

$$\frac{d_4}{d_3} = \frac{1}{2\pi S_t Ma}\sqrt{\frac{d_3}{L_2}}. \tag{13}$$

The above equation is the relationship showing that the Helmholtz cavity diameter (d_4) should be satisfied. When the vortex from d_1 reaches the second injection hole d_2, the pressure signal returns to d_1 after the time $t_L = L_2/c$. If the cavity length L_2 makes the time for the pressure signal to reach d_1 exactly equal to the time required to generate a new vortex, the vibration in the cavity is amplified to produce a strong resonance at the time. So the cavity length L_2 should satisfy the following equation:

$$L_2 = n\lambda - t_L u_c, \tag{14}$$

where n is the number of vortexes and u_c is the vortex convective velocity. Introducing S_t and Ma, then, making $\alpha = u_c/u_1$, it can be derived from the relationship that the cavity length L_2 should satisfy

$$\frac{L_2}{d_3} = \frac{n\alpha}{S_t(1+\alpha\, Ma)}. \tag{15}$$

If the fluid in the organ-pipe oscillation chamber and the Helmholtz oscillation chamber resonates, their natural frequencies must be equal or satisfy the multiple relationship; the following relationship:

$$f_O = N f_H, \tag{16}$$

where f_O is the natural frequency of organ nozzle, f_H is the natural frequency of Helmholtz nozzle, and N is a positive integer.

$$K_N\frac{c}{L_1} = \frac{Nc}{2\pi d_4}\sqrt{\frac{d_3}{L_2}}, \tag{17}$$

where a is the jet velocity, c is the disturbance wave velocity, and K_N is the coefficient which is related to the diameter of pipeline oscillating chamber.

The above Equations (6), (11), (13), (15), and (17) can be applied as a reference for the design of the OH self-excited nozzle.

3. Experimental Setup and Procedures

3.1. Structural Principle of OH Self-Excited Nozzle

The structure diagram of the OH self-excited oscillation nozzle is shown in Figure 3. It consists of two main sections. One is the upstream organ tunnel oscillating chamber and the other is the downstream Helmholtz oscillating chamber. If the two nozzles are compounded by a certain principle, jet pulsation generated in the oscillating chamber of the organ tube is amplified after entering the Helmholtz oscillation chamber. That means that the slight disturbance is amplified twice. Therefore, the nozzle with this structure can produce a pulsation that is more intense than the organ tube nozzle and the Helmholtz nozzle at the same pump pressure.

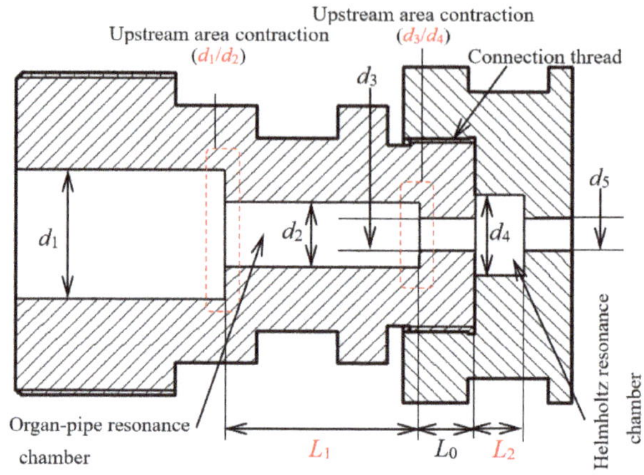

Figure 3. The structure diagram of organ–Helmholtz (OH) self-excited oscillation nozzle.

The key parameters determining the jet performance in the composite nozzle are as follows: d_1/d_2 is called the upper cross-sectional diameter shrink ratio. d_2/d_3 is called a lower cross-sectional diameter shrink ratio. The above two parameters have a great relationship with the performance of organ-pipe resonance chamber. Besides, d_3/d_5 is the ratio of inlet and outlet diameter of the Helmholtz resonance chamber. L_2 and d_4 are the length and diameter of the Helmholtz oscillation chamber, respectively. These three parameters have an important influence on the performance of the Helmholtz resonance chamber. Therefore, designing an OH type compound oscillation nozzle is to continuously adjust and optimize these five parameters so that a better jet oscillation effect can be obtained. In order to process these nozzles more conveniently, the nozzle is machined in two parts during processing. The middle part is threaded and two different structural parts can be freely combined.

Referring to the theoretical analysis and experimental conditions [23], a basic OH nozzle form was designed, and its parameters are shown in Table 1. The nozzles used in the experiments were all obtained by changing the parameters on this nozzle. Except for the altered parameters, the other parameters were the same as those of the basic nozzles. These altered parameters were as follows. d_1/d_2 values were set to 2.2, 2.4, 2.6, 2.8, and 3. Then, d_2/d_3 values were set to 1.5, 2, 2.5, 3, and 3.5, respectively. The values of L_2 were set to 2, 4, 6, 8, and 10 mm, respectively, and values of the cavity diameter d_4 were 6, 8, 10, 12, and 14 mm, respectively. The nozzle inlet and outlet diameter ratios (d_3/d_5) were 0.5, 0.75, 1, 1.25, and 1.5 respectively. Therefore, a total of ten organ nozzles and fifteen Helmholtz nozzles were processed. Their physical diagrams are shown in Figure 4.

Table 1. Parameters of the basic nozzle.

d_1	d_2	d_3	d_4	d_5	L_1	L_2
13 mm	6 mm	2 mm	10 mm	2 mm	21 mm	6 mm

Figure 4. The physical diagram of OH self-excited oscillation nozzle.

3.2. Facilities and Setup

This experiment was carried out on a multi-functional high-pressure water jet test bench. The test bench is composed of four parts, a high-pressure piston pump set, a frequency conversion platform, a test bench, and a control platform. Each component performs unique functions. The experiment used a static pressure sensor to measure the nozzle inlet pressure. Besides, we independently designed a high-pressure water jet pressure test tank to accomplish the experiment. A target disk was mounted on the right side of the water tank, and a measuring hole of 1 mm in diameter was placed in the center of the target disk to measure the striking pressure of the jet center. The connection diagram of the experimental device is shown in Figure 5.

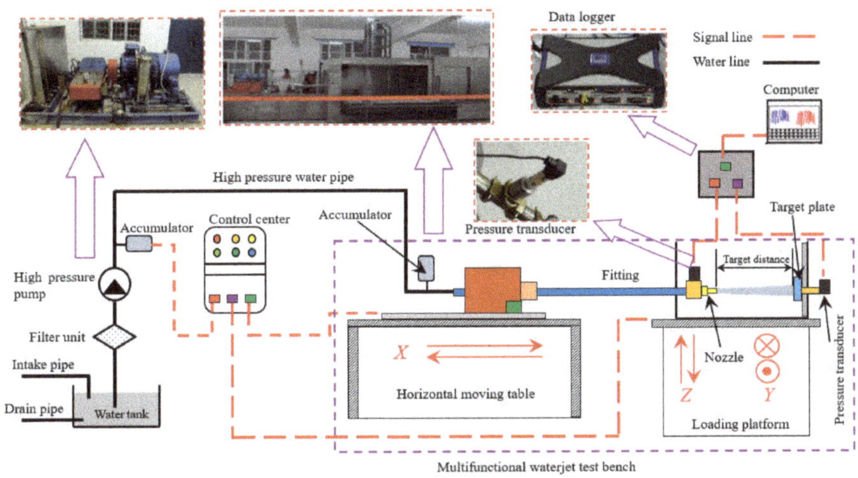

Figure 5. Connection diagram of the device.

The experiment platform works by relying on two key lines, a water flow line and a signal flow line. The water flow line refers to the water passage during the experiment. The water passes through the high-pressure piston pump set through the high-pressure hose and the flow stabilizer to enter the horizontal rigid pipe. Then, it passes through the joint and ejects from the self-excited nozzle. After that, it hits the target on the right side of the tank through the submerged or non-submerged environment. Signal lines refer to the paths through which signals are collected during the experiment. Pressure sensors (Model: JXBS-3001-P304) and the data acquisition instrument (Model: QuantumX MX840B) are the most important components, the latter of which has a sampling frequency of more than 10,000 Hz. When the water flows through the connector and hits the target disk, the upper pressure sensor transmits the pressure signal to the computer through the data acquisition instrument for subsequent analysis and processing.

The following is a brief introduction to the relevant parameters and the working performance of each part. The high-pressure pump's main function is to supply continuous and stable high pressure of water. It works with the frequency conversion platform, whose main function is to control the high-pressure piston pump's speed by changing the frequency of the three-phase current. Therefore, the high-pressure pump was able to achieve smooth pressure conversion and provide specific pressure for the experiment. The test bench was the main operating area for the experiment, which can achieve changes in the target distance both in the horizontal and vertical directions. The movement accuracy is up to 0.1 mm, while the stroke in each direction is greater than 1 m. The control center is the platform for operating the bench. It can realize many functions, such as lifting and lowering the stage, driving the horizontal servo motor, setting the moving speed, and positioning of each moving part, which is the key part of the human–computer interaction of the whole system.

Five pressure drops (10, 15, 20, 25, and 30 MPa) were used to test the performance of the OH type self-excited oscillation nozzle. The experimental pressure range was 10–30 MPa and provided good pressure conditions and regions for optimizing the structural parameters of the oscillation nozzle. At each test, pressure measurements were taken at the nozzle inlet.

The impact pressure of the oscillating nozzle varies at different target distances. The target distance is defined as the distance between the nozzle outlet and the measurement target disk. According to the guiding experiment and the experience of self-excited oscillating jet nozzle research, the target distance of this experiment was set in the range of 10–100 mm, and it was regarded as a group every 10 mm.

3.3. *Perturbation Elimination*

In order to ensure the accuracy of such an experiment, the external influence factors should be reduced during the experiment. For example, in the experiment, the high-pressure piston pump set would inevitably have pulsation of flow and pressure during the working process. This pulsation has a relatively large influence on the pulsation generated by the self-excited nozzle, so two accumulators were provided in the high-pressure line. One of the accumulators is mounted close to the pump set and the other is mounted close to the nozzle. It has been estimated that the installation of the accumulator can reduce the pulsation in the pipeline by 80%.

Water head loss occurs when water passes through a high-pressure hose. In order to reduce the impact caused by head loss, a pressure sensor was provided at the joint where the nozzle was installed to monitor the pressure change of the high-pressure water entering the nozzle in real-time. After the pressure is stabilized, a good experimental result can be obtained.

Before measuring the impact pressure, alignment of the jet axis with the center of the target was performed. In order to make the center of the oscillating nozzle and the center of the target plate overlap as much as possible, the directions of stage y and z needed to be adjusted. After the axis was aligned, the data measured by the pressure sensor were regarded as the striking pressure at the center of the jet. In the experiment, the accuracy of the sensors was ≤0.25% FS (full scale), which meets the experimental requirements.

4. Results and Discussion

The pressure characteristics of the water jet determine its destructiveness and cutting effect. In this experiment, the pressure pulsation peak and amplitude of the OH self-excited oscillation nozzle were measured under different pump pressures. So the effects of different structural parameters on the pressure performance were analyzed.

Figure 6 shows the pressure curve measured by the sensor over a period of time. The experiment was carried out under these conditions: the target distance was 10 mm, pump pressure was 20 MPa, and the nozzle parameter was d_1/d_2 = 2.6. It can be seen that the pressure measured by the sensor changes over time. This indicates that the nozzle created a self-excited oscillation pulsed jet. It can be seen that the peak pressure is not constant during different periods of pressure change. In order to process data efficiently and accurately, the data is divided into series parts by time. Peak pressure is the maximum pressure during each period of time, while trough pressure is the minimum one. The pressure oscillation amplitude is defined as the difference between peak pressure and trough pressure measured at the jet axis in the same cycle during each period time. Then, the average value of the peak pressures for all stable periods of time is taken as the peak pressure of the nozzle. So does the pressure pulsation amplitude. Peak pressure and pressure oscillation amplitude can directly indicate the severity of jet pulsation. Therefore, this article uses these two indicators to evaluate nozzle performance.

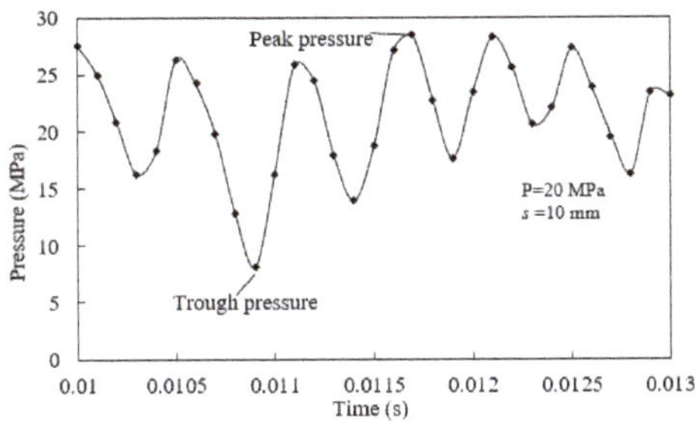

Figure 6. Pressure fluctuations during a period of time.

The performance of the OH nozzle was compared with those of the organ-pipe nozzle and Helmholtz nozzle, and the geometric parameters of the OH nozzle are listed in Table 1. Figure 7 shows the comparison of the peak pressure and the pressure oscillation amplitude of the OH nozzle and the organ nozzle at different target distances under the condition of 30 MPa pressure. Obviously, the OH nozzle performs better than the organ nozzle or Helmholtz nozzle which makes it up. It can be seen that the peak pressure of the optimal OH nozzle would be about 8% higher than that of the organ nozzle or Helmholtz nozzle in the same pump pressure. The pressure pulsation amplitude is about 26% higher. This shows that the OH nozzle does achieve better self-oscillating jet performance.

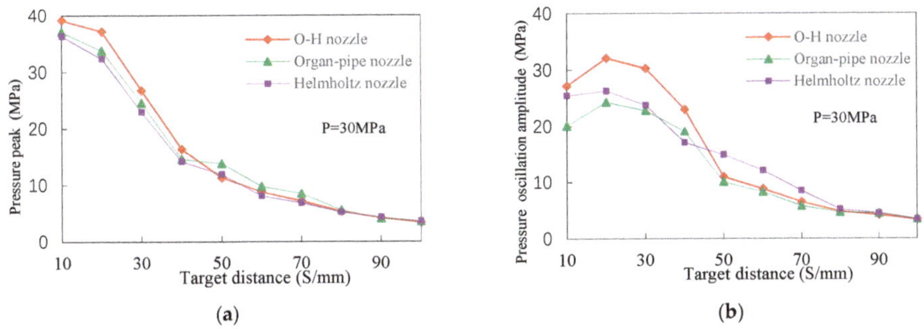

Figure 7. Performance comparison of three nozzles: (**a**). Comparison of pressure peaks at different axis target distances. (**b**). Comparison of pressure oscillation amplitude at different axis target distances.

4.1. *The Effect of d_1/d_2 on Pressure Characteristics*

Figure 8 shows the relationship between the peak pressure and pulsation amplitude on the jet axes of different nozzles as a function of the target distance. It can be seen that peak pressure decreases with increasing jet target distance. The overall trend of decreasing jet peak pressures is accordant, but the decreasing rate and peak value are different for different nozzles. When the pump pressure is 20 MPa, the nozzle with a d_1/d_2 of 2.6 has a peak pressure of 23.3 MPa at a 10 mm target distance.

As the jet target distance increases, the peak pressure value of the pulsation decreases rapidly, while the downward trend slows down and approaches the level with the target distance exceeding 50 mm. Other nozzle peak pressures in the same group have the same trend, but the pressure pulsation peaks are smaller than this.

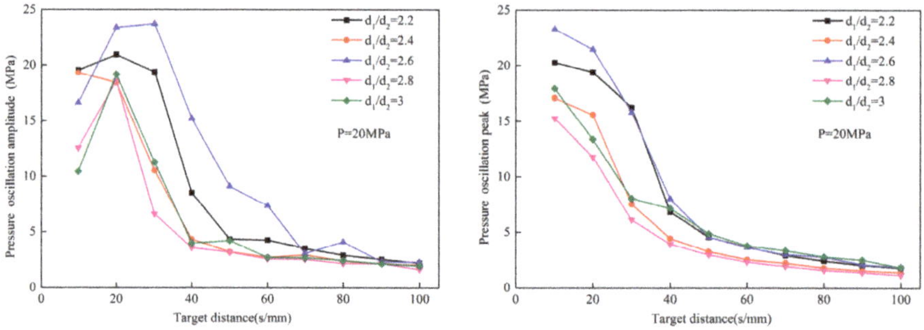

(a) (b)

Figure 8. Effect of d_1/d_2 on nozzle performance at 20 MPa pressure. (**a**). curves of nozzles with different d_1/d_2 on pressure pulsation amplitude at different target distances. (**b**). curves of nozzles with different d_1/d_2 on pressure oscillation peak at different target distances.

It has a complicated relationship between the pressure pulsation amplitude and the jet target distance. When the pressure changes, the change law of peak pressure and pulsation amplitude at different target distances are the same as when the pressure is 20 MPa. The maximal peak pressures of all the nozzles were measured at a 10 mm target distance, which is the smallest target distance measured in the experiment. However, the maximum pressure pulsation amplitudes were measured at a target distance of 20 mm.

The reason is that when the target distance is small, the time when the jet is ejected from the nozzle outlet to the contact with the sensor is too short. It will cause the pressure pulsation in the nozzle to not sufficiently develop. However, peak pressure will also be rapidly reduced when the target distance is too large. The jet is fully developed at the proper target distance, so the pressure amplitude is the largest at a certain target distance.

In fact, the variation law of the peak pressure and pulsation amplitude of different nozzles with the target distance in different pump pressures is similar in this experiment. Generally, the peak pressure is maximal at s = 10 mm and the pressure pulsation amplitude is maximal at s = 20 mm. The effect of the target distance will not be discussed later.

The maximal jet peak pressure and amplitude for different pump pressures are shown in Figure 9. They were measured at the target distance of 20 mm. It can be seen from the two pictures that the best oscillation nozzle structure is d_1/d_2 = 2.6. The reason is that d_1/d_2 of the OH nozzle determines the fluid velocity and reflected wave frequency of the jet entering the oscillating chamber of the organ pipe. Only when the reflected wave frequency is the same or close to the natural frequency of the structure, will the jet produce significant resonance and oscillation effects.

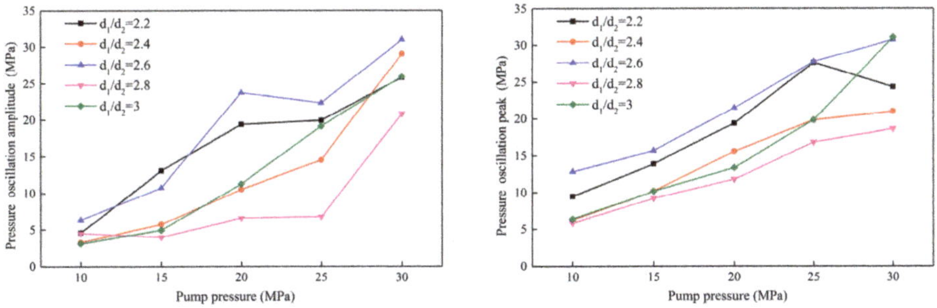

(a) (b)

Figure 9. Performance of different d_1/d_2 nozzles under each pump pressure. (**a**). maximal jet pressure pulsation amplitudes of different d_1/d_2 nozzles. (**b**). maximal jet peak pressures of different d_1/d_2 nozzles at each pump pressure.

4.2. The Effects of d_2/d_3 on Pressure Characteristics

Figure 10 shows the peak pressures and pulsation amplitudes of different d_2/d_3 nozzles at the different target distances. Obviously, the change law here is similar to that discussed previously. It can be seen that the peak pressure is larger when d_2/d_3 is small. When d_2/d_3 increases, the performance of the nozzle will gradually deteriorate. Judging from the peak pressure, the performance of the nozzles is similar when d_2/d_3 is small. However, the variation law of pressure pulsation amplitude is different from that of peak pressure. When d_2/d_3 increases, pressure pulsation amplitude increases first and then decreases. Therefore, in comparison, $d_2/d_3 = 2.5$ is the best nozzle structure. Based on Figure 11, the conclusion that d_2/d_3 has a more pronounced effect on jet pressure pulsation can be drawn. When d_2/d_3 is small, the pressure pulsation amplitude is significantly smaller. The main reason for this phenomenon is that d_2/d_3 will affect the space reflected by the jet after it enters the resonance chamber of the organ pipe and contacts the lower collision wall. This means that inside the nozzle, d_2/d_3 has a major influence on the jet velocity.

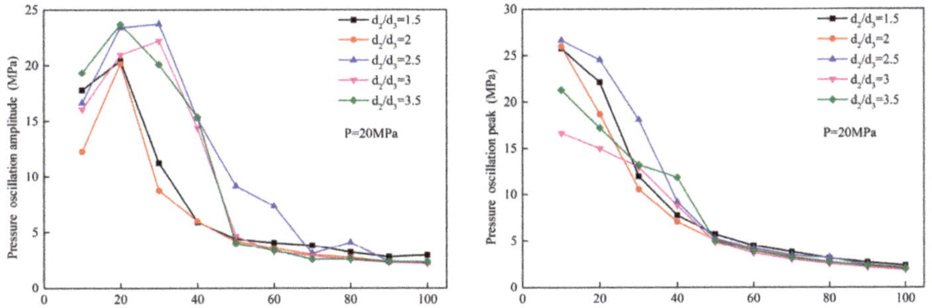

(a) (b)

Figure 10. Effect of d_2/d_3 on nozzle performance at 20 MPa pressure. (**a**). curves of nozzles with different d_2/d_3 on pressure pulsation amplitude at different target distances. (**b**). curves of nozzles with different d_2/d_3 on pressure oscillation peak at different target distances.

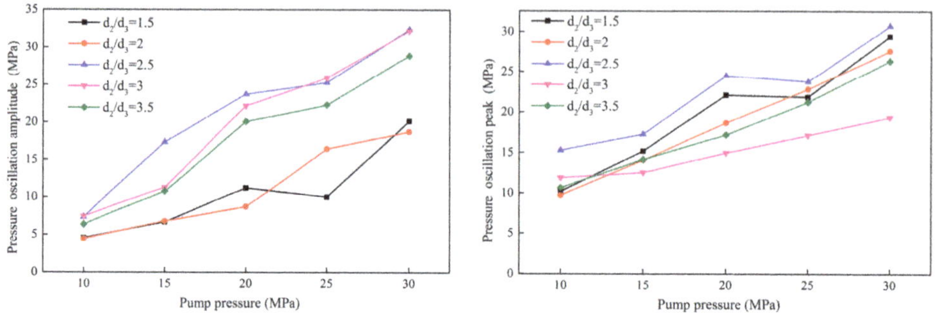

(a) (b)

Figure 11. Performance of different d_2/d_3 nozzles under each pump pressure. (**a**). maximal jet pressure pulsation amplitudes of different d_2/d_3 nozzles. (**b**). maximal jet peak pressures of different d_2/d_3 nozzles at each pump pressure.

4.3. The Effect of L_2 on Pressure Characteristics

Figure 12 shows the peak pressures and pulsation amplitudes of different L_2 nozzles at the different target distances. Their change law stays the same as the previous one. According to Figure 13, nozzle performance first becomes better as L_2 increases, and then it deteriorates. It is obvious that L_2 = 4 mm is the best parameter. Meanwhile, according to the curves, L_2 = 6 mm nozzle also has relatively good performance when pump pressure is low. In fact, the best L_2 will be slightly affected by pump pressure.

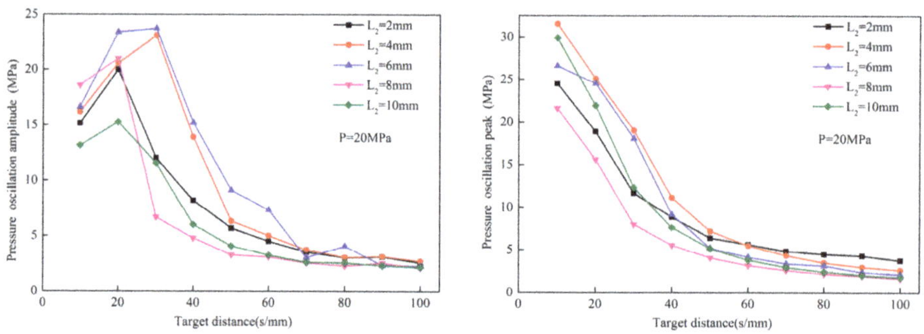

(a) (b)

Figure 12. Effect of L_2 on nozzle performance at 20 MPa pressure. (**a**). curves of nozzles with different L_2 on pressure pulsation amplitude at different target distances. (**b**). curves of nozzles with different L_2 on pressure oscillation peak at different target distances.

Surprisingly, it can be seen that the pressure pulsation amplitude of L_2 = 6mm nozzle is slightly higher than that of 4 mm when the pump pressure is lower. But nozzle performance of L_2 = 4 mm is much better than that of the nozzle of L_2 = 6 mm with the pump pressure increasing. In other words, the comprehensive performance of L_2 = 4 mm nozzle is still the best.

This is because the jet enters into the surrounding fluid in the chamber in the Helmholtz oscillation cavity, and a tiny vortex is continuously generated at the boundary layer. The jet exchanges energy with the environmental fluid, and the degree of turbulence in the cavity increases with the cavity length. At the same time, the feedback generated from the downstream collision wall also exacerbates the attenuation of the axial velocity. When the cavity length is long enough, the ordered vortex ring formed in the jet shear layer has sufficient space to develop. Therefore, the feedback from the downstream collision wall back to the upstream can also be well amplified, which plays an important role in modulating the jet. However, as the cavity length continues to increase, the fluid of the axial center exchanges energy with the surrounding environment, causing a large loss. When this loss is greater than the modulation, the peak pressure and amplitude will decay.

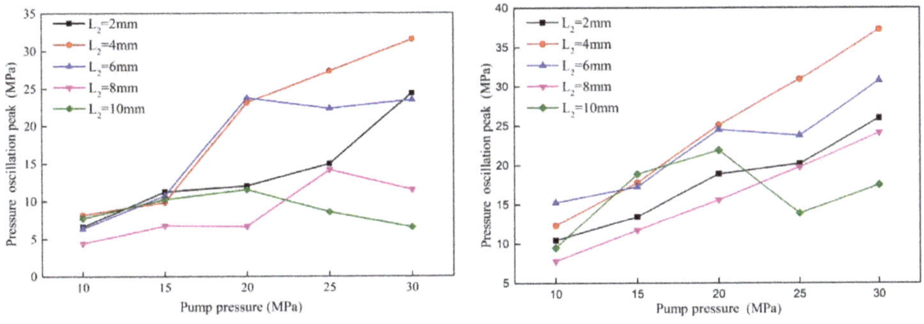

(a) **(b)**

Figure 13. Performance of different L_2 nozzles under each pump pressure. (**a**). maximal jet pressure pulsation amplitudes of different L_2 nozzles. (**b**). maximal jet peak pressures of different L_2 nozzles at each pump pressure.

It can be seen from this that flow in the Helmholtz resonance chamber is more complicated than that in the organ pipe resonance chamber. Therefore, the influence of inlet pressure cannot be ignored when designing the Helmholtz oscillation cavity structure.

4.4. The Effect of d_4 on Pressure Characteristics

Figure 14 shows the peak pressures and pulsation amplitudes of different d_4 nozzles at the different target distances. Figure 15 shows the peak pressures and pulsation amplitudes of different d_4 nozzles at different pressures. From the above peak pressure curves and pressure pulsation curves, it can be seen that the nozzle performance of d_4 = 12 mm is better than that of d_4 = 10mm when the pressure is low. However, with the pump pressure increasing, nozzle with d_4 =10mm has a better performance. This indicates that in the OH type self-excited nozzle, the optimal d_4 structure of the nozzle is extremely sensitive to changes in pressure.

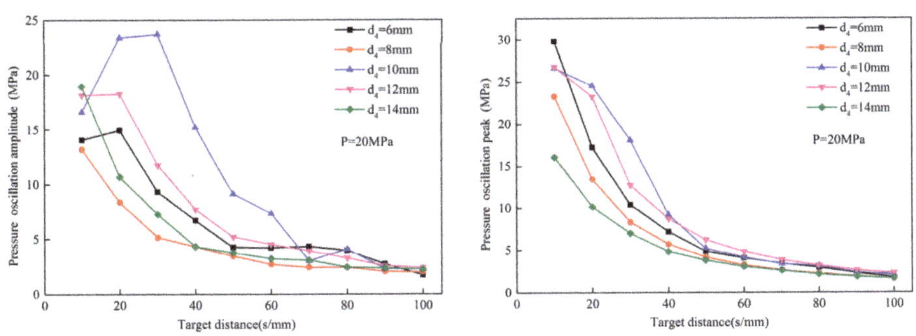

(**a**) (**b**)

Figure 14. Effect of d_4 on nozzle performance at 20 MPa pressure. (**a**). curves of nozzles with different d_4 on pressure pulsation amplitude at different target distances. (**b**). curves of nozzles with different d_4 on pressure oscillation peak at different target distances.

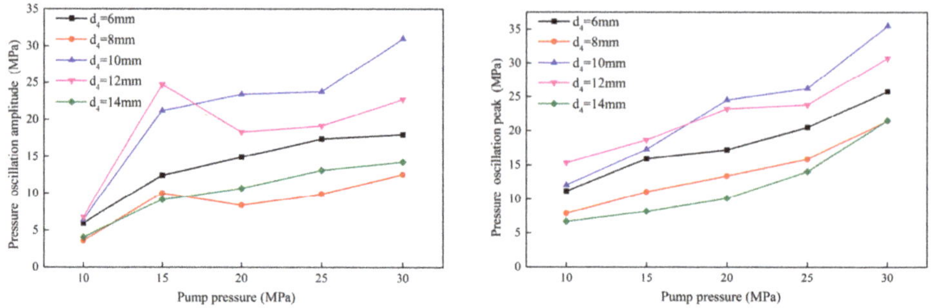

(**a**) (**b**)

Figure 15. Performance of different d_4 nozzles under each pump pressure. (**a**). maximal jet pressure pulsation amplitudes of different d_4 nozzles. (**b**). maximal jet peak pressures of different d_4 nozzles at each pump pressure.

For the Helmholtz nozzle, the vortex ring modulation coupling of the downstream collision wall feedback is mainly affected by d_4. If the cavity diameter is too small, the jet will form a reflux vortex when it engulfs the surrounding fluid in the chamber and cannot be fully developed. This causes the chamber wall to be subjected to limited compression against the backflow of the chamber, resulting in increased energy consumption. If the cavity diameter is too large, the vortex is fully developed and the return flow is away from the jet axis, and the axial jet cannot be effectively modulated and coupled. So d_4 = 12 mm nozzles have better performance when the pressure is lower. When the pressure is 20 MPa or more, the pulsation amplitude of the pressure produces a maximum at d_4 = 10 mm.

4.5. The Effect of d_3/d_5 on Pressure Characteristics

Figure 16 shows the effects of d_3/d_5 on the peak pressure and pressure pulsation amplitude of the OH self-excited pulse nozzle at the different target distances. Figure 17 shows the effect of d_3/d_5 on the peak pressure and pressure pulsation amplitude of the OH self-excited pulse nozzle at different pressures. As can be seen from the two pictures, the jet peak pressure increases as the nozzle diameter increases at first, and then decreases. When the nozzle diameter ratio value is 1.0, the jet pulsation peak is the largest. The pressure pulsation amplitude of the jet shows a more sensitive change. When d_3/d_5 value is 0.5, the pressure pulsation amplitude is almost 5 MPa or less and keeps steady with the pump pressure's increases. It indicates that the fluid does not oscillate when passing through the nozzle at such a time. As d_3/d_5 increases, the oscillation effect generated by the nozzle increases, reaching a maximum value when d_3/d_5 is 1.0.

When d_3/d_5 is small, only the innermost layer of the jet core zone can be ejected outward through the nozzle outlet. Most of the jets will only reciprocate continuously in the oscillating chamber, so it is almost impossible to form an oscillating jet. When d_3/d_5 is too large, the fluid in contact with the collision wall at the periphery of the jet is small, and the oscillation effect generated when propagating upward is insufficient to modulate the core region of the jet. Therefore, no obvious oscillating jets will form in this case.

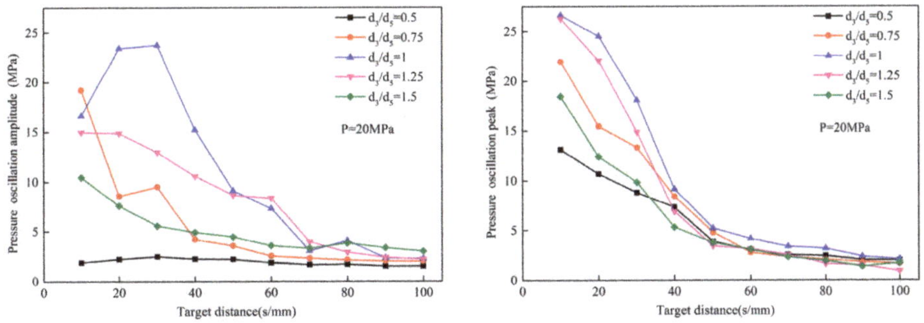

(**a**) (**b**)

Figure 16. Effect of d_3/d_5 on nozzle performance at 20 MPa pressure. (**a**). curves of nozzles with different d_3/d_5 on pressure pulsation amplitude at different target distances. (**b**). curves of nozzles with different d_3/d_5 on pressure oscillation peak at different target distances.

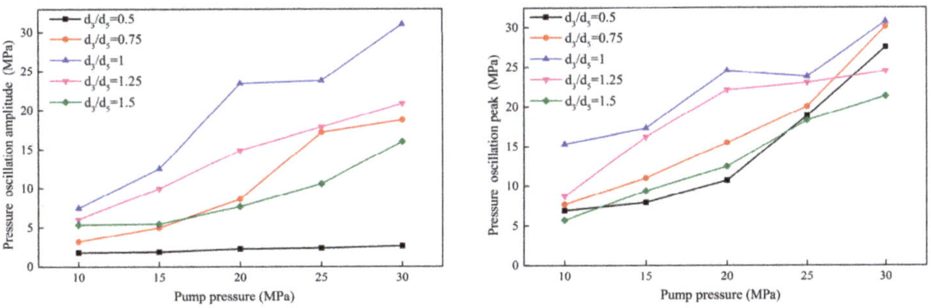

(**a**) (**b**)

Figure 17. Performance of different d_3/d_5 nozzles under each pump pressure. (**a**). maximal jet pressure pulsation amplitudes of different d_3/d_5 nozzles. (**b**). maximal jet peak pressures of different d_3/d_5 nozzles at each pump pressure.

5. Conclusions

OH nozzles (organ–Helmholtz nozzles) are novel self-oscillation nozzles. They do not just connect an organ-pipe nozzle and a Helmholtz nozzle together. They combine the two kinds of ordinary nozzles, and the synergistic effect of OH nozzles makes pressure fluctuations of self-excited jets more obvious. In this paper, the pressure oscillation characteristics of OH nozzles were investigated, and the conclusions were obtained as follows:

(1) Both the jet peak pressure and the pressure pulsation amplitude are positively correlated with the pump pressure. The jet peak pressure is inversely related to the target distance. As the target distance increases, the peak pressure first decreases sharply, and then decreases slowly. As the target distance increases, the pressure pulsation amplitude increases at first, and then decreases. Therefore, there is an optimum target distance, which is 20 mm in the experiment.

(2) Some of the optimal parameters of the OH nozzle are not sensitive to pressure. In this experiment, the performance of that nozzle regarding those parameters was the best regardless of the change in pressure. These parameters may be related to the initial perturbation vortex formation. The optimal parameters are d_1/d_2 = 2.6, d_2/d_3 = 2.5, and d_3/d_5 = 1.

(3) The optimum ranges of L_2 and d_4 are affected by pressure. They are closely related to the development of disturbance waves, and the oscillation pulsed jet can enhance and develop only in appropriate situations. L_2 is relatively lightly affected by pressure. L_2 = 6 mm has good performance

when the pump pressure is low. When the pressure is increased, L_2 = 4 mm performs better. Similarly, d_4 = 12 mm performs well when the pressure is low. When the pressure exceeds 20 MPa, d_4 = 10 mm performs better.

Author Contributions: X.W., Y.H., and X.D. conceived and designed the schemes; Y.L. and M.X. performed the experiments; Y.L. and D.L. analyzed the data; M.X. and Y.L. wrote the paper. All authors have read and agreed to the published version of the manuscript.

Funding: This research was funded by the National Natural Science Foundation of China (number 51804318 and number 51805188), and the National Key Research and Development Program of China (number 2018YFC0808401).

Conflicts of Interest: The authors declare no conflict of interest.

Nomenclature

f	natural frequency of organ-pipe nozzle (Hz)
c	wave velocity (m/s)
St^*	critical Strouhal number
d_3	diameter of organ-pipe nozzle outlet (mm)
Ma	Mach number
d_4	Helmholtz cavity diameter (mm)
V	volume of organ-pipe chamber (m^3)
d_1	diameter of organ-pipe nozzle inlet (mm)
St	Strouhal number
N	mode number
K_N	coefficient
L_1	length of oscillation cavity (mm)
f^*	critical oscillation frequency (Hz)
v	jet velocity (m/s)
S^*_L	Strouhal number based on the cavity length
L_2	cavity length (mm)
d_2	diameter of organ-pipe nozzle resonance chamber (mm)
d_5	diameter of Helmholtz nozzle outlet (mm)
n	number of vortex

References

1. Hu, Y.; Kang, Y.; Wang, X.C.; Li, X.H.; Long, X.P.; Zhai, G.Y.; Huang, M. Mechanism and experimental investigation of ultra high pressure water jet on rubber cutting. *Int. J. Precis. Eng. Manuf.* **2014**, *15*, 1973–1978. [CrossRef]
2. Kohan, O.; Bienen, B.; Gaudin, C.; Cassidy, M.J. The effect of water jetting onspudcan extraction from deep embedment in soft clay. *Ocean Eng.* **2015**, *97*, 90–99. [CrossRef]
3. Soyama, H. Improvement of fatigue strength by using cavitating jets in air and water. *J. Mater. Sci.* **2007**, *42*, 6638–6641. [CrossRef]
4. Lin, B.Q.; Yan, F.Z.; Zhu, C.J.; Zhou, Y.; Zou, Q.L.; Guo, C.; Liu, T. Cross-borehole hydraulic slotting technique for preventing and controlling coal and gas outbursts during coal roadway excavation. *J. Nat. Gas Sci. Eng.* **2015**, *26*, 518–525. [CrossRef]
5. Liu, W.; Kang, Y.; Wang, X.; Yuan, B.; Fang, Z. Forming mechanism and radius prediction of hydraulically generated slots based on a two-walled jet. *J. Nat. Gas Sci. Eng.* **2016**, *36*, 385–401. [CrossRef]
6. Hou, R.; Wang, T.; Lv, Z.; Tian, Y. Investigation of the pulsed waterjet flow field inside and outside of the nozzle excited by ultrasonic vibration. *Int. J. Adv. Manuf. Technol.* **2018**, *99*, 453–460. [CrossRef]
7. Alberdi, A.; Artaza, T.; Suarez, A.; Rivero, A.; Girot, F. An experimental study on abrasive waterjet cutting of CFRP/Ti6Al4V stacks for drilling operations. *Int. J. Adv. Manuf. Technol.* **2016**, *86*, 691–704. [CrossRef]
8. Liu, W.; Kang, Y.; Wang, X.; Liu, Q.; Fang, Z. Integrated CFD-aided theoretical demonstration of cavitation modulation in self-sustained oscillating jets. *Appl. Math. Model.* **2019**. [CrossRef]
9. Bo, Y.; Wang, R.; Zhou, W. Research on the law of rock breaking by rotating jet. *J. Rock Mech. Eng.* **2003**, *22*, 664.

10. Samson, T.; MacDonald, D.; Fernandez, R.; Jodoin, B. Effect of Pulsed Waterjet Surface Preparation on the Adhesion Strength of Cold Gas Dynamic Sprayed Aluminum Coatings. *J. Therm. Spray Technol.* **2015**, *24*, 984–993. [CrossRef]
11. Dehkhoda, S.; Bourne, N.K. Production of a high-velocity water slug using an impacting technique. *Rev. Sci. Instrum.* **2014**, *85*, 025109. [CrossRef] [PubMed]
12. Lehocka, D.; Klich, J.; Foldyna, J.; Hloch, S.; Krolczyk, J.B.; Carach, J.; Krolczyk, G.M. Copper alloys disintegration using pulsating water jet. *Measurement* **2016**, *82*, 375–383. [CrossRef]
13. Xiang, Q.J.; Xue, L.; Kim, K.Y.; Shi, Z.F. Multi-objective optimization of a flow straightener in a large capacity firefighting water cannon. *J. Hydrodyn.* **2019**, *31*, 137–144. [CrossRef]
14. Liao, Z.; Tang, C. Theoretical analysis of self-excited oscillating pulse jet nozzle. *J. Chongqing Univ. (Nat. Sci. Ed.)* **2002**, *02*, 24–27.
15. Ghadi, S.; Esmailpour, K.; Hosseinalipour, S.M.; Mujumdar, A. Experimental study of formation and development of coherent vortical structures in pulsed turbulent impinging jet. *Exp. Therm. Fluid Sci.* **2016**, *74*, 382–389. [CrossRef]
16. Ding, X.; Kang, Y.; Li, D.; Wang, X.; Zeng, D. Experimental investigation on surface quality processed by self-excited oscillation pulsed waterjet peening. *Materials* **2017**, *10*, 989. [CrossRef] [PubMed]
17. Huang, M.; Kang, Y.; Wang, X.; Hu, Y.; Cai, C.; Liu, Y.; Chen, H. Experimental investigation on the rock erosion characteristics of a self-excited oscillation pulsed supercritical CO2 jet. *Appl. Therm. Eng.* **2018**, *139*, 445–455. [CrossRef]
18. Johnson, V.E., Jr.; Lindenmuth, W.T.; Conn, A.F.; Frederick, G.S. *Feasibility Study of Tuned-Resonator, Pulsating Cavitating Water Jet for Deep-Hole Drilling*; SAND 81-7126; Sandia National Laboratories: Albuquerque, NM, USA; Hydronautics, Inc.: Laurel, MD, USA, 1981.
19. Johnson, J.V.E.; Chahine, G.L.; Lindenmuth, W.T.; Conn, A.F.; Frederick, G.S.; Giacchino, J.G.J. Cavitating and Structured Jets for Mechanical Bits to Increase Drilling Rate—Part I: Theory and Concepts. *J. Energy Resour. Technol.* **1984**, *106*, 282–288. [CrossRef]
20. Chahine, G.L.; Johnson, V.E.; Lindenmuth, W.T.; Frederick, G.S. The use of self-resonating cavitating water jets for underwater sound generation. *J. Acoust. Soc. Am.* **1985**, *77*, 113–126. [CrossRef]
21. Chahine, G.; Genoux, P.; Johnson, V.E., Jr.; Frederick, G.S. *Analytical and Experimental Study of the Acoustics and the Flow Field Characteristics of Cavitating Self-Resonating Water Jets*; Tracor Hydronautics, Inc.: Laurel, MD, USA, 1984.
22. Li, D.; Kang, Y.; Ding, X.L.; Wang, X.C.; Fang, Z.L. Effects of area discontinuity at nozzle inlet on the characteristics of high speed self-excited oscillation pulsed waterjets. *Exp. Therm. Fluid Sci.* **2016**, *79*, 254–265. [CrossRef]
23. Li, D.; Kang, Y.; Ding, X.L.; Liu, W.C. Experimental study on the effects of feeding pipe diameter on the cavitation erosion performance of self-resonating cavitating waterjet. *Exp. Therm. Fluid Sci.* **2017**, *82*, 314–325. [CrossRef]
24. Tuna, B.A.; Rockwell, D. Self-sustained oscillations of shallow flow past sequential cavities. *J. Fluid Mech.* **2014**, *758*, 655–685. [CrossRef]
25. Wilson, T.A.; Beavers, G.S.; DeCoster, M.A.; Holger, D.K.; Regenfuss, M.D. Experiments on the Fluid Mechanics of Whistling. *J. Acoust. Soc. Am.* **1971**, *50*, 366–372. [CrossRef]
26. Crow, S.C.; Champagne, F.H. Orderly structure in jet turbulence. *J. Fluid Mech.* **1971**, *48*, 547–591. [CrossRef]
27. Rockwell, D.; Naudascher, E. Self-Sustained Oscillations of Impinging Free Shear Layers. *Annu. Rev. Fluid Mech.* **1979**, *11*, 67–94. [CrossRef]
28. Morel, T. Experimental Study of a Jet-Driven Helmholtz Oscillator. *J. Fluids Eng.* **1979**, *101*, 383–390. [CrossRef]
29. Geveci, M.; Oshkai, P.; Rockwell, D.; Lin, J.C.; Pollack, M. Imaging of the self-excited oscillation of flow past a cavity during generation of a flow tone. *J. Fluids Struct.* **2003**, *18*, 665–694. [CrossRef]
30. Kolsek, T.; Jelic, N.; Duhovnik, J. Numerical study of flow asymmetry and self-sustained jet oscillations in geometrically symmetric cavities. *Appl. Math. Model.* **2007**, *31*, 2355–2373. [CrossRef]
31. Ma, R.L.; Slaboch, P.E.; Morris, S.C. Fluid mechanics of the flow-excited Helmholtz resonator. *J. Fluid Mech.* **2009**, *623*, 1–26. [CrossRef]
32. Hu, D.; Li, X.H.; Tang, C.L.; Kang, Y. Analytical and experimental investigations of the pulsed air-water jet. *J. Fluids Struct.* **2015**, *54*, 88–102. [CrossRef]

33. Hlaváč, L.M. System of chambers for activation of modulation or pulsation in water jets. In Proceedings of the 2007 American WJTA Conference and Expo, Houston, TX, USA, 19–21 August 2007.
34. Mgr. Zuzana Šimonovská, LL.M., advokát. Device to Create and Intensify Modulation of Liquid Flow Velocity. P. Česká Republika Patent 302595, 27 July 2011.
35. Li, G.; Shen, Z. Design principle of self-vibrating cavitation nozzle for organ pipe. *J. Pet. Univ. Nat. Sci. Ed.* **1992**, *5*, 25–29.

energies

MDPI

Article

Numerical Investigation of Very-Large-Scale Motions in a Turbulent Boundary Layer for Different Roughness

Hehe Ren [1,2,3], Shujin Laima [1,2,3,*] and Hui Li [1,2,3]

1 Key Lab of Smart Prevention and Mitigation for Civil Engineering Disasters of the Ministry of Industry and Information, Harbin Institute of Technology, Harbin 150090, China; renhehe@hit.edu.cn (H.R.); lihui@hit.edu.cn (H.L.)
2 Key Lab of Structures Dynamic Behavior and Control of the Ministry of Education, Harbin Institute of Technology, Harbin 150090, China
3 School of Civil Engineering, Harbin Institute of Technology, Harbin 150090, China
* Correspondence: laimashujin@hit.edu.cn

Received: 8 January 2020; Accepted: 26 January 2020; Published: 4 February 2020

Abstract: Wall-model large eddy simulations (WMLES) are conducted to investigate the spatial features of large-scale and very-large-scale motions (LSMs and VLSMs) in turbulent boundary flow in different surface roughnesses at a very high Reynolds number, O (10^6–10^7). The results of the simulation of nearly smooth cases display good agreement with field observations and experimental data, both dimensioned using inner and outer variables. Using pre-multiplied spectral analysis, the size of VLSMs can be reduced or even disappear with increasing roughness, which indirectly supports the concept that the bottom-up mechanism is one of the origins of VLSMs. With increases in height, the power of pre-multiplied spectra at both high and low wavenumber regions decreases, which is consistent with most observational and experimental results. Furthermore, we find that the change in the spectrum scaling law from −1 to −5/3 is a gradual process. Due to the limitations of the computational domain and coarse grid that were adopted, some VLSMs and small-scale turbulence are truncated. However, the size of LSMs is fully accounted for. From the perspective of the spatial correlation of the flow field, the structural characteristics of VLSMs under various surface roughnesses, including three-dimensional length scales and inclination angles, are obtained intuitively, and the conclusions are found to be in good agreement with the velocity spectra. Finally, the generation, development and extinction of three-dimensional VLSMs are analyzed by instantaneous flow and vorticity field, and it shows that the instantaneous flow field gives evidence of low-speed streamwise-elongated flow structures with negative streamwise velocity fluctuation component, and which are flanked on each side by similarly high-speed streamwise-elongated flow structures. Moreover, each of the low-speed streamwise-elongated flow structure lies beneath many vortices.

Keywords: WMLES; VLSMs; LSMs; turbulent boundary flow; roughness

1. Introduction

Large-scale and very-large-scale motions (LSMs and VLSMs) exist in pipe flow, channel flow, and turbulent boundary layer (TBL) flow. LSMs with lengths of $0.3\delta - 3\delta$ (where δ is the boundary layer thickness, pipe radius, or channel half-height) in the streamwise direction has been observed during experiments conducted in turbulent boundary layer flows [1,2], pipe flows [3,4], and channel flows [5,6]. LSMs are packets of hairpin vortices moving upward in the near-wall region. In addition, VLSMs, much longer meandering structures with streamwise lengths larger than 3δ in turbulent boundary layer flows [7–9], pipe flows [3,4,10,11], and channel flows [6] were also observed. Both LSMs and

VLSMs carry a considerable amount of the turbulent kinetic energy (TKE). In contrast with the case of the LSMs, the mechanism of the generation and development of VLSMs remains an open question.

One widespread proposed explanation is the bottom-up mechanism suggested by Kim and Adrian [3], which implies that VLSMs are formed from coherently aligned bottom hairpin packets. This mechanism is approved by Balakumar and Adrian [2] and Lee and Sung [8] for boundary layer flows, by Guala et al. [4] for pipe flows, and by Balakumar and Adrian [2] for channel flows. However, Hutchins and Marusic [7,12], Mathis et al. [13] and Chung and McKeon [14] observed that the interaction of LSMs and VLSMs was not only a mere superposition on near-wall fluctuations, but also that near-wall small-scale structures were subject to an inhibiting modulation effect from the LSMs and VLSMs in the log region. Therefore, Hutchins and Marusic [7] and Hunt and Morrison [15] proposed a new mechanism for the origin of VLSMs, the top-down mechanism, according to which VLSMs may originate in the upper turbulent boundary layer before moving downward and breaking into small-scale structures in the lower boundary layer as a result of shear forces. As we know that the top-down mechanism is at this point, only conjecture for some experimental observations. VLSMs cannot be distinguished at the top of TBL [2]. In addition, Wang and Zheng [9] found that the product of the wavenumber and the power spectra decreases with height on the right side of the maximum value in the pre-multiplied spectra, which can be explained using the bottom-up mechanism. While, on the left side of the peak, the product of the power spectra and the wavenumber increases with height, indicating that the evolution of the VLSMs may not be entirely attributed to the bottom-up mechanism, and other candidate mechanisms (the top-down mechanism) may, therefore, coexist with the bottom-up mechanism.

The simulation study of the characteristics of VLSMs in TBL flow on a rough surface has not yet been performed. A recent review of studies turbulent flows over rough walls was conducted by Jiménez [16]. Turbulence structures on rough-wall boundary layers and spectrum characteristics in rough-wall pipe flow were investigated by Volino et al. [17] and Rosenberg et al. [18], respectively. In particular, although this was not mentioned by Volino et al. [17], Figure 1 in this paper clearly shows the effect of the roughness wall on pre-multiplied spectra, and this specific phenomenon is consistent with results of this study.

In this study, different surface roughness is selected to indirectly study the origin and the mechanism of development of VLSMs. This is because roughness alters near-wall flow structures. Typically, a spacing of about 100 wall units spanwise and extend from the wall about the same distance for near-wall streaks on smooth boundary layers, as was documented by Kline et al. [19]. This size or larger roughness elements undoubtedly disrupt the streaks.

Studies using direct numerical simulation (DNS) have been recently conducted in an attempt to resolve the temporal evolution of spatial structures fully. Jiménez [20] summed up previous works on the subject before identifying the beginning and end of the k_x^{-1} region in pre-multiplied spectra. He further found the VLSMs in the log region to have a streamwise length of $10\delta - 20\delta$ in pipe flow, channel flow, and TBL flow. Del Alamo et al. [21] conducted a channel flow DNS up to friction Reynolds number $\mathrm{Re}_\tau = 1901$ and found evidence for the variant behavior of VLSMs and LSMs in pre-multiplied spectra. Lee and Sung [8] conducted DNS simulations of a TBL flow, with a momentum height Reynolds number of up to $\mathrm{Re}_\theta = 2560$. They showed that the development process of adjacent packet-type structures combines to form VLSMs, and they employed a modified feature-extraction algorithm to identify the properties of the VLSMs. Lee et al. [6] investigated the spatial features of LSMs and VLSMs in a turbulent channel flow with a DNS of $\mathrm{Re}_\tau = 930$, the results indicating that the streamwise length of the VLSMs linearly determines the number of outer LSMs and that the formation of VLSMs possibly comes from the alignment of the positive and negative streamwise-fluctuation structures.

Although DNS is an ideal approach to flow research at low Reynolds numbers, the cost of computation is prohibitively expensive as Reynolds numbers increase. Nowadays, a DNS investigation of these VLSMs at higher Reynolds numbers and with a larger streamwise length domain is not possible. To date, the DNSs of wall-bounded turbulence have been limited to $\mathrm{Re}_\tau = 5000$ [22]. Thus, large-eddy simulation (LES) is very useful for investigating turbulence phenomena in the TBL. This is because LES enables simulations with high Reynolds numbers, and it is computationally more efficient. The horizontally averaged and time-averaged vertical profiles of relevant flow parameters were studied in most LES studies, and hence, they use rather short sizes for the streamwise computational domain [23–26]. Fang and Porté-Agel [27] first used LES to study the VLSMs in the neutral TBL, followed by Salesky and Anderson [28]. As we know, the LES grid is typically constrained by the near-wall region, where turbulent and viscous momentum transport occurs on small scales, especially where the atmospheric surface is covered with roughness elements that could be more expensive to resolve with the grid. Piomelli and Balaras [29] and Smits and Marusic [22] indicated that for wall-bounded flows, the LES computational costs are estimated as $\mathrm{Re}_\tau^{9/5}$, only a small savings over DNS, with its $\mathrm{Re}_\tau^{9/4}$. If, instead, just empirically model the flow that at the near-wall region, the grid requirements would be greatly reduced, which would also significantly reduce computational costs, currently reaching about $\mathrm{Re}_\tau^{1/5}$. Most important is that wall-model LES (WMLES) tends to be more accurate at high Reynolds numbers than at moderate or low ones [29]. The reason for this is that the WMLES approach depends only on large eddies. Consequently, it can effectively simulate high Reynolds-number flows. Chung and McKeon [14] used the WMLES approach to study large-scale structures in a long channel flow, focusing on their amplitude modulation in near-wall small-scale fluctuation. In the present study, LES coupled with a wall model [30] was indirectly used to investigate the origin of VLSMs in TBL, here the word "indirectly" means we cannot consider the top-down mechanism in these idealized simulations. The present WMLES can produce a promising approach to the study of the characteristics of VLSMs.

In this paper, the VLSMs in the TBL under different surface roughnesses were studied progressively, from a single, two-point correlation field analysis to flow field analysis. Our study is arranged as follows. Section 2 introduces our numerical method, wall model, and subgrid-scale (SGS) model. Section 3 shows the validation of the WMLES approach using DNS and field observation data. The simulation results for VLSMs are presented and analyzed in Section 4. Section 5 provides the conclusion.

2. Numerical Method

2.1. Wall Model LES (WMLES)

For incompressible flow, the time-dependent filtered N-S equations are written as follows:

$$\frac{\partial \overline{u}_i}{\partial t} + \frac{\partial}{\partial x_j}(\overline{u}_j\overline{u}_i) = -\frac{\partial}{\partial x_i}\overline{p} - \frac{\partial}{\partial x_j}(\tau_{ij}^D) + f_i^T \tag{1}$$

$$\frac{\partial \overline{u_i}}{\partial x_i} = 0 \tag{2}$$

where x_i are Cartesian coordinates, u_i are corresponding velocity components; p is pressure divided by fluid density; τ_{ij}^D is the total stress; and f_i^T is the other density-normalized forces. The bar on top of the variables represents a filtering operator.

In Equation (1), on the left hand side, the first term represents the time change rate, the second term represents convection, on the right hand side, the first term gives the density normalized pressure gradient, the second term represents the stresses, and the third term is the other density-normalized forces.

SGS model used to represent the effect of unresolved motions on the resolved motions is critical for accurate simulations, especially for the specific purpose of capturing the dynamics of large-scale structures. In the present study, the one-equation of the eddy viscosity model [31] was adopted. The reason for this being that the crucial feature of the Smagorinsky model lies in the assumption of the balance between SGS energy production and dissipation. On using the ensemble-averaging procedure in place of the filtering in LES, the balance of energy production and dissipation holds only in some restricted flow situations. In fact, this balance breaks down in the vicinity of walls in channel and boundary-layer flows, as in jets and wakes. Moreover, for the Smagorinsky model, the eddy-viscosity representation for the Reynolds stress based on the ensemble mean cannot reproduce the anisotropy of turbulence observed in some typical flows like channel flows. However, the one-equation eddy viscosity model is not based on the assumption of SGS energy balance and uses a more accurate representation for SGS Reynolds stress, the deficiency of local balance assumption adopted in algebraic eddy viscosity models can be overcome. At high Reynolds number flows and for coarse grid resolution such a phenomenon may occur, which fits perfectly with the simulation considered in this study.

The SGS stress can be modeled as:

$$\tau_{ij,sgs}^{D} = \frac{2}{3}k_{sgs}\delta_{ij} - 2\upsilon_{sgs}\overline{S}_{ij} \tag{3}$$

where $\tau_{ij,sgs}^{D}$ SGS stress part of the total stress τ_{ij}^{D}. $\overline{S}_{ij}$ resolved-scale strain rate tensor, δ_{ij} Kronecker delta function, k_{sgs} SGS kinetic energy; υ_{sgs} SGS stress eddy viscosity.

Accounting for the historical effect of k_{sgs} because of diffusion, production, and dissipation, a transportation equation is derived:

$$\frac{\partial k_{sgs}}{\partial t} + \frac{\partial \overline{u}_j k_{sgs}}{\partial x_j} = -\tau_{ij,sgs}^{D}\overline{S}_{ij} - C_{\varepsilon}\frac{k_{sgs}^{3/2}}{\Delta} + \frac{\partial}{\partial x_j}(\upsilon_{sgs}\frac{\partial k_{sgs}}{\partial x_j}) \tag{4}$$

where Δ filter scale, C_k and C_ε model constants.

$C_k = 0.0673$ and $C_\varepsilon = 0.93$ were adopted in this study.

The computational requirement of high Reynolds number wall-bounded flow of fully-resolved LES dependents strongly on Reynolds number. Piomelli and Balaras [29] pointed out that at high Reynolds number o (10^6), for smooth boundary layers flow, 99% of the computational mesh grids are used to resolve the inner layer whose thickness is only 10% of the boundary layer height; furthermore, more than 99% of the computational mesh grids is required for the rough boundary layers flow. Consequently, the only economical way to perform LES of high Reynolds number turbulent boundary layer flow is just to compute the outer layer. The outer-flow eddies can determine the grid size in this case, while for the near wall eddies, the grid is too coarse, so must model the wall layer. In particular, near the wall, discrete differentiation cannot evaluate the momentum flux, since the coarse grid cannot resolve the sharp velocity gradient and the quasi-streamwise in this region. Moreover, Piomelli and Balaras concluded that the WMLES tend to be more accurate at high Reynolds number than at moderate or low Reynolds number.

Moreover, for the fully-resolved LES, the SGS model and numerical errors very large in the near-wall region usually; thus, it is conceivable that comparison with fully-resolved LES, using "perfect" wall model can get a more accurate predictions.

Another important point we need to consider involves the case of an atmospheric surface covered with high roughness elements (e.g., sands, rocks, vegetation) that would greatly increase the computation cost when used in conjunction with the fully-resolved LES.

Thus, in the present study, the surface stress model is the following matrix:

$$\tau_{ij}^{D} = \begin{pmatrix} 0 & 0 & \tau_{13}^{tot} \\ 0 & 0 & \tau_{23}^{tot} \\ \tau_{13}^{tot} & \tau_{23}^{tot} & 0 \end{pmatrix} \tag{5}$$

and, the Schumann's model [30] was adopted:

$$\tau_{13}^{tot} = -u_*^2 \frac{(\overline{u}_{1/2} - \langle \overline{u}_{1/2} \rangle)}{(\langle \overline{u}_{1/2} \rangle^2 + \langle \overline{v}_{1/2} \rangle^2)^{1/2}} \tag{6}$$

$$\tau_{23}^{tot} = -u_*^2 \frac{(\overline{v}_{1/2} - \langle \overline{v}_{1/2} \rangle)}{(\langle \overline{u}_{1/2} \rangle^2 + \langle \overline{v}_{1/2} \rangle^2)^{1/2}} \quad (7)$$

where 1/2 is at first level cell centers near the surface; the angle brackets represent horizontal average at a height.

The friction velocity:

$$u_*^2 = (\langle \tau_{13}^{tot} \rangle^2 + \langle \tau_{23}^{tot} \rangle^2)^{1/2} \quad (8)$$

which rough wall log law:

$$\frac{(\langle \overline{u}_{1/2} \rangle + \langle \overline{v}_{1/2} \rangle)^{1/2}}{u_*} = \frac{1}{k} \ln(\frac{z}{z_0} + f(L)) \quad (9)$$

where $f(L)$ atmospheric stability-related function, L Obuhkov length, z_0 aerodynamics roughness height.

In this study, $f(L)$ is zero for neutral stability, z_0 is related to roughness elements shape, height, and distribution.

In this paper, the five different cases of surface roughness are investigated: z_0 = 0.0002 m (case 1, $Re_\tau = 3.538 \times 10^6$), z_0 = 0.002 m (case 2, $Re_\tau = 4.894 \times 10^6$), z_0 = 0.02 m (case 3, $Re_\tau = 6.559 \times 10^6$), z_0 = 0.2 m (case 4, $Re_\tau = 9.927 \times 10^6$), and z_0 = 2.0 m (case 5, $Re_\tau = 1.199 \times 10^7$).

2.2. Simulation Setup

Figure 1 gives the schematic diagram of the computational domain (5000 × 1000 × 300 m for case 1, 5000 × 1000 × 350 m for case 2, 5000 × 1000 × 400 m for case 3, 5000 × 1000 × 500 m for case 4, 5000 × 1000 × 600 m for case 5) adopted in this study. The streamwise, spanwise, and wall-normal coordinates are x, y, and z. At the solid wall, wall model was adopted, and at the upper surface, slip boundary conditions were adopted. For the streamwise and spanwise directions, periodic boundary conditions were applied, same with the LES TBL flow simulation by Fang and Porté-Agel [27], and DNS channel flow simulation conducted by Lee et al. [6]. The logarithmic mean wind-speed profile was used for the inflow, specifically for five cases: Case 1, 250 m, 8 m/s; case 2, 300 m, 8 m/s; case 3, 350 m, 8 m/s; case 4, 450 m, 8 m/s; and case 5, 550 m, 8 m/s.

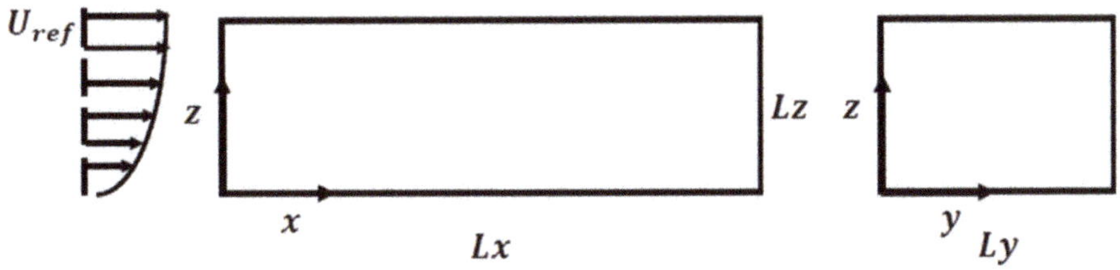

Figure 1. Schematic diagram of the computational domain.

The hexahedral meshes type was adopted in the simulation. The computational domain at the height below 200 m, the mesh grid size is 5 m in each direction. Except for this region, the mesh grid size is 10 m in each direction.

The time step is set to 0.5 s in order to satisfy the Courant number for the stability condition and viscous stability. To guarantee a fully developed flow field and quasi-steady conditions, a spin-up simulation is first made for enough time. To ensure statistical convergence, statistical calculations are then performed over 40,000 timesteps. At 0.015% of domain's height, the initial peak in perturbation magnitude occurs, with both the maximum perturbations of streamwise/spanwise flow near the surface running at 0.25 m/s, a 4.0/20.0 ratio was adopted at the beginning of the simulation for total periods of streamwise/spanwise perturbations.

The simulations were performed using the Simulator for Wind Farm Application (SOWFA) [32], which is developed using OpenFOAM C++ library. The numerical discretization was performed by a finite volume method. A second-order central difference scheme was used for the convection and viscosity terms, and a second-order backward scheme was employed for the unsteady term.

3. Validations

Because the wall layer is modeled in WMLES, the accuracy of the simulation results must be validated before the study can proceed. Here, the simulation results based on non-dimensional inner variables $z^+ = zu_\tau/v$ and u_τ, and outer variables δ and U, are compared with the results of DNS, experiments, and field measurements.

3.1. Validation by Non-Dimensional Inner Variables

To confirm that the WMLES approach is suitable for TBL flows at high Reynolds numbers, variations in mean velocity, Reynolds stress, and turbulent variation with height and Reynolds number are compared with theoretical results and experimental, field-measured data for the TBL flow.

The non-dimensional mean streamwise velocity profiles, shown in Figure 2. The simulated non-dimensional mean velocities $U^+ = U/u_\tau$ are displayed at different heights from case 1. The mean velocity profile agrees well with the logarithmic law and Qingtu Lake Observation Array (QLOA) field-measured data, indicating that the WMLES is highly accurate for TBL flow, despite the down shift in ΔU^+ due to ground roughness.

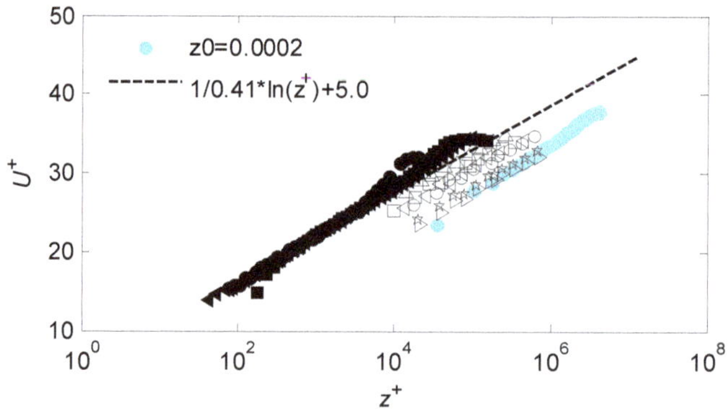

Figure 2. Comparison of mean velocity profiles. The solid colored symbols indicate current simulation data; the black dashed line indicates $U^+ = \ln(z^+)/0.41 + 5.0$ for a smooth-wall wind profile; the black open upward-facing triangles show Surface layer Turbulence and Environmental Science Test (SLTEST) field measured data, $Re_\tau = 6.28 \times 10^5$; the solid black circles indicate the Melbourne wind tunnel experiment data, $Re_\tau = 1.801 \times 10^4$; the solid black squares represent the William B. Morgan Large Cavitation Channel experiment data acquired by the US Navy, $Re_\tau = 6.867 \times 10^4$; the solid black leftward-facing triangles show the Princeton Superpipe experiment data, $Re_\tau = 9.819 \times 10^4$, the observational and experimental data obtained by Marusic et al. [33]; the open symbols present the QLOA field-measured data from Wang and Zheng [9]; the black open squares show $Re_\tau = 1.63 \times 10^6$; the black open circles represent $Re_\tau = 2.19 \times 10^6$; the black open leftward-facing triangles give $Re_\tau = 2.95 \times 10^6$; the black open rightward-facing triangles provide $Re_\tau = 3.14 \times 10^6$; and the black open stars show $Re_\tau = 4.19 \times 10^6$.

Figure 3 shows the non-dimensional Reynolds shear-stress profiles for case 1. These agree well with theoretical predictions [34,35] for canonical TBL flow, the field-observation data of Hutchins et al. [36]

at $Re_\tau = 7.7 \times 10^5$, and the QLOA field measured data with higher Reynolds number of Wang and Zheng [9].

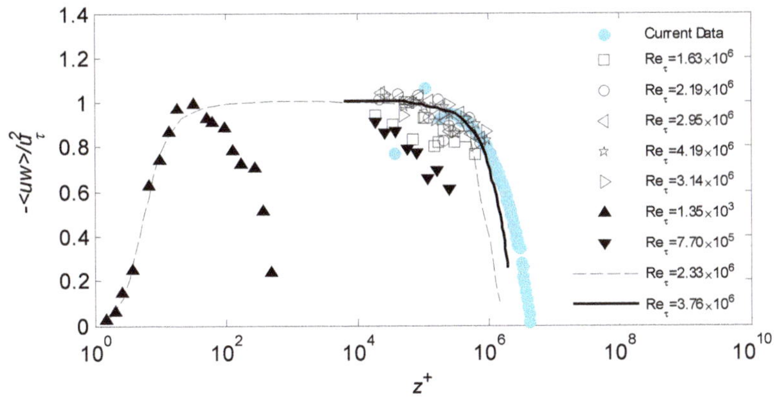

Figure 3. Comparison of Reynolds stress with inner-scaled height: The solid colored symbols are current simulation data, the dashed line is Chauhan [34] for $Re_\tau = 2.33 \times 10^6$, the solid line represents $Re_\tau = 3.76 \times 10^6$, the solid black upward-facing triangles exhibit the wind tunnel results of Graaff and Eaton [37] at $Re_\tau = 1.35 \times 10^3$, the solid black downward-facing triangles give the results of the field observations of Hutchins et al. [36] at $Re_\tau = 7.7 \times 10^5$, the open symbols show the field-measured QLOA data of Wang and Zheng [9], the black open squares represent $Re_\tau = 1.63 \times 10^6$, the black open circles give $Re_\tau = 2.19 \times 10^6$, black open leftward-facing triangles present $Re_\tau = 2.95 \times 10^6$, the black open rightward-facing triangles present $Re_\tau = 3.14 \times 10^6$, and the black open stars are $Re_\tau = 4.19 \times 10^6$.

In addition, $\overline{uu}/u_\tau^2$ is plotted against the inner-scaled height, as shown in Figure 4, including the field measurement data from the SLTEST site [36,38], laboratory data [39,40], and the similarity formulation presented in Marusic and Kunkel [41]. This reasonable agreement suggests that the simulated data agree well with the theoretical formulation [34] and previous measurements at the SLTEST. The magnitude of the normalized streamwise component of normal stress is strongly dependent on the Reynolds number, except for the sublayer. Specifically, the normalized streamwise component of normal stress, simulated at $z/\delta \approx \sim 0.036$, was compared to the results of QLOA [9], SLTEST [36], and wind tunnel experiments [40,42–44] in Figure 5. The current simulation result falls within the QLOA results because the Reynolds number for case 1 is located within the five cases of Wang and Zheng [9], indicating an approximate log-linear relationship between $\overline{uu}/u_\tau^2$ and the Reynolds number.

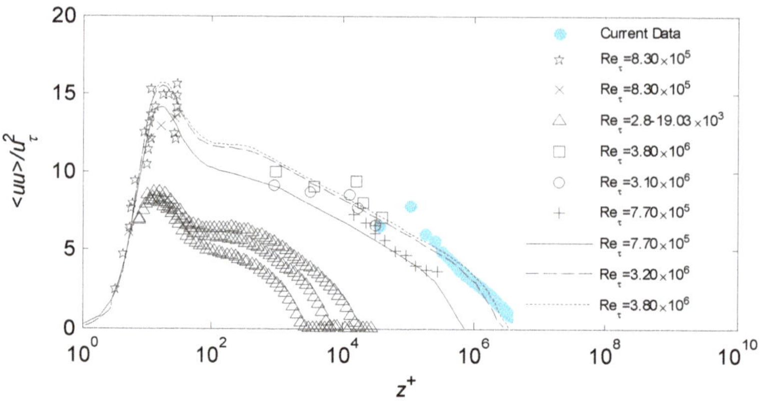

Figure 4. Comparison of $\overline{uu}/u_\tau^2$ in inner-scaled height simulation results with a similarity formulation from Marusic and Kunkel [41]: The solid colored circles are current simulation data at $\mathrm{Re}_\tau = 3.553 \times 10^6$, the open circles and squares are for the result of Kunkel and Marusic [38] at $\mathrm{Re}_\tau \approx 3.1 \times 10^6$ and 3.8×10^6, respectively, the black pluses exhibit the result of Hutchins et al. [36] at $\mathrm{Re}_\tau = 7.7 \times 10^5$, the times and open stars represent Metzger et al. [39] at $\mathrm{Re}_\tau \approx 8.3 \times 10^5$, the open upper triangles show the result of Hutchins et al. [40], $\mathrm{Re}_\tau = 2.8 \times 10^3$, 7.3×10^3, and 1.903×10^4, the lines represent the similarity formulations proposed by Marusic and Kunkel [41] calculated at $\mathrm{Re}_\tau = 7.7 \times 10^5$ (solid black line), $\mathrm{Re}_\tau = 3.2 \times 10^6$ (black dash line), $\mathrm{Re}_\tau = 3.8 \times 10^6$ (black dotted line).

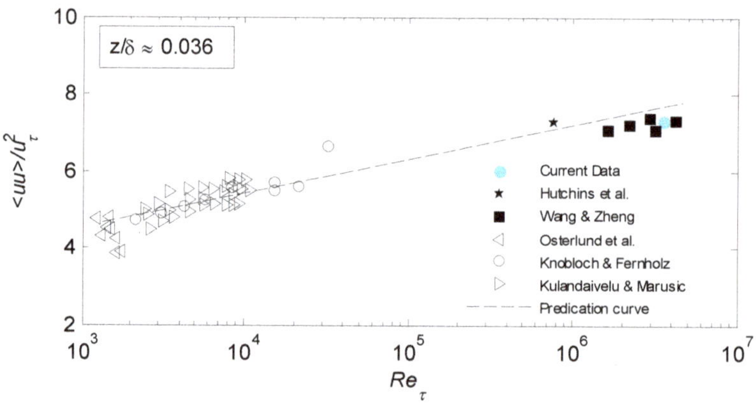

Figure 5. Variation of normalized streamwise component of normal stress $\overline{uu}/u_\tau^2$ with the Reynolds number: Solid colored circles are the current simulation data; the open symbols indicate the wind tunnel results of Österlund et al. [42], Knobloch and Fernholz [43], Hutchins et al. [40], and Kulandaivelu and Marusic [44]; the solid black five-pointed stars represent the field observations of Hutchins et al. [36]; and the solid black squares are the field observations of Wang and Zheng [9].

3.2. Validation by Non-Dimensional Outer Variables

In the outer layer of TBL, the flow is assumed to be independent of viscosity and to depend on the global characteristics of the flow, as represented by δ, U.

The Reynolds shear stress is shown in Figure 6 against the outer-scaled coordinates z/δ. It is clear that the simulated result of case 1 is in a good agreement with the experimental results of Graaff and Eaton [37]; most importantly, around $z/\delta = 0.02$, a trend in the shift of Reynolds shear stress is also captured by this simulation, which indicates the accuracy and high precision of the WMLES adopted in this study.

Normalized streamwise component of normal stress is shown in Figure 7 against outer-scaled coordinates z/δ. The simulated result of case 1 agrees well with the experimental results of Graaff and Eaton [37]; like the Reynolds shear stress, a normalized streamwise component of normal stress also shows a shift in the trend with increasing height.

Figure 8 shows the mean velocity defect profiles according to the outer coordinates. The profile of case 1 is in good agreement with the experiment results of Graaff and Eaton [37] and the DNS results of Lee and Sung [8] throughout the outer region of the boundary layer, albeit at different Reynolds numbers.

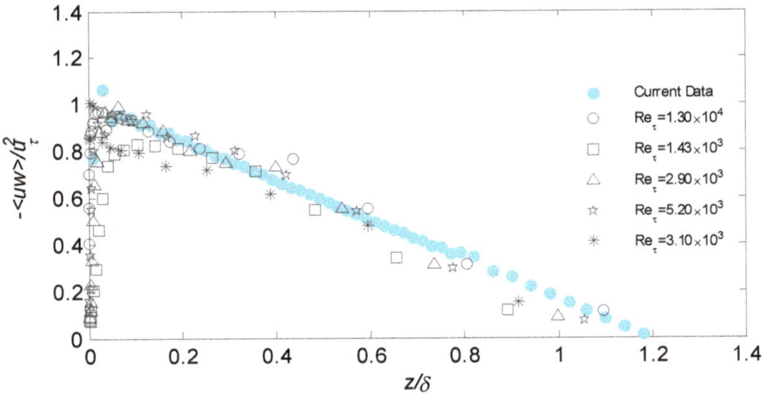

Figure 6. Comparison of $-\overline{uw}/u_\tau^2$ in results of outer-scaled height simulation with the experimental data. The solid colored circles indicate current simulation data, the open symbols are experimental data, the squares show $\text{Re}_\tau = 1.43 \times 10^3$, the upper triangles symbol give $\text{Re}_\tau = 2.9 \times 10^3$, the stars indicate $\text{Re}_\tau = 5.2 \times 10^3$, the circle symbols represent $\text{Re}_\tau = 1.3 \times 10^4$, and the eight-spoked asterisks are $\text{Re}_\tau = 3.1 \times 10^4$.

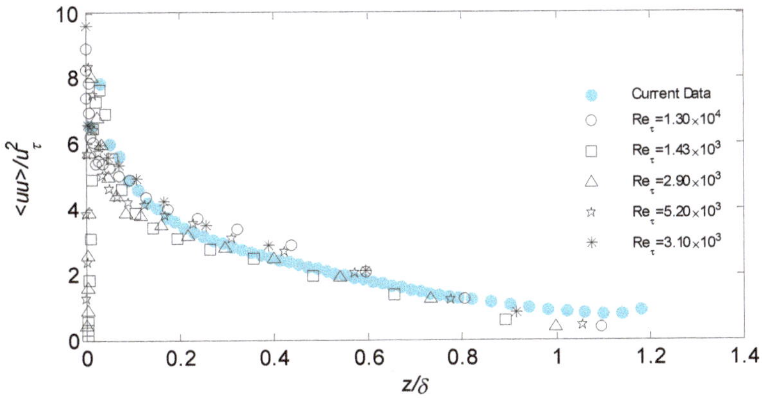

Figure 7. Comparison of $\overline{uu}/u_\tau^2$ in the results of outer-scaled height simulation with the experimental data. The symbols used are the same as those used in Figure 6.

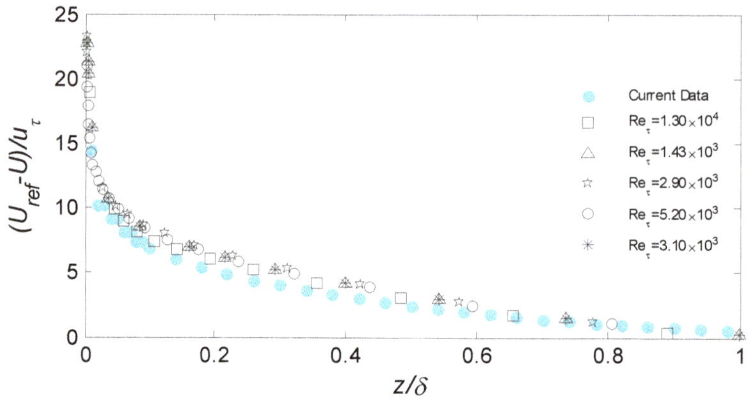

Figure 8. Mean velocity-defect profiles, normalized by friction velocity. The symbols used are the same as those used in Figure 6.

4. Results and Discussion

4.1. Velocity Spectra Characteristics under Different Surface Roughnesses

4.1.1. Streamwise Velocity Spectra

Figure 9 shows the wavenumber-normalized spectra of streamwise velocity $S_u/u_\tau^2\delta$ for all five cases at different heights. Taylor's hypothesis of spatial–temporal transformations was adopted to calculate the wavenumber spectrum, to smooth the spectra, a multi-point averaging method was used, following Balasubramaniam [45]. As Figure 9a–c shows, for $k\delta < 2$ (wavelengths larger than 3δ), each spectral curve has an individual inflection point where the slope of the curves conspicuously changes. The inflection point represents a spectral peak in low-wavenumber region of the pre-multiplied spectra (as shown in Figure 10) associated with the VLSMs [3,9,46]. Thus, it can also be demonstrated that the VLSMs exist in the first three cases for the current TBL study. While, for cases 4 and 5, individual inflection points exist for $k\delta > 2$, which means that VLSMs does not exist in these two cases. Above all, with where roughness increases, the scale of the VLSMs (the width of −1 region) decreases or even disappears. We predict that the reason for this may be associated with roughness elements of similar or larger sizes, which undoubtedly disrupt near-wall quasi-streamwise streaks. This phenomenon is identical to what is shown in Figure 1 from Volino et al. [17], although it is not mentioned as such in the report of that study.

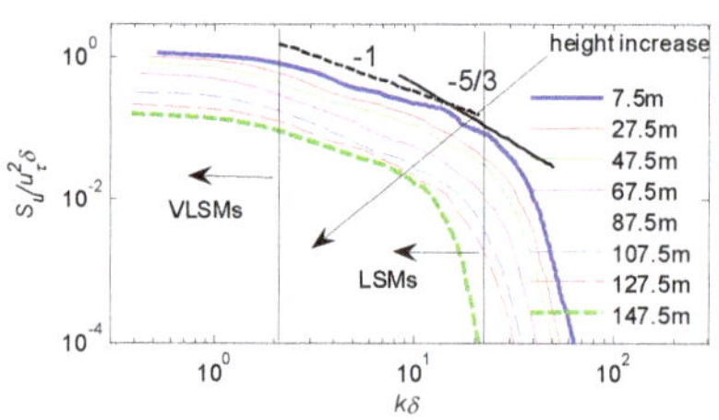

(a)

Figure 9. *Cont.*

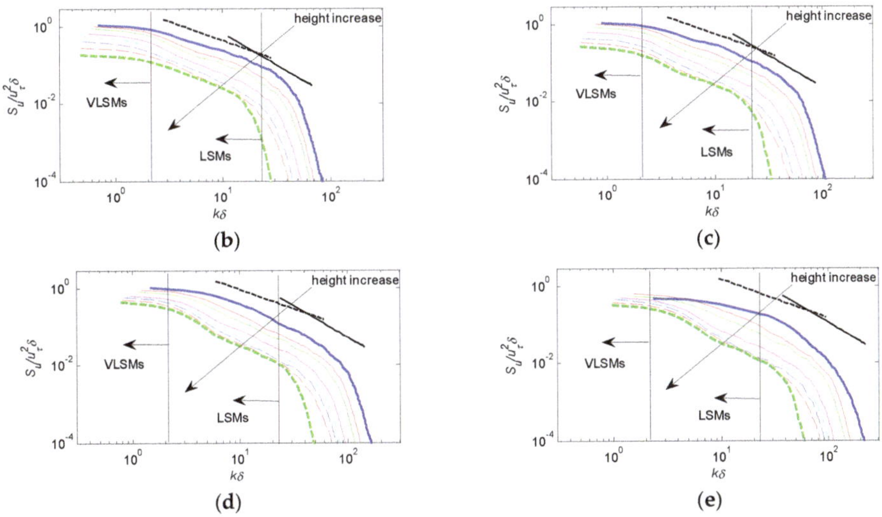

Figure 9. Power spectra of streamwise velocity fluctuations. (**a**) case 1, (**b**) case 2, (**c**) case 3, (**d**) case 4 and (**e**) case 5. The height indicators for each graph are the same as those shown in (**a**), that is, 7.5 m, 27.5 m, 47.5 m, 67.5 m, 87.5 m, 107.5 m, 127.5 m, and 147.5 m. Further, the boundary layer height is 250 m, 300 m, 350 m, 450 m, and 550 m for cases 1, 2, 3, 4, and 5, respectively.

Figure 10 shows the pre-multiplied spectra for all five roughness cases. There is a distinct peak in the spectra for all the heights in Figure 10a–c in the low-wavenumber region that corresponds to VLSMs; another peak appears in high-wavenumber region that corresponds to the LSMs in the spectra for all heights in Figure 10a–e.

In addition, the product of wavenumber and power spectra kS_u decreases with height throughout all of the spectra under various ground roughness, which is consistent with the result of the turbulent-boundary layer [2,46], and the pipe-flow result [4]. This phenomenon can be explained by "bottom-up" mechanism, and VLSMs weaken in the wall-normal direction, due to the TKE produced in near-wall region gradually dissipates.

The slope between the first and second peaks of the streamwise velocity spectra is also analyzed. This is governed by the transformation scaling law of VLSMs and LSMs (cases 1, 2, and 3) or the transformation scaling law of LSMs and the small-scale turbulence vortex (cases 4 and 5); instead of observing jumps from −1 to −5/3, we observe a gradual change as height increases. With increasing roughness, the scaling law grows, and the transformation rate from −1 to −5/3 increases, with the roughness increasing as well, while the VLSMs go extinct. Another interesting phenomenon observed here is the scaling law in cases 4 and 5, showing height increases of −5/3. For other cases, the scaling law is located between the values −1 and −5/3, with the smoother terrain yielding the larger scaling law.

Based on this finding, a new concept could be proposed, according to which the presence or absence of VLSMs cannot simply be defined by the value −1 or −5/3; instead, the scale factor λ/δ (the length of VLSMs and LSMs to the boundary-layer height) and the scaling law should be combined to illustrate the location of the vortex structure in VLSMs, LSMs, or small-scale turbulence vortex regions.

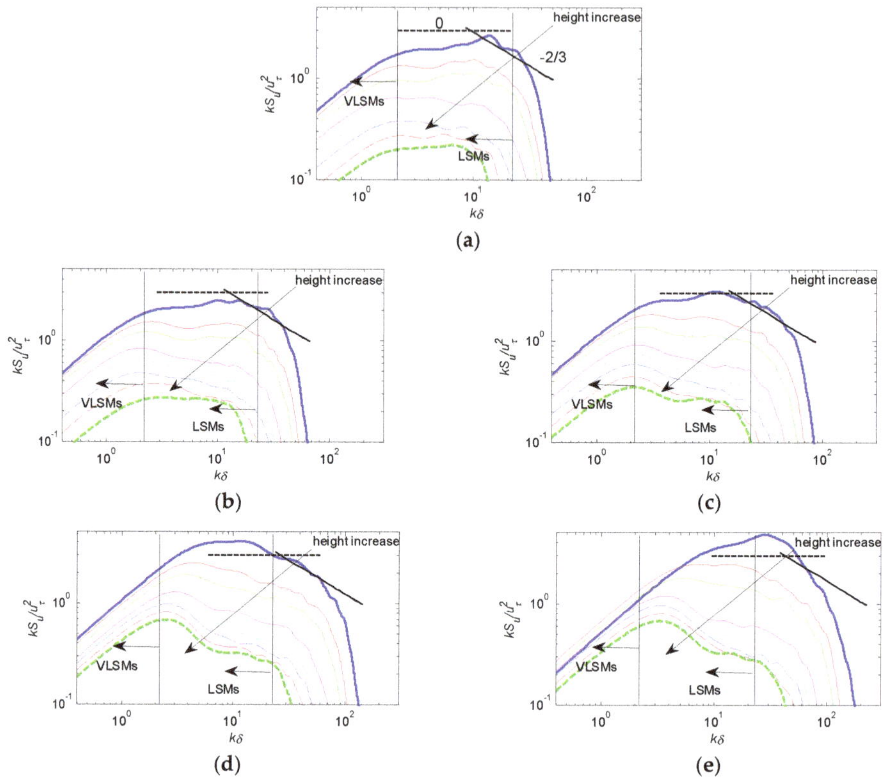

Figure 10. Pre-multiplied spectra for streamwise velocity fluctuation. (**a**) case 1, (**b**) case 2, (**c**) case 3, (**d**) case 4 and (**e**) case 5. The height indicators for each graph are the same as those shown in Figure 9a, that is, 7.5 m, 27.5 m, 47.5 m, 67.5 m, 87.5 m, 107.5 m, 127.5 m, and 147.5 m.

The $k\delta$ decreases at the first peak (meaning $(k\delta)_0$) with increasing height; when the height reaches to a certain extent, $(k\delta)_0$ begins to attain a constant value or even to increase, as determined by the characteristics of the VLSMs and is consistent with the literature [18,46]. These two phenomena can be seen more clearly in Figure 11, which plots the relationship between $(k\delta)_0$ and the dimensionless height z/δ.

In Figure 11, the result of case 1 agrees with the field-measured data, as a result of the similarity of terrain between this and the QLOA site, as shown in Figure 2. With increases in roughness, cases 2 and 3 reach a certain degree of agreement with field data. While, for cases 4 and 5, the results diverge from the field data, implying that the surface roughness has a great influence on the structure of VLSMs. As we know, there is also a certain deviation between the experimental value and the field-measured value, which may be due to the difference in Reynolds number between them.

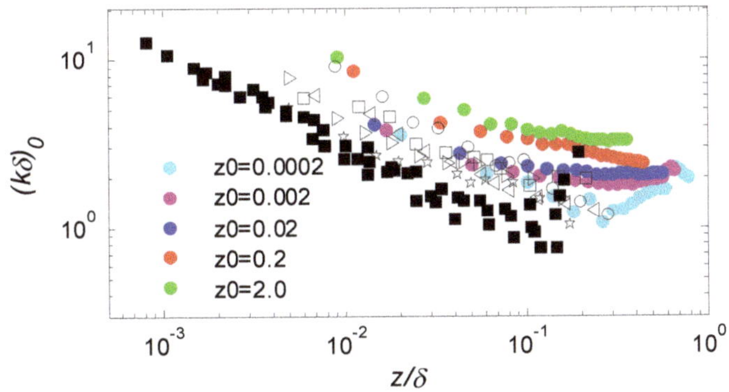

Figure 11. Variation in the peak wavenumber of VLSMs with height. The solid colored circle is the current simulation data, the open symbols show the field-measured data of Wang and Zheng [9], and the solid black symbols represent the experimental data of Valliviki et al. [46].

4.1.2. Wall-Normal Velocity Spectra Characteristics

Wall-normal velocity fluctuation spectra are shown in Figure 12. The trend for wall-normal location is similar to the streamwise velocity fluctuations, with the exception of the crossover phenomenon, which is consistent with the finding of Guala et al. [4]. For the left part of the crossover point, energy density increases from the wall to a wall-normal location, and then it decreases again; for the right part of the crossover point, the energy density always decreases. In addition, it is clear that as the roughness increases, the crossover point moves to a large wavenumber, which indicates that the rougher the terrain, the less the vortex interacts at different heights. Comparison with streamwise velocity spectra, the wall-normal velocity spectra has lower values, which reflects the difference between the two components in the total energy, and the ultimate difference is close to one order of magnitude.

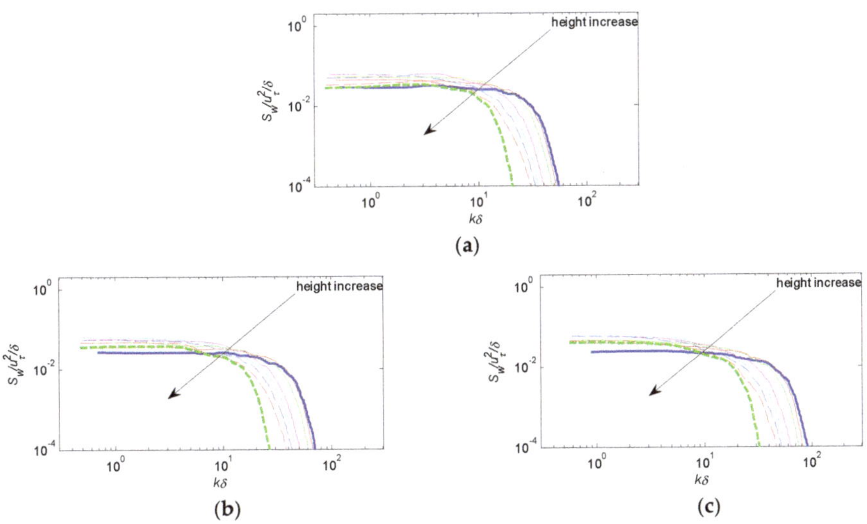

Figure 12. *Cont.*

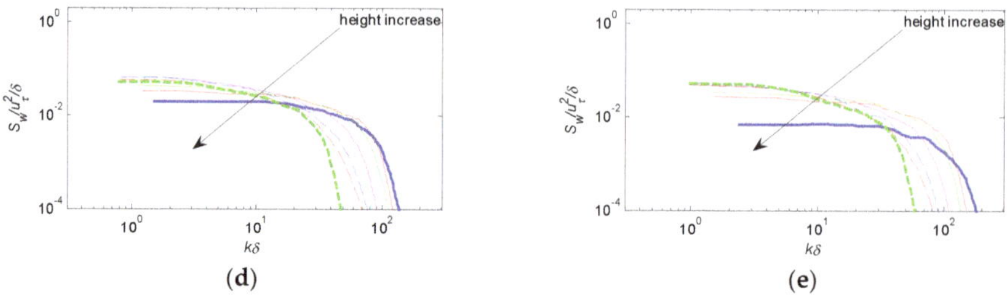

Figure 12. Power spectra of wall-normal velocity fluctuations. (**a**) case 1, (**b**) case 2, (**c**) case 3, (**d**) €e 4 and (**e**) case 5. The height indicators for each graph are the same as those shown in Figure 9a, that is, 7.5 m, 27.5 m, 47.5 m, 67.5 m, 87.5 m, 107.5 m, 127.5 m, and 147.5 m.

4.2. Effects of VLSMs and LSMs on Turbulent Kinetic Energy

One of the principal questions addressed in this study is how far VLSMs and LSMs contribute to TKE. We will examine the distribution of TKE as a function of height.

The TKE of VLSMs and the corresponding energy fraction are plotted against height using a low-pass filter with a cutoff length of 3δ, as shown in Figure 13a,b, respectively. Similarly, band-pass filtering with cutoff lengths of $0.3\delta - 3\delta$ and $\delta - 3\delta$ are used to demonstrate the energy fraction of the LSMs, as shown in Figure 13c,d, respectively. The total kinetic energy of VLSMs of both case 1 in the present study and the experimental data of Balakumar and Adrian [2] slightly decreases with increases in height. However, the total kinetic energy of VLSMs increases with height increases as in Wang and Zheng [9], which may relate to fact that the case investigated by Wang and Zheng [9] found energy increases with height in the low-wavenumber region of the wind spectra, unlike what was found in other studies [2,46]. Case 1 shows a close agreement with the experimental data of Balakumar and Adrian [2], given the increasing ground roughness and nondimensional decreases in TKE. The reason for this is found in the fact that VLSMs decrease or even disappear as roughness increases. It is worth noticing that Figure 13 also shows a result found in cases 4 and 5 that does not exist in VLSMs. It simply is the mathematical expression of the region of VLSMs and LSMs, as shown in Figure 10.

In addition, the energy fraction of the VLSMs in the present study increases with height, which tends to close agreement with the wind-tunnel measurements of Balakumar and Adrian [2], as shown in Figure 13b; however, little deviation from the field-measured data of Wang and Zheng [9] appears, due to the spectrum characteristics mentioned above. The phenomenon, shown in Figure 13 indicates, that contrary to most other studies [4,8,47], which have generally found that VLSMs contribute 40–50% of the total TKE, the present study shows that VLSMs contribute up to 40% only at the highest height, shown in Figure 13b. In Figure 13c, it is seen that in LSMs ($0.3\delta < \lambda < 3\delta$, Guala et al. [4]), the energy fraction is larger than the field-measured data; but in Figure 13d, the energy fraction of the LSMs ($\delta < \lambda < 3\delta$, Lee et al. [48]) is similar to the field measured data. It is clear that the energy fraction of the LSMs is independent of ground roughness, possibly owing to the fact that the size of the element roughness was adopted in the present study only to disrupt the VLSMs, yielding no effect on LSMs, at least from the viewpoint of energy fraction.

The reason why the energy fraction of LSMs is found to be large in our study can be explained as follows. Because the WMLES approach is used, the smallest scale falls in the inertial subrange, while some or even most of the inertial subrange remains unresolved. It can be seen from the wind spectra shown in Figure 10 that it quickly falls into the dissipative region. Therefore, the total energy in this study is small. Due to the relative limitations of the computational domain, the structure of VLSMs is not completely contained, so that the simulation of the energy of VLSMs is less than that of the real atmosphere or the sufficient computational-domain case. This might also explain why the experimental data of Balakumar and Adrian [2] are lower than the field-measured of from Wang and Zheng [9], as shown in Figure 13b, the experimental domain being also limited to some extent.

The scale of LSMs, ranging between $0.3\delta - 3\delta$, can be fully resolved. In summary, the energy of VLSMs is relatively smaller than the field-measured data, while the energy of LSMs is larger.

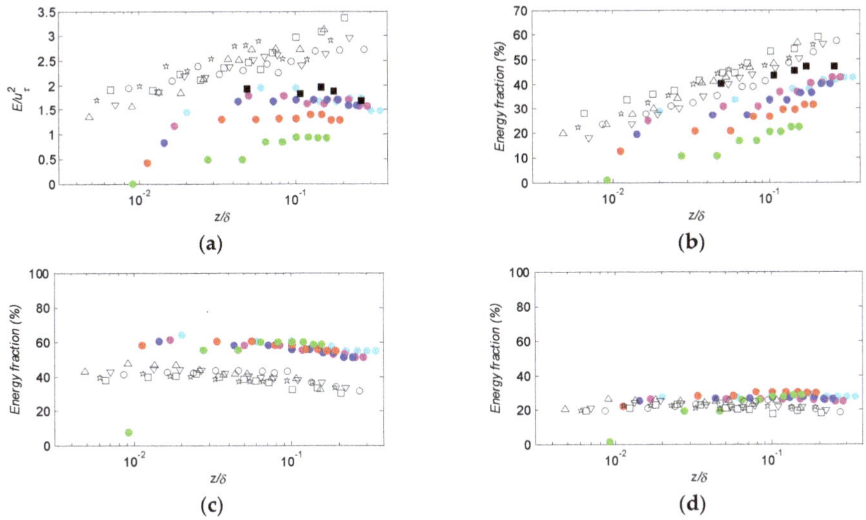

Figure 13. Variations in the total energy (**a**) and energy fraction of VLSMs (**b**) and LSMs (**c**, $0.3\delta < \lambda < 3\delta$; **d**, $\delta < \lambda < 3\delta$) with height. Solid colored circles are current simulation data, the open symbols represent the field-measured data of Wang and Zheng [9], and the solid black symbols give the experimental data of Balakumar and Adrian [2].

4.3. Flow Field Correlation Analysis

To intuitively visualize the structure of VLSMs, the coefficients of spatial correlation of the streamwise fluctuating velocity between the spatial and reference locations ($(x, y, z) = (2500m, 500m, 35m)$) were calculated. Figure 14 shows the contours of the spatial correlation coefficient for five types of roughness cases. The contour level of the correlation coefficient indicating the coherent structure edge was set to be 0.05, following Hutchins and Marusic [7]. As the figure shows, the length scale of the streamwise coherent structure satisfies $\lambda/\delta > 3$ for the first three cases, indicating the existence of the VLSMs. For cases 4 and 5, the $0.3 < \lambda/\delta < 3$ range only bears on the existence of the LSMs. For increasing roughness, the scale of VLSMs and LSMs decreases, indicating that the generation and development of VLSMs and LSMs are affected by the bottom terrain. It also indirectly illustrates that the bottom-up mechanism controls the generation and development of VLSMs.

Along the streamwise direction, the correlation contour lines are tilted, which indicates the inclination of turbulent structures that observed by Marusic and Heuer [49] and Wang and Zheng [9] is also present in the current TBL simulation at higher Reynolds numbers. The inclination angle of the two-point correlation of the fluctuating velocity is related to the average extent of the inclination of the structures of the VLSMs or LSMs. For the present simulation, the inclination angles in Figure 14 are 11.31° for case 1 ($Re_\tau = 3.538 \times 10^6$), 14.04° for case 2 ($Re_\tau = 4.894 \times 10^6$), 16.70° for case 3 ($Re_\tau = 6.559 \times 10^6$), 14.04° for case 4 ($Re_\tau = 9.927 \times 10^6$), and 19.29° for case 5 ($Re_\tau = 1.199 \times 10^7$). The results of case 1 agree closely with those found by Wang and Zheng [9], who indicated that the inclination angles of field-measured data were 12.14° for $Re_\tau = 3.14 \times 10^6$, while for $Re_\tau = 1.63 \times 10^6$, the angle is 20.35°. The difference among all five cases is comparable to the range reported in the literature for smooth- and rough-wall boundary layers. Christensen and Wu [50] found an inclination angle of 11° for their smooth-wall channel flow. Head and Bandyopadhyay [51] observed inclination angles between 15–20°. Christensen and Adrian [52] reported 12–13°. Adrian et al. [53] found an inclination angle of 12°. Tomkins and Adrian [54] found inclination angles between 10–20°. Nakagawa and Hanratty [55] found an angle of 9° on the rough wall in a channel flow, compared to smooth-wall values of 6–8°. In general, the inclination angle of large-scale coherent structures increases with roughness.

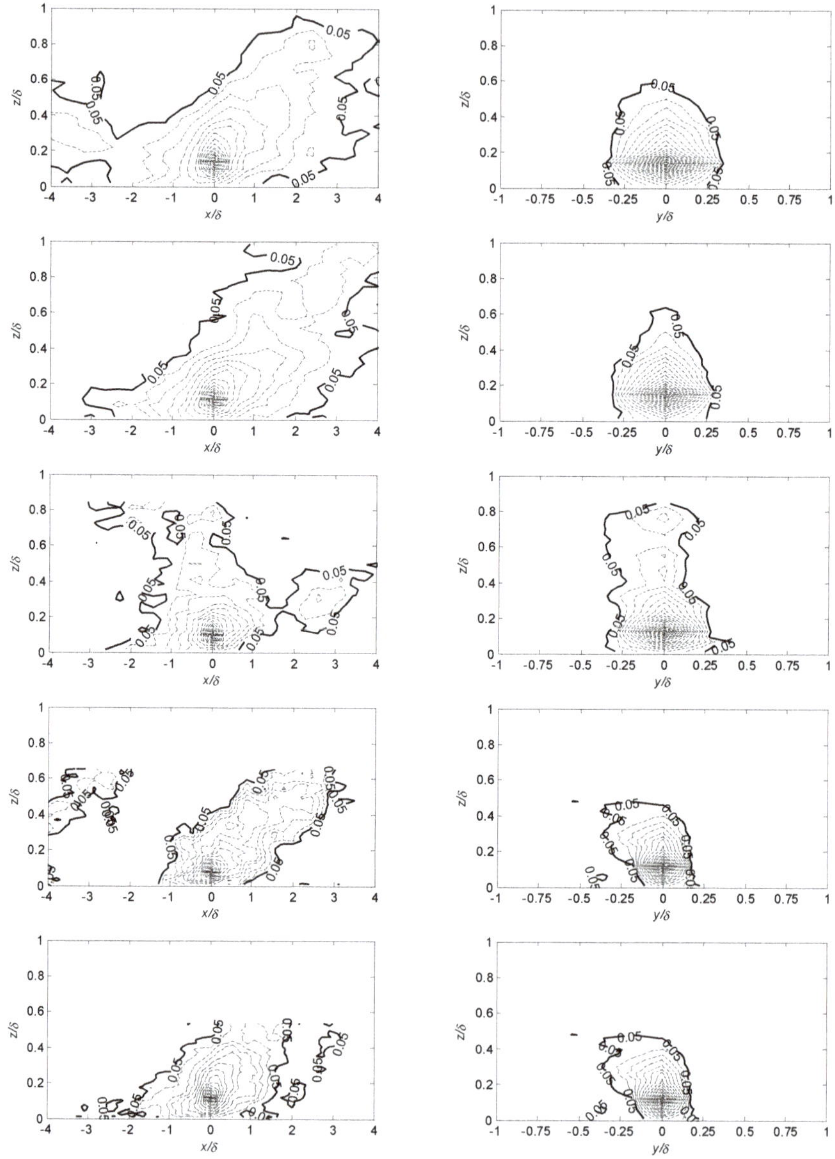

Figure 14. Two-point correlation contours of streamwise-fluctuating velocity.

In addition, the streamwise, wall normal, and spanwise extent of fluctuating-velocity two-point correlations are shown in Figure 15. The distance, Lx_{uu}, twice the distance from the most downstream location to the self-correlation peak on two-point correlations of the fluctuating streamwise velocity equal to 0.5 contour, as defined in Christensen and Wu [50]. Volino et al. [17] found that the results for rough and smooth walls agree well with a value of about $Lx_{uu}/\delta = 0.65$ between $z/\delta = 0.1$ and 0.6

with their experimental data. Close to the wall, some differences are visible. In Figure 15a, the result of case 1 is much smaller than that found by Volino et al. [17]; however, the field observation data of Wang and Zheng [9] with solid symbols are shown in Figure 15a, for $Re_\tau = 3.14 \times 10^6$ measured data, almost same as in case 1 ($Re_\tau = 3.553 \times 10^6$). This indicates that the results of simulation are correct. This can be explained by the fact that field-measured and simulation data have a similarity ratio comparable to the experimental data. In Figure 15a, as roughness increases, Lx_{uu}/δ decreases, which agrees with the results of Krogstad and Antonia [56], who found Lx_{uu}/δ was about 50% lower on their rough wall. The wall-normal extent of the two-point correlations of the fluctuating velocity, Lz_{uu}, is determined based on the wall-normal distance between the points farthest and closest from the surface at a 0.5 contour. As the contours merge with the wall, reliable estimates for Lz_{uu} cannot be obtained for $z/\delta < 0.2$. The conclusion is similar to the Lx_{uu}, as shown in Figure 15b. The spanwise extent Ly_{uu} shown in Figure 15c, with increasing roughness, the Ly_{uu}/δ range from 0.2 to 0.1, which agrees with the DNS result of Lee and Sung [8], who concluded that the streamwise negative motions meander in the downstream direction with a characteristic width of approximately $0.1\delta - 0.2\delta$ in the spanwise direction. Comparing Figure 15a,b, the ratio of Lx_{uu}/Lz_{uu} obtained in Figure 15d, the value is roughly 2.5 for both the experimental data of Volino et al. [17] and the present simulation cases, also consistent with Nakagawa and Hanratty [55], and the smooth-wall results of Krogstad and Antonia [56].

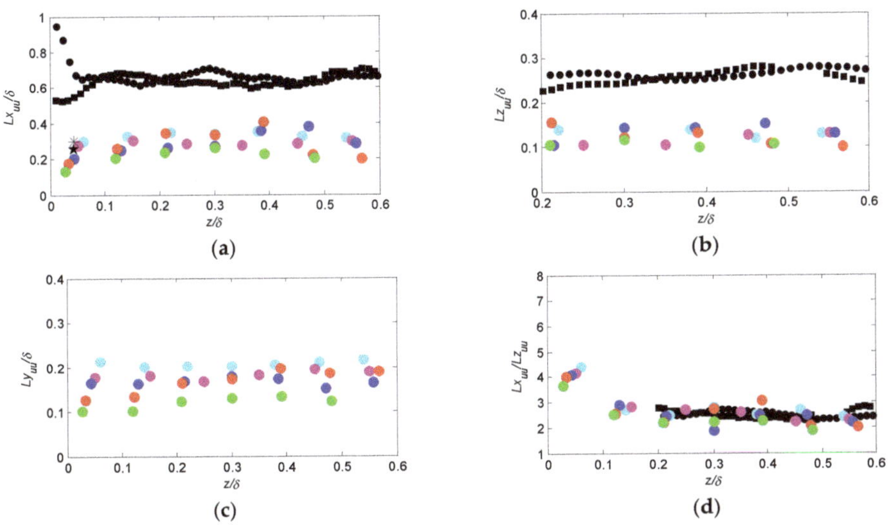

Figure 15. The extent of two-point correlations of the fluctuating velocity, equal to the 0.5 contour as a function of z/δ. (**a**) Streamwise extent, (**b**) wall-normal extent, (**c**) spanwise extent, and (**d**) ratio between streamwise and wall-normal extent. The solid colored symbols are current simulation data, the solid black data are experimental data of Volino et al. [17], the circle symbols represent smooth wall condition, and squares represent roughness wall condition, the field measured data of Wang and Zheng [9], the anise star is $Re_\tau = 3.14 \times 10^6$, and five star is $Re_\tau = 1.63 \times 10^6$ which show in (**a**).

4.4. Instantaneous Flow Field Analysis

Figure 16 shows visualizations of the VLSMs or LSMs structures in the instantaneous flow fields for $z = 30$ m in the xy plane, both the VLSMs and LSMs were visualized based on the streamwise-elongated negative velocity, which is consistent with the results of previous studies [6,8,57–59]. The direction of flow is from left to right. Figure 16a–c show several apparent very long negative (dark color) motions meandering in the streamwise direction, which in the spanwise direction are flanked by positive streamwise fluctuation (white color), and these motions often extend more than 10δ in streamwise direction, and the widths is around $0.1\delta - 0.2\delta$. This phenomenon is in agreement with TBL and turbulent channel DNS studies [7,8]. For Figure 16d,e, there are some intermediate long negative streamwise structures, and the scale characteristics are similar to the LSMs structures.

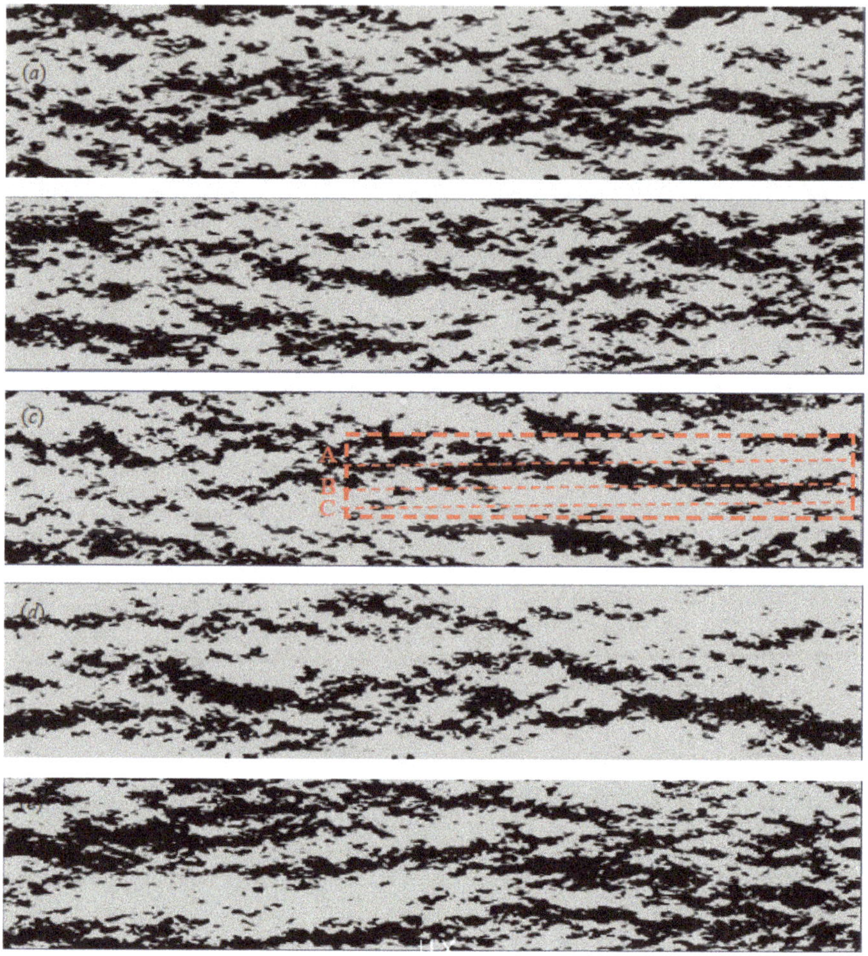

Figure 16. The normalized streamwise velocity fluctuations in xy plane. (**a**) case-1, (**b**) case-2, (**c**) case-3, (**d**) case-4, (**e**) case-5.

Figure 17 shows visualizations of the VLSMs or LSMs structures in the instantaneous flow fields for line A, B and C of Figure 16c in the xz plane, and the dashed line in Figure 17 corresponds to the horizontal plane height of Figure 16. In Figure 16c, line A locates at the very long streamwise structure, line B locates at the intermediate long streamwise structure in order to see the inclination angle of VLSMs, and line C locates at the high-velocity region. Comparing these three lines streamwise fluctuation in Figure 17, there are inclined at angles of 18.26° of line B, very close to the spatial correlation result, 16.70°, which is calculated by the reference point of streamwise direction. Note that the inclination angle is 10°–15° in an almost linear type [8], it also can be seen in the present study, the line A has nearly same inclination angle. Moreover, the flow patterns of these aligned packets are consistent with the observations [53]. For line C, which locates at the high-velocity region, the inclination angles at A and B do not exist at line C, this indicates that LSMs and VLSMs structures located at low momentum region, and comes from the near wall aligned packets, as described by "bottom-up" mechanism.

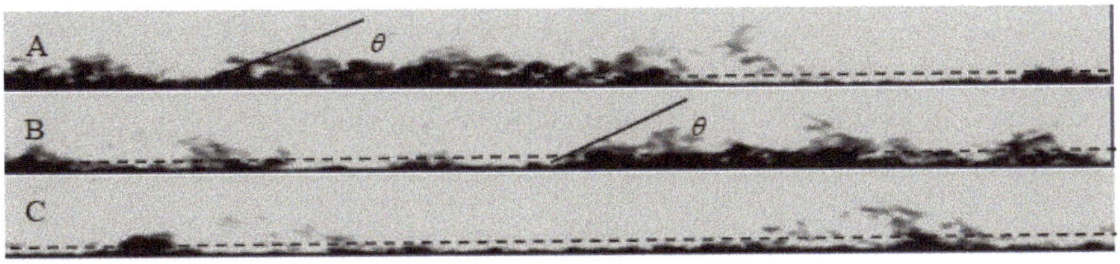

Figure 17. Streamwise velocity fluctuations normalized by the friction velocity in xz plane of case-3. The lines (**A**), (**B**) and (**C**) in the Figure 16c are shown here.

4.5. The Generation and Development of VLSMs

The results of Section 4.4 illustrate the presence of VLSMs and their associate with low-momentum regions. In this section, we address the creation and extinction procession of VLSMs in the flow field. As we know, the hairpin packet model could give a reasonable explanation for logarithmic layer features, for example, quasi-streamwise vortices and bursting processes. An idealized conceptual model Based on the hairpin packet paradigm and in the outer layer, most features of the coherent structures, an idealized conceptual model is built [53]. In contrast, Kim and Adrian [3] suggested that VLSMs is not a new type of turbulent motion. They conjectured that the alignments of hairpin packets could form VLSMs; however, the created and extincted of VLSMs in TBL is little to know. Thus, to obtain spatial information about the formation and disappearance of VLSMs is our object.

The mechanism of Kim and Adrian [3] means that in the streamwise direction continually spawning new hairpins to create packets. After growing, the packets become larger and longer, and then probably merge with adjacent packets or broken up into small structures. In this study, a representative case of merging and extincting of vortex packets is discussed in detail, which VLSMs showed in the dashed red rectangle of Figure 16c, The time evolution of flow field in the xy plane with a time interval of 20 s and the height is 30 m, shown in Figure 18. The dashed lines D, E, F and G were added to show the development of the VLSMs. It is obvious that the VLSMs in Figure 18f originates in several packet motions P1, P2, P3, P4, P5, P6, P7, and P8 present in Figure 18a. For dashed line D1 in Figure 18a, two adjacent low-momentum regions of P1 and P2 are travelling downstream, then they merge together from Figure 18f, the similar phenomenon is observed at dashed line D2. For dashed line E, the P2 and P3 are merge together, as they move downstream, and the low-momentum regions separate in the streamwise direction, as shown in Figure 18j. For dashed line F1, two adjacent low-momentum regions of P3 and P4 are travelling downstream, and the first stage is that they merge together from Figure 18a–e, then separate from each other again in the Figure 18j, the similar phenomenon is observed at dashed line F2. For dashed line G1, as part P5 and P6 move downstream, and the first stage is that they separate from Figure 18c and then merge together again in the Figure 18f, the similar phenomenon is observed at dashed line G2. The four cases of D1, D2, E, F1, F2 and G, represent three typical development process of adjacent hairpin vortices. Due to the low-momentum regions merge in streamwise direction, resulting in a longer structure at the streamwise scale of the flow field in Figure 18f,g. Another phenomenon is that although the streamwise growth of the VLSMs occurs; however, the spanwise length of VLSMs becomes thinner slightly.

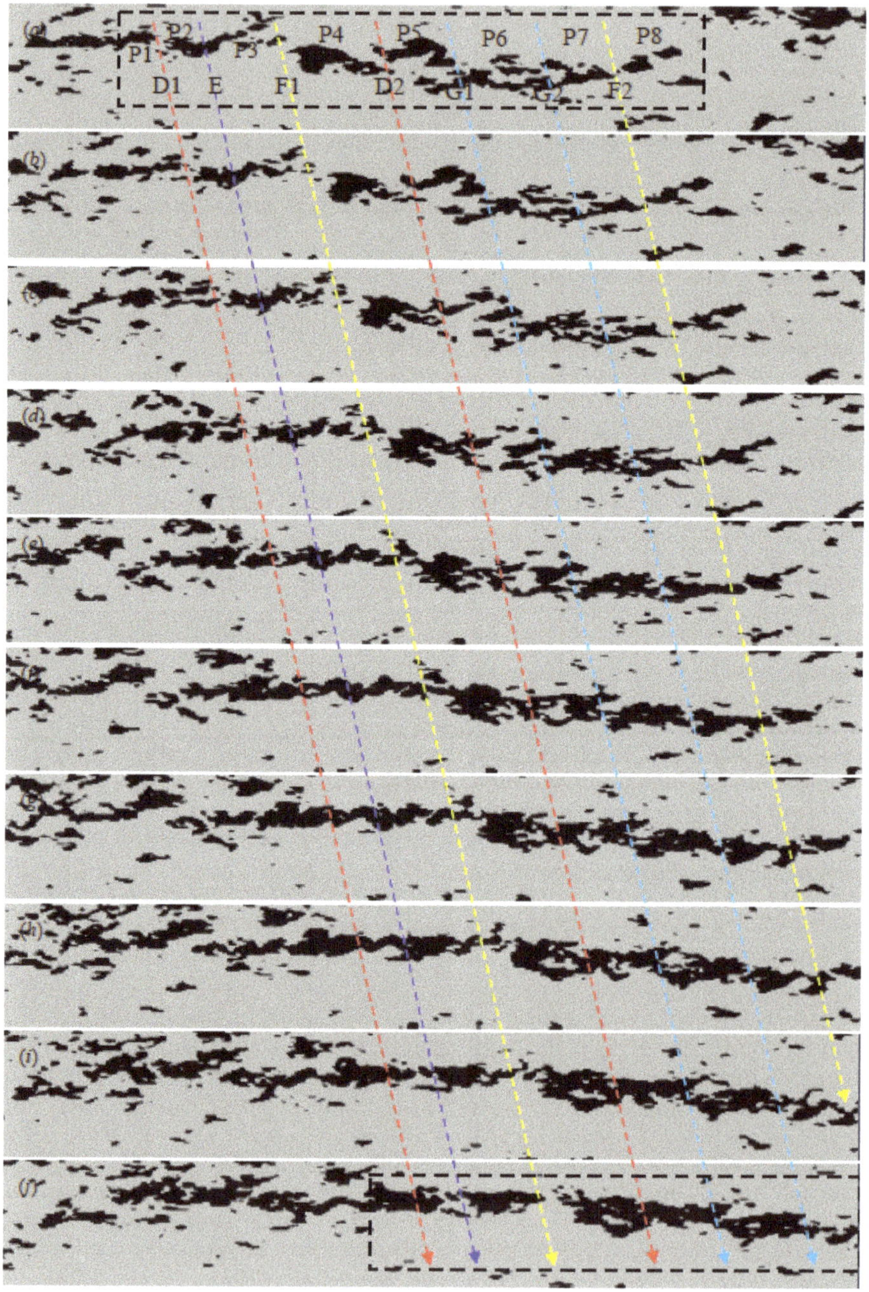

Figure 18. Time evolution of the instantaneous fields at a height of 30 m in xy plane. The dashed line D1, D2 are added to show the growth of the VLSMs, dashed line E is added to show the extinction of the VLSMs, F1, F2 are added to show the growth first then extinction of the VLSMs and G1, G2 are added to show the extinction first then growth of the VLSMs. P-number represents the different part of VLSMs of Figure 18f. Subfigures (**a**–**j**) represent the time evolution of flow field.

Figures 19 and 20 show the associated flow features in the xz plane. The conclusion is consistent with the results show in Figure 18. Combining the Figures 18–20, which indicates that the development and extinction of VLSMs structure is a three-dimensional process. As can be seen in these figures, in the downstream direction, these packets are growing continuously and further away from the wall, part of the upstream packets go through the higher mean flow velocity. The lower parts slower than these part flows, which leads lift vortex away from the wall by a higher velocity, and in the higher part of the upstream packet causes in greater stretching. Consequently, in downstream the adjacent packets are merged to form VLSMs with a shallow angle. These processes are same with hairpin structure dynamics [5] and DNS results [8].

For the formation of VLSMs, present simulations describe one possible mechanism, however, in the future, the analysis of the general mechanism of VLSMs formation is needed, because the trigger of the formation of new packets still no definitive evidence.

Figure 19. Evidence of the streamwise scale growth and extinction in xz plane. Subfigures (**a**–**j**) represent the time evolution of flow field.

Figure 20. Evidence of the streamwise scale growth in xz plane. Subfigures (**a–j**) represent the time evolution of flow field.

4.6. Vortices Field Analysis

The packet merging process is presented schematically in Figure 17 of Tomkins and Adrian [54]. The similar conclusion was obtained in figure 25 of Adrian, Meinhart and Tomkins [53]. This is the reason for the vorticity field analysis, which could show the VLSMs structure is formed by hairpin vortex or hairpin vortex package development.

Figure 21 shows three-dimensional vortical structures that contoured by λ_2 [60] in the horizontal plane. It is clear that several highly elongated low-speed regions are shown in the flow field, and many hairpin-type vortices lie above each of them. Corresponding to the Figure 17, here Figure 22 is shown as the vertical plane to illustrate the negative velocity region is covered by vortices, while positive velocity is not. By comparing A, B, and C, it is clearly shown that vortices are more density for A than B than C, which locates in high-speed region—this is consistent with the preceding results. These flow patterns, together with three-dimensional vortical structures to form longer structures that erupt from the wall and grow towards the outer region, are consist with Lee and Sung [8].

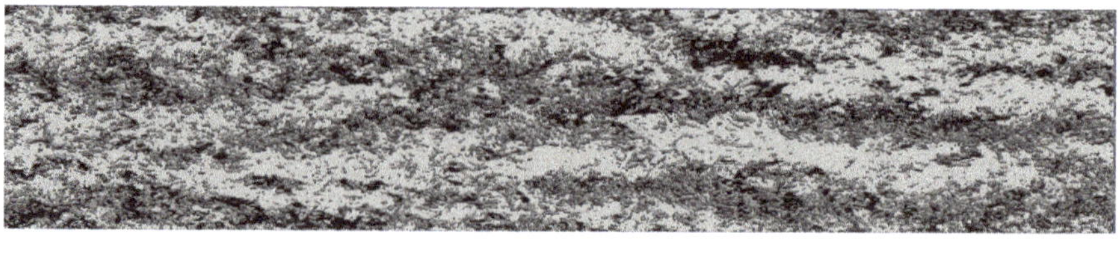

Figure 21. Instantaneous vortical structures visualized with an iso-surface of λ_2 in xy plane.

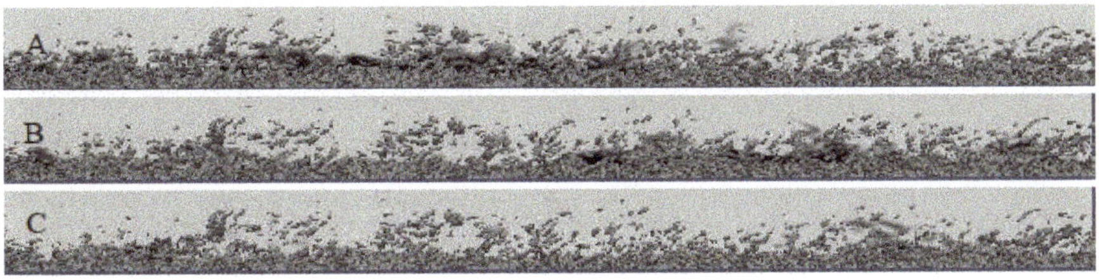

Figure 22. Instantaneous vortical structures visualized with an iso-surface of λ_2 in xz plane, the subfigures (**A**), (**B**) and (**C**) are the same with the Figure 18c, represent the cross section of vortices field.

5. Conclusions

In this paper, the spatial features for LSMs and VLSMs in the TBL flow under different surface roughness at very high Reynolds number, O (10^6–10^7), were investigated using the WMLES approach. The VLSMs in TBL were studied progressively from a single point, in correlation with a field analysis. The statistical properties of the mean and fluctuating velocity for the nearly smooth case were compared with those found in the literature, including field observations, laboratory studies, and DNSs, and a close agreement was found using the inner and outer variables. These results, thus, provide a useful database for the turbulence statistics of TBL.

Our analysis of the streamwise velocity spectra demonstrates the tendency is that with increasing roughness, the scale of VLSMs decreases or even disappears. The $(k\delta)_0$ decreases with height in an approximately log-linear manner. While, when height increases to a certain extent, $(k\delta)_0$ begins to reach a constant value or even to increase. In addition, kS_u decreases with height throughout all spectra.

For wall-normal velocity spectra, the crossover phenomenon occurs, and as roughness increases, the crossover point moves to larger wavenumber.

The VLSMs and LSMs carry a considerable portion of the TKE. While, the energy fraction of VLSMs in the present study is a little smaller than the field-measured data, but it is similar to experimental data, which can be explained by spectral characteristics.

The spatial correlation coefficient pertaining to five different roughness cases were calculated. The results have reached a close agreement with the wind spectra. For the ratio of the streamwise extent and wall-normal extent of the two-point correlations of the fluctuating velocity, the near-constant is 2.5. Moreover, the inclination angle of nearly smooth cases is consistent with field observation data at almost the same Reynolds number. The tendency of the inclination angle of different cases of roughness is similar to what has been shown in other studies.

Then, we conducted the instantaneous flow field analysis to illustrate the generation and development of VLSMs. It is clearly shown that both the VLSMs and LSMs were visualized based on the streamwise-elongated volumes associated with the negative streamwise velocity, and which are flanked on either side by similarly elongated positive streamwise velocity. In addition, the generation and extinction process of VLSMs is seen by three-dimensional flow field analysis.

In the end, Vortices field analysis is conducted to explain further the VLSMs locate in negative streamwise velocity region and each of them lies beneath many hairpin-type vortices.

Above all, in this study, the results at least indirectly support that the bottom-up mechanism is one of the origins of VLSMs. It is confirmed here that the WMLES method has the ability to investigate the spatial features of LSMs and VLSMs in the TBL flow at high Reynolds numbers at neutrally stratified TBL, and the applicability of these results to a real atmospheric boundary layer flow that stability variation will be investigated in the future study.

Author Contributions: Conceptualization, H.R. and S.L.; Methodology, H.R. and S.L.; Software, H.R.; Validation, H.R.; Formal Analysis, H.R.; Investigation, H.R.; Resources, H.L.; Data Curation, H.R.; Writing-Original Draft Preparation, HR.; Writing-Review & Editing, S.L.; Visualization, H.R.; Supervision, S.L. and H.L.; Project

Administration, H.L.; Funding Acquisition, S.L. and H.L. All authors have read and agreed to the published version of the manuscript.

Funding: This study is financially suppborted by the National Key Research and Development Program of China under grant No. 2018YFC0705605, NSFC under grant No. 51878230, the China Postdoctoral Science Foundations (2015M580266 and 2016T90295), the Heilongjiang Postdoctoral Fund (LBH-Z15087), and the Fundamental Research Funds for the Central Universities (HIT. NSRIF. 2017064.)

Conflicts of Interest: The authors declare no conflict of interest.

References

1. Kovasznay, L.S.G.; Kibens, V.; Blackwelder, R.F. Large-scale motion in the intermittent region of a turbulent boundary layer. *J. Fluid Mech.* **1970**, *41*, 283–325. [CrossRef]
2. Balakumar, B.J.; Adrian, R.J. Large-and very-large-scale motions in channel and boundary-layer flows. *Phil. Trans. R. Soc. Lond. A* **2007**, *365*, 665–681. [CrossRef]
3. Kim, K.C.; Adrian, R.J. Very large-scale motion in the outer layer. *Phys. Fluids* **1999**, *11*, 417–422. [CrossRef]
4. Guala, M.; Hommema, S.E.; Adrian, R.J. Large-scale and very-large-scale motions in turbulent pipe flow. *J. Fluid Mech.* **2006**, *554*, 521–542. [CrossRef]
5. Zhou, J.; Adrian, R.J.; Balachandar, S.; Kendall, T.M. Mechanisms for generating coherent packets of hairpin vortices in channel flow. *J. Fluid Mech.* **1999**, *387*, 353–396. [CrossRef]
6. Lee, J.; Lee, J.H.; Choi, J.; Sung, H.J. Spatial organization of large- and very-large-scale motions in a turbulent channel flow. *J. Fluid Mech.* **2014**, *749*, 818–840. [CrossRef]
7. Hutchins, N.; Marusic, I. Evidence of very long meandering features in the logarithmic region of turbulent boundary layers. *J. Fluid Mech.* **2007**, *579*, 1–28. [CrossRef]
8. Lee, J.H.; Sung, H.J. Very-large-scale motions in a turbulent boundary layer. *J. Fluid Mech.* **2011**, *673*, 80–120. [CrossRef]
9. Wang, G.H.; Zheng, X.J. Very large scale motions in the atmospheric surface layer: A field investigation. *J. Fluid Mech.* **2016**, *802*, 464–489. [CrossRef]
10. Monty, J.P.; Stewart, J.A.; Williams, R.C.; Chong, M.S. Large-scale features in turbulent pipe and channel flows. *J. Fluid Mech.* **2007**, *589*, 147–156. [CrossRef]
11. Bailey, S.C.C.; Smits, A.J. Experimental investigation of the structure of large- and very large-scale motions in turbulent pipe flow. *J. Fluid Mech.* **2010**, *651*, 339–356. [CrossRef]
12. Hutchins, N.; Marusic, I. Large-scale influences in near-wall turbulence. *Phil. Trans. R. Soc. Lond. A* **2007**, *365*, 647–664. [CrossRef]
13. Mathis, R.; Hutchins, N.; Marusic, I. Large-scale amplitude modulation of the small-scale structures in turbulent boundary layers. *J. Fluid Mech.* **2009**, *628*, 311–337. [CrossRef]
14. Chung, D.; Mckeon, B.J. Large-eddy simulation of large-scale structures in long channel flow. *J. Fluid Mech.* **2010**, *661*, 341–364. [CrossRef]
15. Hunt, J.C.R.; Morrison, J.F. Eddy structure in turbulent boundary layers. *Eur. J. Mech. B Fluids* **2000**, *19*, 673–694. [CrossRef]
16. Jimenez, J. Turbulence flows over rough walls. *Annu. Rev. Fluid Mech.* **2004**, *36*, 173–196. [CrossRef]
17. Volino, R.J.; Schultz, M.P.; Flack, K.A. Turbulence structure in rough- and smooth-wall boundary layers. *J. Fluid Mech.* **2007**, *592*, 263–293. [CrossRef]
18. Rosenberg, B.J.; Hultmark, M.; Vallikivi, M.; Bailey, S.C.C.; Smits, A.J. Turbulence spectra in smooth- and rough-wall pipe flow at extreme Reynolds numbers. *J. Fluid Mech.* **2013**, *731*, 46–63. [CrossRef]
19. Kline, S.J.; Reynolds, W.C.; Schraub, F.A.; Rundstadler, P.W. The structure of turbulent boundary layers. *J. Fluid Mech.* **1967**, *30*, 741–773. [CrossRef]
20. Jimenez, J. The Largest Scales of Turbulent Wall Flows. 1998. Available online: https://web.stanford.edu/group/ctr/ResBriefs98/jimenez.pdf (accessed on 7 January 2020).
21. Del Alamo, J.C.; Jimenez, J.; Zandonade, P.; Moser, R.D. Scaling of the energy spectra of turbulent channels. *J. Fluid Mech.* **2004**, *500*, 135–144. [CrossRef]
22. Smits, A.J.; Marusic, I. Wall-bounded turbulence. *Physics Today* **2013**, *66*, 25–30. [CrossRef]
23. Moeng, C.H.; Sullivan, P.P. A comparison of shear- and buoyancy-driven planetary boundary layer flows. *J. Atmos Sci.* **1994**, *51*, 999–1022. [CrossRef]

24. Porte-Agel, F.; Meneveau, C.; Parlange, M.B. A scale-dependent dynamic model for large-eddy simulation: Application to a neutral atmospheric boundary layer. *J. Fluid Mech.* **2000**, *415*, 261–284. [CrossRef]
25. Bou-Zeid, E.; Meneveau, C.; Parlange, M.B. A scale-dependent Lagrangian dynamic model for large eddy simulation of complex turbulent flows. *Phys. Fluids* **2005**, *17*, 025105. [CrossRef]
26. Lu, H.; Porte-Agel, F. A modulated gradient model for large-eddy simulation: Application to a neutral atmospheric boundary layer. *Phys. Fluids* **2010**, *22*, 1–12. [CrossRef]
27. Fang, J.N.; Porte-Agel, F. Large-eddy simulation of very-large-scale motions in the neutrally stratified atmospheric boundary layer. *Boundary Layer Meteorol.* **2015**, *155*, 397–416. [CrossRef]
28. Salesky, S.T.; Anderson, W. Buoyancy effects on large-scale motions in convective atmospheric boundary layers: Implications for modulation of near-wall processes. *J. Fluid Mech.* **2018**, *856*, 135–168. [CrossRef]
29. Piomelli, U.; Balaras, E. Wall-layer models for large-eddy simulations. *Annu. Rev. Fluid Mech.* **2002**, *34*, 349–374. [CrossRef]
30. Schumann, U. Subgrid-scale model for finite-difference simulations of turbulent flow in plane channels and annuli. *J. Comput. Phys.* **1975**, *18*, 76–404. [CrossRef]
31. Yoshizawa, A.; Horiuti, K. A statistically-derived subgrid-scale kinetic energy model for the large-eddy simulation of turbulent flows. *J. Phys. Soc. Jpn.* **1985**, *54*, 2834–2839. [CrossRef]
32. Churchfield, M.J.; Moriarty, P.J.; Vijayakumar, G.; Brasseur, J. Wind energy-related atmospheric boundary-layer large-eddy simulation using OpenFOAM. In Proceedings of the 19th Symposium on Boundary Layers and Turbulence, Keystone, CO, USA, 2–6 August 2010.
33. Marusic, I.; Monty, J.P.; Hultmark, M.; Smits, A.J. On the logarithmic region in wall turbulence. *J. Fluid Mech.* **2013**, *716*, R3. [CrossRef]
34. Chauhan, K.A. Study of Canonical Wall-Bounded Turbulent Flows. PhD Thesis, Illinois Institute of Technology, Chicago, IL, USA, January 2007.
35. Nagib, H.M.; Chauhan, K.A. Variations of von Kármán coefficient in canonical flows. *Phys. Fluids* **2008**, *20*, 1518. [CrossRef]
36. Hutchins, N.; Chauhan, K.; Marusic, I.; Monty, J.; Klewicki, J. Towards reconciling the large-scale structure of turbulent boundary layers in the atmosphere and laboratory. *Bound. Layer Meteorol.* **2012**, *145*, 273–306. [CrossRef]
37. De Graaff, D.B.; Eaton, J.K. Reynolds-number scaling of the flat-plate turbulent boundary layer. *J. Fluid Mech.* **2000**, *422*, 319–346. [CrossRef]
38. Kunkel, G.J.; Marusic, I. Study of the near-wall-turbulent region of the high-Reynolds-number boundary layer using an atmospheric flow. *J. Fluid Mech.* **2006**, *548*, 375–402. [CrossRef]
39. Metzger, M.M.; Klewicki, J.C.; Bradshaw, K.L.; Sadr, R. Scaling the near-wall axial turbulent stress in the zero pressure gradient boundary layer. *Phys. Fluids* **2001**, *13*, 1819–1821. [CrossRef]
40. Hutchins, N.; Nickels, T.B.; Marusic, I.; Chong, M.S. Hot-wire spatial resolution issues in wall-bounded turbulence. *J. Fluid Mech.* **2009**, *635*, 103–136. [CrossRef]
41. Marusic, I.; Kunkel, G.J. Streamwise turbulence intensity formulation for flat-plate boundary layers. *Phys. Fluids* **2003**, *15*, 2461–2464. [CrossRef]
42. Österlund, J.M.; Johansson, A.V.; Nagib, H.M.; Hites, M.H. A note on the overlap region in turbulent boundary layers. *Phys. Fluids* **2000**, *12*, 1–4. [CrossRef]
43. Knobloch, K.; Fernholz, H. Statistics, correlations, and scaling in a turbulent boundary layer at $Re_{\delta 2} \leq 1.15 \times 105$. In Proceedings of the IUTAM Symposium on Reynolds Number Scaling in Turbulent Flow, Princeton, NJ, USA, 11–13 September 2002; pp. 11–16.
44. Kulandaivelu, V.; Marusic, I. Evolution of zero pressure gradient turbulent boundary layers. In Proceedings of the 17th Australasian Fluid Mechanics Conference, Auckland, New Zealand, 5–9 December 2010.
45. Balasubramaniam, B.J. Nature of Turbulence in Wall-Bounded Flows. Ph.D. Thesis, University of Illinois at Urbana-Champaign, Urbana/Champaign, IL, USA, January 2005.
46. Vallikivi, M.; Ganapathisubramani, B.; Smits, A.J. Spectral scaling in boundary layers and pipes at very high Reynolds numbers. *J. Fluid Mech.* **2015**, *771*, 303–326. [CrossRef]
47. Morrison, J.F.; Mckeon, B.J.; Jiang, W.; Smits, A.J. Scaling of the streamwise velocity component in turbulent pipe flow. *J. Fluid Mech.* **2004**, *508*, 99–131. [CrossRef]
48. Lee, J.; Ahn, J.; Sung, H.J. Comparison of large- and very-large-scale motions in turbulent pipe and channel flows. *Phys. Fluids* **2015**, *27*, 025101. [CrossRef]

49. Marusic, I.; Heuer, W.D. Reynolds number invariance of the structure inclination angle in wall turbulence. *Phys. Rev. Lett.* **2007**, *99*, 114504. [CrossRef] [PubMed]
50. Christensen, K.T.; Wu, Y. Characteristics of vortex organization in the outer layer of wall turbulence. In Proceedings of the Fourth International Symposium on Turbulence and Shear Flow Phenomena, Williamsburg, VA, USA, 27–29 June 2005; pp. 1025–1030.
51. Head, M.R.; Bandyopadhyay, P. New aspects of turbulent boundary-layer structure. *J. Fluid Mech.* **1981**, *107*, 297–337. [CrossRef]
52. Christensen, K.T.; Adrian, R.J. Statistical evidence of hairpin vortex packets in wall turbulence. *J. Fluid Mech.* **2001**, *431*, 433–443. [CrossRef]
53. Adrian, R.J.; Meinhart, C.D.; Tomkins, C.D. Vortex organization in the outer region of the turbulent boundary layer. *J. Fluid Mech.* **2000**, *422*, 1. [CrossRef]
54. Tomkins, C.D.; Adrian, R.J. Spanwise structure and scale growth in turbulent boundary layers. *J. Fluid Mech.* **2003**, *490*, 37–74. [CrossRef]
55. Nakagawa, S.; Hanratty, T.J. Particle image velocimetry measurements of flow over a wavy wall. *Phys. Fluids* **2001**, *13*, 3504–3507. [CrossRef]
56. Krogstad, P.A.; Antonia, R.A. Structure of turbulent boundary layers on smooth and rough walls. *J. Fluid Mech.* **1994**, *277*, 1–21. [CrossRef]
57. Dennis, D.J.C.; Nickels, T.B. Experimental measurement of large-scale three-dimensional structures in a turbulent boundary layer. Part 1. Vortex packets. *J. Fluid Mech.* **2011**, *673*, 180–217. [CrossRef]
58. Dennis, D.J.C.; Nickels, T.B. Experimental measurement of large-scale three-dimensional structures in a turbulent boundary layer. Part 2. Long structures. *J. Fluid Mech.* **2011**, *673*, 218–244. [CrossRef]
59. Baltzer, J.R.; Adrian, R.J.; Wu, X. Structural organization of large and very large scales in turbulent pipe flow simulation. *J. Fluid Mech.* **2013**, *720*, 236–279. [CrossRef]
60. Jeong, J.; Hussain, F. On the definition of a vortex. *J. Fluid Mech.* **1995**, *285*, 69–94. [CrossRef]

energies

MDPI

Article

Surrogate Model with a Deep Neural Network to Evaluate Gas–Liquid Flow in a Horizontal Pipe

Yongho Seong [1], Changhyup Park [1,*], Jinho Choi [2] and Ilsik Jang [3]

1 Department of Energy and Resources Engineering, Kangwon National University, Chuncheon, Kangwon 24341, Korea; tjddydgh77@naver.com

2 Naval & Energy System R&D Institute, Daewoo Shipbuilding & Marine Engineering Co., Ltd., Siheung, Gyeonggi 15011, Korea; jinhochoi@dsme.co.kr

3 Department of Energy and Resources Engineering, Chosun University, Gwangju 61425, Korea; isjang77@chosun.ac.kr

* Correspondence: changhyup@kangwon.ac.kr; Tel.: +82-33-250-6259

Received: 3 January 2020; Accepted: 20 February 2020; Published: 21 February 2020

Abstract: This study developed a data-driven surrogate model based on a deep neural network (DNN) to evaluate gas–liquid multiphase flow occurring in horizontal pipes. It estimated the liquid holdup and pressure gradient under a slip condition and different flow patterns, i.e., slug, annular, stratified flow, etc. The inputs of the surrogate modelling were related to the fluid properties and the dynamic data, e.g., superficial velocities at the inlet, while the outputs were the liquid holdup and pressure gradient observed at the outlet. The case study determined the optimal number of hidden neurons by considering the processing time and the validation error. A total of 350 experimental data were used: 279 for supervised training, 31 for validating the training performance, and 40 unknown data, not used in training and validation, were examined to forecast the liquid holdup and pressure gradient. The liquid holdups were estimated within less than 8.08% of the mean absolute percentage error, while the error of the pressure gradient was 23.76%. The R^2 values confirmed the reliability of the developed model, showing 0.89 for liquid holdups and 0.98 for pressure gradients. The DNN-based surrogate model can be applicable to estimate liquid holdup and pressure gradients in a more realistic manner with a small amount of computating resources.

Keywords: surrogate model; deep neural network; multiphase flow; horizontal pipe; liquid holdup; pressure gradient

1. Introduction

The accurate evaluation of multiphase flow has been essential not only for optimum facility designs, but also for the estimation of transport features in pre-installed pipes. The oil and natural gas industries require time-consuming experimental analyses, but the flow characteristics, e.g., flow patterns, liquid holdup, superficial velocities, and pressure gradient, are uncertain. These uncertain parameters are important for demonstrating transport phenomena but they are linked nonlinearly. A few analytical models implement closure relationships that depend on experimental data and therefore they have the limitation of the applicable ranges used in the experimental works. The experiments validated the numerical simulations with the closure relationships, as well as the mechanistic modules, e.g., OLGA (Schlumberger, Houston, USA) and Ledaflow (Kongsberg, Norway). However, the detailed designs of these numerical simulations become complicated, e.g., computational fluid dynamics, and require a large amount of computing resources and skillful user interactions [1,2].

Many works have validated their effectiveness using the experimental database, e.g., TUFFPDB (University of Tulsa Fluid Flow Projects Database) and the experimental database of the University of Amsterdam. The empirical interpretations proposed different interpolation rules within the experimental ranges [3–9]. Choi et al. [6] introduced a drift–flux closure relationship to interpolate the experimental results and estimate the liquid holdup. Their model estimated the liquid holdups of experimental data from 0.042 up to 0.156 of the mean absolute error. Lee et al. [7] examined the applicability of the minimum dissipated energy concept to explain the stratified gas–liquid flow in horizontal pipes and confirmed its usefulness by estimating liquid holdups and pressure gradients. Luo et al. [9] proposed an empirical power-law model not subject to the experimental ranges to predict the liquid holdup in vertical gas wells.

Machine learning methods have been actively implemented to predict the multiphase flow characteristics [10–15]: some studies implemented an artificial neural network [10–12] and Mask et al. [13] implemented dimensionless parameters to estimate the flow patterns. Kanin et al. [14] trained a machine learning algorithm to validate the mechanistic models. Mohammadi et al. [15] used a genetic algorithm to select closure relationships in multiphase flow models. These studies have focused on securing the most accurate ways to explain the nonlinearity more effectively between input and output data. Deep neural networks (DNNs), as a class of machine learning techniques, implement a few hierarchical hidden layers of non-linear transforms and improves the performances of a typical artificial neural network with one hidden layer [16–19]. The neural networks normalize the input data such that they can integrate different-scaled data. Added hidden layers reduce the number of calibrating parameters, thereby it can seek the non-linear relationships among the parameters more effectively, i.e., deep learning.

An objective of this paper was to develop a DNN framework, i.e., a data-driven surrogate (proxy) model, to estimate the liquid holdup and pressure gradient for a gas–liquid multiphase flow in horizontal pipes instead of using numerical simulations. DNN with multiple hidden layers was applied to secure an accurate empirical relationship between the experimental input factors and the estimating values. The experimental data were used to validate the reliability, to train the neural network, and to predict the flow-related parameters.

2. Methodology

The representative parameters used to explain gas–liquid flow characteristics are the liquid holdup and pressure gradient; the liquid holdup, i.e., the fraction of a part of pipe occupied by liquid, demonstrates the amount of liquid transported through pipes, while the pressure gradient, i.e., pressure drop divided by the given length of the horizontal pipe, is essential for designing the transport facilities. Various empirical correlations have been proposed to explain these parameters, e.g., Equations (1) and (2) for a horizontal pipe [20–22], but these correlations are challenging to use due to the complexity and the uncertainty of multiphase flow.

$$H_L = \begin{cases} \lambda_L = \frac{A_L}{A_P} = \frac{q_L}{q_L+q_g} = \frac{v_{SL}}{v_{SL}+v_{SG}}, & \text{no–slip condition} \\ \frac{a_1 \lambda_L^{a_2}}{\left(N_{FR}^2\right)^{a_3}}, & \text{slip condition} \end{cases} \quad (1)$$

$$\frac{dP}{dL} = \begin{cases} -\frac{f \rho_n v_m^2}{2d} - \rho_n v_m \frac{dv_m}{dL}, & \text{homogeneous flow} \\ -\frac{\frac{f \rho_n v_m^2}{2d}}{1-E_k}, & \text{non–homogeneous flow} \end{cases} \quad (2)$$

In Equations (1) and (2), the no-slip liquid holdup (λ_L) is the ratio of the liquid volumetric flow rate to the total volumetric flow rate [22]. A_P represents the cross-sectional area of pipe and A_L is the cross-sectional area occupied by liquid in the pipe. q_L is the flow rate of liquid and q_g is the flow rate of gas. $v_{SL} = \frac{q_L}{A_P}$ is the liquid superficial velocity and $v_{SG} = \frac{q_g}{A_P}$ is the gas superficial velocity. a_1, a_2, and a_3 are the coefficients for liquid holdup correlation, which are different for different flow patterns. f is the friction factor related to the pipe diameter (d) and roughness. $N_{FR} = \frac{v_m^2}{gd}$ is the mixture Froude number, where $v_m = v_{SL} + v_{SG}$ (mixture velocity). E_k is a dimensionless kinetic or acceleration energy term. ρ_n is the no-slip density. The no-slip condition occurs where the input phase-superficial velocities are constant in the course of the gas–liquid flow, e.g., a typical stratified flow pattern. A homogeneous two-phase flow is defined as the flow structure in which the two phases travel at the same in-situ velocity while a non-homogeneous flow means the flow structure in which the mixture physical properties vary across the pipe's cross-sectional area [22].

The experimental data were from Gokcal [23,24], which considered a viscous oil–air mixed flow in horizontal pipes, where the viscous oil was Citgo sentry 220 oil with a density range from 833.6 to 884.5 kg/m^3 (Table 1). Gokcal [23] investigated various flow patterns while Gokcal [24] investigated slug flow. The experimental data showed a slip condition since the liquid holdups resulting from the multi-phase flow experiment were not equal to the no-slip liquid holdups (λ_L). The total number of available data was 350: 279 for the supervised training of the DNN, 31 for validating the training performance, and the other 40 data were for testing the predictability. A reason why 310 training data were divided into the training and validation sets was to examine any overfitting or underfitting problems. A data-shuffling process was implemented to split the data. A total of 40 experimental data that were not used in either training or validation were randomly selected for testing the predictability (Table 2).

Table 1. Summary of experimental data for the input and the output parameters in the deep neural network (DNN).

Experimental Data	Parameters [1]	Number of Data Points
Gokcal [23]	Air and oil (Citgo sentry 220 oil) ID = 0.0508 m, T = 20.8–38.1 °C ρ_L = 833.6–884.5 kg/m^3, ρ_G =1.25–4.5 kg/m^3 v_{SL} = 0.01–1.76 m/s, v_{SG} = 0.09–20.3 m/s Annular (33) [2], annular/slug (4), stratified wavy (3), slug (120), elongated bubble (19), dispersed bubble/slug (4)	183
Gokcal [24]	Air and oil (Citgo sentry 220 oil) ID = 0.0508 m, T = 20.8–38.1 °C ρ_L = 833.6–884.5 kg/m^3, ρ_G = 1.12–2.08 kg/m^3 v_{SL} = 0.05–0.8 m/s, v_{SG} = 0.1–2.17 m/s Slug (167)	167

[1] ID: Inner diameter, T: Temperature; [2] (number) indicates the number of available experimental data with the specified flow pattern; The subscripts L, G, SL, SG stand for liquid, gas, liquid superficial velocity, and gas superficial velocity, respectively.

Table 2. Experimental data and their flow patterns used in the DNN model.

	Training Operation		Prediction (Test Set)
	Training	Validation	
Number of Data Points	279	31	40
Flow pattern	Annular (31), annular/slug (3), stratified wavy (3), elongated bubble (14), dispersed bubble/slug (4), slug (257)		Annular (4), annular/slug (1), elongated bubble (5), slug (30)

Figure 1 explains how to establish empirical correlations using a backpropagation algorithm between the input and output data in multi-hidden-layered neural networks (Equation (3); [18,25]).

$$\begin{aligned} a_i^{(1)} &= f_a\left(\sum_{j=1}^{n} w_{i,j}^{(1)} x_j + b_j^{(1)}\right), \quad \forall i = 1, 2, 3, \cdots, m, \\ a_i^{(2)} &= f_a\left(\sum_{j=1}^{m} w_{i,j}^{(2)} a_j^{(1)} + b_j^{(2)}\right), \quad \forall i = 1, 2, 3, \cdots, h, \\ \hat{y}_i &= f_a\left(\sum_{j=1}^{h} w_{i,j}^{(3)} a_j^{(2)} + b_j^{(3)}\right), \quad \forall i = 1, 2, 3, \cdots, l. \end{aligned} \tag{3}$$

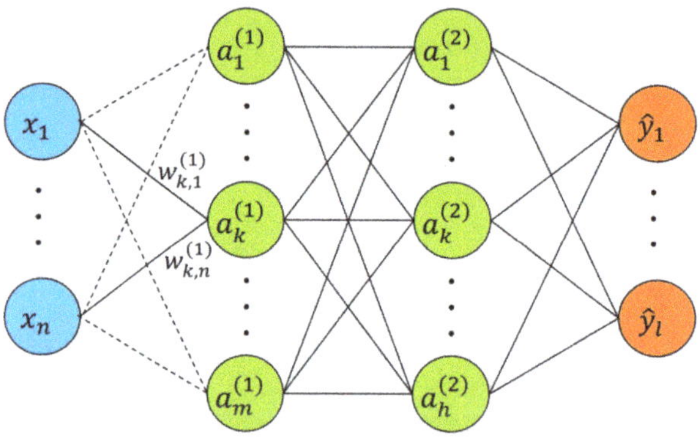

Figure 1. DNN structure to establish the empirical relationship between the input and the output layer.

In Equation (3), a_i^k denotes the value of the ith neuron at the kth hidden layer. f_a is an activation function. w is a weight and b is a bias. x is the value in the input layer and $\hat{y}$ is the estimated value at the output layer. Figure 2 depicts the DNN framework consisting of one input, one output, and two hidden layers in this study. The performance of estimating the output neurons depended on the design of hidden layers, e.g., the number of neurons in the hidden layers, the activation function, and the filters.

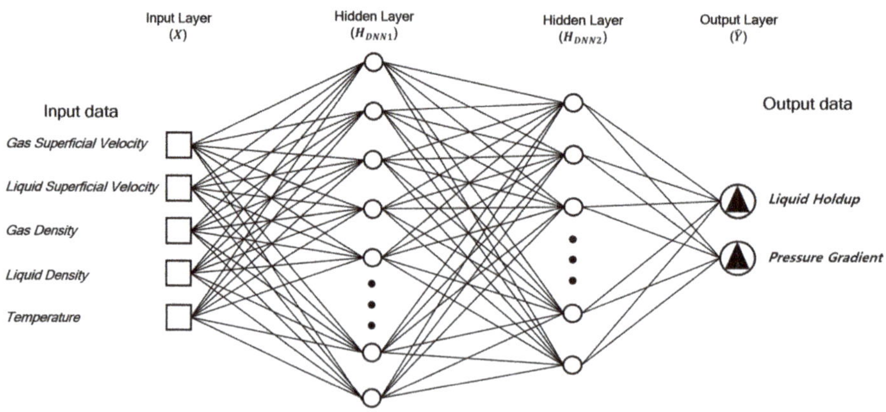

Figure 2. DNN design to estimate the liquid holdup and pressure gradient. One input layer (X), one output layer ($\hat{Y}$), and two hidden layers (H_{DNN1} and H_{DNN2}) were implemented.

The input layer (X) was composed of the normalized properties, such as the liquid superficial velocity, gas superficial velocity, liquid density, gas density, and temperature (see Equation (4)). On the other hand, the output layer ($\hat{Y}$) had the liquid holdup and pressure gradient. Two hidden layers were introduced and the effects of setting the number of neurons were examined as a result of changing the neurons in the hidden layers in terms of the computation cost, i.e., the processing time, and the validation loss, i.e., the mean square error between the estimated values and the true (experimental) data for cost function. The number of nodes in the second hidden layer (H_{DNN2}) was assumed to be half of the node number in the first hidden layer (H_{DNN1}).

$$x_{norm} = \frac{x - x_{MIN}}{x_{MAX} - x_{MIN}} \tag{4}$$

In Equation (4), x_{norm} in X represents the normalized x value, and the subscripts MIN and MAX are the minimum value and the maximum value of x. The Nesterov accelerated gradient was introduced as a momentum optimizer (Equation (5); [26]) and the rectified linear unit (ReLU) was used for the activation function. The training continued until the optimizer converged before 40 epochs. Some methods, e.g., dropout, batch normalization, and weight regularization, was used to try to fix the overfitting matter.

$$\begin{aligned} \theta_t &= \gamma\theta_{t-1} + \eta\nabla_w J(w - \gamma\theta_{t-1}) \\ w &= w - \theta_t \end{aligned} \tag{5}$$

In Equation (5), w is the parameter to update, i.e., weight. θ_t denotes the update vector of the current time, t. γ is the momentum term that this work set to 0.9. J is a loss function, i.e., cost function, and $\nabla_w J$ is a gradient vector of the loss function. η represents the learning rate that was assigned 0.001 in this study. He et al.'s scheme initializes the weights applicable to ReLU (Equation (6); [27]). W is a set of weights. $\sigma^2(W)$ is the variance of weights and n_{in} is the number of nodes in the previous layer.

$$\begin{aligned} W &\sim \mathrm{N}\left(0,\ \sigma^2(\mathrm{W})\right) \\ \sigma^2 &= \tfrac{2}{n_{in}} \end{aligned} \tag{6}$$

Three kinds of errors were applied: the root mean squared error (RMSE; Equation (7)), the percentage error (Equation (8)), and the mean absolute percentage error (MAPE; Equation (9)):

$$\varepsilon_{RMSE} = \sqrt{\frac{1}{n}\sum\nolimits_{i=1}^{n} \left(Y_i - \hat{Y}_i\right)^2}, \tag{7}$$

$$\varepsilon_{PE} = 100\left|\frac{Y_i - \hat{Y}_i}{Y_i}\right|, \tag{8}$$

$$\varepsilon_{MAPE} = \frac{100}{n}\sum_{i=1}^{n}\left|\frac{Y_i - \hat{Y}_i}{Y_i}\right|. \tag{9}$$

In Equations (7)–(9), ε_{RMSE} denotes RMSE, n is the number of output data points, Y_i represents the actual experimental data, and $\hat{Y}_i$ are the values estimated using the DNN. The percentage error (ε_{PE}) measures the difference between the predicted and the experimental value, divided by the experimental value, and expresses the error as a percentage. The MAPE (ε_{MAPE}) is the arithmetic mean value of the percentage errors and is implemented for both the validation and testing sets. The optimum number of nodes in the hidden layers was determined using the results of the processing time, i.e., training time to construct the DNN model, and the MAPEs.

3. Results and Discussion

3.1. Design of the Deep Neural Network: The Number of Hidden Neurons

Both performances of training with validation loss and the predictability with the testing dataset were examined by changing the number of nodes in the hidden layers. Since the node number of the second hidden layer was set as half of that in the first hidden layer, the performances were observed in terms of the number of the first hidden layer from 10 to 1000. The processing time refers to the training time, i.e., time for constructing the DNN using supervised training. It was measured when the optimization scheme, i.e., the Nesterov accelerated gradient, converged until 40 epochs, thereby it can be different with the number of nodes in the hidden layers and the final epoch number. The measuring computer system was as follows: Intel Xeon CPU E5-1620v3 @3.50GHz, 64GB RAM, and NVIDIA GeForce GTX 1060 6GB.

Table 3 summarizes the range of the validation loss to evaluate the DNN training performances. The RMSEs of the validation dataset ranged from 0.0632 to 0.0742 (unitless) for the liquid holdup and from 615.87 to 903.84 Pa/m for the pressure gradient. The predictability of the pressure gradient was less than that of liquid holdup; the reasons for this could be the high nonlinearity and complexity of the pressure gradient. The large range of the pressure gradient, e.g., from 6000 to 8000 Pa/m, caused a high RMSE such that both the RMSE and MAPE should be considered.

Table 3. Summary of the processing time and the errors for validating the DNN training performances.

Number of Nodes [1] (First Hidden Layer)	Processing Time (s)	RMSE	MAPE (%)
10–30	116.02–131.49	0.0648–0.0742 (H_L) 681.04–838.40 (dP/dL)	9.778–11.015 (H_L) 33.564–142.051 (dP/dL)
30–100	128.87–144.19	0.0648–0.0721 (H_L) 618.04–773.73 (dP/dL)	9.791–11.170 (H_L) 45.690–94.514 (dP/dL)
100–1000	138.16–176.02	0.0632–0.0735 (H_L) 615.87–903.84 (dP/dL)	9.376–11.687 (H_L) 47.104–75.831 (dP/dL)

[1] The number of neurons in the second hidden layer assumes the half of given number in the first hidden layer. RMSE: Root Mean Squared Error; MAPE: Mean Absolute Percentage Error. H_L: liquid holdup (unitless); dP/dL: pressure gradient (Pa/m) in RMSE

Figure 3 depicts the processing time and the MAPE (ε_{MAPE} defined in Equation (9)) of liquid holdups estimated using the validation data (31 data not used in training). Figure 4 demonstrates those of the pressure gradient. Both overall trends of processing time increased as the number of nodes in hidden layers increased. The processing time for less than 100 nodes in the first hidden layer was not high, but if applying over 100 nodes, the processing time increased dramatically in the case of predicting the liquid holdup (Figure 3). For the pressure gradient, the trend of processing time was similar to that of the liquid holdup but the errors fluctuated in the range of 30 to 100 for the number of nodes in the first hidden layer. The overall trend of errors, i.e., both for the liquid holdup and pressure gradient, converged after 70 nodes and decreased by up to 30% of the MAPE (Figures 3 and 4). The errors did not necessarily decrease even when the number of neurons in the hidden layer increased. This trend of processing time and error shows that the optimum conditions satisfying the acceptable training time and errors could be contentious because this study assumed several fixed constraints: the neuron number in the second hidden layer was half that in the first hidden layer and the converging criterion was limited until the given epoch. Figures 3 and 4 can be used to recommend 70 nodes in the first hidden layer and 35 nodes in the second hidden layer as the optimum DNN model with the consideration of the processing time and the training accuracy for both the results of validating the liquid holdup and pressure gradient.

(a)

(b)

(c)

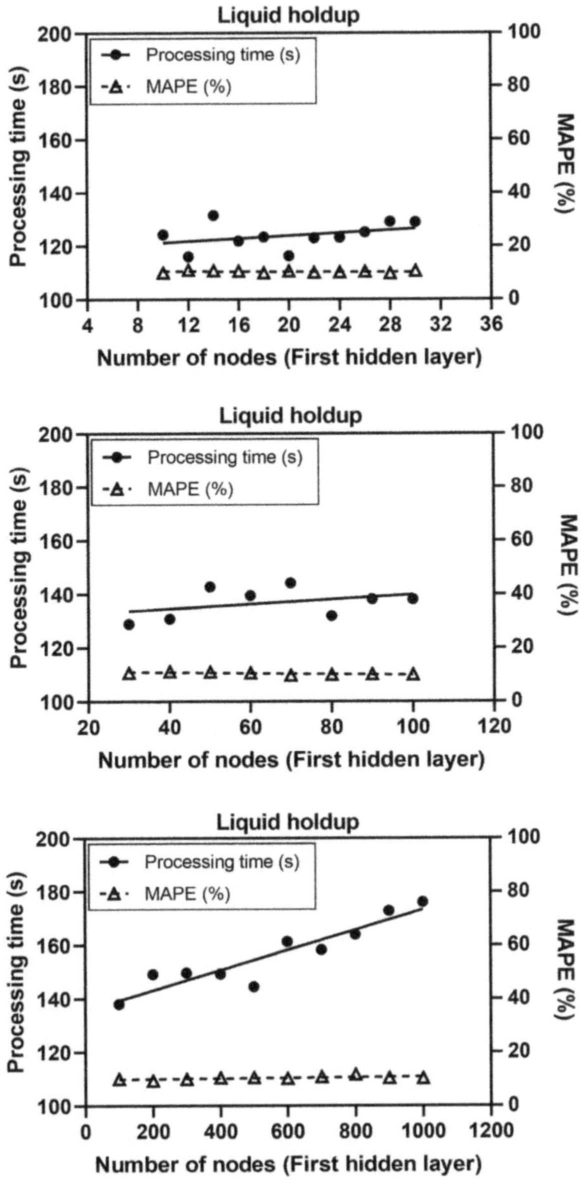

Figure 3. Processing time and training accuracy of the liquid holdup (unitless) for the validation data (31 data points) in relation to the number of nodes in the first hidden layer (H_{DNN1}) in the range of (**a**) 10 to 30, (**b**) 30 to 100, and (**c**) 100 to 1000.

(a)
(b)
(c)

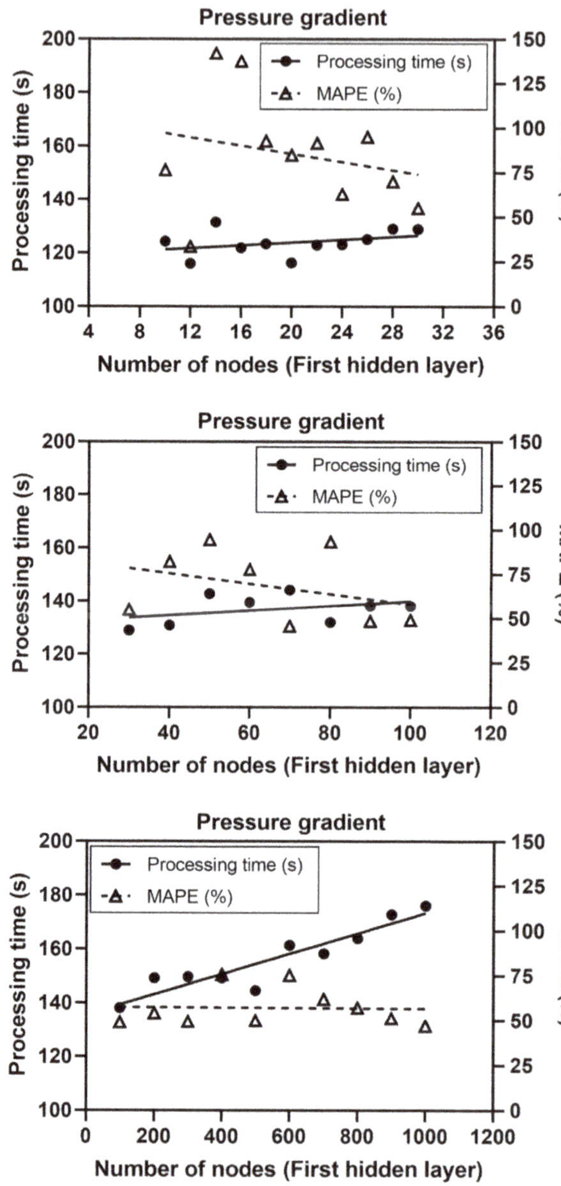

Figure 4. Processing time and training accuracy of pressure gradients (Pa/m) for the validation data (31 data points) in relation to the number of nodes in the first hidden layer (H_{DNN1}) in the range of (**a**) 10 to 30, (**b**) 30 to 100, and (**c**) 100 to 1000.

3.2. Prediction Accuracy of the Liquid Holdups and Pressure Gradients

Figure 5 compares the values predicted using the DNN with those of the experimental results; the proposed DNN implemented 70 nodes in the first hidden layer and 35 nodes in the second hidden layer. The MAPEs were 8.08% for the liquid holdup and 23.76% for the pressure gradient. The R^2 showed a high correlation, i.e., 0.89 for the liquid holdup and 0.98 for the pressure gradient, which was enough to confirm the robustness of the proposed DNN (Table 4). The predictability of the pressure gradients was reliable given the high R^2 value, notwithstanding that they had a wide range from 0 to near 8000 Pa/m in this study. The predictability of the liquid holdup was better or similar to the existing closure relationships [6,8,10,11], e.g., 3.7%–24.5% [6]. The errors of the pressure gradient were not less than the results of the previous works, e.g., around 4%–36% [5,7,8]. Since the available data and the methods in the previous works are different, a direct comparison of the results does not provide a reliable evaluation. Therefore, the errors and the R^2 values, i.e., the correlation between the true and the estimated values, must examine the reliability of the developed model.

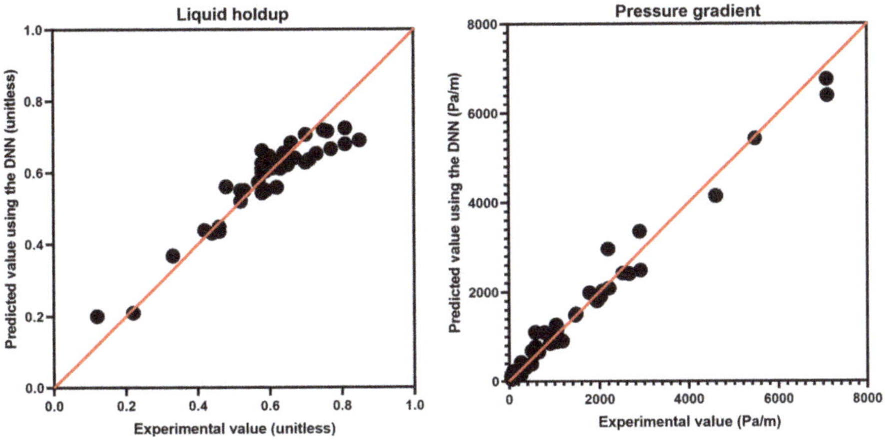

(a) (b)

Figure 5. Prediction accuracy of the proposed DNN model: (**a**) liquid holdup and (**b**) pressure gradient. A total of 40 experimental data points, not from within the training and validation data sets, were used to evaluate the prediction performances.

Table 4. Prediction accuracy of liquid holdups and pressure gradients with the testing dataset (40 data).

Parameter	RMSE	MAPE (%)	R^2
Liquid holdup	0.0056	8.07868	0.8855
Pressure gradient	261.6052	23.7609	0.9802

Figure 6 depicts the percentage errors (Equation (8)) with the values estimated from the DNN. The largest percentage error was 67.23%, for the case of annular flow, with a 0.12 (relatively a small value) liquid holdup, while that of the pressure gradient was 167.0%, observed for slug flow, with 48 Pa/m (a small value). Given the definition of MAPE (Equation (9)), the smaller the value of the denominator, the more the error tends to increase. Five outliers, with percentage errors of 50% or more, influenced the MAPE; the number of outliers when predicting liquid holdups was 1 and that for pressure gradients was 4. All singular values were in the zone where the true values were small. Therefore, Figure 6 proves that the developed model could accurately estimate both the liquid holdup and the pressure gradient, except for the outliers.

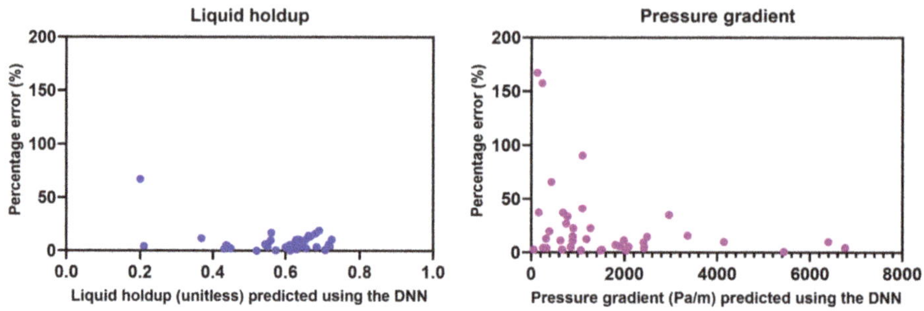

(a) (b)

Figure 6. Scatter plot of the percentage errors using the testing dataset: (**a**) liquid holdup and (**b**) pressure gradient.

A notable result is that the DNN-based model can be applicable to reliably estimate the flow characteristics without time-consuming numerical simulations and user interactions, even though the flow patterns are various in horizontal pipes. The factors influencing the liquid holdup or pressure gradient can be different and thereby the separate model composed of different input parameters can improve the prediction performances. This study discusses the characteristics of gas–liquid multiphase flows that are easily identified using only five simple input parameters. Obtaining enough experimental data to improve the DNN workflow is challenging; the time-consuming work and the complexity of the experimental designs limit the ability to obtain big data suitable for training neural networks and for improving the predictability under various experimental conditions. This paper is limited for the case of horizontal flow and a few flow patterns. If the training and testing data with different flow patterns is enough, a more quantitative analysis would be available that is related to the effects of flow patterns on the prediction accuracy. Constructing the standard forms of neural networks, e.g., determination of the optimum activation functions and elements of the neural networks, is still challenging to obtain to improve the robustness of the deep learning. Enhancement of the neural network requires the intensive examinations of various activation functions and details of designing neural networks to yield more reliable predictions.

4. Conclusions

This study developed a noble machine learning approach based on a deep neural network to evaluate the liquid holdup and pressure gradient of a gas–liquid two-phase flow in a horizontal pipe. The optimal design of two hidden layers in the neural network was obtained from the comparison of the processing time and training accuracy. Experimental results with different flow patterns in a horizontal pipe examined the reliability of the proposed DNN-based surrogate model: the mean absolute percentage errors were 8.08% for the liquid holdup and 23.76% for the pressure gradient, while the R^2 values were 0.89 for the liquid holdup and 0.98 for the pressure gradient. This study developed the data-driven surrogate model to accurately estimate both the liquid holdup and pressure gradient without time-consuming numerical simulations.

Author Contributions: Conceptualization, C.P. and J.C.; formal analysis, C.P. and I.J.; investigation, Y.S. and I.J.; methodology, Y.S. and I.J.; resources, J.C.; software, Y.S.; supervision, C.P.; validation, C.P.; project administration, C.P.; funding acquisition, C.P. All authors have read and agreed to the published version of the manuscript.

Funding: This study was supported by Basic Science Research Program through the National Research Foundation of Korea (NRF) funded by the Ministry of Education (2017R1D1A1B04033060), by the Korea Institute of Energy Technology Evaluation and Planning (KETEP) and the Ministry of Trade, Industry & Energy (MOTIE; Nos. 20172510102150; 20172510102160), and by 2016 Research Grant from Kangwon National University, Korea (No. 520160435).

Acknowledgments: The experimental data were from Gokcal's works [23,24] (The University of Tulsa, Tulsa, OK, USA).

Conflicts of Interest: The authors declare no conflict of interest.

References

1. Choi, J.; Pereyra, E.; Sarica, C.; Lee, H.; Jang, I.; Kang, J. Development of a fast transient simulator for gas–liquid flow in pipes. *J. Petro. Sci. Eng.* **2013**, *102*, 27–35. [CrossRef]
2. López, J.; Pineda, H.; Bello, D.; Ratkovich, N. Study of liquid–gas two–phase flow in horizontal pipes using high speed filming and computational fluid dynamics. *Exp. Therm. Fluid. Sci.* **2016**, *76*, 126–134. [CrossRef]
3. Ottens, M.; Hoefsloot, H.C.J.; Hamersma, P.J. Correlations predicting liquid hold–up and pressure gradient in steady–state (nearly) horizontal co–current gas–liquid pipe flow. *Chem. Eng. Res. Des.* **2001**, *79*, 581–592. [CrossRef]
4. Meng, W.; Chen, X.T.; Kouba, G.E.; Sarica, C.; Brill, J.P. Experimental study of low–liquid–loading gas–liquid flow in near–horizontal pipes. *SPE Prod. Facil.* **2001**, *16*, 240–249. [CrossRef]
5. Vielma, J.C.; Shoham, O.; Mohan, R.S.; Gomez, L.E. Prediction of frictional pressure gradient in horizontal oil/water dispersion flow. *SPE J.* **2011**, *16*, 148–154. [CrossRef]
6. Choi, J.; Pereyra, E.; Sarica, C.; Park, C.; Kang, J. An efficient drift–flux closure relationship to estimate liquid holdups of gas–liquid two–phase flow in pipes. *Energies* **2012**, *5*, 5294–5306. [CrossRef]
7. Lee, H.; Al–Sarkhi, A.; Pereyra, E.; Sarica, C.; Park, C.; Kang, J.; Choi, J. Hydrodynamics model for gas–liquid stratified flow in horizontal pipes using minimum dissipated energy concept. *J. Petro. Sci. Eng.* **2013**, *108*, 336–341. [CrossRef]
8. Xu, D.; Li, X.; Li, Y.; Teng, S. A two–phase flow model to predict liquid holdup and pressure gradient of horizontal well. In Proceedings of the SPE/IATMI Asia Pacific Oil & Gas Conference and Exhibition, Bali, Indonesia, 20–22 October 2015. SPE-176229. [CrossRef]
9. Luo, C.; Zhang, L.; Liu, Y.; Zhao, Y.; Xie, C.; Wang, L.; Wu, P. An improved model to predict liquid holdup in vertical gas wells. *J. Petro. Sci. Eng.* **2020**, *184*, 106491. [CrossRef]
10. Osman, E.A. Artificial neural networks models for identifying flow regimes and predicting liquid holdup in horizontal multiphase flow. In Proceedings of the SPE Middle East Oil Show, Manama, Bahrain, 17–20 March 2001. SPE-68219. [CrossRef]
11. Shippen, M.E.; Scott, S.L. A neural network model for prediction of liquid holdup in two–phase horizontal flow. *SPE Prod. Facil.* **2004**, *19*, 67–76. [CrossRef]
12. Al-Naser, M.; Elshafei, M.; Al-Sarkhi, A. Artificial neural network application for multiphase flow patterns detection: A new approach. *J. Petro. Sci. Eng.* **2016**, *145*, 548–564. [CrossRef]
13. Mask, G.; Wu, X.; Ling, K. An improved model for gas–liquid flow pattern prediction based on machine learning. *J. Petro. Sci. Eng.* **2019**, *183*, 106370. [CrossRef]
14. Kanin, E.A.; Osiptsov, A.A.; Vainshtein, A.L.; Burnaev, E.V. A predictive model for steady–state multiphase pipe flow: Machine learning on lab data. *J. Petro. Sci. Eng.* **2019**, *180*, 727–746. [CrossRef]
15. Mohammadi, S.; Papa, M.; Pereyra, E.; Sarica, C. Genetic algorithm to select a set of closure relationships in multiphase flow models. *J. Petro. Sci. Eng.* **2019**, *181*, 106224. [CrossRef]
16. Reed, R.; Marks, R.J. *Neural Smithing: Supervised Learning in Feedforward Artificial Neural Networks*; MIT Press: London, UK, 1999; ISBN 9780262282215.
17. Hinton, G.; Osindero, S.; Teh, Y. A fast learning algorithm for deep belief nets. *Neural Comput.* **2006**, *18*, 1527–1554. [CrossRef] [PubMed]
18. LeCun, Y.; Bengio, Y.; Hinton, G. Deep learning. *Nature* **2015**, *521*, 436–444. [CrossRef] [PubMed]
19. Mishra, S.; Datta–Gupta, A. *Applied Statistical Modeling and Data Analytics: A Practical Guide for the Petroleum Geosciences*; Elsevier: Amsterdam, The Netherlands, 2018; ISBN 9780128032794. [CrossRef]
20. Beggs, D.H.; Brill, J.P. A study of two–phase flow in inclined pipes. *J. Petrol. Technol.* **1973**, *25*, 607–617. [CrossRef]
21. Shoham, O. *Mechanistic Modeling of Gas–liquid Two–phase Flow in Pipes*; Society of Petroleum Engineers: Richardson, TX, USA, 2006; ISBN 9781555631079.
22. Al-Safran, E.M.; Brill, J.P. *Applied Multiphase Flow in Pipes and Flow Assurance: Oil and Gas Production*; Society of Petroleum Engineers: Richardson, TX, USA, 2017; ISBN 978613994924.
23. Gokcal, B. Effects of High Viscosity on Two-Phase Oil-Gas Flow Behaviour in Horizontal Pipes. Master's Thesis, University of Tulsa, Tulsa, OK, USA, 2005.

24. Gokcal, B. An Experimental and Theoretical Investigation of Slug Flow for High Oil Viscosity in Horizontal Pipes. Ph.D. Thesis, University of Tulsa, Tulsa, OK, USA, 2008.
25. Leung, H.; Haykin, S. The complex backpropagation algorithm. *IEEE T. Signal Process.* **1991**, *39*, 2101–2104. [CrossRef]
26. Ruder, S. An overview of gradient descent optimization algorithms. *arXiv* **2017**, arXiv:1609.04747.
27. He, K.; Zhang, X.; Ren, S.; Sun, J. Delving deep into rectifiers: Surpassing human–level performance on imageNet classification. *arXiv* **2017**, arXiv:1502.01852v1.

energies

MDPI

Article

Numerical Simulation of Coherent Structures in the Turbulent Boundary Layer under Different Stability Conditions

Shujin Laima [1,2,3], Hehe Ren [1,2,3,*], Hui Li [1,2,3] and Jinping Ou [1,2,3]

1 Key Lab of Smart Prevention and Mitigation for Civil Engineering Disasters of the Ministry of Industry and Information, Harbin Institute of Technology, Harbin 150090, China; laimashujin@hit.edu.cn (S.L.); lihui@hit.edu.cn (H.L.); oujinping@hit.edu.cn (J.O.)

2 Key Lab of Structures Dynamic Behavior and Control of the Ministry of Education, Harbin Institute of Technology, Harbin 150090, China

3 School of Civil Engineering, Harbin Institute of Technology, Harbin 150090, China

* Correspondence: renhehe@hit.edu.cn

Received: 7 February 2020; Accepted: 25 February 2020; Published: 1 March 2020

Abstract: Coherent structures in the turbulent boundary layer were investigated under different stability conditions. Qualitative analyses of the flow field, spatial correlation coefficient field and pre-multiplied wind velocity spectrum showed that the dominant turbulent eddy structure changed from small-scale motions to large- and very-large-scale motions and then to thermal plumes as the stability changed from strong stable to neutral and then to strong unstable. A quantitative analysis of the size characteristics of the three-dimensional turbulent eddy structure based on the spatial correlation coefficient field showed that under near-neutral stability, the streamwise, wall-normal and spanwise extents remained constant at approximately 0.3δ, 0.1δ and 0.2δ (δ, boundary layer height), respectively, while for other conditions, the extent in each direction varied in a log-linear manner with stability; only the spanwise extent under stable conditions was also independent of stability. The peak wavenumber of the pre-multiplied wind velocity spectrum moves towards small values from stable conditions to neutral condition and then to unstable conditions; thus, for the wind velocity spectrum, another form is needed that takes account the effects of the stability condition.

Keywords: coherent structures; turbulent boundary layer; stability; pre-multiplied wind velocity spectrum; spatial correlation coefficient field

---

1. Introduction

Over flat and homogeneous terrain, turbulence is a very important factor affecting turbulent motions near the ground. Consequently, the structure of the turbulent boundary layer depends mainly on vertical turbulence transport which has a great influence on mass, momentum, heat, water vapor and energy exchanges [1] with turbulent coherent structures playing an important role [2,3]. Field measurements, laboratory experiments and numerical simulations of the turbulent boundary layer have revealed the existence of a wide variety of coherent structures such as low-speed streaks, hairpin vortices, large-scale motions and very large-scale motions in canonical wall-bounded flows including pipe flows, channel flows and turbulent boundary layer flows [2,4–10]. Actually, coherent structures are characterized by a specific scale and lifetime, the recurrence of characteristic and flow-specific organization and regions of concentrated vorticity. Specifically, Theodorsen [4] proposed the existence of hairpin vortices, and such vortices, together with streamwise packets, have been suggested to be the primary elements forming coherent structures in the neutral turbulent boundary logarithmic layer under experimental conditions [6]. However, most studies have focused on neutral conditions. Under neutral conditions, turbulent structures observed in the field are similar to those found in experiments. Furthermore, Marusic et al. [3] indicated that turbulence and even the pressure field in the turbulent boundary layer are also similar to those in canonical wall-bounded flows, and they suggested that the reason for this is the presence of coherent structures.

The prevalent view regarding the formation of coherent structures in the turbulent surface layer is the "bottom-up" mechanism, in which active structures emerge mainly at the surface and then grow up into the outer layer [8,11–13]. As is well known, the turbulent boundary layer is usually classified as being stable, neutral or unstable on the basis of the dominant mechanism of turbulence generation and stability [14,15]. Differences in stability will affect energy transport and fluxes between the near-surface layer and upper free layer, through the formation and development of coherent structures [16–20]. The stability depends on its ability to resist three-dimensional turbulent motions. In stable conditions, even slight three-dimensional turbulent motions are difficult or even impossible. By contrast, in unstable conditions, slight three-dimensional turbulent motions tend to become stronger, leading to turbulence and convective activity. Li and Bou-Zeid [18] concluded that coherent structures in the atmosphere must be different from those found in experiments, because in the turbulent boundary layer they can be affected by the surface buoyancy flux under stable (nocturnal) or unstable (daytime) conditions which are not taken into account in most experiments. Specifically, unstable turbulent boundary layers include three kinds of coherent structures: under slightly unstable conditions, large-scale horizontal roll vortices occur, owing to a combination of shear and buoyancy [21–24]; under moderately unstable conditions, individual hairpins arise from the ground, leading to an increase in the inclination angle of the vortex packets [25–28]; and under strongly unstable conditions, plume-shaped structures appear, together with streamwise elongated structures, and the magnitude of the vertical turbulent momentum flux increases [17,25,29,30]. Stable turbulent boundary layers can generally be divided into two kinds: weakly stable and strongly stable [31]. Under weakly stable conditions, because turbulence is nearly continuous and quasi-steady, Monin–Obukhov similarity theory applies [32]. Under strongly stable conditions, small eddy motions dominate in the boundary layer, and the layer is much shallower than those under neutral or unstable conditions; however, our understanding of this is still quite limited [31,33,34] partly because of the intrinsic complexity of fluid dynamical phenomena such as intermittency and Kelvin–Helmholtz instability [31,33,35].

In the present study, to obtain a better understanding of turbulent boundary layer flow under different stability conditions, a large eddy simulation (LES) approach was adopted. The LES technique was proposed for meteorological applications by Deardorff [36,37]; since then, it has become an indispensable tool to study the turbulent boundary layer [38–41]. In LES, large-scale structures are resolved explicitly, while the effect of small-scale structures to large-scale structures is not resolved but is parameterized using a subgrid-scale (SGS) model. Thus, because the unstable turbulent boundary layer is dominated by large-scale structures, LES is an appropriate approach [42–44]. Large eddy simulation has also been used successfully for neutral conditions [17,20,42,44,45]. However, simulation of the stable turbulent boundary layer faces a number of challenges. These arise because the way in which the radiative cooling of the ground leads to stable stratification modifying the flow turbulence is still not completely understood [46], and because the characteristic size of eddies decreases with increasing stability. The application of LES to stable conditions has, therefore, been mostly confined to weakly or moderately stable cases [47,48], and this approach has not been considered appropriate for strongly stable conditions [34,49]. However, Churchfield et al. [50] have suggested an approach that is capable of simulating complex terrain and stable conditions using wall model LES (WMLES), and it is this approach that is adopted in the present study.

Turbulent boundary layers have a complex structure and involve some phenomena that are still not completely understood, especially at high Reynolds numbers [9]. Therefore, the objective of this study was to investigate the characteristics of the turbulent boundary layer under different stability conditions and to quantify the variations in the shapes and sizes of eddy structures with stability. The

LES model and simulation setup are described in Section 2, and the model is validated in Section 3. Simulation results are discussed in Section 4. Finally, conclusions are given in Section 5.

2. Simulation Details

The governing equations of the LES model are:

$$\frac{\partial \overline{u}_i}{\partial t} + \frac{\partial}{\partial x_j}(\overline{u}_j\overline{u}_i) = -\frac{\partial}{\partial x_i}\overline{p} - \frac{\partial}{\partial x_j}(\tau_{ij}^D) + f_i^T \tag{1}$$

$$\frac{\partial \overline{u}_i}{\partial x_i} = 0 \tag{2}$$

For potential temperature transport, the governing equation is:

$$\frac{\partial \overline{\theta}}{\partial t} + \frac{\partial}{\partial x_j}(\overline{u}_j\overline{\theta}) = -\frac{\partial}{\partial x_j}(q_j) \tag{3}$$

where x_i represent Cartesian coordinates, u_i represents the corresponding velocity components, p represents the pressure which, divided by the fluid density and τ_{ij}^D, represents the SGS stress, and f_i^T represents the density-normalized forces. The Coriolis force was not considered, the same as in Fang and Porté-Agel [20]. A bar on a variable indicates the mean grid-filtered quantity. $\overline{\theta}$ is the resolved potential temperature, and q_j represents the flux of temperature due to the viscous and sub-filter-scale effects. The details of how the SGS momentum flux τ_{ij}^D was modelled is shown in Ren et al. [51]. The heat flux q_j at the surface was modelled using a similar method [38]:

$$q_j = \begin{bmatrix} 0 \\ 0 \\ q_3^{tot} \end{bmatrix} \tag{4}$$

where, in the simulation, the total average temperature flux is specified, and the fluctuating temperature flux q_3^{tot} is created by the wall model.

For numerical discretization, the finite volume method was used; for the viscous and convective terms, a second-order central difference scheme was adopted; and for the unsteady term, a second-order backward scheme was adopted.

Schumann's model [52] was used as the wall model to model the near wall shear stress, and Moeng's model [38] was adopted for the total temperature flux at the surface. The subgrid stress (SGS) model is the one-equation eddy viscosity model [53]; this is because the one-equation eddy viscosity model does not require the assumption in the algebraic eddy viscosity models, that is, local balance between the production and dissipation of SGS energy. The SGS stress can be modelled as:

$$\tau_{ij,sgs}^D = \frac{2}{3}k_{sgs}\delta_{ij} - 2\upsilon_{sgs}\overline{S}_{ij} \tag{5}$$

where $\tau_{ij,sgs}^D$ is the SGS stress part of the total stress, τ_{ij}^D. $\overline{S}_{ij}$ is the resolved-scale strain rate tensor, δ_{ij} is the Kronecker delta function, k_{sgs} is the SGS kinetic energy; and υ_{sgs} is the SGS stress eddy viscosity.

Accounting for the historic effect of k_{sgs}, because of diffusion, production and dissipation, a transportation equation was derived:

$$\frac{\partial k_{sgs}}{\partial t} + \frac{\partial \overline{u}_j k_{sgs}}{\partial x_j} = -\tau_{ij,sgs}^D\overline{S}_{ij} - C_\varepsilon\frac{k_{sgs}^{3/2}}{\Delta} + \frac{\partial}{\partial x_j}(\upsilon_{sgs}\frac{\partial k_{sgs}}{\partial x_j}) \tag{6}$$

where Δ represents the filter scale, and C_k and C_ε are the model constants.

In the present study, the size of the computational domain was 5 × 1 × 0.3 km for all cases. When the hexahedral mesh was applied, the mesh grid size was 5 m in each direction from the ground surface to 0.2 km, and above 0.2 km, the mesh grid size was 10 m. For the streamwise and spanwise directions, periodic boundary conditions were applied; this agrees well with Fang and Porté-Agel [20] of LES of turbulent boundary layer flow and with Lee et al. [54] of DNS of channel flow; for the upper boundary, the slip boundary condition was adopted; and for the solid wall, the wall model was used. For the initial conditions, the logarithmic mean wind speed profile was used in all cases (gradient wind height of 250 m and a gradient wind speed of 8 m/s).

The computation timestep was 0.5 s which ensures the stability and viscous stability by a given Courant number. First, to obtain a fully developed flow field and quasi-steady state, a spin-up simulation was performed. For the statistical analysis, to ensure statistical convergence, over 40,000 timesteps were simulated. For the perturbation component, the initial peak was set at the height of 0.015% of the computational domain, and for the horizontal direction, the maximum perturbation was 0.25 m/s; at the beginning of the simulation, for the streamwise and spanwise direction, the number of periods of the perturbations were 4.0 and 20.0, respectively.

The aim of this numerical simulation was to calculate the forms of coherent structures under specifically assumed stability conditions and to quantify the variations in the shapes and sizes of eddy structures with stability. The simulation was conducted within a pre-defined stability range. As is well known, stability refers to the air's tendency to either resist vertical movement (stability) or to rise and create storms (instability) which can be expressed in terms of a dimensionless stability parameter, $-z/L$, where z represents the height above the surface ground, and L represents the Obukhov length, given by $L = -u_*^3 T_v/(kgQ_{v0})$, where k represents the von Kármán constant, u_* represents the friction velocity, Q_{v0} is the kinematic virtual temperature flux at the surface, T_v is a reference virtual temperature, and g is the gravitational acceleration. L expresses the relative roles of shear and buoyancy in the production/consumption of turbulence kinetic energy with positive and negative values indicating stable and unstable conditions, respectively, and it approaches zero in the limit of neutral stratification. For $-z/L$, positive and negative values indicate unstable and stable conditions. In this study, 17 different turbulent boundary layer cases were analysed, from case 1 to 17 which range from stable to unstable, with $-z/L = -98.27$ (case 1), −16.19 (case 2), −2.29 (case 3), −1.15 (case 4), −0.84 (case 5), −0.63 (case 6), −0.35 (case 7), −0.05 (case 8), 0 (case 9), 0.54 (case 10), 1.01 (case 11), 1.79 (case 12), 3.16 (case 13), 4.48 (case 14), 8.13 (case 15), 11.31 (case 16), and 31.46 (case 17). It should be noted that to represent the average stability of the turbulent boundary layer, the z used for calculating $-z/L$ here is the boundary layer height δ; this is different from Figures 1 and 2 in which z represents the level height above the ground.

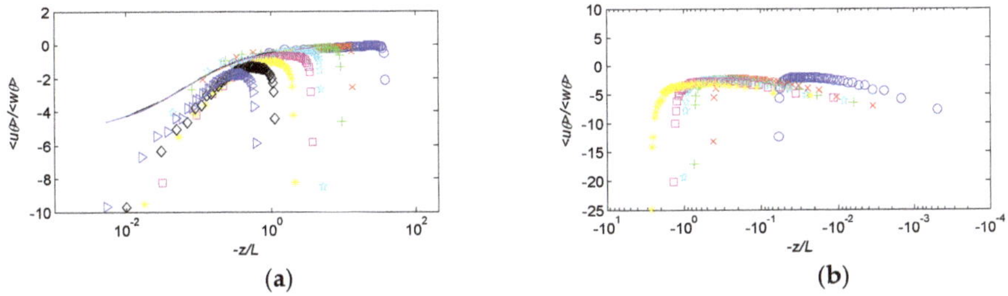

Figure 1. Ratio of horizontal and vertical heat fluxes $\overline{u'\theta'}/\overline{w'\theta'}$ versus $-z/L$. In (**a**), for unstable conditions, the blue circles, red crosses, green plus signs, cyan stars, magenta squares, yellow asterisks, black diamonds and blue rightward-pointing triangles represent cases 17, 16, 15, 14, 13, 12, 11 and 10, respectively, from very to slightly unstable. The curve corresponds to Equation (7). In (**b**), for stable conditions, the blue circles, red crosses, green plus signs, cyan stars, magenta squares, yellow asterisks, black diamonds and blue rightward-pointing triangles represents cases 8, 7, 6, 5, 4, 3, 2, and 1, respectively, going from slightly to very stable.

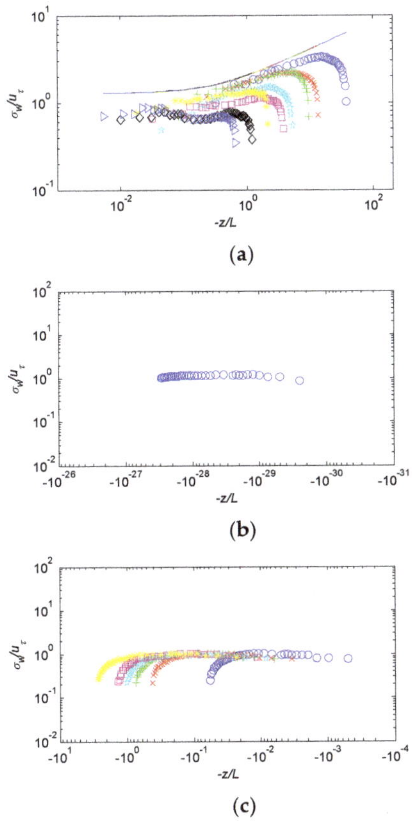

Figure 2. Normalized standard deviation of $w(\sigma_w/u_*)$ versus $-z/L$: (**a**) unstable conditions; (**b**) neutral stability conditions; and (**c**) stable conditions. The symbols are the same as in Figure 1, except that the curve in (**a**) now corresponds to Equation (8).

3. Validation

The basic statistical properties of the velocity under neutral stability conditions were compared with those from previous studies, and good agreement was found when dimensioned both by inner and outer variables; the details are shown in previous studies by Ren et al. [51] and Ren et al. [55].

The ratio of the heat fluxes in the horizontal and vertical directions under unstable conditions is shown in Figure 2. Wyngaard and Cote [56] proposed the following formula for this ratio:

$$\frac{\overline{u'\theta'}}{\overline{w'\theta'}} = a\varphi_m\varphi_h \tag{7}$$

where $\varphi_m = (kz/u_*)(\partial u/\partial z)$ represents the non-dimensional wind shear function, u_* is the surface friction velocity, $\varphi_h = (kz/u_*)(\partial\theta/\partial z)$ is the non-dimensional potential temperature gradient function, and a is a constant. In the present study, $\varphi_m = (1-28z/L)^{-1/4}$, $\varphi_h = (1-20z/L)^{-1/2}$ [57,58], and $a = 5$ [57–59]. In Figure 1, the ratio of the horizontal and vertical heat fluxes decreases with increasing $-z/L$ away from the near-surface region, and $\overline{u'\theta'}/\overline{w'\theta'}$ tends towards 0 which indicates that at higher elevations the shear stress is very small compared with the thermal plume force. The simulation results show good agreement with theory under strongly unstable conditions, but under moderately or slightly unstable conditions, they deviate from the theoretical values. This deviation occurs because under these conditions, the shear stress driving the horizontal motions near the surface is greater than under strongly unstable conditions, while the vertical transport is weaker than under strongly unstable conditions. Therefore, Equation (7) is appropriate for strongly unstable conditions only, and perhaps, after the forms of a, φ_m and φ_h are changed in the present study, it will also prove to be appropriate for moderately or slightly unstable conditions. Another phenomenon revealed in Figure 1 is a shift in the peak of $\overline{w'\theta'}/\overline{w'\theta'}$ towards larger $-z/L$; this is because as conditions become more unstable, thermal plumes reach higher. Similarly, the peak of $\overline{u'\theta'}/\overline{w'\theta'}$ shifts to smaller $-z/L$ as the stability increases under stable conditions. This suggests that as conditions become more unstable, $-z/L$ becomes larger which is related to the upper location of thermal plumes.

Figure 2 shows the normalized standard deviation of the vertical velocity component, σ_w/u_*, which is important for describing stability. The curves in Figure 2 represent the formula proposed by Panofsky et al. [59] in the case of the unstable condition:

$$\sigma_w/u_* = 1.3(1-3z/L)^{1/3} \tag{8}$$

where u_* is the friction velocity, L represents the Obukhov length, and z is the height above the ground. It can be seen that σ_w/u_* remains constant under slightly or moderately unstable conditions but increases with increasing height under strongly unstable conditions. The simulation results show good agreement with Equation (8) which means that σ_w/u_* obeys Monin–Obukhov similarity theory under unstable conditions, fitting well with the 1/3 power law. Under neutral conditions, σ_w/u_* is constant, independent of height and equal to 1.32 (shown in Figure 2b); indeed, a large number of observations show that under neutral conditions, σ_w/u_* should be constant [58,60–64]. In particular, Bowne and Ball [60] and Counihan [61] both concluded that this constant value is 1.3. Roth [64] summarized the values of different researchers, which ranged from 1.1 to 1.7. When the turbulent boundary layer is under stable conditions, the turbulent motions are weakly connected to the ground, and the height is no longer a similarity parameter of this motion, except in the region near the ground. As can be seen from Figure 2c, the normalized standard deviation of the vertical velocity component behaves as expected for stable conditions. Also, similar to the ratio of horizontal and vertical heat fluxes, the peak of σ_w/u_* shifts to smaller $-z/L$ as stability increases, except for neutral conditions; this is related to the upper location of thermal plumes.

4. Results

The above verification analysis has illustrated the feasibility of the WMLES approach and the accuracy of simulation of turbulent boundary layer flow under different stability conditions. In the following, we investigate the characteristics of turbulent boundary layers under different stability conditions and quantify the variations in the shape and size of eddy structures with stability.

4.1. Flow Field

We now turn to an analysis of the flow field with the aim of providing a more intuitive view of its basic characteristics and those of the turbulent eddy scale. We consider a horizontal plane (whole domain) at the height of 0.2δ. Figure 3 shows the flow field for all 17 cases. On the whole, under stable conditions, the dominant turbulent eddy structure was small which is consistent with many previous results [31,33,34]. This is because, under stable conditions, the ground temperature is lower than that of the near-surface flow, and the vertical development of the turbulent eddy structure is limited which suppresses the development of three-dimensional turbulent eddy structure thus small-scale motion is prevalent. The dominant turbulent eddy structure scale increases as the stability decreases, and for the weakly stable condition in case 8, very-large-scale motion structures [8,10,65] even occur, just as under neutral conditions. It is obvious under neutral conditions that both large-scale and very-large-scale motion structures are visualized, based on the streamwise-elongated volumes associated with negative streamwise velocity, which is consistent with the results of Dennis and Nickels [66,67], Lee and Sung [65], Baltzer et al. [68] and Lee et al. [54]. Under slightly and moderately unstable conditions, obvious large-scale horizontal roll vortices form which are usually attributed to the combined effects of shear and buoyancy [21,22,24]. These can be seen as corresponding to the extended turbulent eddy structure that appears under neutral stability conditions. Under strongly unstable conditions, the effects of buoyancy dominate those of shear, so there is an obvious distinction between low-velocity and high-velocity regions which extends over the whole spanwise direction. This is consistent with the experimental results of Hommema and Adrian [25].

In Figure 3, going from the strongly stable condition to the neutral stable condition to the strongly unstable condition, some interesting phenomena can be seen. First, the form and shape of the dominant turbulent eddy structure changed from small-scale motions to large-scale and very-large-scale motions and then to regional motions as the turbulent boundary layer became unstable. Second, the dominant size of the turbulent eddy structure varied as the stability changed; a quantitative analysis of this is presented in Section 4.2 based on the three-dimensional spatial correlation coefficient field.

4.2. Spatial Correlation Coefficient Field

To quantify the size characteristics of the three-dimensional turbulent eddy structure under different stability conditions, a flow field correlation analysis was performed with reference location $(x, y, z) = (2500\,\text{m}, 500\,\text{m}, 35\,\text{m})$. Figure 4 presents contours of the spatial correlation coefficient for all 17 cases. The plots in the left column are for the xz plane, and those in the right column are for the yz plane; thus, it is possible to visualize the three-dimensional turbulent eddy structure. The conclusion is similar to that from the flow field analysis, namely, that with increasing stability, the dominant size of the turbulent eddy decreases. This clearly shows that under weakly stable, neutral and slightly and moderately unstable conditions, very-large-scale motion structures appeared, while strongly unstable conditions led to thermal plumes. Meanwhile, an interesting phenomenon was that the spanwise size of the dominant vortices remained almost constant, except under strongly unstable conditions which may be due to the fact that the effect of the shear force was mainly in the streamwise direction and that of the buoyancy force was mainly in the wall-normal direction; however, under strongly unstable conditions, the temperature of the ground was higher than that of the adjacent near-surface air, and so the buoyancy force had a positive effect on the upward motion of air near the ground in the whole spanwise direction. The spanwise and wall-normal velocity correlations were different from the streamwise velocity correlation. For the spanwise velocity correlation under stable conditions, the spanwise size of the correlation structure was smaller than the streamwise velocity correlation; moreover, the degree of tilt was smaller than for the streamwise velocity correlation. Under unstable conditions, even light unstable conditions, the plume-shaped structure occurred, meaning that this structure form occurred earlier than the streamwise velocity correlation. For the wall-normal velocity correlation, from stable to unstable conditions, the spanwise and wall-normal sizes of the correlation structure became wide and deep which is very different to the other two direction velocity correlations. These results are not shown here.

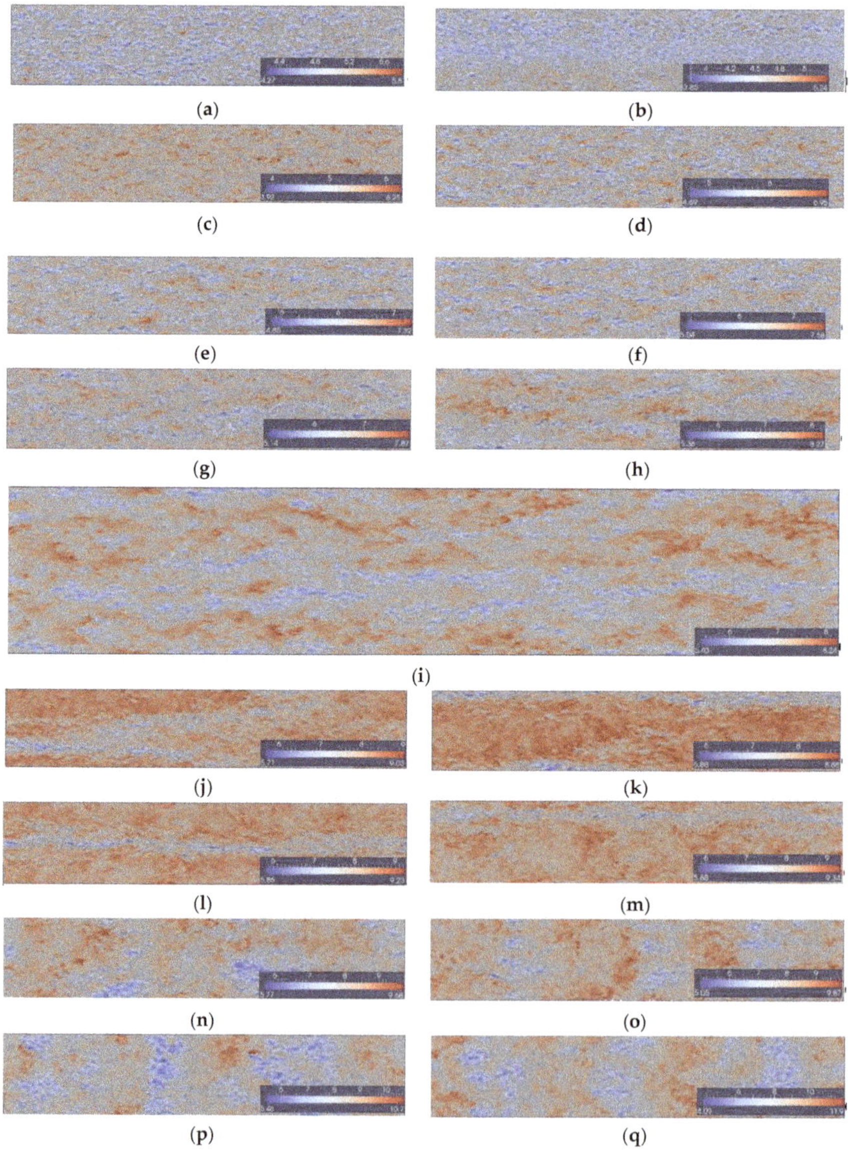

Figure 3. Streamwise velocity in the xy plane (whole domain) at a height of 0.2δ for each case. (**a**) to (**q**) represent cases 1 to 17, respectively, going from very stable, to neutral and, finally, very unstable conditions.

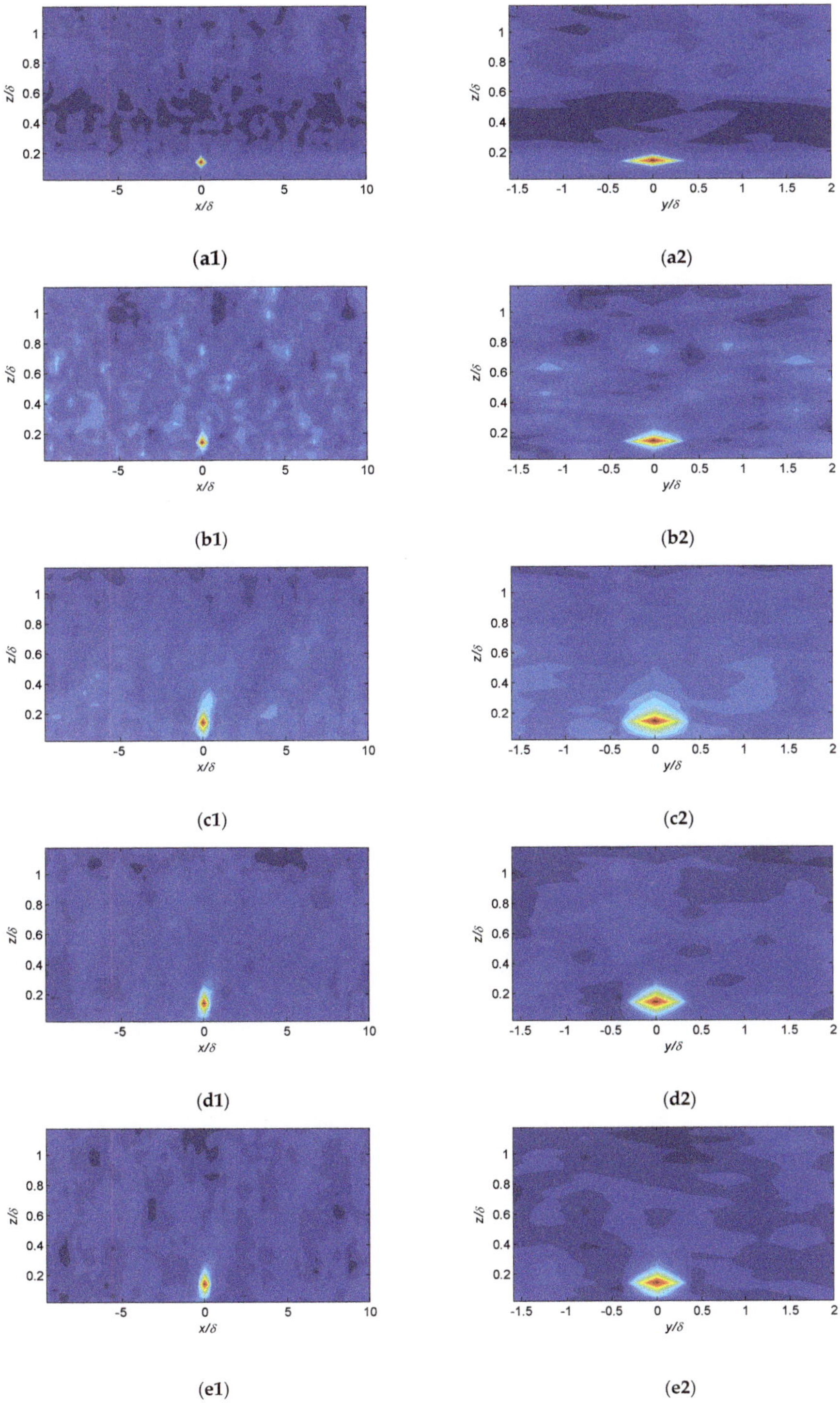

Figure 4. *Cont.*

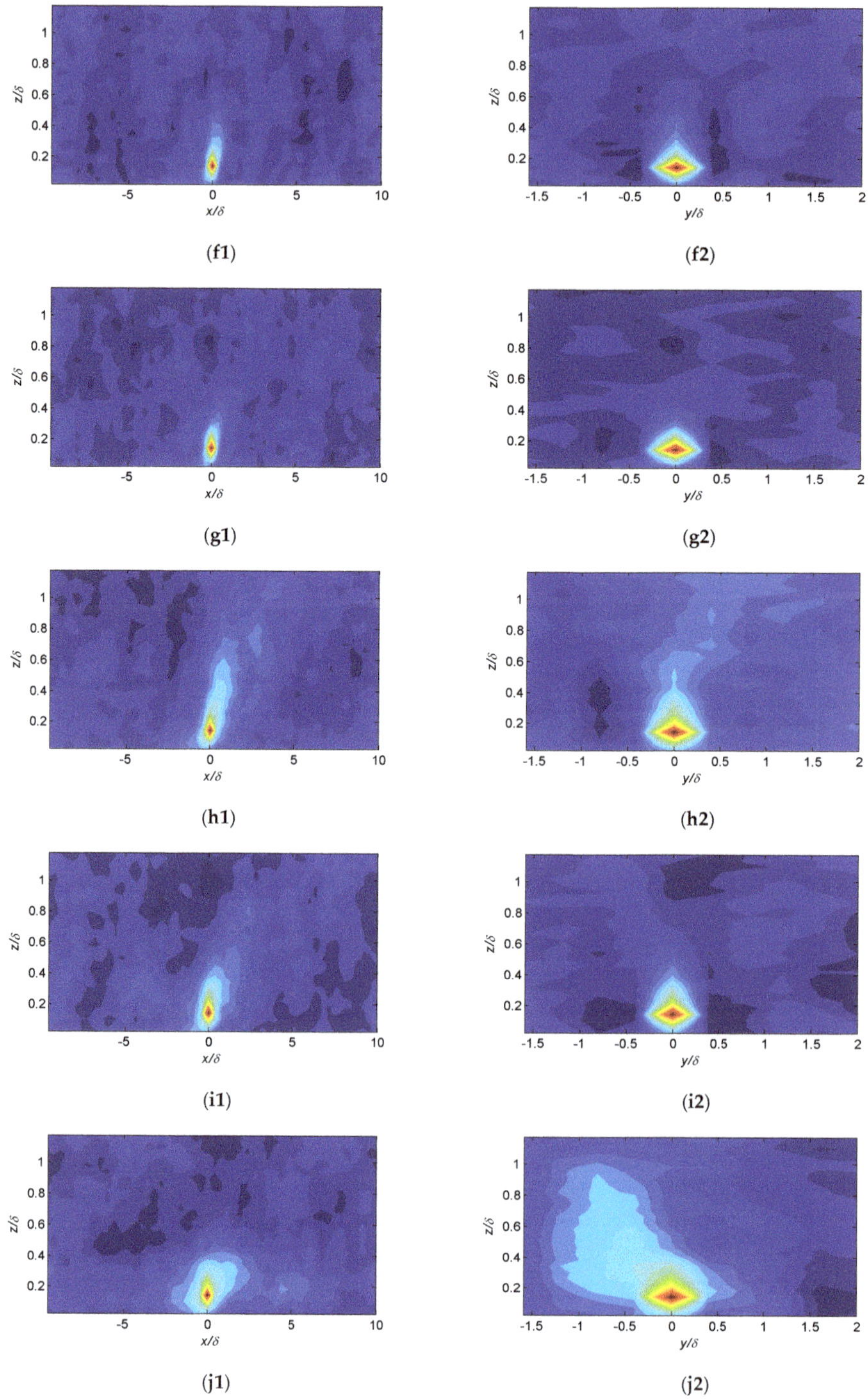

Figure 4. *Cont.*

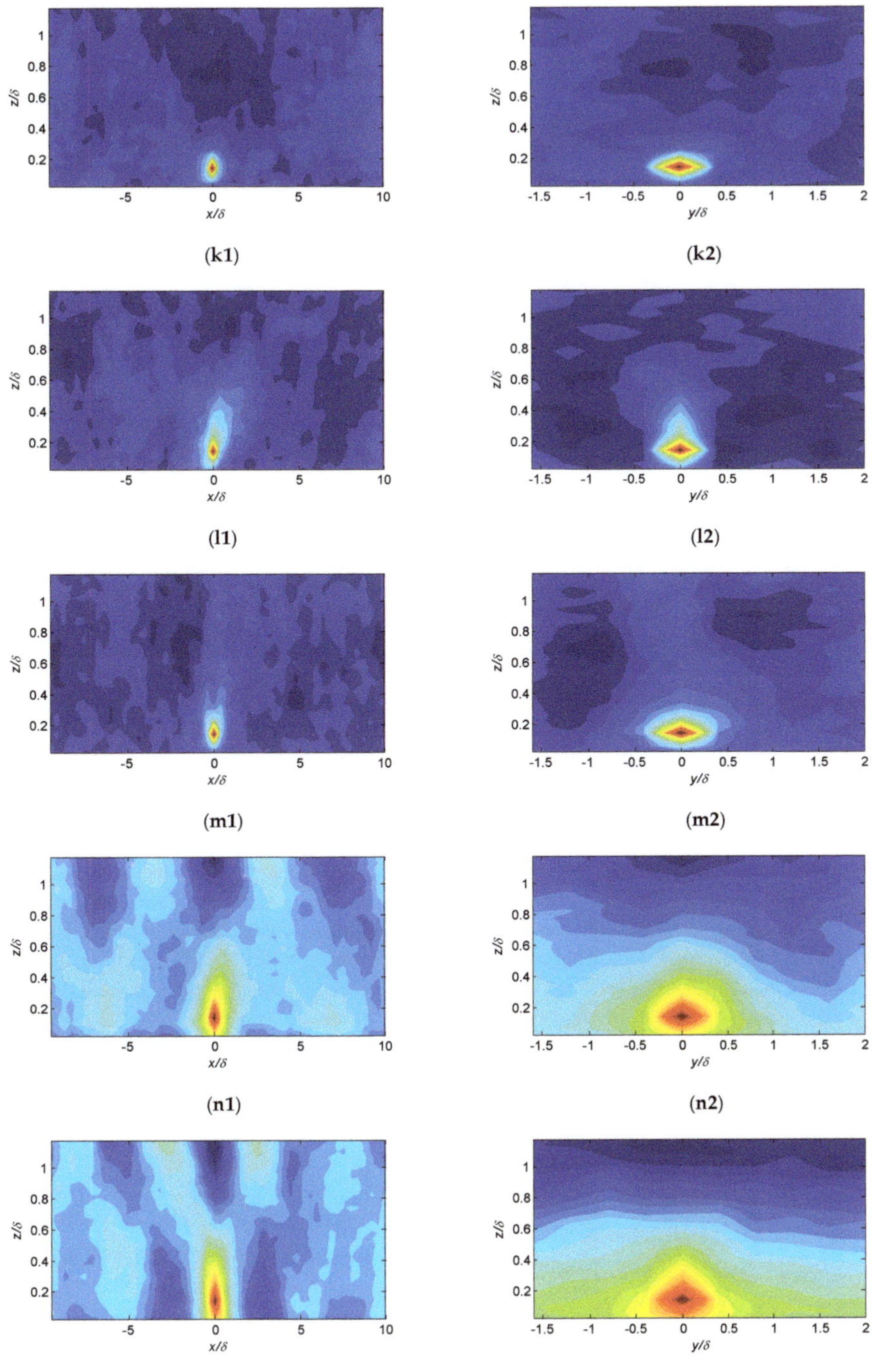

Figure 4. *Cont.*

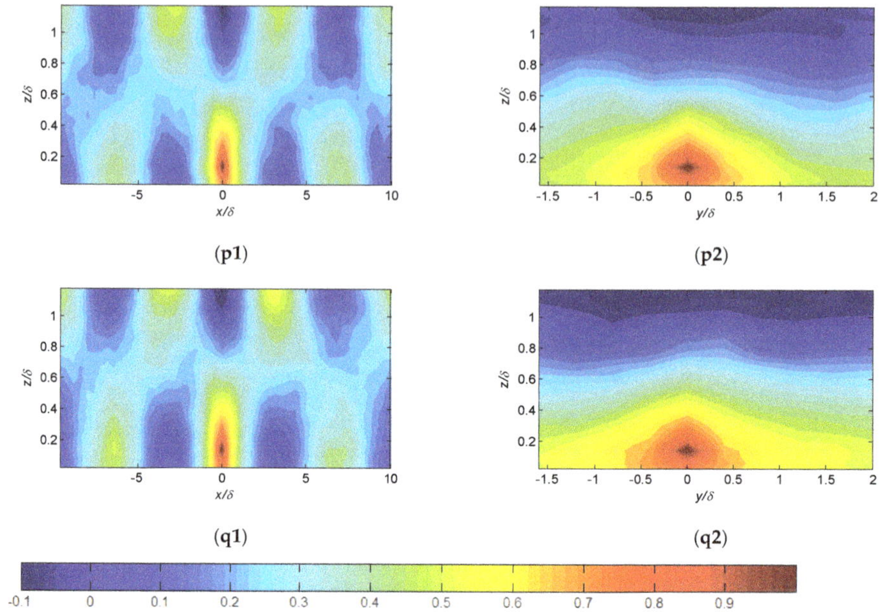

Figure 4. Contours of the two-point correlation of streamwise velocity fluctuations. Subfigures (**a1**) to (**q2**) means: each row represents case 1 to 17, from very stable, to neutral, and then very unstable conditions, with the left and right columns showing the contours in the xz and yz planes, respectively.

Figure 5 shows the size of the turbulent eddy structure under different stability conditions at a height of 0.2δ. The distances Lx_{uu}, Ly_{uu} and Lz_{uu} are defined in Christensen and Wu [69] and Volino et al. [70] as twice the distances from the self-correlation peak to the most distant locations on the 0.5 contour of the two-point correlation of the fluctuating velocity in the downstream, spanwise and wall-normal directions, respectively. We can see that under strongly and moderately stable conditions, the streamwise extent Lx_{uu} increased in a log-linear manner as $-z/L$ increased, and that it behaved similarly under strongly unstable conditions. However, under near-neutral stability conditions, Lx_{uu} was independent of $-z/L$ and remained constant at approximately 0.3δ. Similar behaviour was found for the wall-normal extent Lz_{uu} which remained constant at approximately 0.1δ. From strongly stable through neutral to slightly unstable conditions, the spanwise extent Ly_{uu} remained constant at approximately 0.2δ, while under strongly unstable conditions, Ly_{uu} also increased in a log-linear manner with increasing $-z/L$. This corresponds to the behaviour of the spatial correlation coefficient field.

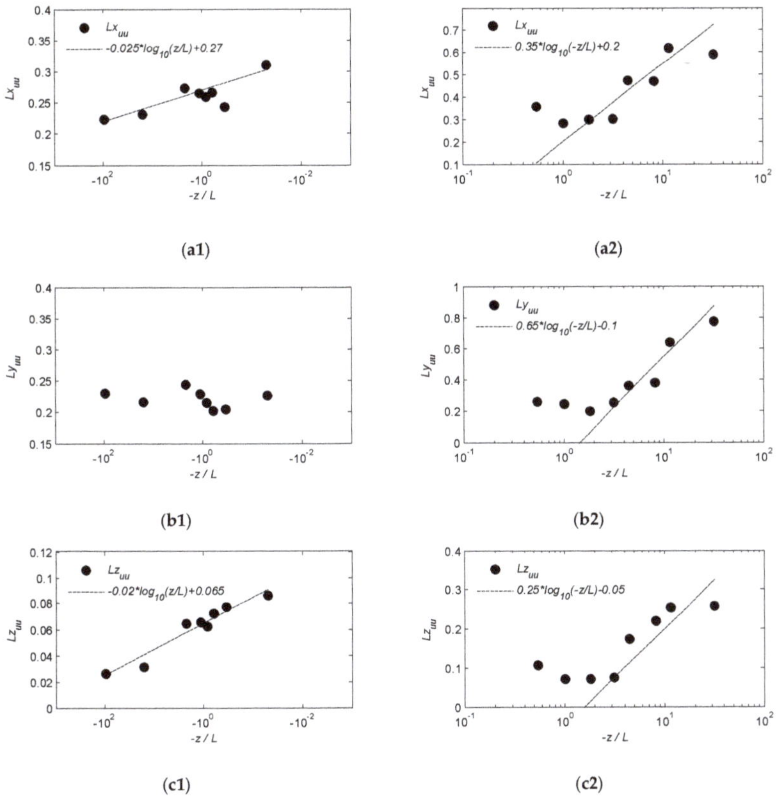

Figure 5. Extents of the 0.5 contours of the two-point correlations of the fluctuating velocity as functions of $-z/L$ at a height 0.2δ: (**a**) streamwise extent; (**b**) spanwise extent; and (**c**) wall-normal extent. The left and right columns are for stable and unstable conditions, respectively. For the neutral stability condition, $Lx_{uu} = 0.32\delta$, $Ly_{uu} = 0.20\delta$ and $Lz_{uu} = 0.12\delta$.

4.3. Pre-Multiplied Spectral Analysis

Pre-multiplied spectral analysis is a popular approach for determining the distribution characteristics of vortices of different scales in canonical wall-bounded flows [8,10,71,72]. Figure 6 shows the multipoint averaged [73] pre-multiplied wind velocity spectra for all 17 cases. As we know, the most energetic point in the pre-multiplied wind velocity spectrum represents the dominant turbulent eddy structure. Therefore, under stable conditions, the peak wavenumber of the wind velocity spectrum is located in the high-wavenumber region which indicates that the dominant vortices are small. With decreasing stability, the peak wavenumber moved to smaller values which means that the size of the dominant turbulent eddy structure increased and very-large-scale motion structures occurred under the slightly stable conditions of case 8. Here, it should be noted that very-large-scale motion structures are those with turbulent eddy size larger than 3δ. Under near-neutral stability conditions (slightly stable, neutral and slightly unstable conditions), the stationary stage of the pre-multiplied wind velocity spectrum has a large width, because the stationary stage corresponds to the very-large-scale motion structure [6,8,10,12]. Under strongly unstable conditions, the peak wavenumber is located in the small wavenumber region which means that the dominant turbulent eddy structure is very large, and the shape of the wind velocity spectrum is similar to that of a typhoon wind velocity spectrum near the eye-wall region. This may be because the eye-wall region of a typhoon is also characterized by strongly convective conditions, accompanied by a strong updraft. For spanwise and wall-normal pre-multiplied wind velocity spectra, the maximum energy points are consistent with the streamwise pre-multiplied wind velocity spectrum, and as the instability increases, it moves to a small wavenumber. The difference is that the steady stage is more obvious in the streamwise pre-multiplied wind velocity spectrum than the other two directions, and that the spanwise and wall-normal pre-multiplied wind velocity spectra have obvious cross-over phenomenon at different heights, which does not occur for streamwise pre-multiplied wind velocity spectrum. These results are not shown here.

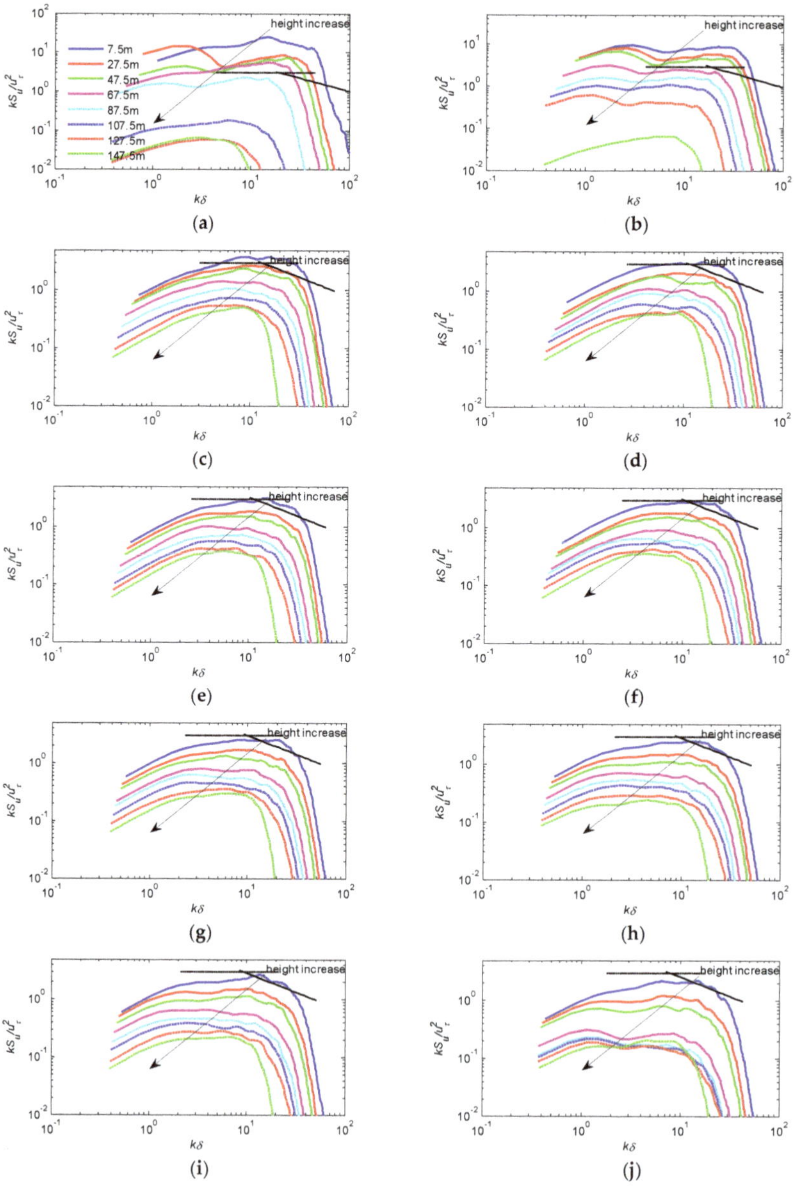

Figure 6. *Cont.*

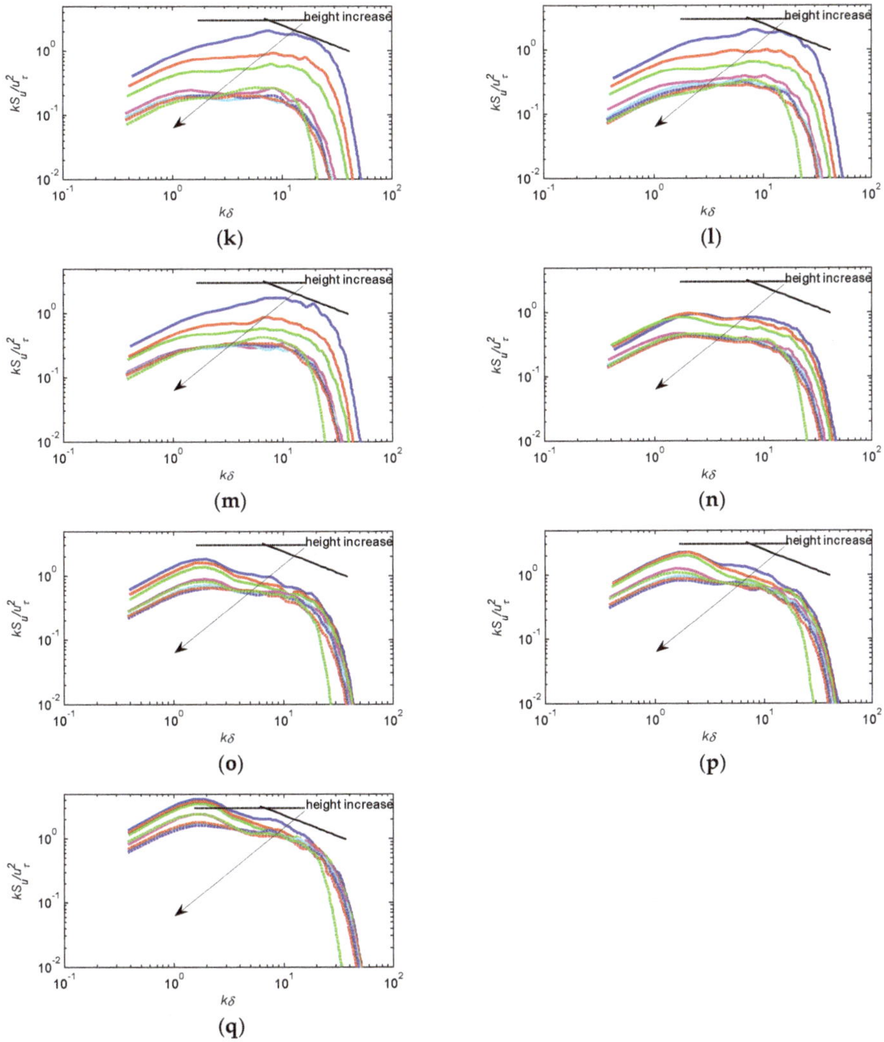

Figure 6. Pre-multiplied wind velocity spectra of the streamwise velocity fluctuation for each case at different heights: (**a**) to (**q**) are for cases 1 to 17, respectively. The dashed and solid black lines represent 0 and −2/3 slope, respectively.

It can be seen from Figure 6 that as the stability conditions changed from stable to neutral to unstable, the peak wavenumber of the pre-multiplied wind velocity spectrum moved towards smaller values which is consistent with the above analyses of the flow field and the spatial correlation coefficient field, and the stationary stage of the wind velocity spectrum existed only under near-neutral conditions. Under strongly stable or unstable conditions, the small-scale and large-scale turbulent eddy structures were the dominant structures, respectively, and only one dominant peak existed in the pre-multiplied wind velocity spectrum.

The canonical wind velocity spectral styles are based on the neutral stability condition, but in this paper, we found that the different stability conditions gave different wind velocity spectral forms. Therefore, in future studies, we should try to find a stability parameter that represents the differences in the wind velocity spectra and allows us to obtain a universal wind velocity spectrum that is suitable for all stability conditions.

5. Conclusions

We investigated the coherent structures in the turbulent boundary layer under different stability conditions and performed qualitative and quantitative analyses of the shape and size of eddy structures.

First, the WMLES approach and the basic variables of the simulation results were validated by comparing these results with theoretical predictions, observed data, experimental data and other simulation results.

Then, based on a flow field analysis, it was found that the form and shape of the dominant turbulent eddy structure changed from small-scale motions to large-scale and very-large-scale motions and then to thermal plumes as the stability condition changed from strong stable conditions to neutral stability conditions and then, to strongly unstable conditions. A spatial correlation analysis was performed, and the results led to similar conclusions as those obtained from the flow field analysis. An interesting phenomenon was also found, namely, that the spanwise size of the dominant vortices remained almost constant, except under strongly unstable conditions which may be because the main effect of shear is in the streamwise direction while that of buoyancy is in the wall-normal direction. From an analysis of the pre-multiplied wind velocity spectrum, it was found that as the stability condition goes from stable to neutral to unstable, the peak wavenumber of the pre-multiplied wind velocity spectrum moves towards smaller values, and the stationary stage of wind velocity spectrum only exists under near-neutral stability, whereas under strongly stable or unstable conditions, the small- and large-scale turbulent eddy structures have an overwhelming advantage, so only one dominant peak exists in the pre-multiplied wind velocity spectrum. In future work, we need to consider a new style for the wind velocity spectrum that takes account of the effects of stability.

Finally, a quantitative analysis of the size characteristics of the three-dimensional turbulent eddy structure under different stability conditions based on the spatial correlation coefficient field showed that for near-neutral stability, the streamwise, wall-normal and spanwise extents remain constant at approximately 0.3δ, 0.1δ and 0.2δ, respectively. For other conditions, it was found that the streamwise, wall-normal and spanwise extents increase in a log-linear manner with increasing stability, except for the spanwise extent under stable conditions which is independent of stability.

Author Contributions: Conceptualization, H.R., S.L., H.L. and J.O.; Methodology, H.R. and S.L.; Software, H.R.; Validation, H.R.; Formal Analysis, H.R.; Investigation, H.R. and S.L.; Resources, S.L.; Data Curation, H.R.; Writing–Original Draft Preparation, H.R.; Writing–Review and Editing, S.L.; Visualization, H.R.; Supervision, H.L. and J.O.; Project Administration, H.L.; Funding Acquisition, H.L. All authors have read and agreed to the published version of the manuscript.

Funding: This study was financially supported by the National Key Research and Development Program of China under grant no. 2018YFC0705605, NSFC under grant no. 51878230, the China Postdoctoral Science Foundations (2015M580266 and 2016T90295), and the Heilongjiang Postdoctoral Fund (LBH-Z15087).

Conflicts of Interest: The authors declare no conflict of interest.

References

1. Monin, A.S.; Obukhov, A.M. Basic laws of turbulent mixing in the surface layer of the atmosphere. *Contrib. Geophys. Inst. Acad. Sci. USSR* **1954**, *24*, 163–187.
2. Robinson, S.K. Coherent motions in the turbulent boundary-layer. *Annu. Rev. Fluid Mech.* **1991**, *23*, 601–639. [CrossRef]
3. Marusic, I.; Mathis, R.; Hutchins, N. Predictive model for wall-bounded turbulent flow. *Science* **2010**, *329*, 193–196. [CrossRef] [PubMed]
4. Theodorsen, T. Mechanism of turbulence. In Proceedings of the Second Midwestern Conference on Fluid Mechanics, Ohio State University, Columbus, OH, USA, 17–19 March 1952; pp. 1–19.
5. Kline, S.J.; Reynolds, W.C.; Schraub, F.A.; Runstadl, P.W. Structure of turbulent boundary layers. *J. Fluid Mech.* **1967**, *30*, 741–773. [CrossRef]

6. Adrian, R.J. Hairpin vortex organization in wall turbulence. *Phys Fluids* **2007**, *19*, 041301. [CrossRef]
7. Zhou, J.; Adrian, R.J.; Balachandar, S.; Kendall, T.M. Mechanisms for generating coherent packets of hairpin vortices in channel flow. *J. Fluid Mech.* **1999**, *387*, 353–396. [CrossRef]
8. Kim, K.C.; Adrian, R.J. Very large-scale motion in the outer layer. *Phys Fluids* **1999**, *11*, 417–422. [CrossRef]
9. Hunt, J.C.R.; Morrison, J.F. Eddy structure in turbulent boundary layers. *Eur. J. Mech. B* **2000**, *19*, 673–694. [CrossRef]
10. Wang, G.H.; Zheng, X.J. Very large scale motions in the atmospheric surface layer: A field investigation. *J. Fluid Mech.* **2016**, *802*, 464–489. [CrossRef]
11. Adrian, R.J.; Meinhart, C.D.; Tomkins, C.D. Vortex organization in the outer region of the turbulent boundary layer. *J. Fluid Mech.* **2000**, *422*, 1–54. [CrossRef]
12. Guala, M.; Hommema, S.E.; Adrian, R.J. Large-scale and very-large-scale motions in turbulent pipe flow. *J. Fluid Mech.* **2006**, *554*, 521–542. [CrossRef]
13. Balakumar, B.J.; Adrian, R.J. Large-and very-large-scale motions in channel and boundary-layer flows. *Philos. Trans. R. Soc. Lond. A* **2007**, *365*, 665–681. [CrossRef] [PubMed]
14. Stull, R.B. *An Introduction to Boundary Layer Meteorology*; Kluwer Academic: Dodrecht, The Netherlands, 1988.
15. Arya, S.P. *Introduction to Micrometeorology*; Academic Press: New York, NY, USA, 2001.
16. Kumar, V.; Kleissl, J.; Meneveau, C.; Parlange, M.B. Large-eddy simulation of a diurnal cycle of the atmospheric boundary layer: Atmospheric stability and scaling issues. *Water Resour. Res.* **2006**, *42*, 650–664. [CrossRef]
17. Park, S.B.; Baik, J.J. A large-eddy simulation study of thermal effects on turbulence coherent structures in and above a building array. *J. Appl. Meteorol. Climatol.* **2013**, *52*, 1348–1365. [CrossRef]
18. Li, D.; Bou-Zeid, E. Coherent structures and the dissimilarity of turbulent transport of momentum and scalars in the unstable atmospheric surface layer. *Bound. Layer Meteorol.* **2011**, *140*, 243–262. [CrossRef]
19. Basu, S.; Porté-Agel, F. Large-eddy simulation of stably stratified atmospheric boundary layer turbulence: A scale-dependent dynamic modeling approach. *J. Atmos. Sci.* **2005**, *63*, 2074–2091. [CrossRef]
20. Fang, J.N.; Porté-Agel, F. Large-eddy simulation of very-large-scale motions in the neutrally stratified atmospheric boundary layer. *Bound. Layer Meteorol.* **2015**, *155*, 397–416. [CrossRef]
21. Mason, P.J.; Sykes, R.I. A two-dimensional numerical study of horizontal roll vortices in the neutral atmospheric boundary-layer. *Q. J. R. Meteorol. Soc.* **1980**, *106*, 351–366. [CrossRef]
22. Etling, D.; Brown, R.A. Roll vortices in the planetary boundary-layer—A review. *Bound. Layer Meteorol.* **1993**, *65*, 215–248. [CrossRef]
23. Khanna, S.; Brasseur, J.G. Three-dimensional buoyancy-and shear-induced local structure of the atmospheric boundary layer. *J. Atmos. Sci.* **1998**, *55*, 710–743. [CrossRef]
24. Boppe, R.S.; Neu, W.L.; Shuai, H. Large-scale motions in the marine atmospheric surface layer. *Bound. Layer Meteorol.* **1999**, *92*, 165–183. [CrossRef]
25. Hommema, S.E.; Adrian, R.J. Packet structure of surface eddies in the atmospheric boundary layer. *Bound. Layer Meteorol.* **2003**, *106*, 147–170. [CrossRef]
26. Carper, M.A.; Porté-Agel, F. The role of coherent structures in subfilter-scale dissipation of turbulence measured in the atmospheric surface layer. *J. Turbul.* **2004**, *5*, 32. [CrossRef]
27. Chauhan, K.; Hutchins, N.; Monty, J.; Marusic, I. Structure Inclination Angles in the Convective Atmospheric Surface Layer. *Bound. Layer Meteorol.* **2013**, *147*, 41–50. [CrossRef]
28. Salesky, S.T.; Anderson, W. Buoyancy effects on large-scale motions in convective atmospheric boundary layers: Implications for modulation of near-wall processes. *J. Fluid Mech.* **2018**, *856*, 135–168. [CrossRef]
29. Cortese, T.; Balachandar, S. Vortical nature of thermal plumes in turbulent convection. *Phys. Fluids* **1993**, *5*, 3226–3232. [CrossRef]
30. Zhou, S.Q.; Xia, K.Q. Plume statistics in thermal turbulence: Mixing of an active scalar. *Phys. Rev. Lett.* **2002**, *89*, 184502. [CrossRef]
31. Mahrt, L. Stratified atmospheric boundary layers and breakdown of models. *Theor. Comput. Fluid Dyn.* **1998**, *11*, 263–279. [CrossRef]
32. Monin, A.S. Atmospheric boundary layer. *Annu. Rev. Fluid Mech.* **1970**, *2*, 225. [CrossRef]
33. Derbyshire, S.H. Stable boundary-layer modelling: Established approaches and beyond. *Bound. Layer Meteorol.* **1999**, *90*, 423–446. [CrossRef]

34. Holtslag, A.A.M. *GABLS Initiates Intercomparison for Stable Boundary Layer Case*; GEWEX News, No. 13.; International GEWEX Project Office: Silver Spring, MD, USA, 2003; pp. 7–8.
35. Hunt, J.C.R.; Shutts, G.J.; Derbyshire, S.H. Stably stratified flows in meteorology. *Dyn. Atmos. Oceans* **1996**, *23*, 63–79. [CrossRef]
36. Deardorff, J.W. Theoretical expression for the countergradient vertical heat flux. *J. Geophys. Res.* **1972**, *77*, 5900–5904. [CrossRef]
37. Deardorff, J.W. Three-dimensional numerical study of turbulence in an entraining mixed layer. *Bound. Layer Meteorol.* **1974**, *7*, 199–226. [CrossRef]
38. Moeng, C.H. A large-eddy simulation model for the study of planetary boundary-layer turbulence. *J. Atmos. Sci.* **1984**, *41*, 2052–2062. [CrossRef]
39. Sullivan, P.P.; McWilliams, J.C.; Moeng, C.H. A subgrid-scale model for large-eddy simulation of planetary boundary-layer flows. *Bound. Layer Meteorol.* **1994**, *71*, 247–276. [CrossRef]
40. Porté-Agel, F.; Meneveau, C.; Parlange, M.B. A scale-dependent dynamic model for large-eddy simulations: Application to a neutral atmospheric boundary layer. *J. Fluid Mech.* **2000**, *415*, 261–284. [CrossRef]
41. Beare, R.J.; Macvean, M.K.; Holtslag, A.A.M.; Cuxart, J.; Esau, I.; Golaz, J.C.; Jimenez, M.A.; Khairoutdinov, M.; Kosovic, B.; Lewellen, D.; et al. An intercomparison of large-eddy simulations of the stable boundary layer. *Bound. Layer Meteorol.* **2006**, *118*, 247–272. [CrossRef]
42. Schmidt, H.; Schumann, U. Coherent structure of the convective boundary layer derived from large-eddy simulations. *J. Fluid Mech.* **1989**, *200*, 511–562. [CrossRef]
43. Nieuwstadt, F.T.M.; Mason, P.J.; Moeng, C.H.; Schumann, U. Large-eddy simulation of the convective boundary layer: A comparison of four computer codes. In *Turbulent Shear Flows 8*; Durst, F., Ed.; Springer: Berlin/Heidelberg, Germany, 1991; pp. 343–367.
44. Churchfield, M.J.; Moriarty, P.J.; Vijayakumar, G.; Brasseur, J. *Wind Energy-Related Atmospheric Boundary-Layer Large-Eddy Simulation Using OpenFOAM*; NREL/CP-500-48905; National Renewable Energy Lab.: Golden, CO, USA, 2010.
45. Andren, H. Effects of Habitat Fragmentation on Birds and Mammals in Landscapes with Different Proportions of Suitable Habitat: A Review. *Oikos* **1994**, *71*, 355–366. [CrossRef]
46. Flores, O.; Riley, J.J. Analysis of turbulence collapse in the stably stratified surface layer using direct numerical simulation. *Bound. Layer Meteorol.* **2011**, *139*, 241–259. [CrossRef]
47. Mason, P.J.; Derbyshire, S.H. Large-eddy simulation of the stably stratified atmospheric boundary layer. *Bound. Layer Meteorol.* **1990**, *53*, 117–162. [CrossRef]
48. Brown, A.; Derbyshire, S.H.; Mason, P.J. Large-eddy simulation of stable atmospheric boundary layers with a revised stochastic subgrid model. *Q. J. R. Meteorol. Soc.* **1994**, *120*, 1485–1512. [CrossRef]
49. Saiki, E.M.; Moeng, C.H.; Sullivan, P.P. Large-eddy simulation of the stably stratified planetary boundary layer. *Bound. Layer Meteorol.* **2000**, *95*, 1–30. [CrossRef]
50. Churchfield, M.J.; Sang, L.; Moriarty, P.J. *Adding Complex Terrain and Stable Atmospheric Condition Capability to the OpenFOAM-Based Flow Solver of the Simulator for on/Offshore wind Farm Applications (SOWFA)*; National Renewable Energy Lab.: Golden, CO, USA, 2013; Volume 2.
51. Ren, H.; Laima, S.; Li, H. Numerical study of amplitude modulation in the atmospheric boundary layer at very high Reynolds number. *AIP Adv.* **2019**, *9*, 105112. [CrossRef]
52. Schumann, U. Subgrid-scale model for finite-difference simulations of turbulent flow in plane channels and annuli. *J. Comput. Phys.* **1975**, *18*, 376–404. [CrossRef]
53. Yoshizawa, A.; Horiuti, K. A statistically-derived subgrid-scale kinetic energy model for the large-eddy simulation of turbulent flows. *J. Phys. Soc. Jpn.* **1985**, *54*, 2834–2839. [CrossRef]
54. Lee, J.; Lee, J.H.; Choi, J.; Sung, H.J. Spatial organization of large- and very-large-scale motions in a turbulent channel flow. *J. Fluid Mech.* **2014**, *749*, 818–840. [CrossRef]
55. Ren, H.; Laima, S.; Li, H. Numerical Investigation of Very-Large-Scale Motions in a Turbulent Boundary Layer for Different Roughness. *Energies* **2020**, *13*, 659. [CrossRef]
56. Wyngaard, J.C.; Cote, O.R. Cospectral similarity in the atmospheric surface layer. *Q. J. R. Meteorol. Soc.* **1972**, *98*, 590–603. [CrossRef]
57. Zhang, H.S.; Chen, J.Y.; Zhang, A.C.; Wang, J.M.; Mitsuta, Y. An experiment and the results on flux-gradient relationships in the atmospheric surface over Gobi Desert Surface. In Proceedings of the International Symposium on HEIFE, Kyoto, Japan, 1–5 March 1993; Volume IV-12, pp. 349–362.

58. Zhang, H.; Chen, J.Y.; Park, S.U. Turbulence structure in unstable conditions over various surfaces. *Bound. Layer Meteorol.* **2001**, *100*, 243–261. [CrossRef]
59. Panofsky, H.A.; Tennekes, H.; Lenschow, D.H.; Wyngaard, J.C. The characteristics of turbulent velocity components in the surface layer under convective conditions. *Bound. Layer Meteorol.* **1977**, *11*, 355–361. [CrossRef]
60. Bowne, N.E.; Ball, J.T. Observational comparison of rural and urban boundary-layer turbulence. *J. Appl. Meteorol.* **1970**, *9*, 862–873. [CrossRef]
61. Counihan, J. Adiabatic atmospheric boundary layers: A review and analysis of data from the period 1880–1972. *Atmos. Environ.* **1975**, *19*, 871–905. [CrossRef]
62. Steyn, D.G. Turbulence in an unstable surface layer over suburban terrain. *Bound. Layer Meteorol.* **1982**, *22*, 183–191. [CrossRef]
63. Rotach, M.W. *Turbulence Within and Above an Urban Canopy*; ETH: Zurich, Switzerland, 1991.
64. Roth, M. urbulent transfer relationships over an urban surface. II: Integral statistics. *Q. J. R. Meteorol. Soc.* **1993**, *119*, 1105–1120. [CrossRef]
65. Lee, J.H.; Sung, H.J. Very-large-scale motions in a turbulent boundary layer. *J. Fluid Mech.* **2011**, *673*, 80–120. [CrossRef]
66. Dennis, D.J.C.; Nickels, T.B. Experimental measurement of large-scale three-dimensional structures in a turbulent boundary layer. Part 1. Vortex packets. *J. Fluid Mech.* **2011**, *673*, 180–217. [CrossRef]
67. Dennis, D.J.C.; Nickels, T.B. Experimental measurement of large-scale three-dimensional structures in a turbulent boundary layer. Part 2. Long structures. *J. Fluid Mech.* **2011**, *673*, 218–244. [CrossRef]
68. Baltzer, J.R.; Adrian, R.J.; Wu, X. Structural organization of large and very large scales in turbulent pipe flow simulation. *J. Fluid Mech.* **2013**, *720*, 236–279. [CrossRef]
69. Christensen, K.T.; Wu, Y. Characteristics of vortex organization in the outer layer of wall turbulence. In Proceedings of the Fourth International Symposium on Turbulence and Shear Flow Phenomena, Williamsburg, VA, USA, 27–29 June 2005; Volume 3, pp. 1025–1030.
70. Volino, R.J.; Schultz, M.P.; Flack, K.A. Turbulence structure in rough- and smooth-wall boundary layers. *J. Fluid Mech.* **2007**, *592*, 263–293. [CrossRef]
71. Hutchins, N.; Marusic, I. Evidence of very long meandering features in the logarithmic region of turbulent boundary layers. *J. Fluid Mech.* **2007**, *579*, 1–28. [CrossRef]
72. Mathis, R.; Hutchins, N.; Marusic, I. Large-scale amplitude modulation of the small-scale structures in turbulent boundary layers. *J. Fluid Mech.* **2009**, *628*, 311–337. [CrossRef]
73. Balasubramaniam, B.J. Nature of turbulence in wall-bounded flows. Ph.D. Thesis, University of Illinois at Urbana-Champaign, Urbana, IL, USA, 2005.

energies

MDPI

Article

Numerical Study of the Effects of the Jet Fan Speed, Heat Release Rate and Aspect Ratio on Smoke Movement in Tunnel Fires

Ha Thien Khieu, Young Man Lee, Ji Tae Kim and Hong Sun Ryou *

School of Mechanical Engineering, Chung-Ang University, 84, Heukseok-ro, Dongjak-gu, Seoul 06974, Korea; hathienkhieu@gmail.com (H.T.K.); ymlee@alllitelife.com (Y.M.L.); sdd322@naver.com (J.T.K.)

* Correspondence: cfdmec@cau.ac.kr; Tel.: +82-2-820-5280

Received: 4 February 2020; Accepted: 3 March 2020; Published: 5 March 2020

Abstract: In this study, the effects of the jet fan speed, heat release rate and aspect ratio on smoke movement in tunnel fires have been investigated. The jet fan speed was changed from 6.25 (25%) to 12.5 m/s (50%), 18.75 m/s (75%), and 25 m/s (100%). The heat release rate was set up from 3.9 to 6 MW and 16 MW, the aspect ratio was changed from 0.6 to 1 and 1.5, respectively. The lower the jet fan speed is, the longer the smoke back-layering length is. With a higher velocity, the smoke tends to move out of the tunnel quickly; however, smoke stratification also occurs, and this reduces visibility. This could make it difficult for people to evacuate. With a higher heat release rate, the smoke tends to move far away from the fires quickly when compared with other cases. Additionally, the higher the heat release is, the longer the smoke back-layering is. Finally, with a higher aspect ratio, the smoke back layering length in the tunnel is also longer. The smoke layer thickness is also larger than in other cases. The correlation of velocity, heat release rate and aspect ratio has been investigated to avoid the smoke back layer length in tunnel fires.

Keywords: tunnel fires; jet fan speed; heat release rate; aspect ratio; smoke movement; visibility; smoke layer thickness; smoke stratification

1. Introduction

Recently, there have been many tunnels under construction worldwide. A tunnel is an underground passageway commonly used in transportation, and studying fires in tunnels is very critical for human safety. When a fire occurs, smoke and toxic gases will be accumulated in the tunnel due to the nature of a tunnel structure. Two parts of typical smoke are CO_2 and CO, which are poisonous to humans. As smoke spreads across the tunnel, the smoke prohibits the evacuation of occupants and prevents firefighters from extinguishing the fire.

Many fire accidents have occurred in tunnels. These typically cause significant damage in both human and economic terms, causing many people to be killed and injured and many facilities destroyed; for these reasons, investigating tunnel fires is very important.

In order to prevent the smoke from spreading through an entire tunnel, mechanical ventilation systems are used to transport smoke from the tunnel to the outside. A jet fan is one of the main methods for ventilation in a tunnel; a jet fan can quickly move smoke from vehicles and fires outside of the tunnel. Correspondingly, investigating the jet fan speed on smoke movement in tunnels is necessary for decreasing human casualties.

João Carlos Viegas (2009) [1] investigated the interaction between a fire ceiling jet and the flow driven by the jet fan. In that paper, the use of impulse ventilation for smoke control in underground car parks was performed by using Computational Fluid Dynamics (CFD) simulations. M.G.L.C et al. (2010) [2] performed modeling and simulation of a jet fan for controlling airflow in large enclosures. A relatively simple CFD model of the free jet fan was proposed and validated. Moreover, the different series modes of jet fans in a longitudinal tunnel ventilation system were investigated by Guihong Pei and Jie Pan (2014) [3]. When most operational fans are located near the tunnel entrance, it is more conducive to increase the velocity and pressure inside the tunnel. Beyer M et al. (2016) [4] has considered the evaluation of the jet fan performance in tunnels. In a tunnel, due to the lack of space, the jet fan is usually installed close to the wall. However, this increases the wall shear stress around the jet, and thus, the effectiveness is significantly reduced.

An experimental study of the effect of the aspect ratio on the critical velocity in longitudinal ventilation tunnel fires was studied by Ryou et al. (2003) [5]. In that study, the aspect ratios of the tunnel were changed from 0.5, 0.667, 1.0, 1.5 and 2.0. The results showed that the aspect ratio affects the growth of smoke fires in the tunnel. The critical velocity increases with the aspect ratio; it was confirmed that high-temperature smoke moved along the ceiling and low-temperature air moved along the lower part of the tunnel.

Moreover, Haukur Ingason et al. (2008) [6] investigated the effect of air velocity on the heat release rate (HRR) and fire characteristics during a fire in the tunnel. The results showed that when increasing the ventilation rate, causing the maximum heat, the release rate increases. The effect of the cross-section on the maximum heat release rates in the tunnel fires was studied by Li et al. (2016) [7]. In that study, they made a model scale tunnel fire with heptane pool fires. For heptane pool fires, decreasing the tunnel height results in an increase of the heat release rate; the heat release rate is about three times that in a free burn test. However, in this research, there is a lack of information about the effect of the aspect ratio on smoke movement and stratification. On the other hand an experimental study on temperature evolution inside the compartment with fire growth and flame ejection through an opening under external wind was considered by Hu, Longhua, et al [8]. They also considered the characterization of buoyant flow stratification and flame extension lengths in the tunnel fires [9,10]. Wan, Huaxian, et al. [11] investigated a numerical study on the smoke back-layering length and inlet air velocity of fires in an inclined tunnel under natural ventilation with a vertical shaft. The tunnel inlet air velocity increases when increasing the tunnel slope and HRR. Besides, Ji, Jie, et al. (2019) [12] also performed a numerical investigation on the influence of the length–width ratio of the fire source on the smoke movement and temperature distribution in tunnel fires. Furthermore, many researches relating to fires or fans were carried out in literature [13–37].

Although there are many kinds of research regarding the jet fan operation and efficiency, there is no research relating to the effect of jet fan speed on smoke movement or stratification in tunnel fires. The purpose of this study is to analyze the effect of the jet fan speed, heat release rate and aspect ratio on smoke movement in a tunnel fire. For this purpose, a numerical study of the effect of the jet fan speed, heat release rate and aspect ratio in the tunnel fires has been carried out by using CFD simulation, as shown in Figure 1. The smoke movement, temperature, velocity and visibility are also investigated in this paper to clarify the impact of the jet fan location and velocity in the tunnel.

Figure 1. Numerical analysis objective.

2. Theoretical Analysis

2.1. *Jet Fan Speed*

The jet fan centerline velocity can be calculated as follows:

Awbi (2003) [38] and Baturin (1972) [39] provide the equation regarding the centerline velocity and distance for the experiment. The equation includes the centerline velocity, u_m, the supply velocity, u_0, distance, X, supply diameter of the fan, d_0, and constant, a (from 0.076 to 0.080 for cylindrical tubes).

With a square duct, Kümmel et al. (2007) [40] provided another equation of jet fan experimental data. It includes the length of the potential core, x_0, the side length of the square duct, h, and constant, m (from 0.12 to 0.2).

From these equations, the jet fan centerline velocity can be calculated, as shown in the Figure 2, with the supply velocity at 25 m/s, and the side length of the square duct at 0.91 m.

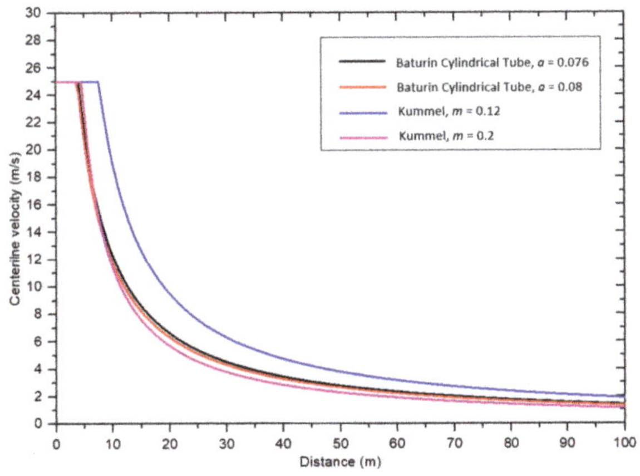

Figure 2. Jet fan centerline velocity.

2.2. *Heat Release Rate*

In this study, the heat release rate was obtained from our experimental study; one car was assumed to be fired in the tunnel. The maximum heat release rate was 3896 kW at 1551 seconds as shown below Figure 3. Besides, the heat release rate was around 6 MW, and 16 MW for two vehicles [41] and truck fires, respectively.

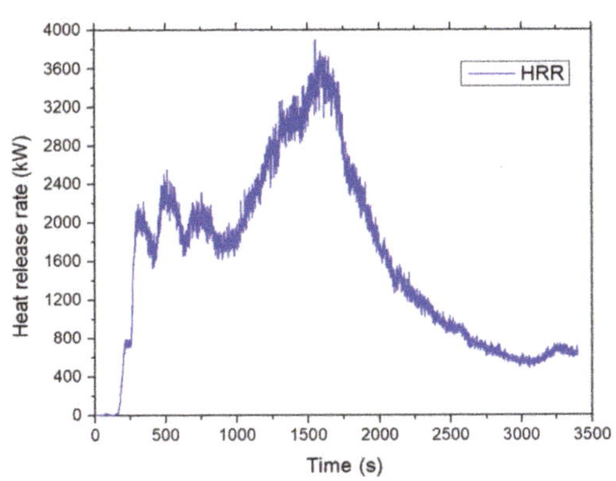

Figure 3. The heat release rate from an experimental study (one vehicle) by Park et al. (2019) [41].

2.3. Smoke Layer Height and Thickness

There are several methods to determine the smoke layer height and thickness. These include the *N*-percentage rule and integral ratio method. The *N* percentage rule was developed by Cooper et al. [42]. It is the simplest way to define the smoke layer. This method has been used widely in a much different application, however, it has a limitation. Depending on every region, other *N* values will be used; this may cause confusion when using *N* values.

He et al. (1998) [43] proposed another method, which was an integral ratio method. This method divided the temperature profile into two regions, which are the smoke layer and cold air layer. The integral ratio of the smoke layer is r_u and the air layer is r_l. The sum of the integral ratio is r_t; when r_t reaches the minimum value, the interface height can be calculated. As this method is not based on any external reference parameter, it is adopted to obtain a more accurate estimation of the smoke layer thickness in this study.

3. Numerical Analysis

3.1. Computational Domain for Numerical Study

For this numerical study, the effect of the jet fan speed, heat release rate and aspect ratio on the smoke movement and visibility in tunnel fires would be considered. The geometry of the tunnel and the computation flow domain is presented in Figure 4. The computational domain was extended over the exit and entrance surface, in order to eliminate the influence of the tunnel exit and entrance conditions. Moreover, the computational domain was expanded to the place where the pressure boundary condition was applied. Twenty-four cases have been investigated in this simulation. The parameters of the jet fan velocity, heat release rate and aspect ratio are shown in Table 1.

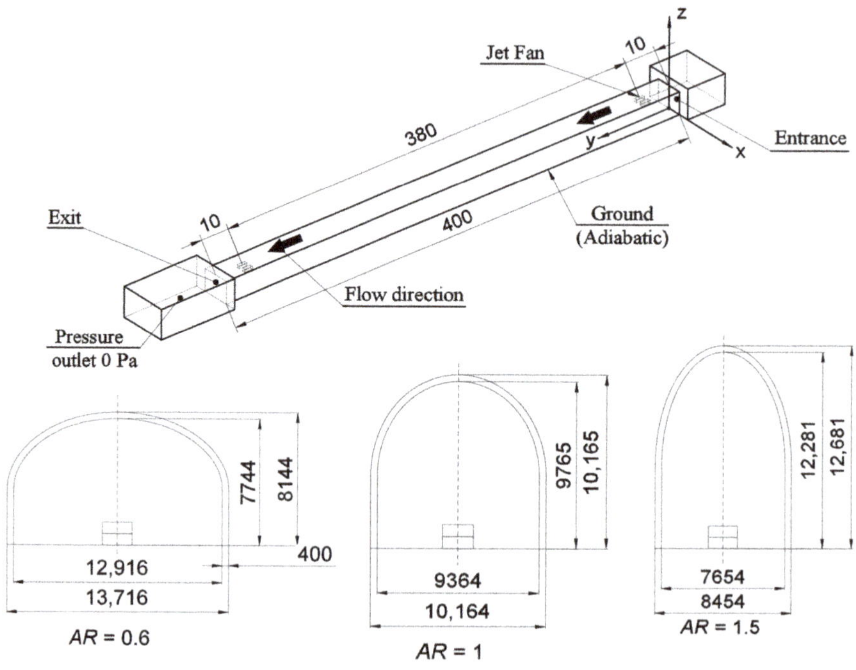

Figure 4. The geometry of the tunnel.

Table 1. The numerical case studies.

Case	HRR (kW)	Jet Fan Velocity (m/s)	Aspect Ratio
1	3896	25	0.6
2	3896	18.75	0.6
3	3896	12.5	0.6
4	3896	6.25	0.6
5	6000	25	0.6
6	6000	12.5	0.6
7	16,000	25	0.6
8	16,000	12.5	0.6
9	3896	25	1
10	3896	25	1.5
11	3896	12.5	1
12	3896	12.5	1.5
13	3896	6.25	1
14	3896	6.25	1.5
15	6000	18.75	0.6
16	6000	6.25	0.6
17	16,000	18.75	0.6
18	16,000	6.25	0.6
19	6000	25	1
20	6000	25	1.5
21	16,000	25	1
22	16,000	25	1.5
23	3896	18.75	1
24	3896	18.75	1.5

3.2. Numerical Method and Boundary Conditions

3.2.1. Numerical Method

The Fire Dynamics Simulator (FDS) is a computational fluid dynamics software developed by the National Institute of Standards and Technology (NIST) of the United States Department of Commerce, in cooperation with the VTT (Valtion Teknillinen Tutkimuskeskus) Technical Research Centre of Finland. This computer program numerically solves large-eddy simulations using Navier–Stokes equations, which are suitable for low-speed flow, thermally driven flow, with an emphasis on smoke and heat transport from the fire to describe the progression of the fire.

The governing equations of FDS can be described as in [44].

3.2.2. Numerical Condition

In order to simulate the fire in the tunnel, the initial condition and boundary condition were set up as follows. The ground of the tunnel is adiabatic, the tunnel structure is also adiabatic and the computation domain at the entrance and exit of the tunnel was set up as a zero Pascal pressure gauge.

For the initial condition, the value was set up, as follows in Table 2. The velocity in three directions was 0 m/s, the temperature was 293.15 K and the gauge pressure was 0 Pa.

Table 2. The initial conditions.

Physics	Fluid Flow
Cell zone	Gauge pressure = 0 Pa x, y, z velocity = 0 m/s Temperature: 293.15 K

3.3. *Grid Independence Test*

For a numerical study, the independent grid test was performed to get accurate analysis results and the proper size. The number of grids was increased from 915,000 to 2,300,000, and a total of three cases, of a number of grids, were selected as Table 3.

Table 3. The number of grid cells.

Case Index	Number of Grid Cells
Case 1	2,300,000
Case 2	1,275,000
Case 3	915,000

From the above Figure 5, when the grid size was 0.1 m, the results showed the best accuracy, so in this case, the grid size of 0.1 m was used for a nearby fire source, and the grid size of 0.4 m was used for a location that was far away from the fire source. As it is less critical in this region, the coarse mesh was applied.

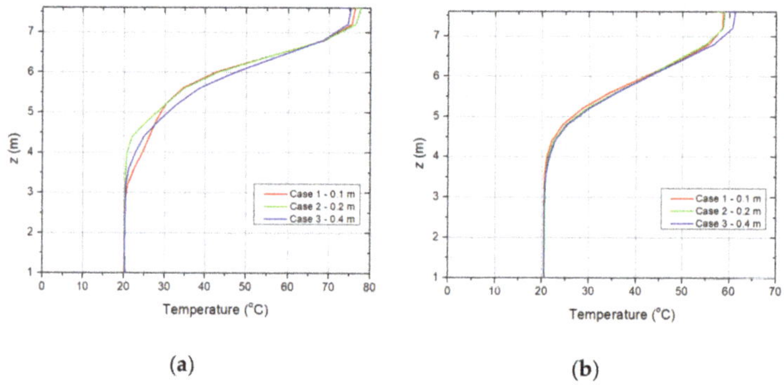

Figure 5. The vertical temperature distribution: (**a**) Location A ($Y = 120$) and (**b**) Location B ($Y = 300$).

4. Results and Discussion

In this study, a numerical study was conducted to investigate the effect of jet fan speed, heat release rate and aspect ratio on the smoke movement in the tunnel fire for optimizing the ventilation system in the tunnel. For this purpose, the separate effects will be analyzed as follows.

4.1. *Effect of Jet Fan Speed*

To consider the effect of the jet fan speed on smoke movement, the heat release rate and aspect ratio were fixed, as below in Figure 6. The smoke movement is shown through visibility because it is directly related to visibility.

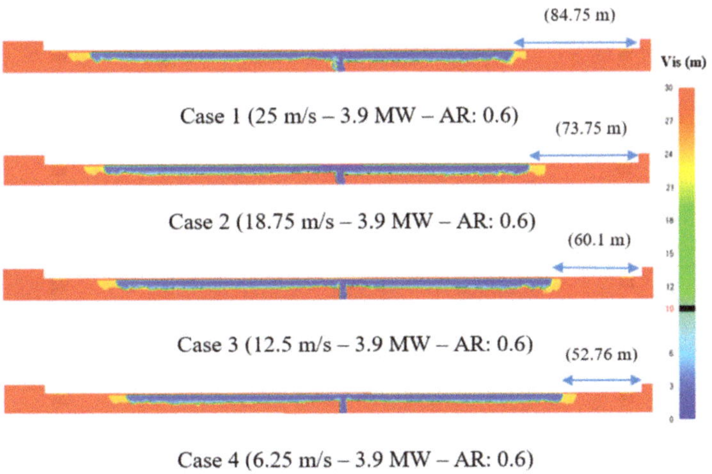

Figure 6. Visibility in the tunnel when changing the jet fan speed.

When the jet fan operates with a high velocity, the smoke tends to move to the left side (the exit direction), as there is a significant amount flow rate of the air entrained into the tunnel from the entrance. However, when the jet fan speed is decreased, the smoke tends to move in both directions; and the visibility between the left side and right side had the same distribution.

When changing the jet fan speed, the distances between the smoke and entrance position for case 1, 2, 3 and 4 were 84.75 m, 73.75 m, 60.1 m and 52.76 m, respectively. This indicates that the higher the jet fan speed is, the lower the smoke back layer length is. At location $Y = 300$ m, the smoke layer thicknesses for case 1, 2, 3 and 4 were 2.3 m, 2.1 m, 2 m and 1.8 m, respectively. From this result, when decreasing the jet fan speed, the smoke layer thickness decreased. Moreover, the temperature in the tunnel on the tunnel center's plane was also affected by the jet fan speed. In this case, it was confirmed that high-temperature smoke moved along the ceiling and low-temperature air moved along the lower part of the tunnel.

For case 1 in Figure 7b, when the height (z) was from 1 to 5.4 m, the velocity was a minus value, indicating that, at this region, the flow (air) moved into the tunnel (moved from the right side to the left side). When the height (z) was from 5.4 to 7.6 m, the velocity was a positive value, and in this region, the flow (smoke) from the fire in the tunnel moved outside (moved from the left side to right side). Similarly, from case 2 to case 3 and 4, the temperature and velocity distribution were also different between these cases. In case 2, when the height was from 1 to 5.4 m, the air moved into the tunnel. When the height was from 5.4 to 7.6 m, the smoke from the fire in the tunnel moved outside. For cases 3 and 4, when the height was from 1 to 5 m, the air moved into the tunnel. When the height was from 5 to 7.6 m, the smoke from the fire in the tunnel moved outside. The effect of the jet fan speed on flow can be obtained from Figure 7, as well. In the lower part, as the flow was air induced, the absolute velocity value of case 1 (highest speed) was larger than the other cases. However, at the upper part, which was near the ceiling, because all flow was smoke, the absolute velocity value of case 1 (the highest speed) was lower than for the other cases.

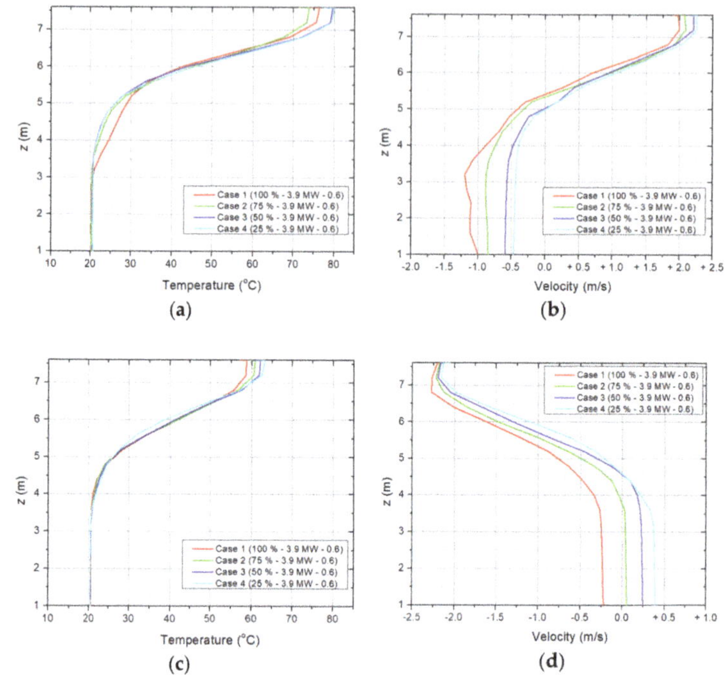

Figure 7. Temperature and velocity distribution in the tunnel: (**a**) temperature distribution at $Y = 120$; (**b**) velocity distribution at $Y = 120$; (**c**) temperature distribution at $Y = 300$ and (**d**) velocity distribution at $Y = 300$.

4.2. Effect of Heat Release Rate

First, the value of the jet fan speed was fixed at 25 m/s, the aspect ratio was 0.6 and the heat release rate was changed. The smoke movement in these cases on the tunnel center's plane is shown in Figure 8. In case 1, the velocity of the jet fan was 25 m/s (100%), and the smoke movement was totally different between the left side and right side. The smoke tended to move to the left side (exit direction). As, in this case, the jet fan operated with a high velocity, there was a significant amount of flow rate of the air coming inside the tunnel from the entrance. Therefore, it pushed the smoke to move to the exit direction. Moreover, the heat release rate was 3.9 MW, thus the amount of smoke was not high. However, when increasing the heat release rate, the amount of smoke was more significant. In case 7, with a 16 MW heat release rate, the smoke moved in both directions. In the region near the entrance, due to the effect of a jet fan, stratification occurred. The visibility was lower when compared with other cases, as a result, and it was difficult for evacuation.

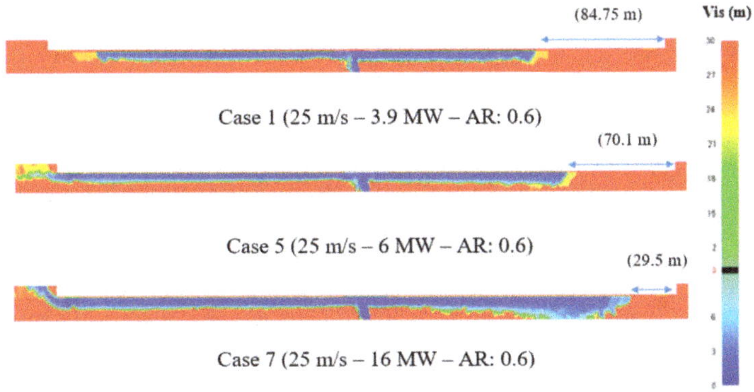

Figure 8. Visibility in the tunnel fire when changing the HRR.

Moreover, when changing the heat release rate, the length of the smoke back layer was quite different. The distances between the smoke and entrance position for case 1, case 5 and case 7 were 84.75 m, 70.1 m and 29.5 m, respectively. This indicates that the smoke back layer length was affected by the heat release rate. With the higher heat release rate, the smoke back layer length was higher. Additionally, at location $Y = 300$ m, the smoke layer thicknesses for case 1, 5 and 7 were 2.3 m, 2.4 m and 2.7 m, respectively. Thus, the larger the heat release rate is, the thicker the smoke layer is.

The velocity vector at the center plane for cases 1, 5 and 7 are shown in Figure 9. In case 7, at the distance 29.5 m from the entrance, the smoke and air from the jet fan were mixed. The velocity vector of the air from jet fan tended to move down. The stratification occurred in this region.

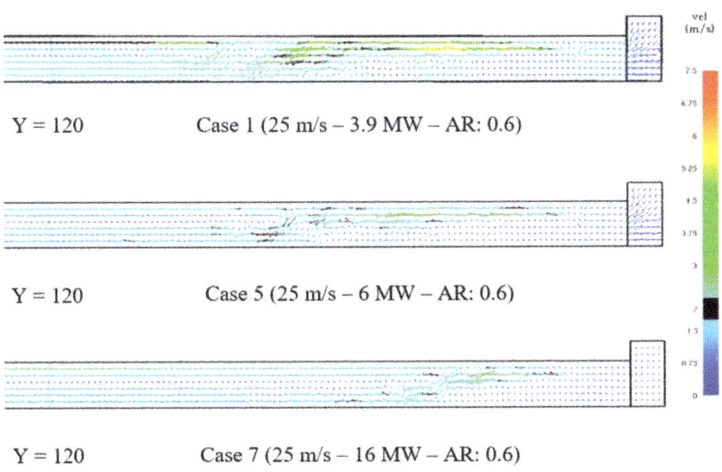

Figure 9. Velocity vector in the tunnel fire when changing the HRR.

In Figure 10a at the upper part, the larger the heat release rate was, the higher the temperature was. Besides, the velocity was also larger when the heat release rate was larger. In the lower part, because all flow was air so that the absolute velocity value of case 1 (lowest heat release rate) was lower than other cases. When the heat release rate was low, it means the amount of air needed for fire combustion was also lower, so that the air coming inside the tunnel was low (case 1), and the smoke movement was quite weak when compared with other cases. However, with a higher heat release rate, the amount of air needed for the combustion process was larger and the smoke generated by the combustion process was also larger, so the velocity value, in this case, was greater than in other cases.

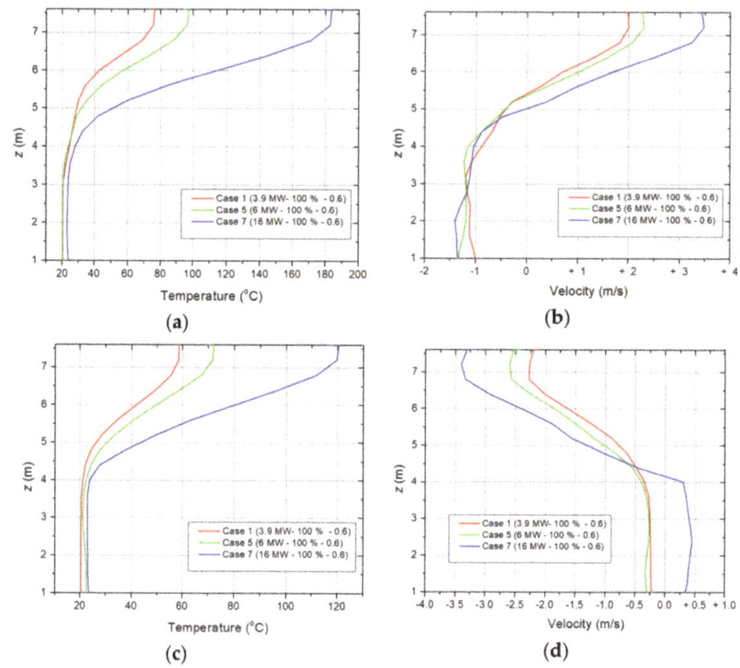

Figure 10. Temperature and velocity distribution in the tunnel: (**a**) temperature distribution at $Y = 120$; (**b**) velocity distribution at $Y = 120$; (**c**) temperature distribution at $Y = 300$ and (**d**) velocity distribution at $Y = 300$.

4.3. Effect of Aspect Ratio

In order to analyze the effect of aspect ratio on smoke movement, the numerical study was investigated by fixing the value of the jet fan speed and the heat release rate, and the aspect ratio was changed from 0.6 to 1 and 1.5.

First, the value of the jet fan speed was 25 m/s, and the heat release rate was 3.9 MW, and the aspect ratio was changed. The smoke movement in these cases was shown in the figures below, through visibility.

For case 1 in Figure 11, the velocity of the jet fan was 25 m/s, and the smoke movement was completely different between the left side and right side. In this case, the smoke tended to move to the left side (exit direction). As, in case 1, the jet fan operated with a high velocity, thus there was a tremendous flow rate of the air-entrained inside the tunnel. Therefore, it pushed the smoke to move to the exit direction. The heat release rate was 3.9 MW; thus, the amount of smoke was not high. However, when increasing the aspect ratio, the smoke could move outside quickly. In the region near the entrance, due to the effect of the jet fan, stratification occurred with the largest aspect ratio (case 10). The visibility was lower when compared with other cases, and as a result, it was difficult for the evacuation to proceed.

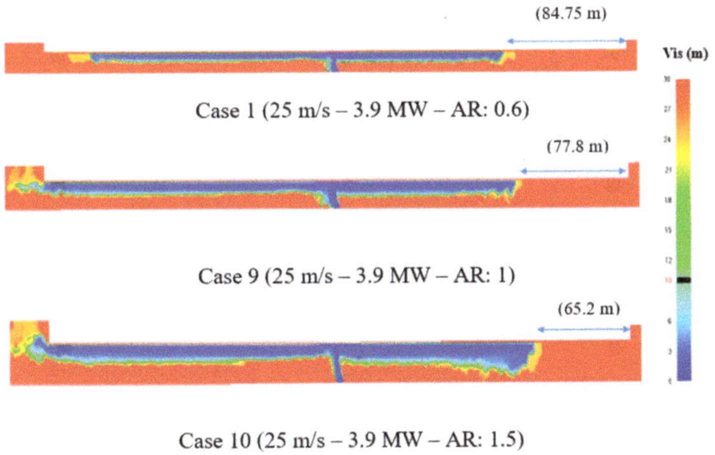

Figure 11. Visibility in the tunnel fire when changing the aspect ratio.

Furthermore, when changing the aspect ratio, the distance between the smoke and entrance position from case 1, case 9 and case 10 were 84.75 m, 77.8 m and 65.2 m, respectively. This indicates that the smoke back layer length was affected by the aspect ratio. With the high aspect ratio, the smoke back layer length was higher.

In Figure 12a, at the upper part, the lower the aspect ratio was, the larger the temperature was. When the aspect ratio increased, the velocity increased. This is because the width of the tunnel decreased when increasing the aspect ratio. In the lower part, as all flow was air, the absolute velocity value of case 1 (lowest aspect ratio) was lower than in other cases.

As shown in the previous sections, when increasing the heat release rate, the smoke layer length was also increased. The reason for this is that when the heat release rate of fire was larger, the amount of smoke generated was also higher. This caused the smoke to move quickly, so the smoke layer length was also increased. In contrast, when the jet fan speed was increased, the smoke layer length was decreased.

The result shows that when the aspect ratio increased, the smoke back layer length also increased. When the aspect ratio was larger, this indicates that the height of the tunnel was higher. For that reason, the smoke was affected by buoyancy force, leading to the smoke movement being larger, and the smoke layer length being longer. The higher the jet fan speed was, the shorter the smoke back layer length was.

Furthermore, when changing both the heat release rate and aspect ratio, with a higher heat release rate or aspect ratio, the smoke layer length was also larger. This was in agreement with the above conclusion in previous figures.

In the tunnel fires, the ceiling jet velocity is calculated by the following equation [45]

$$\left(\frac{V_{\max}}{\sqrt{gH}}\right)/Q_c^{*1/3} = 0,5812\left(\frac{l_b}{H}\right)^{-1/3}\exp\left[-1.567.St.\frac{x}{H}\left(\frac{l_b}{H}\right)^{1/3}\right] \tag{1}$$

where V_{max} is maximum ceiling jet velocity, H is tunnel height, $Q_c{}^*$ dimensionless heat release rate = $Q_c/(\rho_\infty C_p T_\infty g^{1/2} H^{5/2})$, l_b is half-width of tunnel = $W/2$, St is Stanton number and x is the horizontal distance from the center of fire source (m).

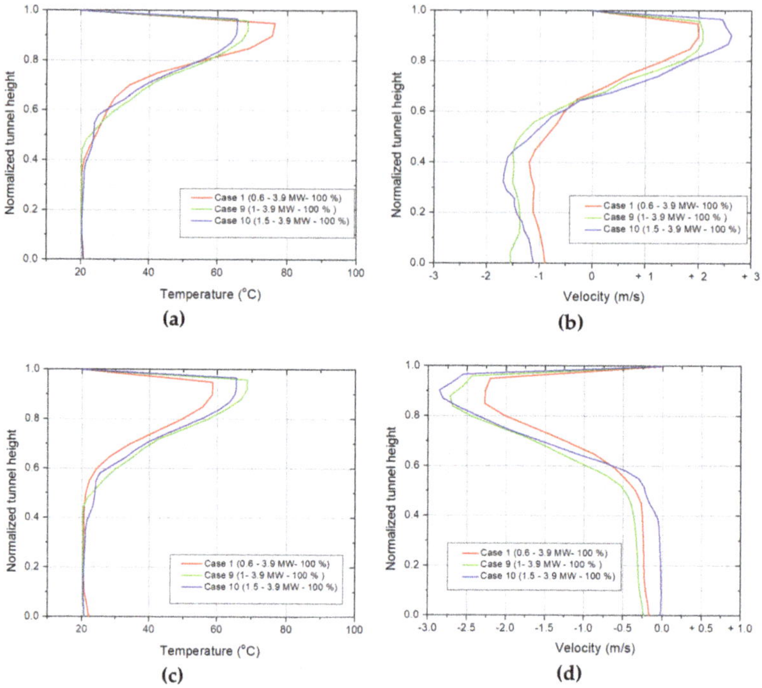

Figure 12. Temperature and velocity distribution in the tunnel: (**a**) temperature distribution at $Y = 120$; (**b**) velocity distribution at $Y = 120$; (**c**) temperature distribution at $Y = 300$ and (**d**) velocity distribution at $Y = 300$.

Based on the above equation, the following equation can be derived.

$$V_{\max} = \sqrt{g.AR.2l_b}.Q_c^{*1/3}.0,5812\left(\frac{1}{2.AR}\right)^{-1/3}\exp\left[-1.567.St.\frac{x}{AR.l_b}\left(\frac{1}{2.AR}\right)^{1/3}\right] \tag{2}$$

where AR is aspect ratio = $H/2l_b$

On the other hand, the jet fan velocity can be applied by the equation as in Section 2.1.

$$U_m = \frac{0.48U_0}{(aX/d_0 + 0.145)} \tag{3}$$

In order to avoid the smoke back layer length, the velocity of the jet fan at distance X (U_m) has to be larger than the ceiling jet velocity (V_{max}). So, that the following equation can be derived from Equations (2) and (3)

$$U_0 > \frac{(a(L-x)/d_0 + 0.145)}{0.48}.\sqrt{g.AR.2l_b}.Q_c^{*1/3}.0,5812\left(\frac{1}{2.AR}\right)^{-1/3}\exp\left[-1.567.St.\frac{x}{AR.l_b}\left(\frac{1}{2.AR}\right)^{1/3}\right] \tag{4}$$

where L is the distance from the center of fire source to a jet fan location.

Based on Equation (4), as the left term and right term are equal, it means that the smoke movement keeps stable. From that point, the distance x is the place in which the ceiling jet velocity and jet fan velocity are equal. So, x (smoke back layer length) can be derived from Equation (4). The results of Equation (4) are shown in Figures 13 and 14.

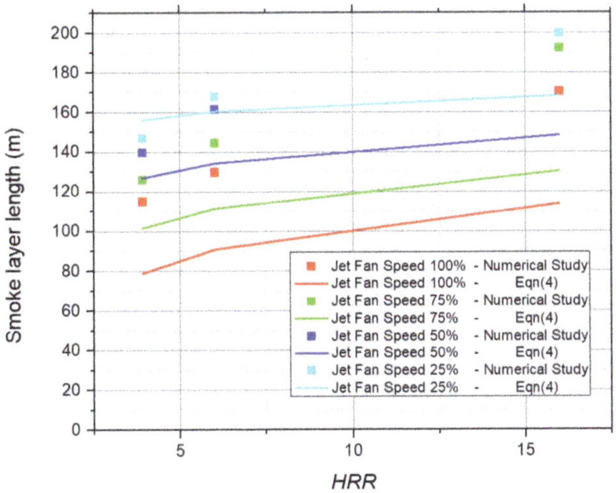

Figure 13. Smoke back layer length in the relationship between *HRR* and jet fan speed.

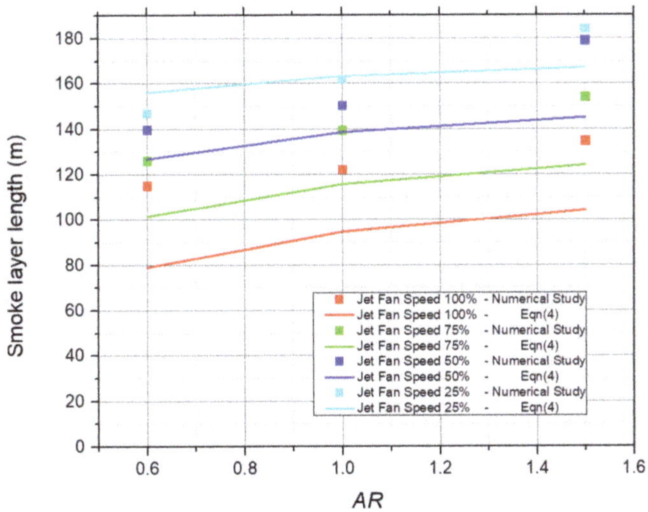

Figure 14. Smoke back layer length in relationship between *AR* and jet fan speed.

The smoke back layer length in the correlation between the heat release rate and jet fan speed is shown in Figure 13, the dots represent the results obtained by numerical study, and the line is the results from Equation (4). The figure shows that when increasing the heat release rate, the smoke layer length was also increased for both the numerical study and Equation (4). The reason is that when the heat release rate of fire was larger, the amount of smoke generated was also higher, it tended to cause the smoke to move quickly, the result was that the smoke layer length was also increased. In contrast, when the jet fan speed increased, the smoke layer length decreased, as this depends on a large amount of air that has been transported at high speed. Similarly, in Figure 14, when increasing the aspect ratio, the smoke layer length was also increased for both the numerical study and Equation (4), because when the aspect ratio was larger, it means the height of the tunnel was higher. For that reason, the smoke affected by buoyancy force led to the smoke movement being larger and the smoke layer length being longer, obviously. In this figure, the effect of the jet fan speed on the smoke layer length was also the same as the previous figure. The higher the jet fan speed was, the shorter the smoke layer length was. In Equation (4), there are some coefficients representing the temperature and velocity attenuation. Besides, because the two jet fans operated, it caused discrepancy in the coefficient. For that reason, the results had a little bit of error between the equation and numerical study. The future work will be investigated to find out that correction.

5. Conclusions

In this study, the numerical study of the effect of jet fan speed, heat release rate and aspect ratio on smoke movement in tunnel fires was investigated. The effect of the jet fan speed, heat release rate and aspect ratio on smoke movement and visibility and stratification in tunnel fires were drawn as follows:

The lower the jet fan speed is, the longer the smoke back-layering is, as this depends on a small amount of air that has been transported at low speed. The smoke tends to spread in both directions.

When changing the jet fan's speed to a high velocity, the smoke tends to move to the exit direction, because the flow rate of air that enters the tunnel is higher, thus, the smoke can quickly move outside. However, smoke stratification also occurs and the smoke and air are mixed. This reduces visibility, and it is difficult for people to evacuate.

The lower the heat release is, the lower the smoke back-layering is. With a higher heat release rate, the smoke tends to move far away from the fires when compared with other cases. With a high heat release rate, the stratification of smoke occurs quite strongly, as it is affected by both the heat release rate and jet fan. This causes difficulty for people evacuating; thus, it is important to optimize the jet fan speed to avoid stratification occurring in the tunnel.

The higher the aspect ratio is, the higher the smoke back-layering is. With a higher aspect ratio, the smoke can move outside quickly. On the other hand, stratification also occurs strongly, as compared with other cases.

The jet fan velocity has been investigated to avoid the smoke back layering as Equation (4).

For these reasons, it is necessary to consider the effects of jet fan speed, heat release rate and aspect ratio when designing mechanical ventilation systems in tunnels.

Author Contributions: Writing—original draft preparation, H.T.K. and Y.M.L.; formal analysis, H.T.K. and Y.M.L. and J.T.K.; writing—review and editing, H.S.R. All authors have read and agreed to the published version of the manuscript.

Funding: This research was funded by NRF, grant number 2019R1F1A1061741 and Chung-Ang University Young Scientist Scholarship (CAYSS) Program 2018.

Acknowledgments: This research was supported by Chung-Ang University Young Scientist Scholarship (CAYSS) Program.

Conflicts of Interest: The authors declare no conflict of interest.

References

1. Viegas, J.C. The use of impulse ventilation for smoke control in underground car parks. *Tunn. Undergr. Space Technol.* **2010**, *25*, 42–53. [CrossRef]
2. Vd Giesen, B.J.M.; Penders, S.H.A.; Loomans, M.G.L.C.; Rutten, P.G.S.; Hensen, J.L.M. Modelling and simulation of a jet fan for controlled airflow in large enclosures. *Environ. Model. Softw.* **2011**, *26*, 191–200. [CrossRef]
3. Pei, G.; Pan, J. Numerical study on different series modes of the jet fan in a longitudinal tunnel ventilation system. *Math. Probl. Eng.* **2014**, *2014*, 194125. [CrossRef]
4. Beyer, M.; Sturm, P.J.; Saurwein, M.; Bacher, M. Evaluation of Jet Fan Performance in Tunnels. In Proceedings of the 8th International Conference Tunnel Safety and Ventilation, Graz, Austria, 25–26 April 2016.
5. Lee, S.R.; Ryou, H.S. An experimental study of the effect of the aspect ratio on the critical velocity in longitudinal ventilation tunnel fires. *J. Fire Sci.* **2005**, *23*, 119–138. [CrossRef]
6. Lönnermark, A.; Ingason, H. The effect of air velocity on heat release rate and fire development during fires in tunnels. *Fire Saf. Sci.* **2008**, *9*, 701–712. [CrossRef]

7. Li, Y.Z.; Fan, C.G.; Ingason, H.; Lönnermark, A.; Ji, J. Effect of cross section and ventilation on heat release rates in tunnel fires. *Tunn. Undergr. Space Technol.* **2016**, *51*, 414–423. [CrossRef]
8. Hu, L.; Ren, F.; Hu, K.; Tang, F.; Lu, K. An experimental study on temperature evolution inside compartment with fire growth and flame ejection through an opening under external wind. *Proc. Combust. Inst.* **2017**, *36*, 2955–2962. [CrossRef]
9. Tang, F.; Li, L.J.; Dong, M.S.; Wang, Q.; Mei, F.Z.; Hu, L.H. Characterization of buoyant flow stratification behaviors by Richardson (Froude) number in a tunnel fire with complex combination of longitudinal ventilation and ceiling extraction. *Appl. Therm. Eng.* **2017**, *110*, 1021–1028. [CrossRef]
10. Qiu, A.; Hu, L.; Chen, L.; Carvel, R.O. Flame extension lengths beneath a confined ceiling induced by fire in a channel with longitudinal air flow. *Fire Saf. J.* **2018**, *97*, 29–43. [CrossRef]
11. Wan, H.; Gao, Z.; Han, J.; Ji, J.; Ye, M.; Zhang, Y. A numerical study on smoke back-layering length and inlet air velocity of fires in an inclined tunnel under natural ventilation with a vertical shaft. *Int. J. Therm. Sci.* **2019**, *138*, 293–303. [CrossRef]
12. Ji, J.; Tan, T.; Gao, Z.; Wan, H.; Zhu, J.; Ding, L. Numerical investigation on the influence of length–width ratio of fire source on the smoke movement and temperature distribution in tunnel fires. *Fire Technol.* **2019**, *55*, 963–979. [CrossRef]
13. Khattri, S.K.; Log, T.; Kraaijeveld, A. Tunnel Fire Dynamics as a Function of Longitudinal Ventilation Air Oxygen Content. *Sustainability* **2019**, *11*, 203. [CrossRef]
14. Khattri, S.K. From small-scale tunnel fire simulations to predicting fire dynamics in realistic tunnels. *Tunn. Undergr. Space Technol.* **2017**, *61*, 198–204. [CrossRef]
15. Ji, J.; Guo, F.; Gao, Z.; Zhu, J. Effects of ambient pressure on transport characteristics of thermal-driven smoke flow in a tunnel. *Int. J. Therm. Sci.* **2018**, *125*, 210–217. [CrossRef]
16. Gao, Z.; Jie, J.; Wan, H.; Zhu, J.; Sun, J. Experimental investigation on transverse ceiling flame length and temperature distribution of sidewall confined tunnel fire. *Fire Saf. J.* **2017**, *91*, 371–379. [CrossRef]
17. Carvel, R.O.; Beard, A.N.; Jowitt, P.W.; Drysdale, D.D. Fire size and fire spread in tunnels with longitudinal ventilation systems. *J. Fire Sci.* **2005**, *23*, 485–518. [CrossRef]
18. Kim, J.H.; Kim, J.H.; Kim, K.Y.; Yoon, J.Y.; Yang, S.H.; Choi, Y.S. High-efficiency design of a tunnel ventilation jet fan through numerical optimization techniques. *J. Mech. Sci. Technol.* **2012**, *26*, 1793–1800. [CrossRef]
19. Lee, S.R.; Ryou, H.S. A numerical study on smoke movement in longitudinal ventilation tunnel fires for different aspect ratio. *Build. Environ.* **2006**, *41*, 719–725. [CrossRef]
20. Janssens, M.; Tran, H.C. Data reduction of room tests for zone model validation. *J. Fire Sci.* **1992**, *10*, 528–555. [CrossRef]
21. Muckett, M.; Furness, A. *Introduction to Fire Safety Management*; Routledge: Abingdon upon Thames, UK, 2007.
22. Šekularac, M.B. Experimental determination of tunnel ventilation axial ducted fan performance. *Therm. Sci.* **2016**, *20*, 209–221. [CrossRef]
23. Dahm, W.J.A.; Dimotakis, P.E. Measurements of entrainment and mixing in turbulent jets. *AIAA J.* **1987**, *25*, 1216–1223. [CrossRef]
24. Lönnermark, A.; Ingason, H. *Fire Spread in Large Industrial Premises and Warehouse*; SP Swedish National Testing and Research Institute: Borås, Sweden, 2005.
25. Grant, G.; Drysdale, D. Numerical modelling of early flame spread in warehouse fires. *Fire Saf. J.* **1995**, *24*, 247–278. [CrossRef]
26. Drysdale, D. *An Introduction to Fire Dynamics*; John Wiley and Sons: Hoboken, NJ, USA, 1986.
27. Jun, Y.J. Deduction of Key Items and Task Improvement for Quality Test of Curtain Wall. Master's Thesis, University of Seoul, Seoul, Korea, 2012.
28. Kweon, O.S.; Yoo, Y.H.; Kim, H.Y.; Kim, J.H. Real Scale Experiment for Assessing Fire Model of Office Compartment. *J. Korean Soc. Hazard Mitig.* **2012**, *12*, 61–65. [CrossRef]
29. Lee, J.S.; Yim, H.C.; Kim, H.J.; Kim, H.Y.; Cho, B.H. An Experimental Study on the Evaluation of Fire Resistance Performance of Curtain-Wall. *J. Archit. Inst. Korea* **2011**, *27*, 141–148.
30. Mowrer, F.W. Window Breakage Induced by Exterior Fires. In Proceedings of the Second International Conference on Fire Research and Engineering (ICFRE2), Maui, HI, USA, 3–9 May 1998; pp. 404–415.
31. Pagni, P.J.; Joshi, A.A. Glass Breaking in Fires. *Fire Saf. Sci.* **1991**, *3*, 791–802. [CrossRef]
32. Rahkonen, O.K. Breaking of Window Glass Close to Fire. *Fire Mater.* **1988**, *12*, 60–69.

33. Roh, J.S.; Yang, S.S.; Ryou, H.S. Tunnel fires: Experiments on Critical Velocity and Burning Rate in Pool Fire During Longitudinal Ventilation. *J. Fire Sci.* **2005**, *25*, 161–176. [CrossRef]
34. Shields, T.J.; Silcock, G.W.H.; Flood, M. Performance of a Single Glazing Assembly Exposed to a fire in the Centre of and Enclosure. *Fire Mater.* **2002**, *26*, 51–75. [CrossRef]
35. Shields, T.J.; Silcock, G.W.H.; Flood, M. Behavior of Double Glazing in Corner Fires. *Fire Technol.* **2005**, *41*, 37–65. [CrossRef]
36. Wu, C.W.; Lin, T.H.; Lei, M.Y.; Chung, T.H.; Huang, C.C.; Chiang, W.T. Fire Resistance Tests of a Glass Pane with Down-flowing Water Film. *J. Chin. Inst. Eng.* **2008**, *31*, 737–744. [CrossRef]
37. Xie, Q.; Zhang, H.; Wan, Y.; Zhan, Q.; Cheng, X. Full-scale experimental study on crack and fallout of toughened glass with different thicknesses. *Fire Mater.* **2008**, *32*, 293–306. [CrossRef]
38. Awbi, H.B. *Ventilation of Buildings*, 2nd ed.; Spon Press: London, UK, 2003.
39. Baturin, V.V. *Fundamentals of Industrial Ventilation*; Pergamon: Oxford, UK, 1972.
40. Kümmel, W. *Technische Strömungsmechanik*, 3rd ed.; Technical Report; Teubner, B.G., Ed.; Teubner: Wiesbaden, Germany, 2007; p. 25.
41. Park, Y.; Ryu, J.; Ryou, H.S. Experimental Study on the Fire-Spreading Characteristics and Heat Release Rates of Burning Vehicles Using a Large-Scale Calorimeter. *Energies* **2019**, *12*, 1465. [CrossRef]
42. Cooper, L.Y.; Harkleroad, M.; Quintiere, J.; Rinkinen, W. An experimental study of upper hot layer stratification in full-scale multiroom fire scenarios. *J. Heat Transfer* **1982**, *104*, 741–749. [CrossRef]
43. He, Y.; Fernando, A.; Luo, M. Determination of interface height from measured parameter profile in enclosure fire experiment. *Fire Saf. J.* **1998**, *31*, 19–38. [CrossRef]
44. Kevin, M.; Simo, H.; Jason, F. Fire dynamics simulator (version 5) technical reference guide. *NIST Spec. Publ.* **2007**, *1018*, 5.
45. Oka, Y.; Oka, H. Velocity and temperature attenuation of a ceiling-jet along a horizontal tunnel with a flat ceiling and natural ventilation. *Tunn. Undergr. Space Technol.* **2016**, *56*, 79–89. [CrossRef]

energies

MDPI

Article

Numerical Investigation of Vertical Crossflow Jets with Various Orifice Shapes Discharged in Rectangular Open Channel

Hao Yuan [1], Ruichang Hu [2], Xiaoming Xu [3,*], Liang Chen [1], Yongqin Peng [1] and Jiawan Tan [4]

1 Southwest Research Institute for Water Transport Engineering, Chongqing Jiaotong University, Chongqing 400074, China; yuanhao@cqjtu.edu.cn (H.Y.); Liangchenchm315@163.com (L.C.); pyqin_xks@163.com (Y.P.)

2 State Key Laboratory of Hydraulics and Mountain River Engineering, Sichuan University, Chengdu 610065, China; hurrayjy@163.com

3 Changjiang Chongqing Waterway Bureau, Chongqing 400074, China

4 School of Shipping and Naval Architecture, Chongqing Jiaotong University, Chongqing 400074, China; taninfo@cqjtu.edu.cn

* Correspondence: hurray1@sina.com

Received: 6 February 2020; Accepted: 15 March 2020; Published: 22 March 2020

Abstract: Vertical jet in flowing water is a common phenomenon in daily life. To study the flow and turbulent characteristics of different jet orifice shapes and under different velocity ratios, the realizable k-ε turbulent model was adopted to analyze the three-dimensional (3D) flow, turbulence, and vortex characteristics using circular, square, and rectangular jet orifices and velocity ratios of 2, 5, 10, and 15. The following conclusions were drawn: The flow trajectory of the vertical jet in the channel exhibits remarkable 3D characteristics, and the jet orifice and velocity ratio have a significant influence on the flow characteristics of the channel. The heights at which the spiral deflection and maximum turbulent kinetic energy (TKE) occur for the circular jet are the smallest, while those for square jets are the largest. As the shape of the jet orifice changes from a circle to a square and then to a rectangle, the shape formed by the plane of the kidney vortices and the region above it gradually changes from a circle to a pentagon. With the increase in the velocity ratio, the 3D characteristics, maximum TKE, and kidney vortex coverage of the flow all gradually increase.

Keywords: orifice shape; vertical jet; velocity ratio; numerical investigation; hydraulic characteristics

1. Introduction

In real life, it is a typical vertical jet that pollutants such as waste gas and wastewater enter the river through the riverbed. After entering the water, the pollutants can have a good or bad impact on the water environment. The study of vertical jets is of great significance for pollutant diffusion, sediment transport, and water temperature regulation.

When a high-velocity flow is ejected from a pipeline into a low-velocity water environment, a vertical jet is formed. In recent years, different researchers have conducted numerous studies on the flow characteristics and pollutant diffusion characteristics when there is a vertical jet in flowing water [1–5]. Scorer [6] is arguably the first person who noticed the presence of a counterrotating vortex in a vertical jet. He pointed out that a pair of counterrotating kidney vortices exist in the channel when a small circular orifice emits a high-velocity flow. In recent years, many scholars have conducted extensive research on vertical jets under different conditions from physical experiments [7,8] and numerical simulations [9–11], but the existing research is mainly focused on circular jets. Smith and Mungal [12] employed a planar laser-induced fluorescence technology (PLIF) technique to divide the flow field under the condition of a circular hole jet in a wide range of 5 to 25 into three parts and analyzed the kidney vortices and their attenuation characteristics. Based on the relevant experimental data of momentum transport and the turbulence characteristics of a circular vertical jet, Wang and Law [13] proposed a second-order integral model to analyze the turbulence characteristics of a vertical jet. The data calculated by Huai et al. [14] using the k-ε turbulent model are in good agreement with Andreopolous and Rodi [15] test data. Zhu and Shin [16] used different turbulent models to analyze the flow characteristics of water flow under co-current jet conditions and found that the simulation results of renormalization group (RNG) k-ε and realize k-ε are significantly better than the standard k-ε, and the difference between the calculation results of k-ε and realize k-ε is small. Zeng and Huai [17] analyzed the flow field structure and bifurcation phenomenon of a thermal buoyant jet under the condition of a circular hole vertical jet using the finite analysis method based on the RNG k-ε turbulence model. Large eddy simulation (LES) [18] and Scale adaptive simulation (SAS) [19] can also realize relevant research of the vertical jet.

Although circular orifice jets are the most common in practice, they are not unique. The change in the shape of the jet orifice can affect the flow trajectory and energy conversion of the jet. Therefore, it is necessary to study the shape of the jet orifice, but there are few studies on this aspect. Haven and Kuroska [20] adopted the laser-induced fluorescence (LIF) technique and particle image velocimetry (PIV) in a top-down jet experiment with a small velocity ratio range of 0.4 to 2.0 and found that the square and rectangular jets can also produce kidney vortices, and the change in the jet orifice shape affects the local shape of downstream vortices. Due to the limitation of the accuracy of the physical experiment, the trajectory and turbulent characteristics of the jet in the water have not been thoroughly studied due to the change in the shape of the jet orifice. However, the vortex structure due to buoyancy in the process of jet from top to bottom and jet from bottom to top is also different.

The trajectory of the jet deflects, which moves along the direction of the open channel while rising, due to the influence of the original flowing water in the open channel when the jet flows from the riverbed into the open channel environment. The velocity of the jet beam in the channel is a combination of its vertical upward velocity and the original horizontal velocity of the water in the channel. The trajectories and other effects of the jet beam in the closed and open channel are different due to the different velocity distributions along the water depth, especially at a large velocity ratio. In this paper, we choose a smaller range of velocity ratio to study the vertical jet in an open channel.

In order to further investigate the jet trajectory and turbulent characteristics with various jet orifice shapes, this paper analyzes the 3D flow structure, turbulent kinetic energy, and vortex of circular, square, and rectangular jet orifices in a wide range of velocity ratios of 2, 5, 10, and 15, respectively. This research complements the lack of research on different jet orifice shapes under vertical jet conditions, and can provide basic guidance for pollutant diffusion, water temperature regulation, and sediment transport in the river.

2. Materials and Methods

2.1. Model Layout

The model built in present work is a straight channel with a rectangular cross-section, the length (L), width (B), and height of which are 3 m, 0.8 m, and 0.3 m, respectively (shown in Figure 1). The water depth (H) at the inlet is 0.26 m and the distance between the center of the jet outlet, placed at the bottom of the model, to the inlet is 1 m. To study the influence of the jet orifice shape on the flow and turbulence characteristics of the straight channel, the shape of the jet orifice was set to be circular, square, and rectangular while maintaining the same jet orifice area and average flow velocity. The circular diameter is D = 5 mm, the square side length is a = 4.43 mm, and the rectangular outlet has a side length ratio of b/c = 2, where b and c are the dimensions along the width and length directions respectively, of the channel. The average velocity, u_0, at the inlet is 0.0826 m/s, the average velocity at the jet outlet is u_j, and the velocity ratio is calculated using Equation (1). Different velocity ratios of 2, 5, 10, and 15 are adopted to analyze the influence of the jet orifice shape on the channel under different velocity ratios (listed in Table 1). Re_j, calculated using Equation (2), is the Reynolds number of the jet hole. In the present work, the center of the jet orifice is set as the origin of the coordinates, the flow direction is the positive direction of the Y axis, the positive direction of the X axis is determined using the right-hand rule, and the positive direction of the Z axis is vertical upward.

$$r = u_j/u_0 \tag{1}$$

where r is the velocity ratio.

$$Re_j = \rho u_j D/u_0 \tag{2}$$

where Re_j, u_j, and D are the Reynolds number, velocity, and diameter of the jet hole respectively, and ρ and u_0 are the density and dynamic viscosity of the water, respectively.

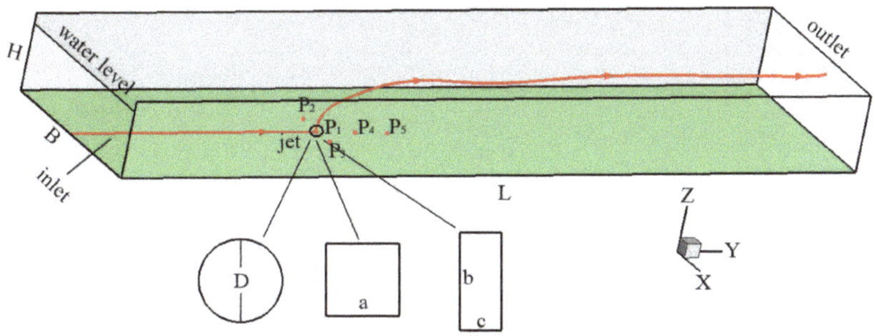

Figure 1. Three-dimensional (3D) view of the model layout.

Table 1. Parameters of different cases.

Series	Shape	B (m)	H (m)	L (m)	u_0 (m/s)	D (mm)	a (mm)	b (mm)	c (mm)	u_j (m/s)	r	Re_j	Cells
Case 1	Circular									0.165	2	924	703,080
Case 2	Circular					5	-	-	-	0.413	5	2312	703,080
Case 3	Circular									0.826	10	4625	703,080
Case 4	Circular									1.23	15	6887	703,080
Case 5	Square									0.165	2	924	703,080
Case 6	Square	0.8	0.3	3	0.0826	-	4.43	-	-	0.413	5	2312	703,080
Case 7	Square									0.826	10	4625	703,080
Case 8	Square									1.23	15	6887	703,080
Case 9	Rectangular									0.165	2	924	698,040
Case 10	Rectangular					-	-	6.26	3.13	0.413	5	2312	698,040
Case 11	Rectangular									0.826	10	4625	698,040
Case 12	Rectangular									1.23	15	6887	698,040

2.2. Governing Equations

The realizable k-ε turbulent model proposed by Shih et al. [21] is adopted in this paper. This model is optimized on the basis of the standard k-ε turbulent model and can accurately simulate the jet flow, especially in terms of strong streamtrace bending, vortex, and rotation. The governing equations are as follows:

Continuity equation:

$$\frac{\partial \rho}{\partial t} + \frac{\partial(\rho u_i)}{\partial x_i} = 0 \tag{3}$$

Momentum equation:

$$\frac{\partial(\rho u_i)}{\partial t} + \frac{\partial(\rho u_i u_j)}{\partial x_j} = -\frac{\partial p}{\partial x_i} + \frac{\partial}{\partial x_j}\left[(\mu + \mu_t) \times (\frac{\partial u_i}{\partial x_j} + \frac{\partial u_j}{\partial x_i})\right] \tag{4}$$

Turbulent kinetic energy (k) equation:

$$\frac{\partial(\rho k)}{\partial t}+\frac{\partial}{\partial x_j}\left(\rho u_j k\right) = \frac{\partial}{\partial x_j}\left[\left(\mu+\frac{\mu_t}{\sigma_k}\right)\frac{\partial k}{\partial x_j}\right]+G_k+G_b-\rho\varepsilon-Y_M+S_k \tag{5}$$

Turbulent dissipation rate (ε) equation:

$$\frac{\partial(\rho\varepsilon)}{\partial t}+\frac{\partial}{\partial x_j}\left(\rho u_j\varepsilon\right) = \frac{\partial}{\partial x_j}\left[\left(\mu+\frac{\mu_t}{\sigma_\varepsilon}\right)\frac{\partial\varepsilon}{\partial x_j}\right]+\rho C_1 S_\varepsilon-\rho C_2\frac{\varepsilon^2}{k+\sqrt{v\varepsilon}}+C_{1\varepsilon}\frac{\varepsilon}{k}C_{2\varepsilon}G_b+S_\varepsilon \tag{6}$$

$$C_1 = max\left[0.43,\frac{\eta}{\eta+5}\right] \tag{7}$$

$$\eta = \frac{Sk}{\varepsilon} \tag{8}$$

$$S = \sqrt{2S_{ij}S_{ij}} \tag{9}$$

$$S_{ij} = 0.5(\frac{\partial u_j}{\partial x_i}+\frac{\partial u_i}{\partial x_j}) \tag{10}$$

$$\mu_t = \frac{\rho C_\mu k^2}{\varepsilon} \tag{11}$$

where G_k and G_b are the turbulent kinetic energy values caused by the average velocity gradient and buoyancy lift, respectively. Y_M is the contribution of the compressible turbulent wave expansion to the overall dissipation rate, ρ is the water density, t is time, v is the turbulent kinematic viscosity coefficient, u_i is the average velocity in the direction of i, μ_t is the turbulent viscosity, C_μ, $C_{1\varepsilon}$, $C_{2\varepsilon}$, σ_k, and σ_ε are model constants with values of $C_\mu = 0.09$, $C_{1\varepsilon} = 1.44$, $C_{2\varepsilon} = 1.92$, $\sigma_k = 1.0$, and $\sigma_\varepsilon = 1.3$, and S_k and S_ε are user-defined source items.

2.3. Solver Settings and Boundary Conditions

Fluent software is adopted and the volume of fluid (VOF) method proposed by Hirt and Nichols [22] in 1981 is used to solve the free surface tracking problem in multiphase flow. This method uses a volume fraction function, F (x, y, z), to describe the position and volume ratio of the gas and liquid phases. F (x, y, z, t) = 1 indicates that the calculation unit is full of liquid, and F (x, y, z, t) = 0 indicates that the calculation unit is full of gas. Therefore, when 0 < F (x, y, z, t) < 1, there must exist an interface between the water and gas phases. Accordingly, a difference method can be used to calculate the interface position and solve for the free surface. The equations for calculating the volume fraction function in the VOF method are as follows:

$$\frac{\partial F}{\partial t}+u_i\frac{\partial F}{\partial x_i} = 0 \tag{12}$$

$$\rho = F_w\rho_w+(1-F_w)\rho_a \tag{13}$$

$$\mu = F_w\mu_w+(1-F_w)\mu_a \tag{14}$$

where, ρ_w and ρ_a are the densities of water and gas respectively, μ_w and μ_a are the viscosity coefficients of water and gas respectively, and F_w is the volume fraction of the liquid phase.

A transient solver is used for the calculation. Pressure-velocity coupling is calculated by adopting the SIMPLE pressure correction algorithm proposed by Patankar [23], and open channel flow is used for the VOF sub-models.

To save computation time and maintain calculation accuracy, the model is divided into five sections (shown in Figure 2). A fine mesh is used near the jet orifice location, and a relatively coarse mesh is used for locations near the inlet and outlet of the channel in the Y direction. The number of grids in the jet orifice is 4, and the total number of grids of the channel is 84 in the X direction. The number of grids is 30 in the Z direction. So, the layouts of the mesh of different jet orifices are (100 + 4 + 50 + 75 + 50) × 84 × 30, (100 + 4 + 50 + 75 + 50) × 84 × 30, and (100 + 2 + 50 + 75 + 50) × 84 × 30, respectively. The red line in Figure 2 is the grid of the jet orifice. The numbers of grids in different cases are shown in Table 1.

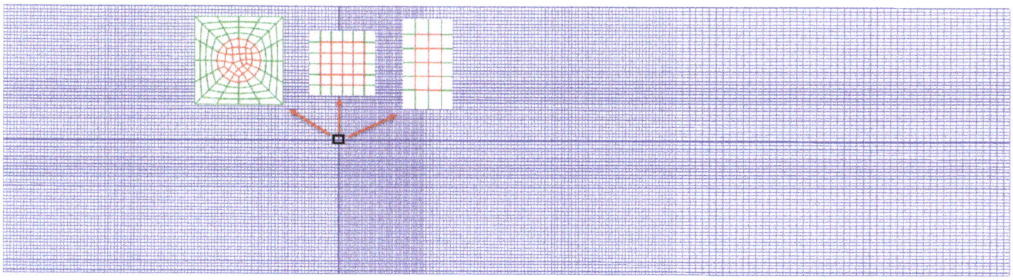

Figure 2. The grid of the numerical model.

The velocity-inlet boundary or pressure-inlet boundary are adopted to the water part or the air part of the channel, respectively. The inlet velocity of the water in the channel is set to the average and the magnitude is 0.0826 m/s. The pressure outlet is set at the outlet and the air–water interface is standard atmospheric pressure. The material of side walls of the sink in Reference [24] are glass, which is set to a non-slip boundary condition and a standard wall function is used near the sidewalls.

2.4. Convergence Stability and Mesh Sensitivity Analysis

The flow in the straight channel without a vertical jet is first calculated until it is steady, then its flow field is taken as the initial flow field for the channel with a vertical jet to carry out the relevant calculation. In the calculation, the mass flow rate error (MFRE) for the inlet and outlet of the channel are traced and the calculation is considered to be steady when the MFRE is small enough. To ensure the accuracy of the calculation, the vertical distribution of the velocity at the center of the jet orifice is compared at different times when the MFRE is relatively small. If the velocities are almost consistent, the calculation is considered to reach the final steady state required in this study. The time-history curve of the variation in the MFRE with the calculation time is shown in Figure 3, and the MFRE values of different cases are shown in Table 2.

Table 2. MFRE values under different cases in steady state.

Title 1	Case 1	Case 2	Case 3	Case 4	Case 5	Case 6	Case 7	Case 8	Case 9	Case 10	Case 11	Case 12
MFRE (kg/s)	−0.006	−0.006	0.00002	0.007	0.002	0.01	−0.006	0.0004	−0.002	0.006	0.002	−0.001
Relative mass flow rate (%)	−0.36	−0.36	0.00	0.42	0.12	0.58	−0.35	0.02	−0.12	0.35	0.12	−0.06

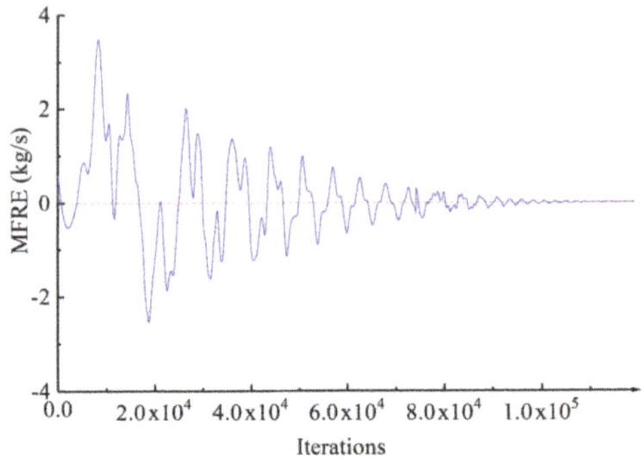

Figure 3. Mass flow rate error (MFRE) time-history curve.

For case 4, the grid convergence index (GCI) [25] of the velocity distributed vertically at the center of the jet orifice is calculated with the number of grids approximately equal to 0.54 million, 0.7 million, and 1.08 million to verify that the calculation in this study is independent of the meshing. The GCI is calculated as follows:

$$GCI = \frac{1.25\left|\frac{\phi_1-\phi_2}{\phi_1}\right|}{(h_2/h_1)^P-1} \tag{15}$$

$$P = \frac{1}{\ln(h_2/h_1)}\left|\ln\left|(\phi_3-\phi_2)/(\phi_2-\phi_1)\right| + \ln\left(\frac{(h_2/h_1)^P - \mathrm{sgn}((\phi_3-\phi_2)/(\phi_2-\phi_1))}{(h_3/h_2)^P - \mathrm{sgn}((\phi_3-\phi_2)/(\phi_2-\phi_1))}\right)\right| \tag{16}$$

where ϕ_i is the calculated value for the ith numerical grid with i = 1, 2, and 3, and h_i is the height of the first-layer grid of the ith numerical grid and satisfies $h_1 < h_2 < h_3$.

The variation in the vertically distributed GCI of the vertical velocity at the center of the jet orifice is shown in Figure 4, where ω is the vertical velocity. As shown in Figure 4, the maximum value of GCI is only 2.3, which indicates that the number of grids used in the present study (approximately 0.7 million) is appropriate.

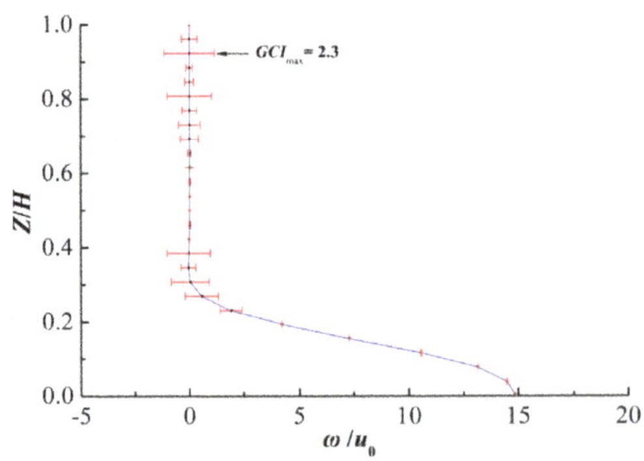

Figure 4. Vertical distribution of the grid convergence index (GCI) in jet orifice centerline.

2.5. Numerical Model Validation

Before carrying out the study, it is necessary to conduct an accuracy analysis on the chosen numerical model to verify that it can meet the needs of this work. The calculation model selected in this study is from Reference [24], and the experimental water tank has a length of 15 m and a width of 4 m. The calculation and analysis in the present study show that the flow velocity of the water tank calculated using a 1:5 scale model is almost the same as that of the prototype. Therefore, to save calculation time, this scale is used for the width and length of the channel but the water depth in the original experiment and numerical model along the height direction is maintained at 0.26 m, the average velocity at the inlet of the straight channel is 0.0826 m/s, the jet outlet has a circular shape with a diameter of D = 5 mm, and the jet velocity ratio is 15. Using the center of the jet orifice as the coordinate origin, the numerical calculation results and the experimental data of the vertical velocities at the intersecting lines between the central plane face of the channel and the sections of Y = 2D, 4D, 12D, and 16D along the flow direction are selected and compared in Figure 5.

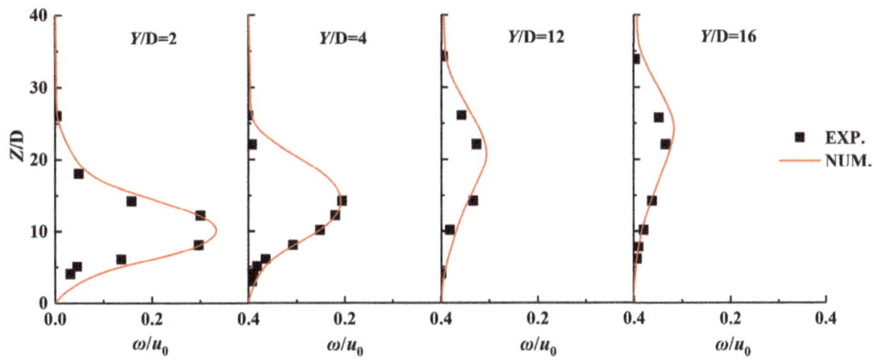

Figure 5. Distribution of the vertical flow velocity along the water depth at different locations.

The comparison shows that the numerical calculation data are slightly larger than the experimental data at different positions. Considering that there may be a slight deviation in the positions in the actual measurement process of the physical experiment and that the vertical velocity at the middle sections is greater than that on the two sides, the numerical calculation results are considered to be in good agreement with the experimental results and can thus be used for further in-depth investigation in this study.

3. Results and Discussion

3.1. 3D Structure of the Flow Field

In the absence of the free surface flow, the jet behaves as a simple jet, with the fluid decelerating as it moves away from the jet exit. But the trajectory of the jet deflects, which moves along the direction of the open channel while rising, due to the influence of the original flowing water when the jet flows from the riverbed into the water environment flowing in the open channel. The jet beam successively undergoes three processes: linear jetting, spiral motion, and downstream steady linear motion (shown in Figure 6). When the flow is ejected rapidly from the orifice, the jet beam moves linearly. During the upward motion, the jet beam gradually increases due to the effect of gravity, the original flow field, and the absence of orifice constraints. When developed to a certain stage, the water with a high velocity in the middle of the jet orifice deflects significantly to move outward, which simultaneously drives the deflection of the surrounding water and forms a spiral flow due to the influence of the original water movement. After the energy is dissipated over a sufficiently long distance, the vertical velocity and spanwise velocity of the jet flow gradually decrease and eventually move downstream in line with the mainstream of the channel.

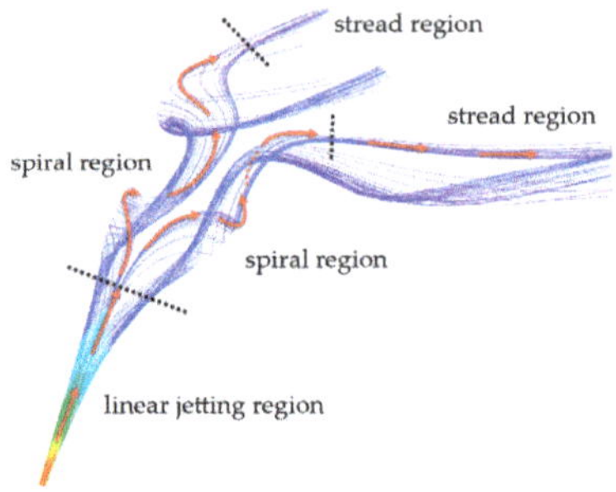

Figure 6. 3D structure of the jet streamtrace.

In this study, the trajectories of the jets with different orifice shapes and different velocity ratios at Y = 0 were analyzed, as shown in Figure 7.

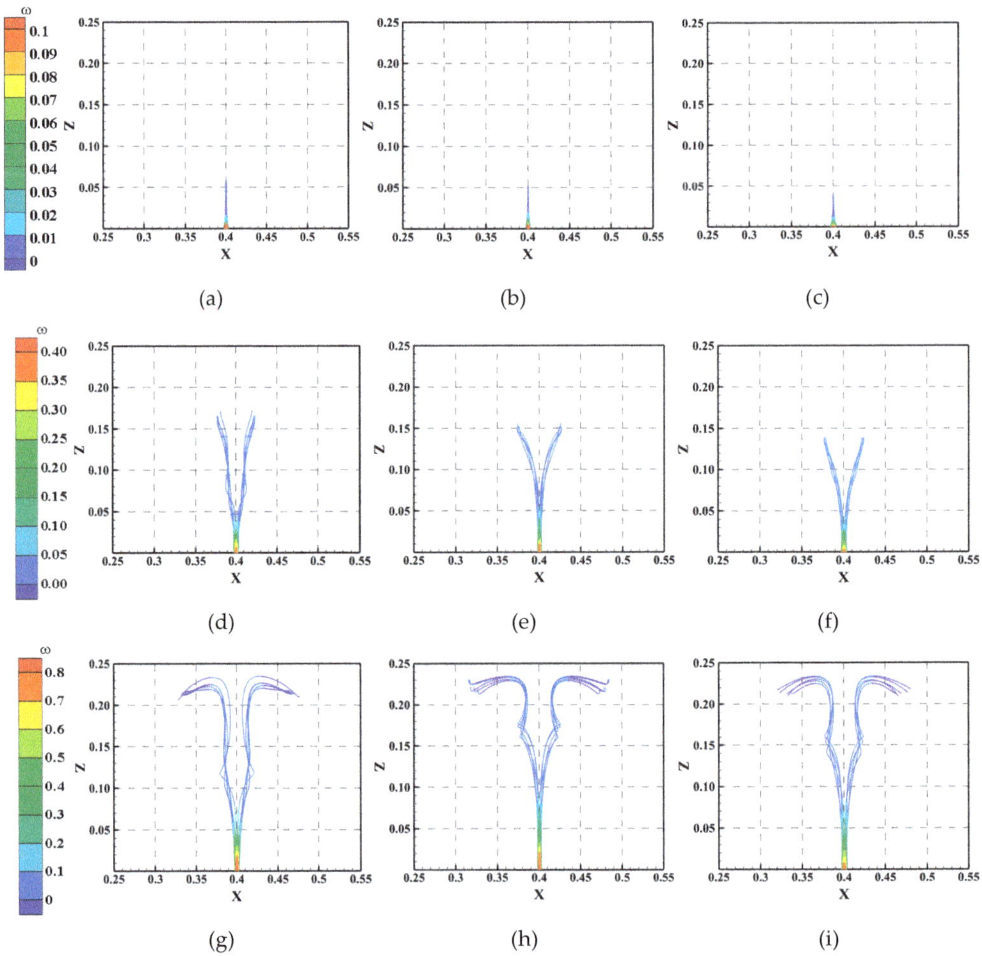

Figure 7. *Cont.*

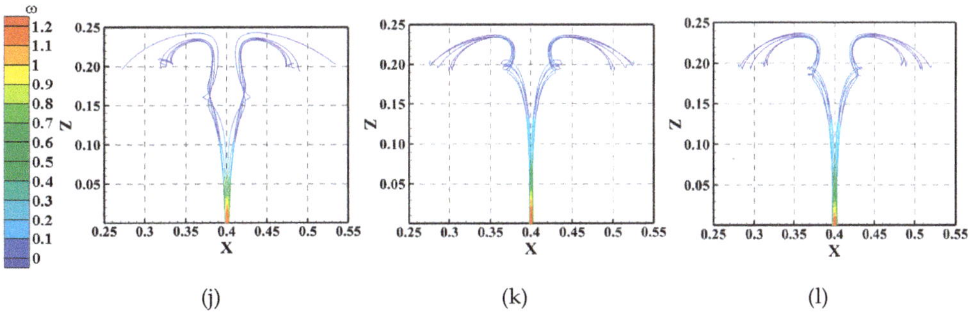

Figure 7. Variations of streamtraces under different cases: (a) case 1, (b) case 5, (c) case 9, (d) case 2, (e) case 6, (f) case 10, (g) case 3, (h) case 7, (i) case 11, (j) case 4, (k) case 8, (l) case 12.

As shown in Figure 7, when the velocity ratio is 2, the water is ejected from the orifice and then gradually draws closer toward the middle in the vertical direction, eventually forming a straight line without bifurcation. The shape of the jet orifice has a notable influence on the jet height and the jet height is the largest for the circular jet, followed by the square jet and then the rectangular jet. When the velocity ratio is 5, the jet beam bifurcates. The jet beams with circular and rectangular orifices bifurcate even when Z is less than 0.05 m, while the height at which the jet beam of the square orifice starts to bifurcate is relatively large, at more than 0.05 m. Further analysis reveals that the streamtrace line of the circular jet beam at the center of the orifice exhibits significant spiral deflection, while no apparent spiral deflection occurs in the jet beams with the other jet orifice shapes. When the velocity ratio is 10, there are two significant jet deflections in all the jet beams regardless of the orifice shape. The height at which the jet beam exhibits the first notable spiral deflection is the smallest for the circular jet beam, followed by the rectangular jet and then the square jet. For each of the three orifice shapes, the amplitude of the second deflection of the jet is significantly larger than that of the first deflection. After the second deflection, the streamtrace lines of the jets from the circular and rectangular orifices move vertically downward and then tend to stabilize and move downstream. After the second deflection, the jet beam with the square orifice exhibits a third deflection but with a smaller amplitude, and then tends to stabilize and move downstream. When the velocity ratio is 15, the variation in the jet beam trajectory for each of the three orifice shapes is similar to that with a velocity ratio of 10, although the amplitude of the second deflection has a large decrease in the vertical direction and all three exhibit third deflections. Based on the analysis of the influence in the Y direction, the circular jet beam has the largest influence range, while the rectangular jet beam has the smallest influence range.

An analysis of the cases with different velocity ratios shows that as the velocity ratio increases, the jets on the two sides first gather toward the center and gradually bifurcate; then, a spiral deflection occurs, and the number of occurrences continuously increases. The circular jet has the largest jet height, the smallest vertical height at which a spiral deflection occurs, and the largest influence range. The square jet has the largest vertical height at which the spiral deformation occurs and the rectangular jet has the smallest jet height and the smallest influence range.

3.2. The Analysis of Turbulent Kinetic Energy (TKE)

When the flow velocity is relatively low, the flow in the straight channel is laminar, and the TKE (calculated by Equation (17)) near the sidewall is relatively large due to the influence of the flow velocity fluctuation. When the jet beam enters the water, the flow fluctuation increases, and the TKE increases as the jet beam passes. In this work, several TKE distributed along the water depth at several locations are selected to analyze the influence of the jet orifice shape and velocity ratio on the distribution of TKE in the channel, as shown in Figure 8. As shown in Figure 1, P_1 is the center of the jet orifice, the distances of P_2 and P_3 to P_1 are both X/D = 4, and the distances of P_4 and P_5 to P_1 are *Y*/D = 10 and 20, respectively.

$$TKE = \frac{1}{2}[(u')^2 + (v')^2 + (\omega')^2] \tag{17}$$

where u', v', and ω' represent the fluctuating flow velocity in the X, Y, and Z directions, respectively.

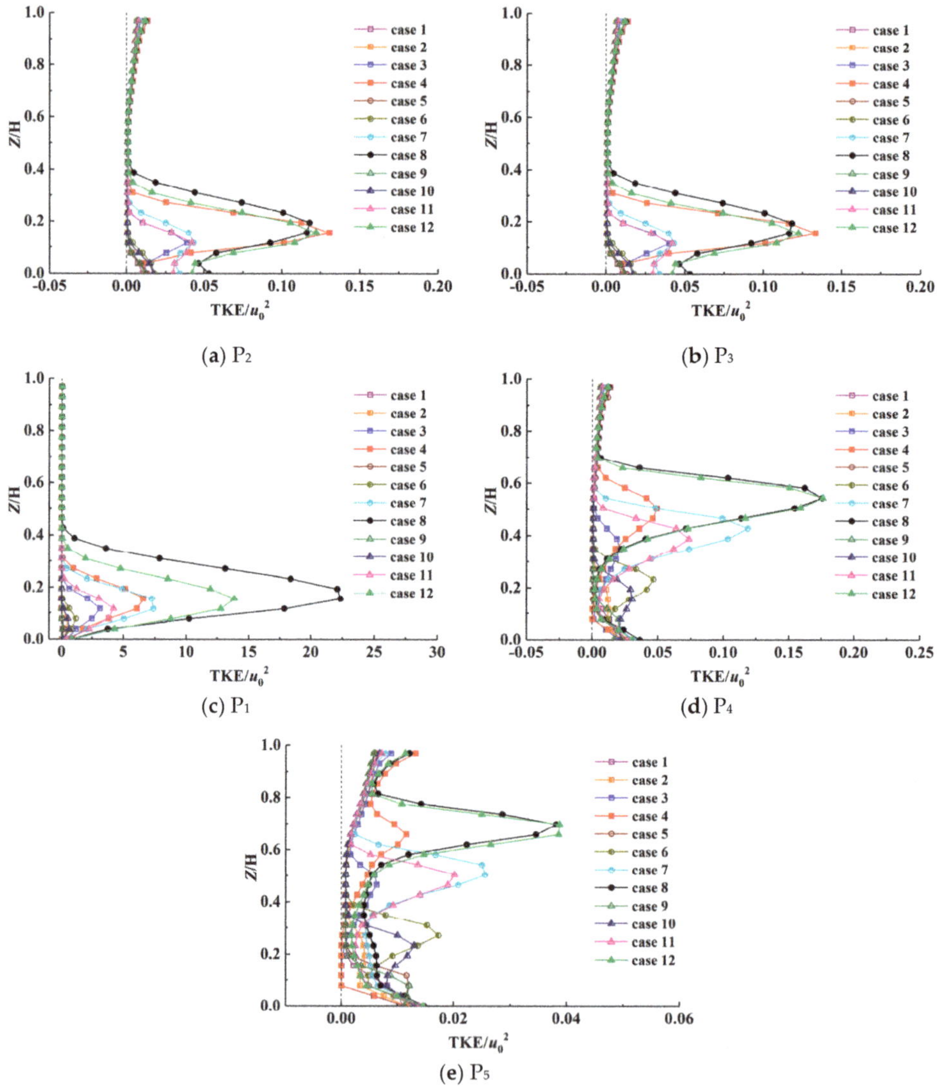

Figure 8. Distribution of the turbulent kinetic energy (*TKE*) values at different locations along the water depth: (**a**) P_2, (**b**) P_3, (**c**) P_1, (**d**) P_4, (**e**) P_5.

It can be seen from Figure 8a,b that the TKE value along the water depth is distributed symmetrically under different cases at symmetrical positions on both sides of the center of the jet orifice. Overall, the maximum TKE value gradually increases as the velocity ratio increases, and the influence of the different jet orifice shapes is not significant. When the flow velocity is small (r = 2 and 5), the maximum

TKE values under different cases are relatively close to each other and adjacent to the channel. When the flow velocity is high (r = 10 and 15), the maximum TKE value significantly increases, and the height of the corresponding occurrences also increases significantly. In both situations, the height at which the maximum TKE appears is Z/H = 0.16. As shown in Figure 8c, the variation in the TKE value at the center of the jet orifice along the water depth under different cases has a notable pattern. As the velocity ratio increases, the maximum TKE increases gradually as does the corresponding height of the occurrences. With different jet orifice shapes, the heights where the maximum TKE value occurs under the same velocity ratio are basically the same, but the corresponding numerical values are significantly different: the maximum TKE value of the square jet is the largest, and the maximum TKE value of the round jet is the smallest.

According to Figure 8d, at the position Y/D = 10 away from the jet orifice downstream of the channel, the maximum TKE value gradually increases as the velocity ratio increases, as does the corresponding height of the occurrences. The maximum TKE value of the circular jet occurs at the smallest height and its value is also the smallest. The maximum TKE value of the square jet occurs at the largest height, and its value is also the largest. When the velocity ratio is small (r = 2 and 5), the maximum TKE is close to the channel, and when the velocity ratio is high (r = 10 and 15), the heights at which the maximum TKE values occur for different jet orifice shapes differ greatly. When r = 10, the maximum TKE values corresponding to the circular orifice, square orifice, and rectangular orifice are Z/H = 0.35, 0.49, and 0.43 respectively, and when r = 15, the maximum TKE values are Z/H = 0.50, 0.54, and 0.54, respectively. Combined with the earlier analysis of Figure 7, it is believed that the TKE value is highly related to the spiral deflection of the jet beam. The height at which the spiral deflection occurs is the largest for the square jet, followed by the rectangular jet and then the round jet; thus, the height corresponding to the maximum TKE value also exhibits a similar distribution pattern. According to Figure 8e, the variation in the TKE distribution along the water depth at a distance Y/D = 20 from the jet outlet is relatively complex. Further analysis shows that the variation pattern of the *TKE* value is the same as that at Y/D = 10, but the maximum TKE value is significantly reduced, and the positions at which the maximum TKE value appears under different cases increase remarkably.

Analysis of Figure 8a–e shows that the abscissa of Figure 8c is significantly larger than that of the other figures. This is because the turbulence is relatively strong at the center of the jet orifice due to the sharp collision between the jet and the original water in the open channel. Then, the TKE gradually decreases as the distance to the jet hole increases. The four points (P_2, P_3, P_4, and P_5) selected in this paper are far from the jet orifice (P_1) and close to the edge of the area affected by the jet. Therefore, the turbulent kinetic energy at these four positions is significantly smaller than that at the position of the jet orifice.

3.3. The Analysis of Vortex

After the jet beam enters the channel, the velocity of the original water changes significantly, and vortices can be found near the jet beam. In this study, the vortex characteristics at the cross-section Y = 0 under different cases are analyzed, as shown in Figures 9–11 (where the legend in the cloud plots of the velocity is consistent with that in Figure 7).

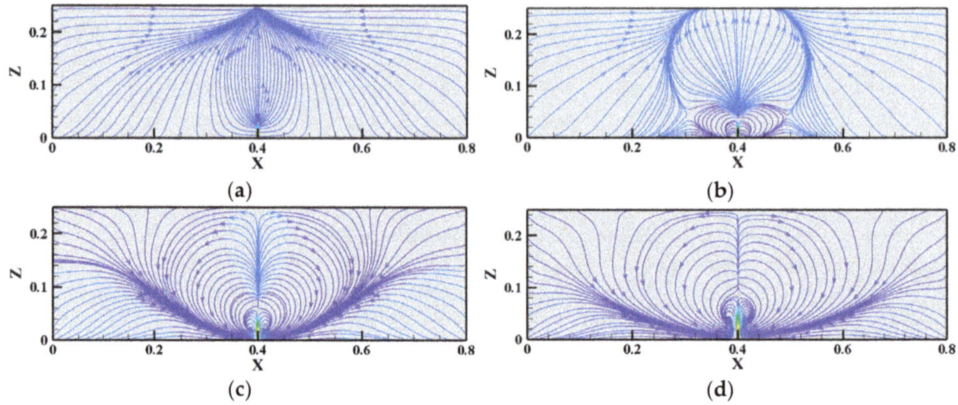

Figure 9. Vortex of the circular jet at cross-section Y = 0: (**a**) case 1, (**b**) case 2, (**c**) case 3, (**d**) case 4.

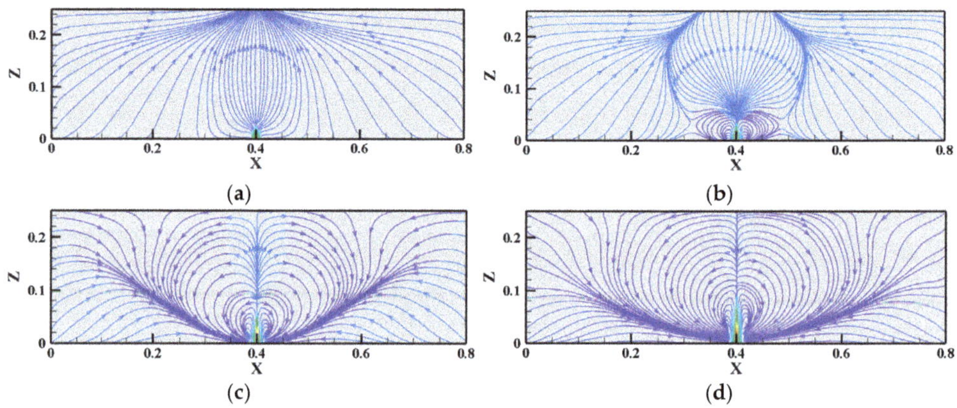

Figure 10. Vortex of the square jet at cross-section Y = 0: (**a**) case 5, (**b**) case 6, (**c**) case 7, (**d**) case 8.

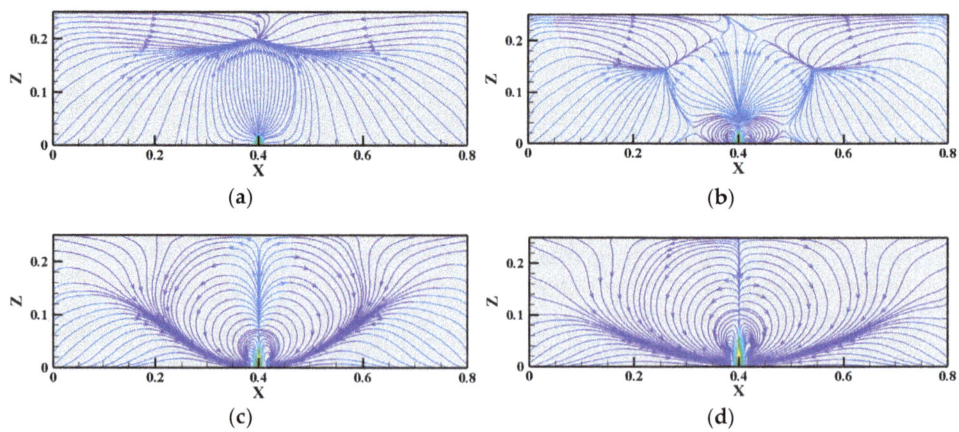

Figure 11. Vortex of the rectangular jet at cross-section Y = 0: (**a**) case 9, (**b**) case 10, (**c**) case 11, (**d**) case 12.

Figure 9 shows that there is a very small vortex at the deflection position of the cross-sectional jet orifice when the circular jet has a velocity ratio of 2 because the jet velocity is too small, which causes the jet beam to deflect along the Y direction near the channel while only slightly deflecting in the X direction. When the velocity ratio is 5, there is a notable kidney vortex near the jet orifice, the water above the kidney vortex moves upward, the water on both sides deflects on the upper side of the vortex, and the kidney vortex and the upper streamtrace lines form a circular distribution on the plane. When the velocity ratio is 10, the range of the kidney vortex is further expanded. The region above the jet orifice is within the range of the kidney vortex, and both sides are affected by the kidney vortex, resulting in a notable streamtrace line convergence near the lower side of the kidney vortex. When the velocity ratio is 15, the plane streamtrace line distribution is basically consistent with that when the velocity ratio is 10, with the main difference being the further expanded range of the kidney vortex. An analysis of Figures 10 and 11 shows that as the velocity ratio increases, the range of the kidney vortex increases continuously.

The cross-sectional streamtrace distribution under different orifice shapes and the same velocity ratio as shown in Figures 9–11 was analyzed. When the velocity ratio is 2, the shapes of the streamtrace line distribution of the circular and square jet outlets are basically the same, and the height of the converged ellipse-like streamtrace lines above the section of the rectangular jet is reduced significantly. When the velocity ratio is 5, as the shape of the jet orifice gradually changes from circular to rectangular, the kidney vortex and the shapes formed in the upper convergence region gradually develop from circular to pentagonal. It is believed that, with a relatively small velocity ratio, the influence height corresponding to the circular, square, and rectangular jet orifices gradually decrease, while different jet orifice shapes result in different coverages and thus different streamtrace line distributions. When the velocity ratios are 10 and 15, there is no significant difference in the sectional distribution of the kidney vortex with different jet outlet shapes.

4. Conclusions

In this study, the realizable k-ε turbulent model was used to analyze the 3D flow structure, TKE, and vortex structure of water under jets with different orifice shapes and velocity ratios in straight channels. The conclusions are as follows:

(1) The flow trajectory of the vertical jet in the channel exhibited notable 3D characteristics, and the jet orifice shape and velocity ratio significantly affected the spiral characteristics of the flow. As the velocity ratio increased, the number of spiral deflections of the jet beam increased gradually, and the magnitude became larger. The circular jet beam had the largest height and the smallest vertical height of the spiral deflection, the square jet had the largest vertical height at which the spiral deformation occurs, and the rectangular jet had the smallest jet height.

(2) As the velocity ratio increased, the maximum TKE value gradually increased, and the height corresponding to the maximum value also gradually increased. The maximum TKE value of the circular jet was the smallest and occurred at the smallest height, the maximum TKE value of the square jet was the largest and occurred at the largest height. As the downstream distance increased, the maximum TKE value gradually decreased, and the height at which the maximum TKE value occurs gradually increased.

(3) As the velocity ratio increased, the influence range of the kidney vortex increased. When the shape of the jet orifice gradually changed from circular to square and rectangular, the shape formed by the kidney vortex and the region above it gradually changed from circular to pentagonal, indicating that the shape of the jet orifice affects the shape distribution of the cross-sectional vortex.

Author Contributions: Conceptualization, H.Y., R.H., and X.X.; validation, H.Y., R.H., and X.X.; formal analysis, H.Y., X.X., and J.T.; investigation, H.Y., Y.P., and R.H.; resources, H.Y. and Y.P.; writing—original draft preparation, H.Y. and R.H.; writing—review and editing, H.Y., R.H., L.C., J.T., and X.X. H.Y. and R.H. contributed to the work equally and R.H. should be regarded as co-first authors. All authors have read and agree to the published version of the manuscript.

Funding: This research was funded by the National Natural Science Foundation of China (Grant No. 51709293) and the China National Key Research and Development Plan (GrantNo.2016YFC0402006).

Conflicts of Interest: The authors declare no conflict of interest.

References

1. Papanicolaou, P.; List, E. Investigation of round vertical turbulent buoyant jets. *J. Fluid Mech.* **1988**, *195*, 341–391. [CrossRef]
2. Peterson, J.; Bayazitoglu, Y. Measurements of velocity and turbulence in vertical axisymmetric isothermal and buoyant jets. *J. Heat Trans.* **1992**, *114*, 135–142. [CrossRef]
3. Mossa, M.; De Serio, F. Rethinking the process of detrainment: Jets in obstructed natural flows. *Sci. Rep.* **2016**, *6*, 39103. [CrossRef] [PubMed]
4. Eroglu, A.; Breidenthal, R.E. Exponentially accelerating jet in crossflow. *AIAA J.* **1998**, *36*, 1002–1009. [CrossRef]
5. Morton, B.R.; Ibbetson, A. Jets deflected in a crossflow. *Exp. Therm. Fluid Sci.* **1996**, *12*, 112–133. [CrossRef]
6. Scorer, R.S. *Natural Aerodynamics*; Pergamon: Oxford, UK, 1958.
7. Kelso, R.M.; Lims, T.T.; Perry, A.E. An experimental study of round jets in cross-flow. *J. Fluid Mech.* **1996**, *306*, 111–144. [CrossRef]
8. Yuan, H.; Xu, W.L.; Li, R.; Feng, Y.Z.; Hao, Y.F. Spatial Distribution Characteristics of Rainfall for Two-Jet Collisions in air. *Water* **2018**, *10*, 1600. [CrossRef]
9. Gao, M.; Huai, W.X.; Xiao, Y.Z.; Yang, Z.H.; Jin, B. Large eddy simulation of a vertical buoyant jet in a vegetated channel. *Int. J. Heat Fluid Flow* **2018**, *70*, 114–124. [CrossRef]
10. Chai, X.; Iyer, P.S.; Mahesh, K. Numerical study of high speed jets in crossflow. *J. Fluid Mech.* **2015**, *785*, 152–188. [CrossRef]
11. Guan, H.; Wu, C. Large-eddy simulations and vortex structures of turbulent jets in crossflow. *Sci. China Phys. Mech.* **2007**, *50*, 118–132. [CrossRef]
12. Smith, S.H.; Mungal, M.G. Mixing, structure and scaling of the jet in crossflow. *J. Fluid Mech.* **1998**, *357*, 83–122. [CrossRef]
13. Wang, H.W.; Law, A.W.K. Second-order integral model for a round turbulent buoyant jet. *J. Fluid Mech.* **2002**, *459*, 397–428. [CrossRef]
14. Huai, W.X.; Li, W.; Peng, W.Q. Calculation and behavior analysis on turbulent jets in cross-flow. *J. Hydrau. Eng.* **1998**, *4*, 7–14. (In Chinese)
15. Andreopolous, J.; Rodi, W. Experimental investigation of jets in a cross-flow. *J. Fluid Mech.* **1984**, *138*, 43–127.
16. Zhu, J.; Shih, T.H. Computation of confined crossflow jets with three turbulence models. *Int. J. Numer. Methods Fluids* **1994**, *19*, 939–956. [CrossRef]
17. Zeng, Y.H.; Huai, W.X. Characteristics of round thermal discharging in a flowing environment. *J. Hydro-Environ. Res.* **2009**, *2*, 164–171. [CrossRef]
18. Ma, F.; Satish, M.; Islam, M.R. Large eddy simulation of thermal jets in cross flow. *Eng. Appl. Comput. Fluid Mech.* **2007**, *1*, 25–35. [CrossRef]
19. Rogowski, K.; Hansen, M.O.L.; Maroński, R.; Lichota, P. Scale Adaptive Simulation Model for the Darrieus Wind Turbine. *J. Phys. Conf. Ser.* **2016**, *753*, 022050. [CrossRef]
20. Haven, B.A.; Kurosaka, M. Kidney and anti-kidney vortices in crossflow jets. *J. Fluid Mech.* **1997**, *352*, 27–64. [CrossRef]
21. Shih, T.H.; Liou, W.W.; Shabbir, A.; Yang, Z.; Zhu, J. A new k-ε eddy viscosity model for high reynolds number turbulent flows. *Comput. Fluids* **1995**, *24*, 227–238. [CrossRef]
22. Hirt, C.W.; Nichols, B.D. Volume of fluid (VOF) method for the dynamics of free boundaries. *J. Comput. Phys.* **1981**, *39*, 201–225. [CrossRef]
23. Patankar, S.V.; Spalding, D.B. A calculation procedure for heat, mass and momentum transfer in three-dimensional parabolic flows. *Int. J. Hear Mass Transf.* **1972**, *15*, 1787–1806. [CrossRef]

24. Malcangio, D.; Mossa, M. A laboratory investigation into the influence of a rigid vegetation on the evolution of a round turbulent jet discharged within a cross flow. *J. Environ. Manag.* **2016**, *173*, 105–120. [CrossRef] [PubMed]
25. Celik, I.B.; Ghia, U.; Roache, P.J.; Freitas, C.J. Procedure of estimation and reporting of uncertainty due to discretization in CFD applications. *J. Fluids Eng.* **2008**, *130*, 078001.

energies

MDPI

Article

Numerical Study of the Normal Impinging Water Jet at Different Impinging Height, Based on Wray–Agarwal Turbulence Model

Hongliang Wang [1,2], Zhongdong Qian [1], Di Zhang [3], Tao Wang [4] and Chuan Wang [3,*]

1 State Key Laboratory of Water Resources and Hydropower Engineering Science, Wuhan University, Wuhan 430072, China; wanghl@czu.cn (H.W.); zdqian@whu.edu.cn (Z.Q.)

2 School of Aerospace and Mechanical Engineering/Flight College, Changzhou Institute of Technology, Changzhou 213032, China

3 College of Hydraulic Science and Engineering, Yangzhou University, Yangzhou 225009, China; dx120180054@yzu.edu.cn

4 Key Laboratory of Fluid and Power Machinery, Ministry of Education, School of Energy & Power Engineering, Xihua University, Chengdu 610039, China; mailtowangtao@163.com

* Correspondence: wangchuan@ujs.edu.cn

Received: 9 March 2020; Accepted: 3 April 2020; Published: 5 April 2020

Abstract: As a kind of water jet technology with strong impinging force and simple structure, the submerged impinging water jet can produce strong scouring action on subaqueous sediments. In order to investigate the flow field characteristics and impinging pressure of submerged impinging water jets at different impinging heights, the Wray-Agarwal (W-A) turbulence model is used for calculation. The velocity distribution and flow field structure at different impinging heights ($1 \leq H/D \leq 8$), and the impinging pressure distribution at the impingement plate under different Reynolds numbers ($11,700 \leq Re \leq 35100$) are studied. The results show that with the increase of the impinging height, the diffusion degree increases and the velocity decreases gradually when the jet reaches the impingement region. The fluid accelerates first and then decelerates near the stagnation point. The maximum impinging pressure and the impinging pressure coefficient decrease with the increase of the impinging height, but the effective impinging pressure range remains unchanged. In this paper, the distribution characteristics of the impinging pressure in the region of the impingement plate at different heights are clarified, which provides theoretical support for the prediction method of the impinging pressure.

Keywords: impinging water jet; impinging height; numerical calculation

1. Introduction

Impinging jets are widely used in many fields, such as waste treatment, irrigation and drainage, cooling and heating, chemical vapor deposition and so on. The flow characteristics of the impinging jet depend on the nozzle shape, the Reynolds number at the exit of the jet, the distance from the nozzle to the impingement plate, and the impinging angle (the angle between the center axis of the jet nozzle and the impingement plate), etc. In order to study the flow structure of impinging jet, domestic and foreign scholars have used hot wire anemometer, laser Doppler velocimeter, particle image velocimetry and other technologies for experimental research [1–5], and different turbulence models are also used for numerical calculation [6–8]. According to the impinging angle, the submerged water jet flow can be divided into vertical impinging jet ($\theta = 90°$) and oblique impinging jet ($0° < \theta < 90°$). The impinging jet can be divided into three regions: free-jet region, impingement region and wall-jet region [9]. In the free-jet region, the flow characteristics are similar to that of the free jet. When the jet fluid moves towards the impingement plate, the surrounding fluid is sucked in and the overall velocity is attenuated; when the jet fluid approaches the impingement plate, the axial velocity decreases gradually, and the direction changes in the impingement region $1D$ to $2D$ away from the impingement plate. In the downstream of the impingement region, the jet fluid diffuses outward, almost parallel to the impingement plate; as the jet fluid exchanges momentum with the ambient fluid, the impinging jet eventually develops into a wall jet.

With the rapid development of computational fluid dynamics (CFD) technology, many scholars have applied CFD to solve complex problems encountered in engineering such as turbomachinery [10–14]. At the same time, some scholars have conducted a lot of research on vertical submerged impinging water jets through CFD. Chen et al. [15] studied the evolution mechanism and characteristics of the submerged laminar round jet in a viscous homogenous shallow water layer through computational modeling. In order to visualize the formation and evolution of the flow pattern, the volume of fluid (VOF) method was adopted, to simulate the free surface of the water layer below the air and to trace the jet fluid. The results show that the jet forms a class of quasi-two-dimensional (Q2D) vortex structures in the ambient fluid, with unequal influence from the bottom wall and free surface. Amamou et al. [16] computationally investigated a turbulent round jet issuing into a uniform counterflow stream. The simulation was carried out using the Reynolds stress model. Numerical results agree well with experimental results and the penetration and spreading of the jet are studied. Chen et al. [17] used a SIMPLE algorithm and RNG k-ε turbulence model to analyze the flow field of the plane jet in the semi-closed space with inclined impinging plate (θ = 30°, 60°, 90°), the results show that the change of impingement angle has a significant effect on the flow field structure. Guo et al. [18] studied the three-dimensional numerical simulation of the flow field of the oblique jet, by using the two turbulence models of k-ε and large eddy simulation, the results show that the large eddy simulation can accurately reveal the evolution process of the oblique submerged jet flow field and the shear and mixing mechanism of energy dissipation. Afroz et al. [19] presented the results of a numerical investigation of fluid flow and heat transfer due to a turbulent annular jet impingement on an isothermally heated flat surface. Annular impinging jets enhance the heat transfer and spread it more uniformly over the impingement surface compared to the round impinging jet, and have one noteworthy characteristic of forming a reverse stagnation flow. Huang et al. [20] used the numerical tool of arbitrary Lagrangian–Eulerian formulations to model the arc-curved jet impacting these different solid surface types, and a double/multiple-peaked pressure structure is observed for the cases of the water jet impacting the concave and convex solid surfaces. Stahl et al. [21] used large-eddy simulations to examine the primary flow phenomena for two adjacently placed round jets, with identical and dissimilar exit conditions, impinging on a ground plane. Particular emphasis was placed on the dynamics of the fountain flow between the jets, entrainment, vortex tube formation and acoustic feedback mechanisms. Battistin et al. [22] established a numerical model to describe the impinging water jet, and incorporated it into the solver to evaluate the model capability. After a careful verification, the proposed model is validated through comparisons with the similarity solution of the wedge impact with constant entry velocity. Singh et al. [23] numerically studied the flow characteristics of a turbulent offset jet impinging on a wavy wall surface. The variation in integral constant of momentum flux, wall shear stress, and pressure along the wall is presented and compared. Wienand et al. [24] made numerical results of a turbulent impinging jet on a flat plate, and four different jet-to-plate distances from H/D = 2 to 14 had been considered at a jet Reynolds number of 23,000. The results show the influence of the dimensionless distance of the first node near the wall on flow and heat transfer characteristics. In summary, the existing numerical studies on impinging jets are based on common commercial turbulence models.

In this paper, the W-A (Wray–Agarwal) turbulence model was used to calculate the submerged impinging jet at different impinging heights ($1 \leq H/D \leq 8$), and the influence of impinging heights on the vertical jet flow field was studied. In this study, a common circular jet was selected, and the jet was a fully-developed round jet.

2. Computational Model

2.1. Geometric Model and Boundary Conditions

The vertical jet fluid is sprayed into a square water tank through a round copper pipe perpendicular to the bottom wall surface, as shown in Figure 1. The jet fluid flows in from the copper pipe, and the inlet of the copper pipe is the Inlet surface; the outflow from the left and right side walls is the Outlet surface; the top surface of the water tank contacts with the air, which is the free surface; the other surfaces are the wall surface. D is the inner diameter of the copper pipe, which is selected as 20 mm. H is the vertical distance from the end face of the vertical jet nozzle to the bottom surface. In this paper, the impinging height H/D = 1–8 was set. As shown in the figure, in order to better describe the motion in the jet flow field, the intersection of the extension line of the nozzle axis and the impingement plate was taken as the origin of the coordinate system, and the rectangular coordinate system oxyz was established.

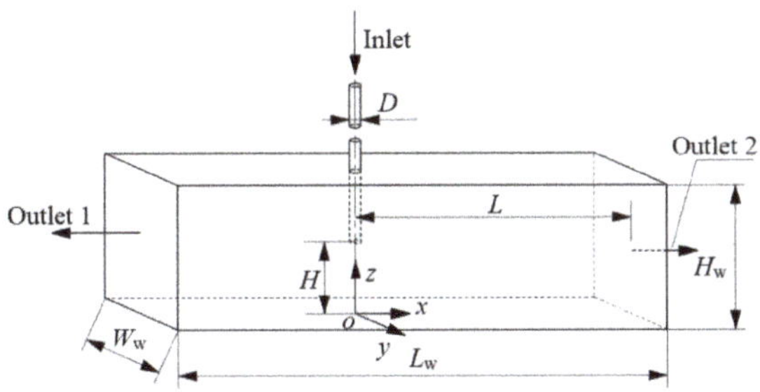

Figure 1. Three-dimensional submerged vertical jet.

Setting of boundary conditions: (1) inlet: velocity inlet was adopted, and the velocity was evenly distributed, different inlet conditions are shown in Table 1; (2) outlet: pressure outlet, and the pressure value was set as 0 Pa; (3) free surface: the rigid-lid assumption was adopted, and the symmetrical surface was set; (4) wall surface: fixed wall surface with no slip; (5) reference pressure was set to 0 Pa.

Table 1. Computational conditions of submerged impinging jet.

Parameters	Condition 1	Condition 2	Condition 3
Inlet velocity(m/s)	0.585	1.17	1.76
Re	11,700	23,400	35,100

2.2. Mesh Information

The calculation domain should be discretized before simulation. In this study, the whole calculation domain was divided into structural mesh by ICEM software, to shorten the calculating time. The total number of elements is 950,8963, which is near ten million. Due to the fact that the number of elements is rather large, the working capacity of the common computer, which has a CPU with 10 cores and RAM with 64G, has reached a limit. The greater the number of elements, the greater the calculation accuracy, therefore, it's not necessary to prove the mesh independence for the model. The value of the mesh quality is greater than 0.67, which fully meets the need of calculation accuracy. Moreover, the time of every calculation case will last for 5–7 days under the premise that 8 cores are used. The mesh can be shown in Figure 2.

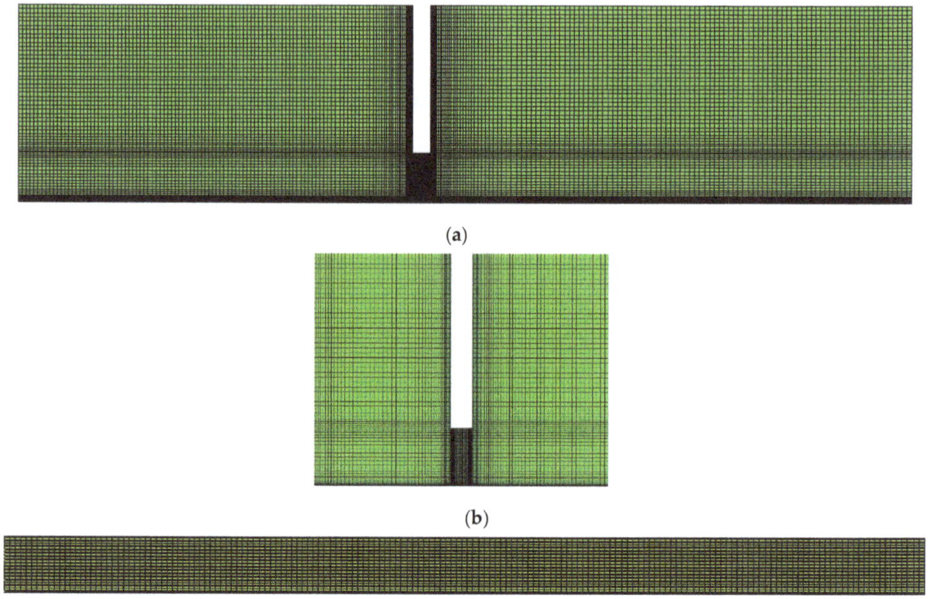

(a)

(b)

(c)

Figure 2. Middle section of the mesh of the impinging jet model: (**a**) *oxz* plane without jet; (**b**) *oyz* plane without jet; (**c**) enlarged jet rotating 90 degrees clockwise.

2.3. Wray–Agarwal Turbulence Model

The Wray–Agarwal (W-A) turbulence model is a single-equation model developed based on the k - ω turbulence model. Wray et al. [25] used Wray-Agarwal (W-A), SA and SST k-ω turbulence model to simulate several classical separation flows, and compared with the experimental results, it was found that the W-A model predicted the separation and reattachment characteristics of the boundary layer more accurately.

From our previous studies on the oblique impinging jet with the impinging angle of $\theta = 45°$ and impinging height of $H/D = 3$, the Wray-Agarwal, standard k-ε, RNG k-ε, realizable k-ε, standard k-ω, and SST k-ω turbulence models were used for numerical calculations, which were also compared with the experimental data of PIV.

Figure 3a shows the comparison between the numerical velocity V/V_{max} with the empirical formula $V/V_{\text{max}} = (1-2r/D)^{1/n}$ for the fully developed circular jet. Where, V is the velocity at any position of the jet exit; V_{max} is the maximum velocity at the jet exit; V_{b} represents the bulk velocity at the jet exit, which can be defined as $V_{\text{b}} = 4Q/\pi D^2$ (Q is the flow rate of the jet); r represents the radial direction. When $n = 7$, the formula is considered to be approximately consistent with the fully developed velocity distribution of the circular jet [26]. It can be seen from the figure that the numerical results by applying different turbulence models are in good agreements with the empirical formula, indicating that the flow at the jet exit is a fully developed jet flow. Among them, the result calculated by the W-A turbulence model is closest to the empirical formula.

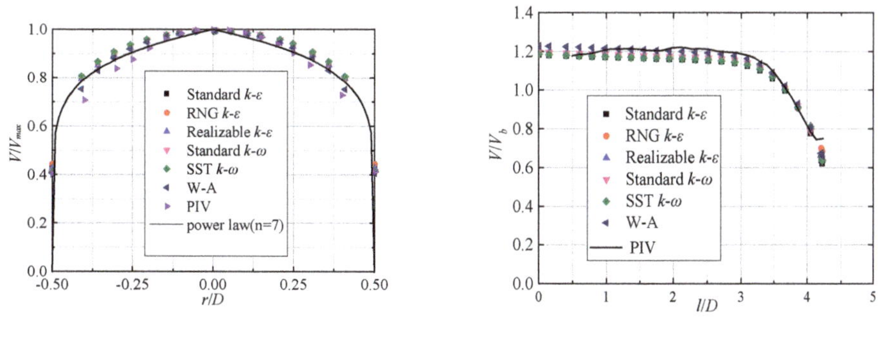

(a) (b)

Figure 3. Verification of Wray-Agarwal turbulence model: (**a**) Profiles of the normalized axial velocity V/V_{max} near the jet exit; (**b**) Normalized mean axial velocity V/V_{max} profiles along the centerline of the impinging jet.

Figure 3b illustrates the axial velocity V/V_b in the centerline of the jet, with different turbulence models. It can be seen that the axial velocity remains basically unchanged in the free jet region (about $0 \leq l/D \leq 3$), and decreases rapidly in the impinging region (about $3 < l/D \leq 4.24$). In the free jet region, the axial velocity calculated by the W-A turbulence model is greater than that by other models, and the velocity in the impinging region is approximately the same. In the initial stage of the free jet region (about $0 \leq l/D \leq 0.8$), the experimental data are smaller than the calculation result with the W-A turbulence model, and the experimental data in the region ($1 \leq l/D \leq 1.5$) agree well with the calculation results. Finally, it is found that, compared with other turbulence models, the calculation results with the W-A turbulence model agree well with the experimental data.

The flow angle φ of the impinging water jet, which is defined as φ = arctan (V_z/V_x), can represent the flow direction at any point in the middle section (*oxz* section) of the jet. Figure 4 shows the distribution of the flow angle on the centerline of the jet. In the free jet region (about $0 \leq l/D \leq 3$), the flow angle is maintained at about 45 degrees, while it rapidly decreases to zero in the impinging region (about $3 < l/D \leq 4.24$). Except for the near-wall region, the flow angles calculated by different turbulence models are in good agreement with the experimental results.

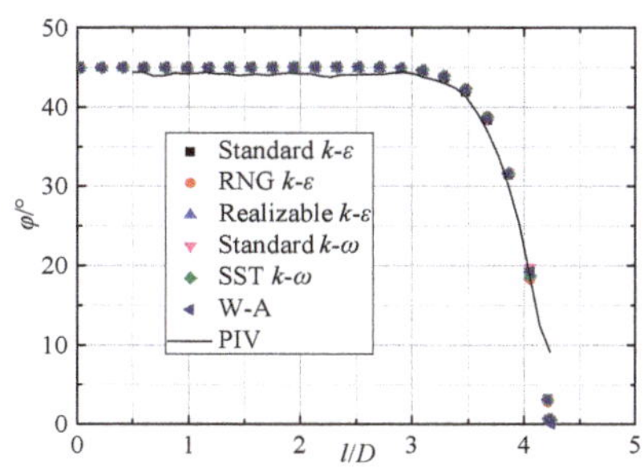

Figure 4. Flow angle φ along the jet centerline.

3. Results and Discussion

3.1. Analysis of Vertical Submerged Jet Flow Field under Different Impinging Heights

Figure 5 shows the mid-section ($y/D = 0$) V/V_b contour and streamline (Re = 35,100) of the jet at various impinging heights. When $H/D = 1$, a pair of vortices of similar size appear in the jet flow field; When $H/D \leq 3$, the size of the vortex is almost unchanged; thereafter the range of influence of the vortex becomes larger as the impinging height increases. When H/D = 8, the vortex position is close to the left and right exit surfaces.

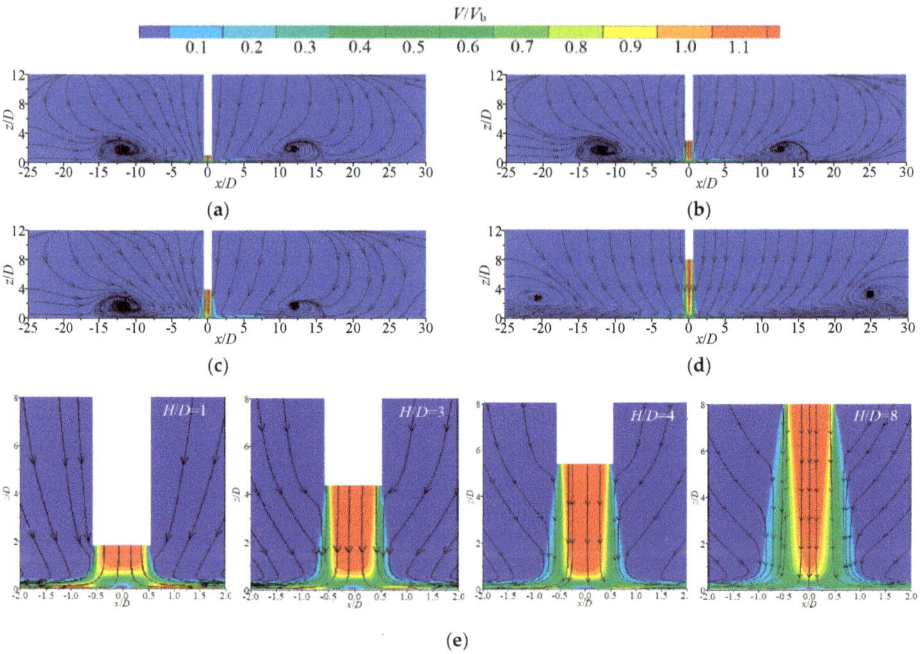

Figure 5. Normalized mean velocity V/V_b contour and streamline of mid-section on submerged impinging jet at various impinging heights:(**a**) H/D = 1; (**b**) H/D = 3; (**c**) H/D = 4; (**d**) H/D = 8; (**e**) enlarged view on the impinging region.

Figure 6 shows the cross section (z/D = 1, 0.5, 0.1) velocity V/V_b contour and streamline (Re = 35,100) of the jet at various impinging heights. When H/D = 1, the velocity contour and streamline of the jet have good symmetry with respect to the section in the jet; There are two pairs of vortices near the left and right exit of z/D = 1 section; At the z/D = 0.5 section, the jet velocity contour is distributed in a ring shape, and a ring with zero velocity appears. The outer boundary of the jet at the z/D = 0.5 section is affected by the front and rear wall surfaces, and the velocity contour is square outside and circular inside. With the increase of the impinging height, the area of the ring with a zero velocity at z/D = 0.5 gradually decreases, indicating that the thickness of the jet from the impingement region to the wall-jet region gradually increases. It can be seen from the figure that the velocity distribution and streamline of the jet at different heights at the section with z/D = 0.1 are similar.

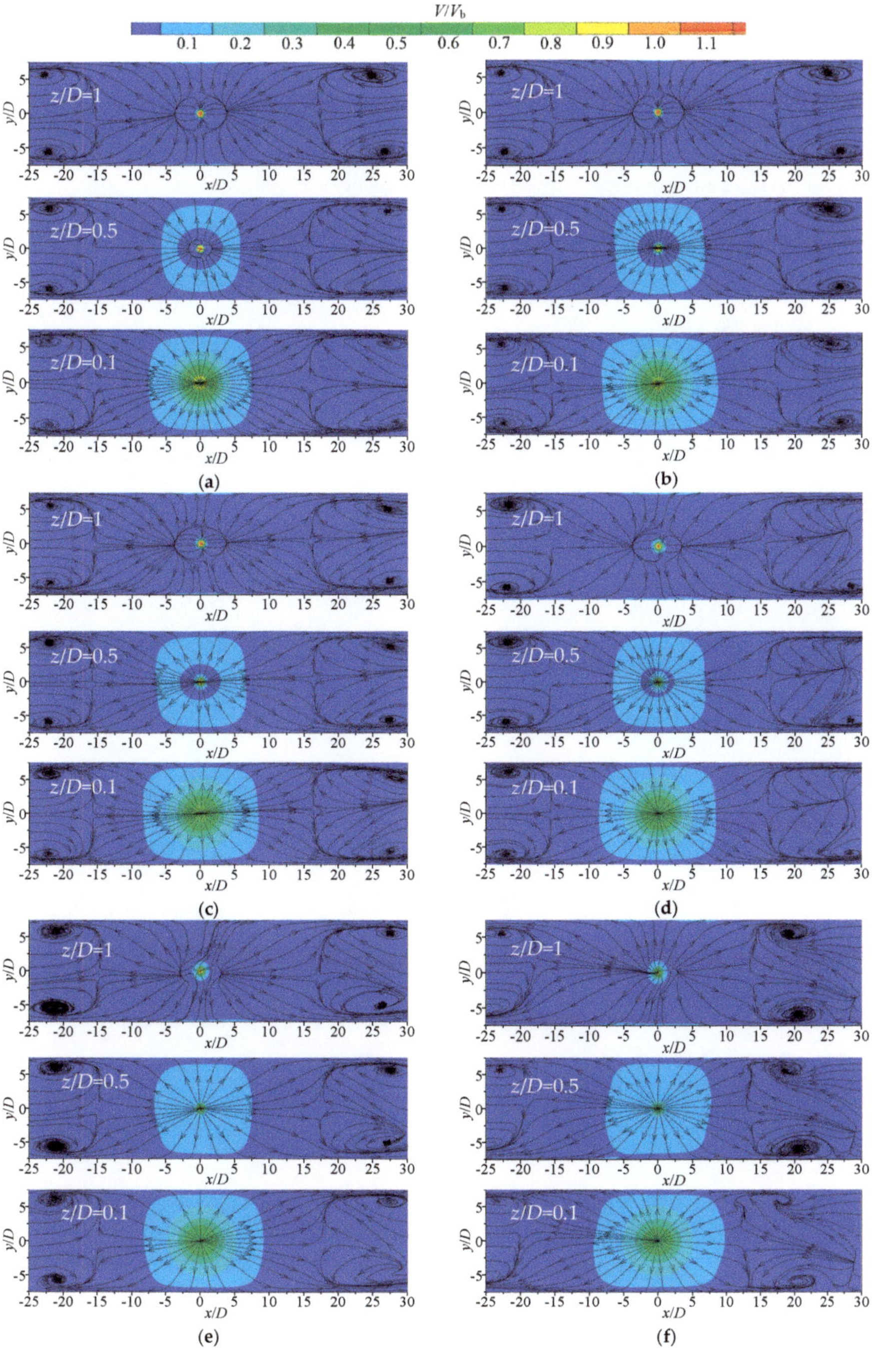

Figure 6. Normalized mean velocity V/V_b contour and streamline of different cross sections on submerged impinging jet at various impinging heights: (**a**) H/D = 1; (**b**) H/D = 2; (**c**) H/D = 3; (**d**) H/D = 4; (**e**) H/D = 6; (**f**) H/D = 8.

Figure 7 shows the vertical cross section (x/D = 0, 1, 5) velocity V/V_b contour and streamline (Re = 35100) of the jet at various impinging heights. The contour of V/V_b in the vertical cross section can reflect the lateral (y-direction) diffusion of the wall-jet region, and the streamline can reflect the development of vortex. When H/D = 1, the thickness of the wall jet increases as x increases; on the x/D = 0 section near the two side wall surfaces, a pair of vortices are generated symmetrically; the thickness of the wall jet decreases first and then increases. When H/D = 3, the thickness of the wall jet increases and the vortex size is almost the same at the vertical cross section of the jet at the same position. When H/D = 6, the vortex is asymmetric in the flow field.

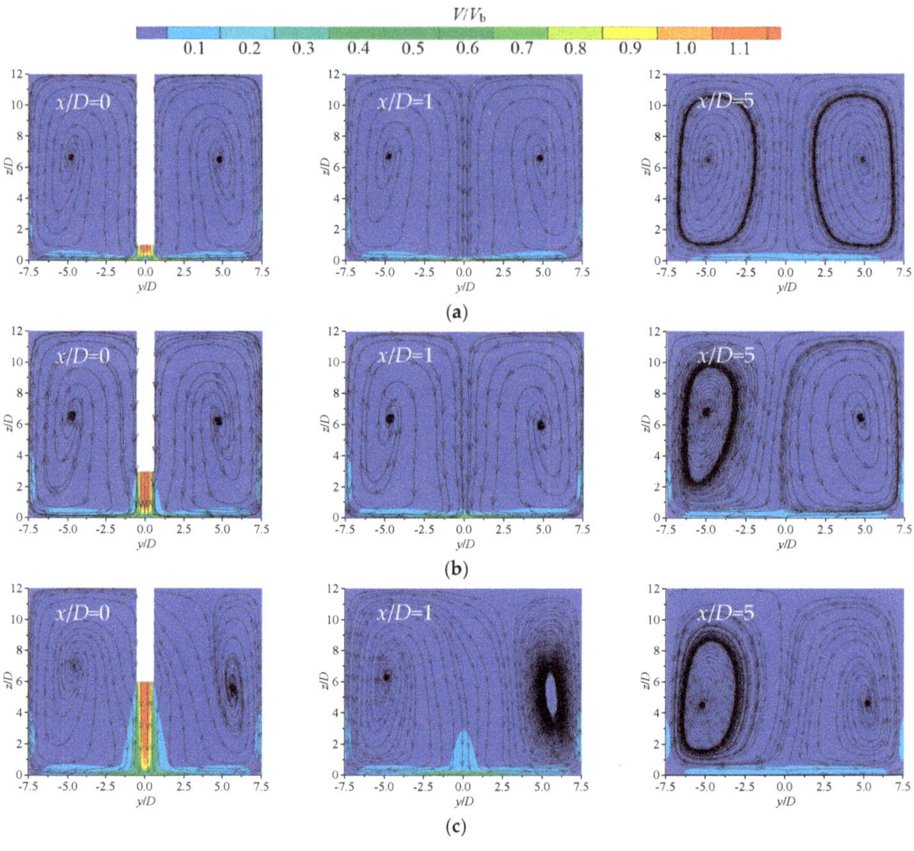

Figure 7. Normalized mean velocity V/V_b contour and streamline of different vertical cross sections on submerged impinging jet at various impinging height: (**a**) H/D = 1; (**b**) H/D = 3; (**c**) H/D = 6.

Figure 8 shows the radial distribution of the axial velocity V/V_b of the jet at various impinging heights (Re = 35,100). When H/D = 1, the velocity distribution has a good symmetry; when $z/D \leq 0.25$, the jet velocity presents a double hump distribution, and two velocity peaks are observed at different x values; with the decrease of z, the velocity on the axis decreases gradually, but the velocity peak increases gradually. When H/D = 2, the velocity distribution remains the same in the region of $1 \leq z/D \leq 1.5$, and the jet is in the core region. With the increase of the impinging height, the range of the jet core region gradually increased, and the position of the double hump velocity near the impingement plate gradually.

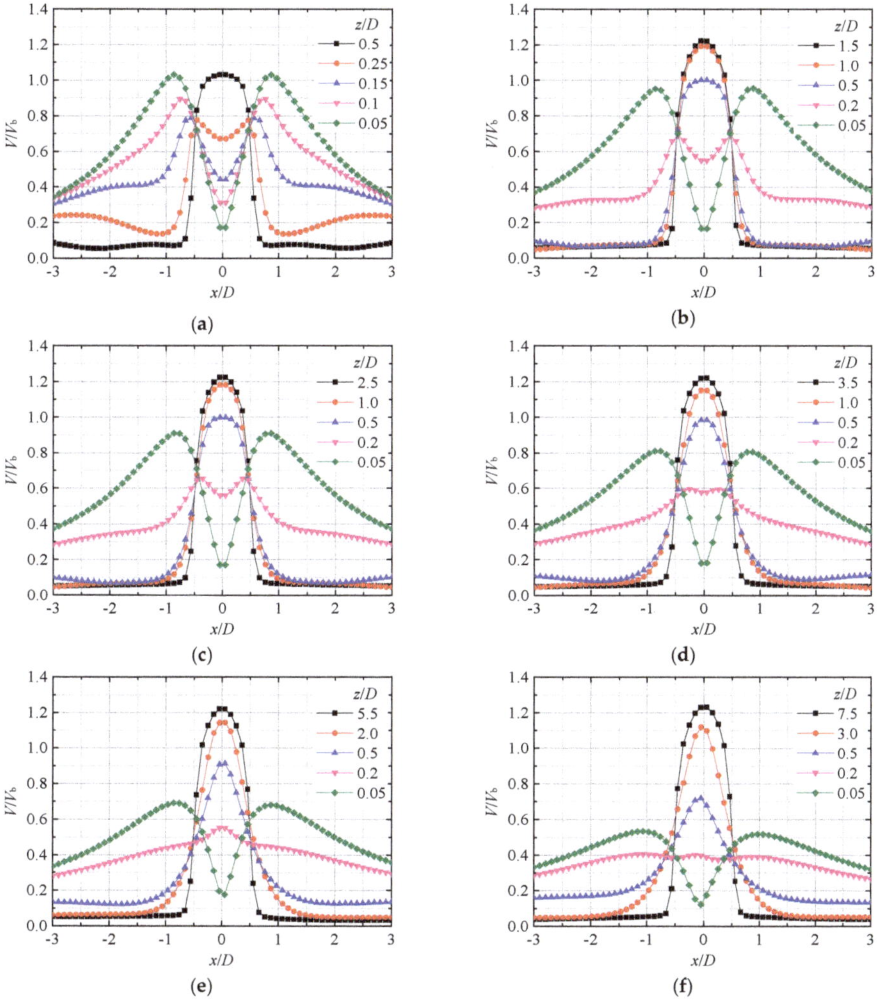

Figure 8. Radial profile development of the normalized mean axial velocity V/V_b at various impinging height: (**a**) $H/D = 1$; (**b**) $H/D = 2$; (**c**) $H/D = 3$; (**d**) $H/D = 4$; (**e**) $H/D = 6$; (**f**) $H/D = 8$.

Figure 9 shows the velocity V/V_b distribution along the jet centerline (Re = 35,100). It can be seen that the velocity in the core region of the jet remains almost unchanged, and the velocity in the impinging region decreases rapidly. The height of the impinging region is about $1D$, and the velocity on the impingement plate is 0. When $H/D \geq 6$, the jet has a transition zone, and the velocity decreases when the fluid reaches the impinging region.

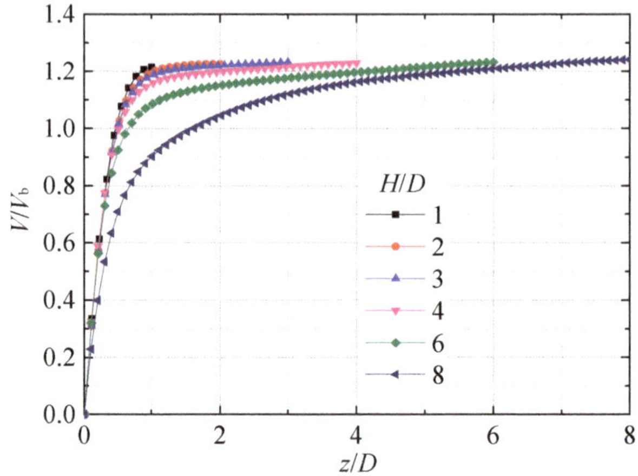

Figure 9. Normalized mean axial velocity V/V_b along the jet centerline.

Figures 10 and 11 show the dimensionless horizontal velocity component and vertical velocity component distributions at different x positions (Re = 35,100). It can be seen from Figure 9 that when H/D = 1, the horizontal velocity component reaches a maximum value at x/D = 0.8, which is about 1.05; when H/D = 3, the maximum horizontal velocity component is about 0.95. When H/D = 6, the horizontal velocity component reaches the maximum value at x/D = 1, which is about 0.72. As shown in Figure 10, as x increases, the vertical velocity component gradually decreases.

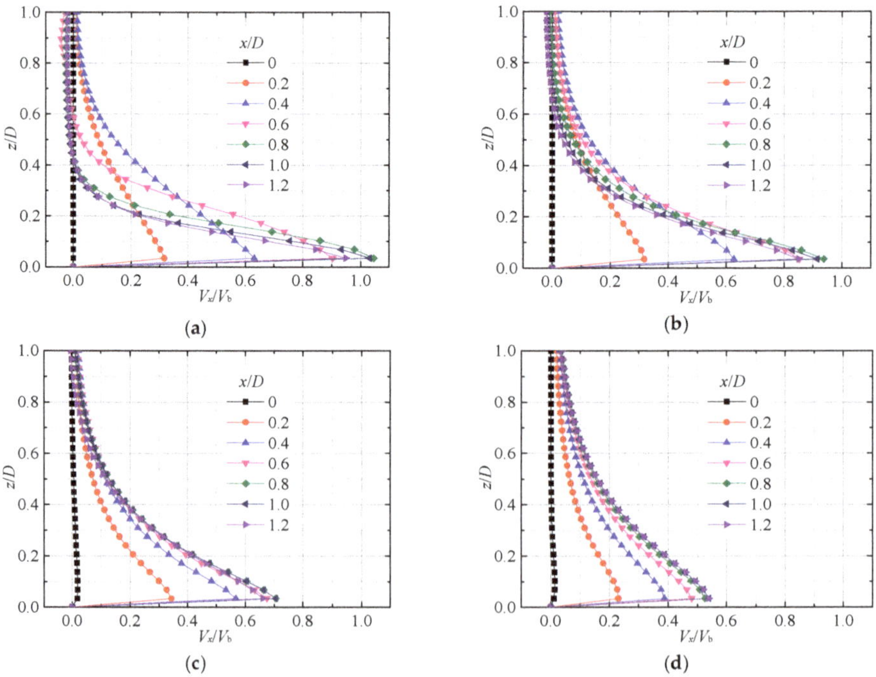

Figure 10. Development of mean horizontal velocity (V_x/V_b) along the horizontal direction at different vertical locations on the near-wall region:(**a**) H/D = 1; (**b**) H/D = 3; (**c**) H/D = 6; (**d**) H/D = 8.

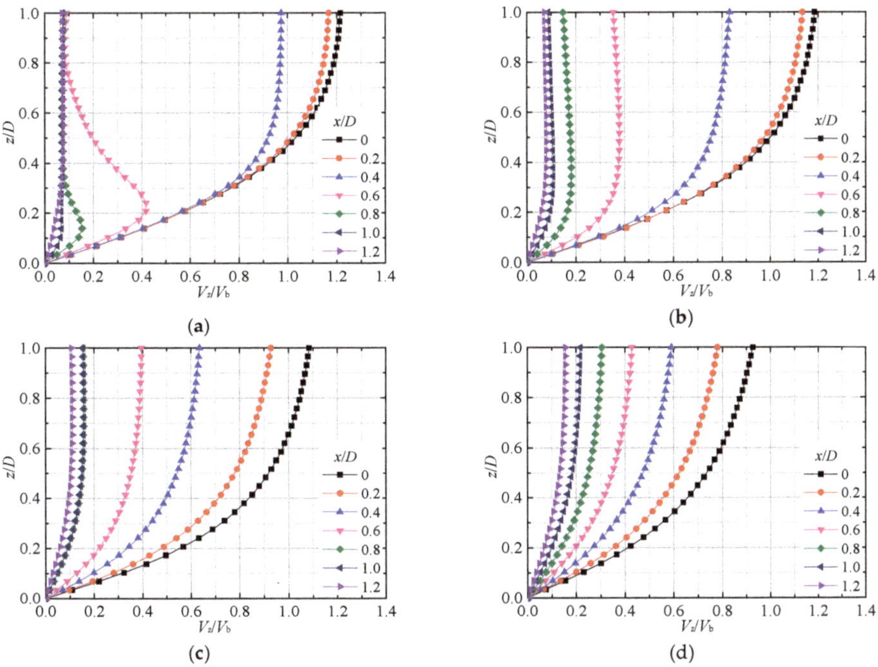

Figure 11. Development of mean vertical velocity (V_z/V_b) along the horizontal direction at different vertical locations on the near-wall region: (**a**) $H/D = 1$; (**b**) $H/D = 3$; (**c**) $H/D = 6$; (**d**) $H/D = 8$.

Figure 12 shows the axial distribution of the average velocity V/V_b at different x positions ($H/D = 3$, $Re = 35{,}100$). It can be seen that, with the development of the wall jet, the local maximum velocity decreases gradually, and the thickness of the wall jet increases gradually. Figure 13 shows the maximum velocity V_m/V_b distribution of the wall jet in the forward flow direction ($Re = 35{,}100$). As shown in the figure, when $H/D \leq 8$, there is a large difference in the value of V_m/V_b at different impinging heights in the region of $x/D \leq 2$, and a small difference in the region of $x/D > 2$, and the fitting formula $V_m/V_b = (x/D)^{-1}$ is very consistent.

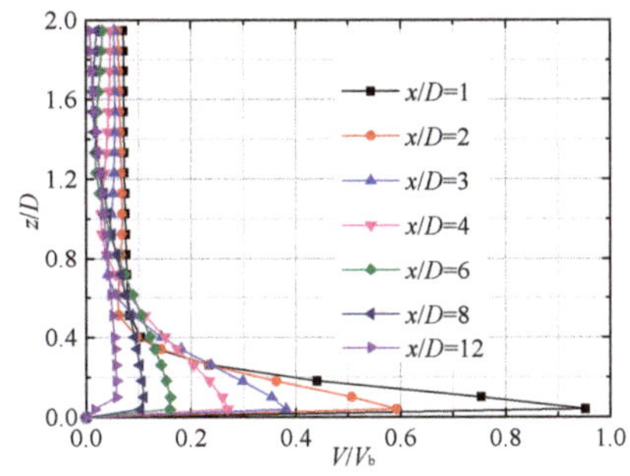

Figure 12. Variation of normalized average velocity V/V_b.

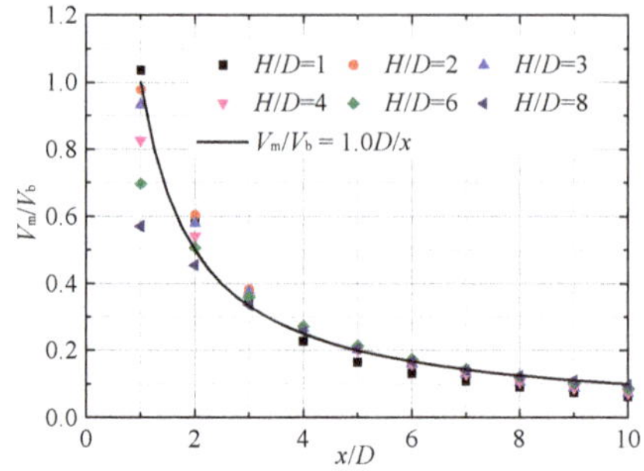

Figure 13. Variation of normalized maximum velocity V_m/V_b along the x direction.

3.2. Time-Averaged Impinging Pressure Distribution

Figure 14 shows the pressure contour (Re = 35,100) of the impingement plane (oxy plane) with different impinging heights. When H/D = 1, the pressure on the impingement plate is circular and symmetrical, and the maximum pressure is obtained at the impinging point. As the impinging height increases, the maximum impinging pressure on the impingement plane gradually decreases.

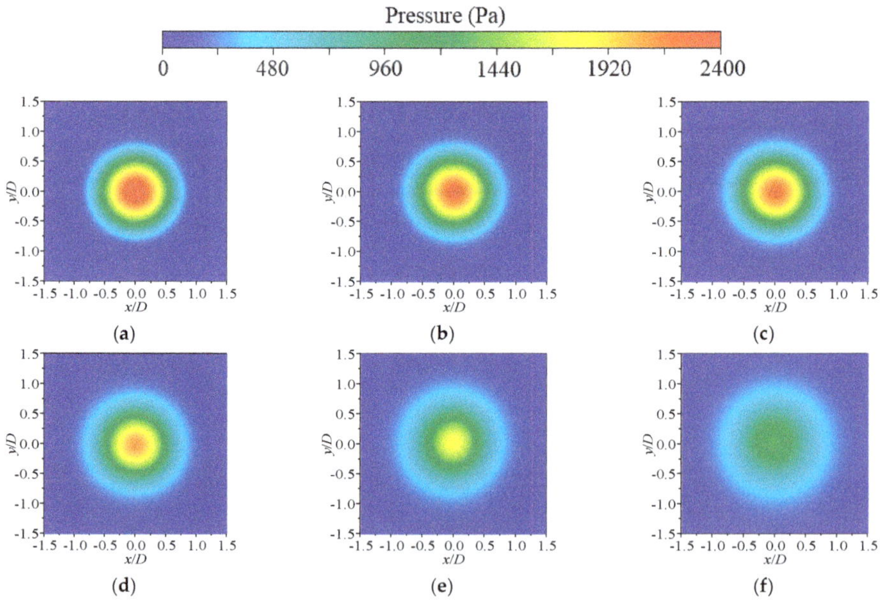

Figure 14. Pressure contour of the plane (oxy plane) of jet at various impinging height: (**a**) H/D = 1; (**b**) H/D = 2; (**c**) H/D = 3; (**d**) H/D = 4; (**e**) H/D = 6; (**f**) H/D = 8.

Figure 15 shows the change of the maximum pressure coefficient C_{pmax} with H/D, and the numerical results are fitted with a straight line. The fitting formula is:

$$C_{pmax} = \frac{P_{max} - P_0}{\frac{1}{2}\rho V_b^2} = -0.012\left(\frac{H}{D}\right)^2 + 0.011\frac{H}{D} + 1.539 \quad (1)$$

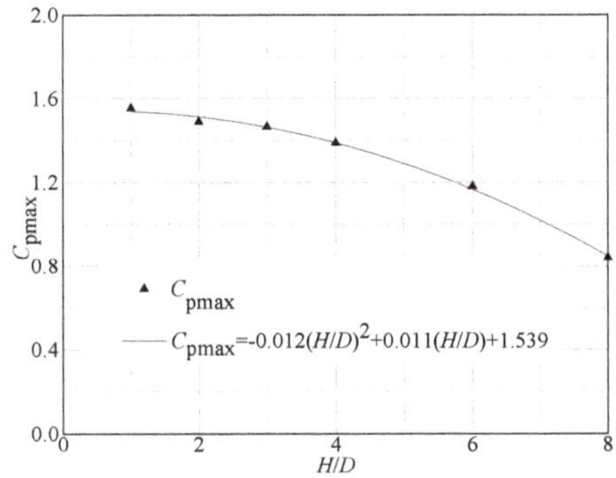

Figure 15. Variation of C_{pmax} as a function of *H/D*.

The applicable range of formula (1) is $\theta = 90°$, $11{,}700 \le Re \le 35{,}100$. It can be seen that the decreasing rate of C_{pmax} gradually increases with *H/D*.

Figure 16 shows the distribution of the pressure coefficient C_p on the *x*-axis at different heights. It can be seen that C_p increases first and then decreases with the increase of *x*. The effective impinging pressure range is within $-2 \le x/D \le 2$, which is independent of the impinging height. The impinging pressure is dimensionless treated with P/P_{max} as the ordinate and *x/D* as the abscissa. The pressure distribution under different impinging heights is shown in Figure 17. It can be seen that the jet pressure distributions at different impinging heights are similar. When $H/D \le 6$, the dimensionless pressure distribution is almost the same. When $H/D = 8$, the dimensionless pressure distribution range increases, and the pressure concentration degree decreases.

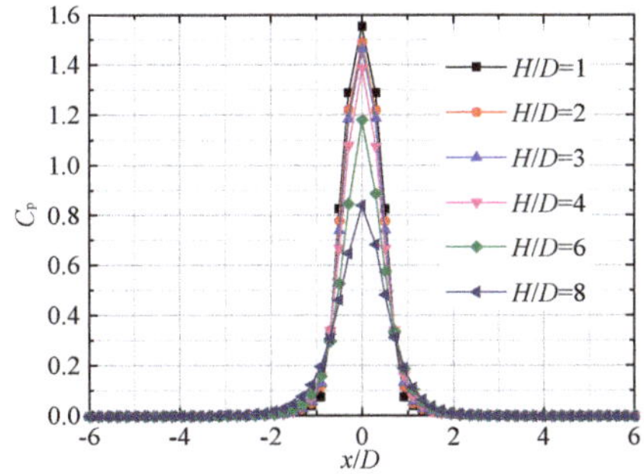

Figure 16. Variation of the pressure coefficient C_p along the *x* axis in the flow direction.

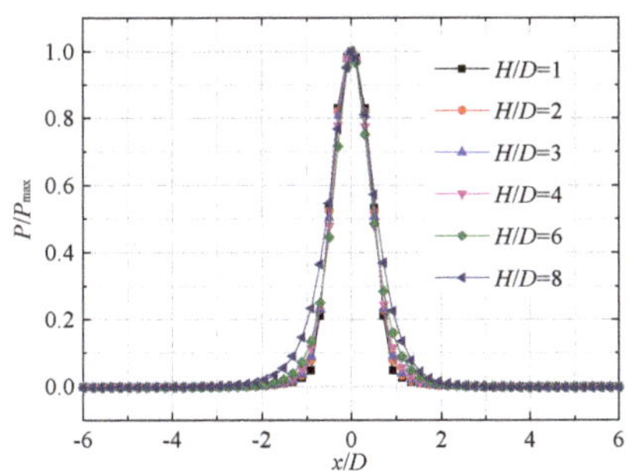

Figure 17. Variation of the pressure P/P_{max} along the *x* axis in the flow direction.

4. Conclusions

In this manuscript, submerged impinging water jets with different impinging heights were investigated in depth, by using CFD based on the W-A turbulence model, and the conclusions can be shown as follows:

(1) The jet flow is highly dependent on the impinging height H, but is relatively insensitive to Re. The normalized mean velocity V/V_b along the jet centerline remains constant in the potential-core region, and decreases rapidly in the transition zone and the impinging region. With the increase of impinging height, the diffusion degree of the jet flow reaching the impinging region increases and the velocity decreases gradually. Near the stagnation point, the fluid velocity increases first and then decreases.

(2) In the impinging region, the jet decelerates rapidly and starts to flow along the wall in the transverse direction. With the increase of impinging height, the value of the maximum axial velocity V_m/V_b decreases rapidly in the region of $x/D \leq 2$ (impinging region), while it remains basically unchanged in the region of $x/D > 2$ (wall-jet region), which also keeps consistent with x/D. It shows that the velocity distribution in the wall-jet region is relatively insensitive to impinging height.

(3) The pressure coefficient C_p along the plane mainly concentrates on the impinging region in the range $-2 \leq x/D \leq 2$, with the maximum value being observed at the stagnant point. The maximum impinging pressure coefficient at the impinging plate decreases with the impinging height, and the decreasing rate increases constantly.

Author Contributions: Data curation, H.W.; Formal analysis, Z.Q.; Writing—original draft, C.W. and D.Z.; Writing—review and editing, T.W. All authors have read and agreed to the published version of the manuscript.

Funding: This research was funded by the National Natural Science Foundation of China (No. 51979240 and 51609105), Priority Academic Program Development of Jiangsu Higher Education Institutions (grant number PAPD), Open Research Subject of Key Laboratory of Fluid and Power Machinery (Xihua University), Ministry of Education (szjj2016-061), Changzhou Sci&Tech Program (Grant No. CJ20190048), Visiting Researcher Fund Program of State Key Laboratory of Water Resources and Hydropower Engineering Science (2017SDG02).

Conflicts of Interest: The authors declare no conflicts of interest.

References

1. Cooper, D.; Jackson, D.C.; Launder, B.E.; Liao, G.X. Impinging jet studies for turbulence model assessment-I. Flow-field experiments. *Int. J. Heat Mass Transfer.* **1993**, *36*, 2675–2684. [CrossRef]
2. Naib, S.A.; Sanders, J. Oblique and Vertical Jet Dispersion in Channels. *J. Hydraul. Eng.* **1997**, *123*, 456–462. [CrossRef]
3. Wang, C.; Wang, X.; Shi, W.; Lu, W.; Ling, Z. Experimental Investigation on Impingement of a Submerged Circular Water Jet at Varying Impinging Angles and Reynolds Numbers. *Exp. Therm. Fluid Sci.* **2017**, *89*, 189–198. [CrossRef]
4. Krishnan, G.; Mohseni, K. An experimental study of a radial wall jet formed by the normal impingement of a round synthetic jet. *Eur. J. Mech. B-Fluid.* **2010**, *29*, 269–277. [CrossRef]
5. Shestakov, M.V.; Dulin, V.M.; Tokarev, M.P.; Sikovsky, D.P.; Markovich, D.M. PIV study of large-scale flow organisation in slot jets. *Int. J. Heat Fluid Flow.* **2015**, *51*, 335–352. [CrossRef]
6. Lu, Z.; Lu, Y.; Xia, B.; Liu, Y.; Zhao, L.; Ge, Z.; Zuo, W. Numerical simulation on hydrodynamic characteristics of percussion pulsed jet. *J. China Univ. Pet. Ed. Nat. Sci.* **2013**, *37*, 104–108.
7. Craft, T.J.; Graham, L.J.W.; Launder, B.E. Impinging jet studies for turbulence model assessment—II. An examination of the performance of four turbulence models. *Int. J. Heat Mass Transfer.* **1993**, *36*, 2685–2697. [CrossRef]
8. Baughn, J.W.; Shimizu, S. Heat transfer measurements from a surface with uniform heat flux and an impinging jet. *J. Heat Transfer.* **1989**, *111*, 1096–1098. [CrossRef]
9. Gutmark, E.; Wolfshtein, M.; Wygnanski, I. The plane turbulent impinging jet. *J. Fluid Mech.* **1978**, *88*, 737–756. [CrossRef]

10. Wang, C.; Chen, X.; Qiu, N.; Zhu, Y.; Shi, W. Numerical and experimental study on the pressure fluctuation, vibration, and noise of multistage pump with radial diffuser. *J. Braz. Soc. Mech. Sci. Eng.* **2018**, *40*, 481. [CrossRef]
11. Wang, H.; Long, B.; Wang, C.; Han, C.; Li, L. Effects of the Impeller Blade with a Slot Structure on the Centrifugal Pump Performance. *Energies* **2020**, *13*, 1628. [CrossRef]
12. Shi, L.; Zhu, J.; Tang, F.; Wang, C. Multi-disciplinary optimization design of axial-flow pump impellers based on the approximation model. *Energies* **2020**, *13*, 779. [CrossRef]
13. He, X.; Zhang, Y.; Wang, C.; Zhang, C.; Cheng, L.; Chen, K.; Hu, B. Influence of critical wall roughness on the performance of double-channel sewage pump. *Energies* **2020**, *13*, 464. [CrossRef]
14. Wang, H.; Long, B.; Yang, Y.; Xiao, Y.; Wang, C. Modelling the influence of inlet angle change on the performance of submersible well pumps. *Int. J. Simul. Model.* **2020**, *19*, 100–111. [CrossRef]
15. Chen, K.; Zhao, K.; You, Y. Numerical study on flow structure of a shallow laminar round jet. *J. Shanghai Jiaotong Univ.* **2017**, *22*, 257–264. [CrossRef]
16. Amamou, A.; Habli, S.; Saïd, N.M.; Bournot, P.; Le Palec, G. Numerical study of turbulent round jet in a uniform counterflow using a second order Reynolds Stress Model. *J. Hydro-Environ. Res.* **2015**, *9*, 482–495. [CrossRef]
17. Chen, Q.; Xu, Z.; Wu, Y.; Zhang, Y. Numerical analysis of a planar oblique impinging jet flow field. *J. Eng. Thermophys.* **2005**, *02*, 237–239.
18. Guo, W.; Li, L.; Liu, C.; Li, N. Study on the Flow Field of a Submerged Jet. *Adv. Eng. Sci.* **2017**, *49*, 35–43.
19. Afroz, F.; Sharif, M.A.R. Numerical study of turbulent annular impinging jet flow and heat transfer from a flat surface. *Appl. Therm. Eng.* **2018**, *138*, 154–172. [CrossRef]
20. Huang, F.; Li, S.; Zhao, Y.; Liu, Y. A numerical study on the transient impulsive pressure of a water jet impacting nonplanar solid surfaces. *J. Mech. Sci. Technol.* **2018**, *32*, 4209–4221. [CrossRef]
21. Stahl, S.L.; Gaitonde, D.V. Computational Investigation of Dual Impinging Jet Dynamics at Mixed Operating Conditions. In Proceedings of the AIAA Scitech 2020 Forum, Reno, NV, USA, 15–19 June 2020; p. 0818.
22. Battistin, D.; Iafrati, A. A numerical model for the jet flow generated by water impact. *J. Eng. Math.* **2004**, *48*, 353–374. [CrossRef]
23. Singh, T.P.; Kumar, A.; Satapathy, A.K. Fluid flow analysis of a turbulent offset jet impinging on a wavy wall surface, Part C. *J. Mech. Eng. Sci.* **2020**, *234*, 544–563. [CrossRef]
24. Wienand, J.; Riedelsheimer, A.; Weigand, B. Numerical study of a turbulent impinging jet for different jet-to-plate distances using two-equation turbulence models. *Eur. J. Mech. B/Fluids* **2017**, *61*, 210–217. [CrossRef]
25. Han, X.; Wray, T.J.; Fiola, C.; Agarwal, R.K. Computation of Flow in S Ducts With Wray–Agarwal One-Equation Turbulence Model. *J. Propuls. Power* **2015**, *31*, 1338–1349. [CrossRef]
26. Fairweather, M.; Hargrave, G. Experimental investigation of an axisymmetric, impinging turbulent jet. 1. Velocity field. *Exp. Fluids* **2002**, *33*, 464–471. [CrossRef]

energies MDPI

Article

Optimized Design of a Swirler for a Combustion Chamber of Non-Premixed Flame Using Genetic Algorithms

Daniel Alejandro Zavaleta-Luna [1], Marco Osvaldo Vigueras-Zúñiga [1,2], Agustín L. Herrera-May [1,3], Sergio Aurelio Zamora-Castro [1] and María Elena Tejeda-del-Cueto [1,2,*]

1 Master in Applied Engineering, Veracruzana University, 94294 Veracruz, Mexico; zs17024372@estudiantes.uv.mx (D.A.Z.-L.); mvigueras@uv.mx (M.O.V.-Z.); leherrera@uv.mx (A.L.H.-M.); szamora@uv.mx (S.A.Z.-C.)

2 Mechanical Engineering Department, Veracruzana University, 94294 Veracruz, Mexico

3 Micro and Nanotechnology Research Center, Veracruzana University, 94294 Veracruz, Mexico

* Correspondence: etejeda@uv.mx

Received: 13 March 2020; Accepted: 26 April 2020; Published: 3 May 2020

Abstract: Recirculation in a combustion chamber is required for stabilizing the flame and reducing pollutants. The swirlers can generate recirculation in a combustion chamber, inducing a swirling flow that breaks vorticity and improves the mixing of air and fuel. The swirl number (S_n) is related to the formation of recirculation in conditions of high-intensity flows with $S_n > 0.6$. Thus, the optimized design of a swirler is necessary to generate enough turbulence that keeps the flame stable. We present the optimized design of a swirler considering the main parameters for a non-premixed combustion chamber. This optimization is made with genetic algorithms to ensure the generation of a recirculation zone in the combustion chamber. This recirculation phenomenon is simulated using computational fluid dynamics (CFD) models and applying the renormalization group (RNG) k-ε turbulence method. The chemistry is parameterized as a function of the mixture fraction and dissipation rate. A CFD comparison of a baseline swirler model and the proposed optimized swirler model shows that a recirculation zone with high intensity and longer length is generated in the primary zone of the combustion chamber when the optimized model is used. Furthermore, the CFD models depict swirling effects in the turbulent non-premixed flame, in which the stabilization is sensitive to the recirculation zone. The temperature results obtained with the CFD models agree well with the experimental results. The proposed design can help designers enhance the performance of combustion chambers and decrease the generation of CO and NO_x.

Keywords: swirler; optimized; genetic algorithms; recirculation; combustion; CFD; experimental validation

1. Introduction

Recirculation has an important role in most combustion systems. Several equipments such as combustion chambers, gas turbines, and industrial furnaces need recirculation to increase the combustion efficiency and keep the flame stable [1–3]. Recirculation is generated by a reversible toroidal flow that causes a low-pressure zone and a break in the vorticity [4]. This allows efficient mixing between hot gaseous fuel and fresh air entering the combustor [5,6]. Lefebvre and Ballal [7] showed that the process of mixing and combustion occurs inside the primary zone. This primary zone provides an anchor to flame and supplies the temperature, time, and turbulence necessary to achieve complete combustion [8]. The formation of a recirculation zone depends on the swirl number, which is an important parameter in swirling flows [9]. It allows the generation of toroidal recirculation zone to burn the fuel and prevent the formation of nitrogen oxide (NO_x). Mundus and Kwark [10,11] reported that the emissions generated during the combustion can be reduced keeping a flow with rotation and controlling the flame intensity. The flame instability can increase the CO and NO_x emissions; therefore, swirl flows with a sufficient amount of turbulence are required to induce a reversal flow and define a recirculation zone [12–17]. Tuttle [18] described the importance of mechanical components to generate a recirculation zone inside a combustion chamber. Tejeda-del-Cueto [19] indicated that flame stability could be achieved through three means: the generation of turbulence using a bluff body, swirling the flow or a combination with a bluff body to improve mixing of the reactants, and finally the optimization of air injection. Furthermore, Kilik [20] studied high-intensity turbulence in the primary zone using perforated plates with spiral-shaped strips inside a tube or by swirler, which was the most efficient method to inject tangential flows.

A swirler has a set of directed vanes to generate a centrifugal force in the flow, which moves the fluid with a swirling shape within the walls of the flame tube (liner). Valera-Medina [21] observed that flow with high turbulence can produce constant temperature during combustion, avoiding the formation of hot spots. The vane angle of a swirler influences the flame behavior and the pollutant emissions [21–23]. For the swirler design, it is necessary to develop analytical and numerical methods to optimize their performance. Several experimental studies have determined the rotatory flows generated by swirlers and their effect on combustion [24–28]. In addition, computational fluid dynamics (CFD) models have been used to predict the flow behavior in function of the swirler geometry [29–34]. The swirler designs are based on preliminary combustion chamber models [11,29,35,36]. Swirler geometry is defined by variables such as the swirl number, pressure drop due to mass flow, and the area of the swirler. These variables influence the formation of a recirculation zone at the primary area inside the combustion chamber as well as flame stability [35].

Genetic algorithm (GA) is a method that is used to solve constrained and unconstrained optimization problems considering natural selection and genetics [37]. Hiroyasu [38] used genetic algorithms to improve fuel efficiency and reduce the generation of pollutants during the combustion of a diesel engine. By using this method, the operating costs and NO_x generation were reduced. On the other hand, Seneca et al. [39] and Liu et al. [40] used genetic algorithms to determine the optimal geometry of a combustion chamber, reducing the fuel consumption and pollutant emissions. They observed that NO_x generation and fuel consumption are reduced by varying the parameters of fuel injection and the geometry of the combustion chamber. In order to improve the performance of a combustion chamber, we propone the optimal design of a swirler using genetic algorithms. This design considers the main parameters of a swirler and the physical limitations of the combustor, as well as the conditions to enhance the recirculation in the combustion chamber. In addition, CFD models are developed to study the recirculation zone generated by the swirler and the temperature behavior of the flame under real operating conditions. The results of the CFD simulations show suitable conditions for a complete combustion. Furthermore, the optimized swirler generates high-intensity turbulence that can decrease the formation of CO and NO_x. The response of the temperature behavior obtained using CFD models agrees well with the experimental results.

2. Design

A swirler induces a flow with rotation to generate axisymmetric stagnation in the center of the flame tube, improving the flame stability [41]. Nowadays, there are two types of swirlers [42]: annular and axial. In this work, we present the optimized design of an axial swirler. Figure 1 depicts a schematic view of an axial swirler, in which the flow is axially applied.

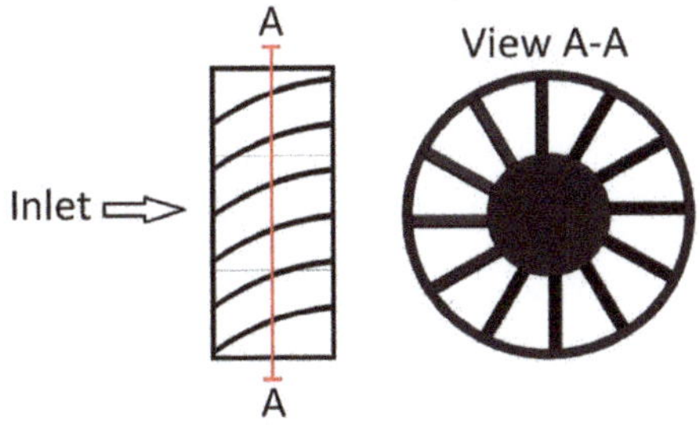

Figure 1. Schematic view of an axial swirler.

The vanes of an axial swirler have an outlet angle used to direct the air flow. This swirler generates a low-pressure drop, since vanes do not have a sharp change of direction. Keshtkar [43] demonstrated that the variation of vane angle can affect the size of the flame and decrease the pollutants generated during combustion. These swirler vanes can be designed with two shape types: flat or curved [44]. Thus, Beer and Syred [45] evaluated the performance of curved and flat vanes of swirlers. They reported that curved vanes achieve higher efficiency (about 70%) and generate less pressure drop compared with flat vanes. Lefebvre and Ballal [7] described a conventional notation for elements of an axial swirler, as shown in Figure 2.

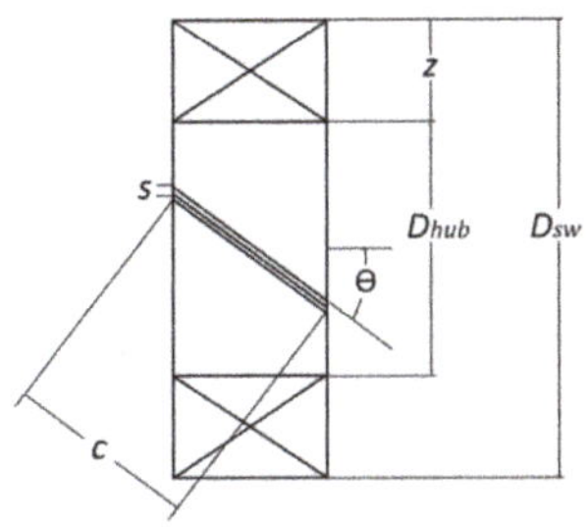

Figure 2. Geometrical parameters of the main elements of an axial swirler. Where D_{sw} is the swirler diameter, θ is the vane angle, D_{hub} is the swirler hub diameter, c is the vane chord, s is the vane space, and z is the annular space length formed by the hub diameter and the swirler diameter.

3. Design of Swirler

The design procedure presented in this work follows steps as shown in Figure 3. The operating conditions and physical dimensions of the combustion chamber defined the starting point for optimization. Deep research was done to determine the variables, objective (fitness) function, limits, and restrictions for running the genetic algorithm. The design procedure finished when the swirl number is maximized, keeping variables within the established limits.

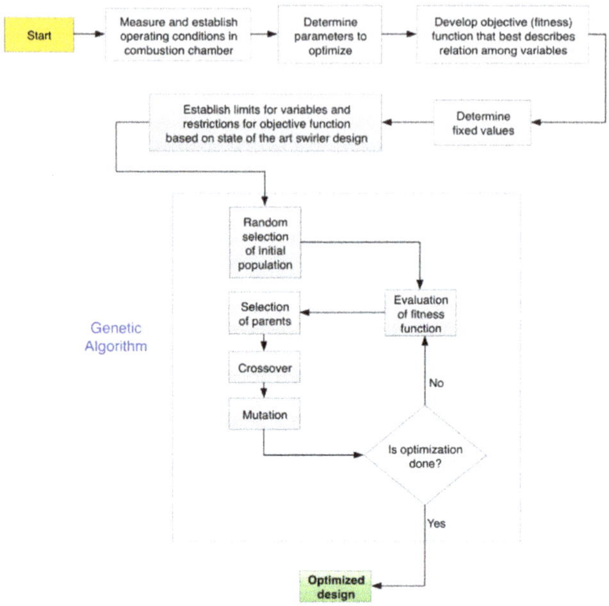

Figure 3. Flow chart of design procedure.

Generally, the design of swirlers is based on three variables: swirl number, mass flow, and swirler area [7].

3.1. Swirl Number

The swirl number (S_n) is the main variable considered in swirler design. The swirl number is the ratio of the axial flux of angular momentum (G_t) to the axial flux of the axial momentum (G_x). The swirl number can be determined by [43]:

$$S_n = \frac{G_t}{G_x r_a} \tag{1}$$

where r_a is the hydraulic radius.

Beer and Syred [45] proposed the following equation to determine the swirl number (S_n) in an annular swirler with a constant vane angle:

$$S_n = \frac{2}{3}\left(\frac{1-\left(\frac{D_{hub}}{D_{sw}}\right)^3}{1-\left(\frac{D_{hub}}{D_{sw}}\right)^2}\right)\tan(\theta). \tag{2}$$

3.2. Mass Flow

Knight and Walker [46] reported that the mass flow ($\dot{m}_{sw}$) of air entering the combustion chamber is a requirement in the swirler design. The $\dot{m}_{sw}$ can be expressed as:

$$\dot{m}_{sw} = \left\{\frac{2\rho\Delta P_{comb}}{k_{sw}\left[\frac{\sec(\theta)}{A_{sw}}\right]^2 - \frac{1}{A_L^2}}\right\}^{\frac{1}{2}} \tag{3}$$

where A_{sw} is the swirler frontal area, A_L is the liner area, ΔP_{comb} is the pressure drop through combustor, k_{sw} is the vane shape factor, and ρ is the air density.

The pressure drop (ΔP_{sw}) in the combustor due to the mass flow in a swirler must be minimal.

3.3. Swirler Area

The frontal area (A_{sw}) of the swirler represents the relationship between the frontal annular area and the area occupied by the vane profiles. This parameter can be calculated by [47]:

$$A_{sw} = \frac{\pi}{4}\left(D_{sw}^2 - D_{hub}^2\right) - \frac{n_v t_v}{2}\left(D_{sw} - D_{hub}\right) \quad (4)$$

where n_v is the number of swirler vanes and t_v is the thickness of the swirler vane.

4. Genetic Algorithm

The genetic algorithm (GA) of MATLAB (R2015b, The MathWorks, Inc., Natick, USA) is a tool that enables defining the value of the parameters involved in an optimization process. These parameters are restricted by limits that allow the values to be optimized efficiently [37]. The genetic algorithm used in this work is based on generating a set of individuals called a population of possible solutions; these individuals are manipulated through the intervention of the following operators: selection, crossover, and mutation to generate a new generation of individuals. When the new generation is created, the new individuals are analyzed, and their properties are evaluated to determine if convergence criteria are satisfied or not. If the convergence criteria are satisfactory, the process stops, but if the convergence criteria are not satisfactory, the process continues until a convergence solution is found [48,49].

4.1. Objective Function

In the design of swirlers, we consider three main parameters: the swirl number, the mass flow, and the swirler area. We propone the swirl number as objective function, which will be maximized to achieve a stability in the flame. This is due to the generation of a recirculation zone in the combustion chamber. The objective function of swirl number (S_n) can be represented by:

$$S_n = \frac{2}{3}\frac{\left(D_{sw}^2 + D_{sw}D_{hub} + D_{hub}^2\right)(\tan\theta)}{\left[\left(\frac{4D_{sw}}{\pi(D_{sw}-D_{hub})}\right)\left(\sqrt{\frac{\sec^2\theta}{\left(\frac{2\rho\Delta P_{comb}}{k_{sw}\dot{m}_{sw}^2}\right)}} + \left(\frac{2n_v t_v D_{sw}}{\pi}\right)\right)\right]} \quad (5)$$

where ΔP_{sw} is the swirler pressure drop, n_v is the number of vanes, $\dot{m}_{sw}$ is the mass air flow, and t_v is the thickness.

4.2. Limits

Table 1 shows the values of k_{sw}, n_v, t_v, and θ recommended by [46,50] for swirler design. These values can allow the generation of a recirculation zone in the combustion chamber. However, Keshtkar [43] and Beer and Syred [45] reported that a similar behavior of the recirculation zone is generated starting with a vane angle of 10°. Based on these research studies, the minimum angle value is modified. In addition, the shape of the curved vane is considered. Table 2 depicts the range of values of the main geometrical parameters used by the genetic algorithm in the swirler design.

Table 1. Recommended values of geometrical parameters for the swirler design.

Geometrical Parameter	Minimum	Maximum
θ	30°	60°
t_v	0.7 mm	1.5 mm
n_v	8	16
k_{sw}	1.15 (curved vane)	1.3 (flat vane)

Table 2. Limits of the geometrical parameters used in the genetic algorithms for the swirler design.

Variable	Minimum	Maximum
θ	10°	60°
t_v	0.7 mm	1.5 mm
n_v	8	16

4.3. Restriction Parameters

The swirl number directly influences the formation of the recirculation zone in the combustion chamber. However, it is necessary for a higher turbulence intensity of the flow. For this, the swirl number must increase to generate a backward flow. Beer and Syred [45] indicated that flows with high-intensity turbulence have a range of swirl numbers between 0.6 and 2.5. Thus, the recirculation zone increases, which establishes the main restriction of the objective function in the genetic algorithms.

Another parameter is the ratio of the hub diameter (D_{hub}) to the swirler diameter (D_{sw}). This ratio allows the reduction of the pressure drop in the flow, which is established by the ratio $D_{\text{hub}}/D_{\text{sw}}$. This ratio has a recommended value to reduce losses of $D_{\text{hub}}/D_{\text{sw}} = 0.5$. Nevertheless, Beer and Syred [45] expressed that recirculation zones with similar behavior can be generated when the ratio $D_{\text{hub}}/D_{\text{sw}}$ has values from 0.4 to 0.6. Based on this restriction and considering $D_{sw} = 9.8$ cm, D_{hub} is measured from the ratios $D_{\text{hub}}/D_{\text{ sw}} = 0.4$ and $D_{\text{hub}}/D_{\text{sw}} = 0.6$. Thus, the maximum and minimum dimensions of D_{hub} are 5.88 cm and 3.92 cm, respectively. Another restriction is added regarding the need to insert the fuel injector at the center of the swirler. It is established that the minimum value of D_{hub} is 5 cm. Table 3 shows the values of the restriction parameters used in genetic algorithms.

Table 3. Range of restrictions for swirler design using genetic algorithms.

Restrictions	Minimum	Maximum
S_n	0.6	2.5
$D_{\text{hub}}/D_{\text{sw}}$	0.4	0.6
D_{hub}	0.05 m	0.0588 m

4.4. Constant Values

Figure 4 shows the different components of the combustion chamber design. In this design, the diffuser has a length of 5.6 cm, an inlet diameter of 6 cm, and an outlet diameter of 15.70 cm. The flame tube has a cylindrical shape with an internal diameter and length of 10.70 cm and 26.10 cm, respectively. It has 5 secondary holes with diameters of 0.54 cm around the flame tube, which are located 12 cm from the air inlet. Moreover, it has 5 tertiary holes with diameters of 1.20 cm that are located 18 cm from the air inlet. The fuel injector has a diameter of 1.30 cm and a length of 11.50 cm. Combustion exhaust gases leave the chamber through a nozzle, which has a length of 5.60 cm and interior and outer diameters of 15.70 cm and 6.0 cm, respectively. The combustion chamber has two peepholes of 2.54 cm diameter. The first and second peephole are located 20.60 cm and 26.60 cm from the air inlet, respectively. Table 4 shows the values of different parameters used in the objective function of the genetic algorithm.

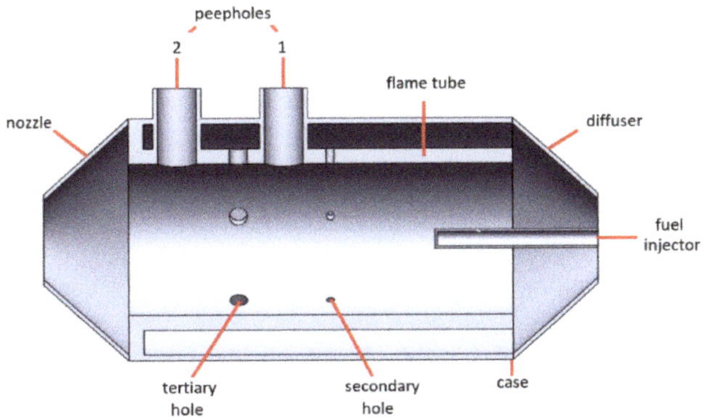

Figure 4. Diffuse flame combustion chamber.

Table 4. Values of different parameters considered in the objective function of the swirl number.

Parameters	Value
Internal diameter of the case	0.157 m
Diameter of combustor	0.107 m
Minimum diameter of diffuser	0.06 m
Air density	1.16 kg/m^3
Vane shape factor	1.15
Air mass flow	4.398×10^{-2} kg/s
Percentage of air entering the combustor	70%
Combustor pressure drop relative to dynamic pressure	20 Pa
Ratio of the combustor pressure drop to the combustor inlet	0.06
Gas constant	286.9 N m/kg
Temperature of incoming air	303.15 K

4.5. Optimal Values

The swirler depth is determined in proportion to D_{sw}, considering 40% of the swirler diameter. Table 5 shows the values obtained by the genetic algorithm. In addition, Figure 5 depicts the final design of the optimized swirler with the coupled injector.

Table 5. Values of optimum swirler design obtained by genetic algorithm.

θ	n_v	t_v	D_{hub}	Depth
18°	8	0.001 m	0.0576 m	0.039 m

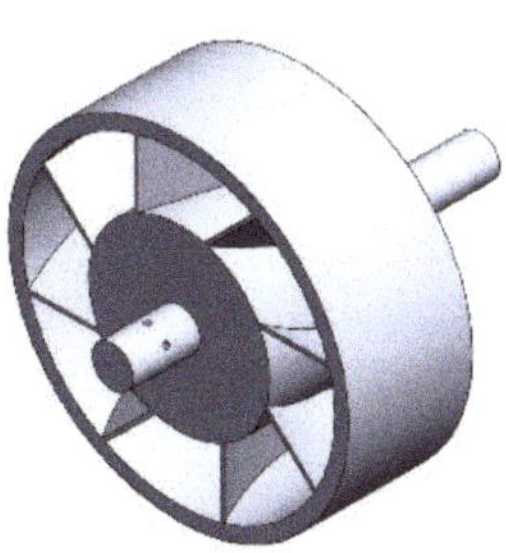

Figure 5. Optimized design of swirler with injector.

The solidity is a parameter that intervenes in swirler design and is defined as the ratio of vane chord length to vane pitch. Martin [51] described that for an axial swirler, the solidity as a function of the vane angle is defined by the following equation:

$$\sigma = \frac{c \sin\theta}{s} \tag{6}$$

where σ is the solidity of the vane, c is the chord, θ is the vane angle, and s is the gap between adjacent vanes.

Considering this, the solidity of the vanes of the designed swirler has a value of 0.51, which indicates that the flow area increases, reducing the pressure drop of the element. In addition to this, since it is a value less than 1, the generation of a high-intensity swirl number is favored.

5. Numerical Models

5.1. CFD Model

5.1.1. Turbulence Model

The RNG k-ε turbulence model is a variation of the standard k-ε turbulence model derived by Yakhot and Orszag, and it is based on the infinite scale expansion in η [52]. This replaces the constant $C_{\varepsilon 1}$ by the function $C_{\varepsilon 1RNG}$ within the transport equation for turbulence dissipation and is based on the theory of renormalization of groups through the solution of transport equations for the generation and dissipation of turbulence [53]. The turbulence model used for the analysis of recirculation zone and the flame behavior generated by proposed the swirler have been used in several research studies related to rotational flows, providing correct results in the turbulence developed in flows with high rotation and also good results in predicting the behavior of the steady-state combustion [54–58]. This RNG k-ε turbulence model can be represented as:

$$\frac{\partial(\rho k)}{\partial t} + \frac{\partial(\rho u_j k)}{\partial x_j} = \frac{\partial}{\partial x_j}\left[\left(\mu + \frac{\mu_t}{\sigma_k}\right)\frac{\partial k}{\partial x_j}\right] + \mu_t\left(\frac{\partial u_i}{\partial x_j} + \frac{\partial u_j}{\partial x_i}\right)\frac{\partial u_i}{\partial x_j} - \rho\varepsilon \tag{7}$$

$$\frac{\partial(\rho \varepsilon)}{\partial t} + \frac{\partial(\rho u_j \varepsilon)}{\partial x_j} = \frac{\partial}{\partial x_j}\left[\left(\mu + \frac{\mu_t}{\sigma_\varepsilon}\right)\frac{\partial \varepsilon}{\partial x_j}\right] + C^*_{\varepsilon 1}\mu_t\frac{\varepsilon}{k}\left(\frac{\partial u_i}{\partial x_j} + \frac{\partial u_j}{\partial x_i}\right)\frac{\partial u_i}{\partial x_j} - C_{\varepsilon 2}\rho\frac{\varepsilon^2}{k} \tag{8}$$

with

$$\mu_t = \rho C_\mu \frac{k^2}{\varepsilon} \tag{9}$$

$$C_{\varepsilon 1RNG} = 1.42 - \frac{\eta\left(1 - \frac{\eta}{\eta_0}\right)}{(1 + \beta\eta^3)} \tag{10}$$

$$\eta = \sqrt{\mu_t\left(\frac{\partial u_i}{\partial x_j} + \frac{\partial u_j}{\partial x_i}\right)\left(\frac{\partial u_i}{\partial x_j}/\rho C_\mu \varepsilon\right)} \tag{11}$$

where the term μ_t represents the turbulent viscosity and the value of the constants are $C_\mu = 0.085$, $C_{\varepsilon 2} = 1.68$, $\sigma_\varepsilon = 1.3$, $\sigma_k = 1.0$, $\beta = 0.015$, and $\eta_0 = 4.38$.

5.1.2. Laminar Flamelet Model

The laminar flamelet model is based on the temperature and species mass fractions transport equations. These equations are standardized in terms of two variables: scalar dissipation (χ) and mixture fraction (Z) [59,60], and they can be represented by:

$$\rho\frac{\partial Y_i}{\partial t} = \rho\frac{\chi}{2}\frac{\partial^2 Y_i}{\partial Z^2} + w_i - \rho\left(\frac{D\chi}{2}\right)^{1/2} k\frac{\partial Y_i}{\partial Z} \tag{12}$$

$$\rho\frac{\partial T}{\partial t} = \rho\frac{\chi}{2}\frac{\partial^2 T}{\partial Z^2} + w_T - \rho\left(\frac{D\chi}{2}\right)^{1/2} k\frac{\partial Y}{\partial Z} \quad (13)$$

where w_T and w_i are the chemical sources for temperature and species, Y_i is the mass fraction of the specie i, T is the temperature, k is the curvature of a mixture fraction iso-line, and t represents the time.

The laminar flamelet model uses the mixture fraction (Z) to incorporate chemical reactions into the flame during turbulent combustion [61]. To quantify the proximity to balance, this model employs scalar dissipation (χ), which represents the stretching of the flame [60]. These parameters are expressed by:

$$\chi = 2D|\nabla Z|^2 \quad (14)$$

$$\rho\frac{\partial Z}{\partial t} + \rho v_k\frac{\partial Z}{\partial \chi} = \frac{\partial}{\partial \chi}\left(\rho D\frac{\partial Z}{\partial \chi}\right) \quad (15)$$

where D is the diffusion coefficient and v_k is the velocity.

Combustion is studied using the laminar flamelet model with a chemical kinetics mechanism [62], which incorporates 58 pollutants (including CO and CO_2) and 270 reactions, but the chemical kinetics mechanism for NOx production is not included in this model.

5.1.3. Numerical Modeling

Meshing and Boundary Conditions

A CFD model of the combustion chamber is developed to study its performance using the finite volume method in Fluent ANSYS. Combustion is modeled using liquefied petroleum gas (LPG) as fuel, which is a mixture of 80% propane and 20% butane on a volumetric basis. This concentration can decrease the contaminant emissions [63]. Figure 6 shows the longitudinal section of the combustion chamber mesh.

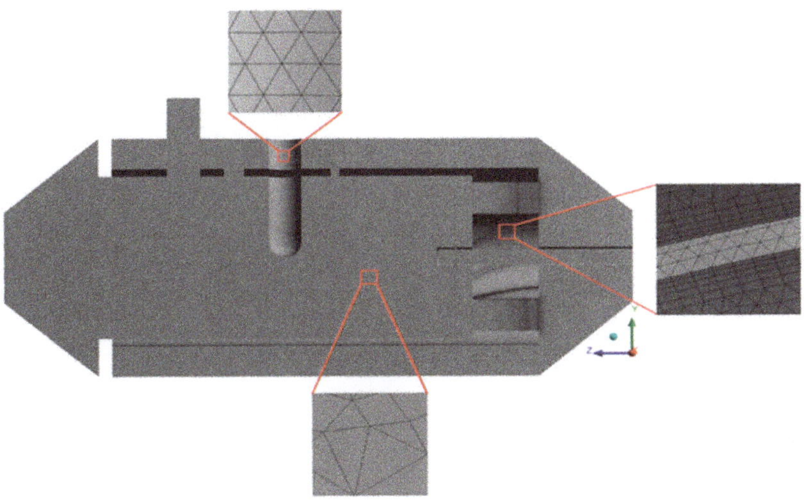

Figure 6. Mesh used in the computational fluid dynamics (CFD) model of the combustion chamber, including a view of the boundary layer on the swirler vanes (right), the boundary of the thermocouple (top), and the elements of the burner (bottom).

The mesh used represents the entire volume of the combustion chamber, and it is set to simulate the fluid that constitutes the flow along the combustor. The mesh is of the tetrahedral type and the use of a boundary layer is established in the swirler vanes, with local refinement on the internal and external surfaces of the vanes. The mesh has 7,555,837 elements and 1,565,195 nodes with an approximate quality of 84%. The residual values of continuity, momentum, turbulent kinetic energy, mixture fraction, and variance of mixture fraction were 10-4, while for energy, it was 10-7.

Table 6 indicates the boundary conditions used in the CFD model of the combustion chamber.

Table 6. Boundary conditions for numerical analysis.

Variable	Value
Air inlet pressure	2100 (Pa)
Mass air flow	4.398×10^{-2} (kg/s)
Air temperature	303.15 (K)
Fuel inlet pressure	2100 (Pa)
Mass fuel flow	2.95×10^{-4} (kg/s)
Fuel temperature	300.15 (K)
Combustion chamber outlet pressure	375 (kg/s)
Wall temperature	300 (K)

Given that the model is based on density terms, the second-order upwind discretization scheme is used. This scheme provides a higher-order precision at the cell faces using Taylor series expansion of the cell-centered solution at its centroid [64,65].

Cold Air Flow

The recirculation zone generated by the optimized swirler was compared to a baseline swirler under the same boundary conditions. Figure 7 shows the views of the optimized swirler and the baseline swirler. Figure 7a,b shows isometric and front views of the baseline swirler, while Figure 7c,d depicts the isometric and front views of the optimized swirler.

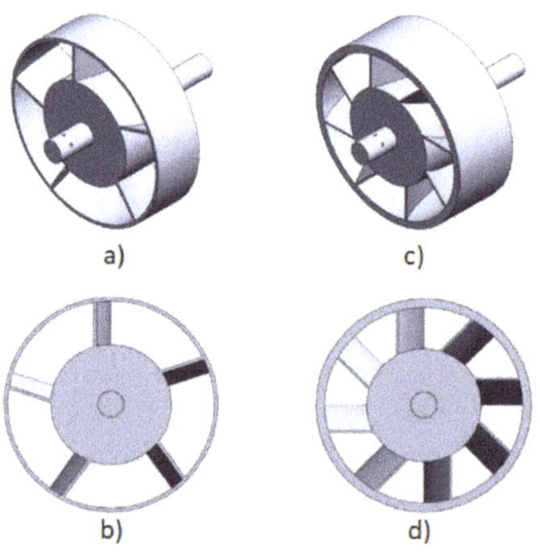

Figure 7. Swirlers compared in numerical analysis. (**a**) baseline swirler, (**b**) front view of the baseline swirler, (**c**) proposed optimized swirler, (**d**) front view of the proposed optimized swirler.

Table 7 shows the comparison of the design parameters for both an optimized swirler and baseline swirler.

Table 7. Comparative table of the design parameters of the swirlers.

	θ	n_v	t_v	D_{hub}	Depth	S_n
Optimized swirler model	18°	8	0.001 m	0.0576 m	0.039 m	2.48
Baseline swirler model	12°	5	0.0015 m	0.06 m	0.03 m	0.18

Figure 8 shows the behavior of the flow circulating in the combustion chamber. Figure 8a represents the velocity path lines generated by the optimized swirler. On the other hand, Figure 8b shows the velocity path lines generated by the baseline swirler. It is observed that the recirculation zone generated by the optimized swirler has a longer length than the baseline swirler and the velocities developed reach up to 5 m/s; it is also observed that a high-intensity turbulence zone is developed at the primary zone. Additionally, a low-speed region is found at the combustion chamber core, ending close to the secondary zone. The recirculation zone obtained with the baseline swirler is wider and shorter than its optimized counterpart. The recirculation zone ends at the primary zone and is bounded by the hub of the swirler. Velocities developed with the baseline swirler are smaller, having average values of 3 m/s.

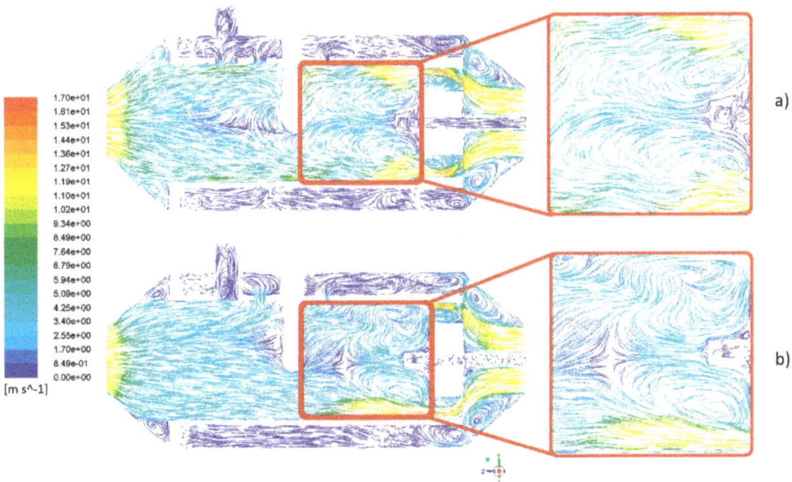

Figure 8. Path lines of velocity along the *yz* plane of a CFD model of a combustion chamber. (**a**) Recirculation zone generated by the optimized swirler model, (**b**) recirculation zone generated by the baseline swirler model.

As the optimized swirler shows better behavior, further analysis is performed regarding cold air flow conditions. Figure 9 shows the air velocity streamlines in the *zx* plane of the combustion chamber using the optimized swirler. Figure 9a depicts a recirculation core formed by air inside the primary zone, which benefits the mixing process by retaining the particles in that zone and delays the process in order to burn properly. In addition, Figure 9b illustrates the rotating flow along the combustor.

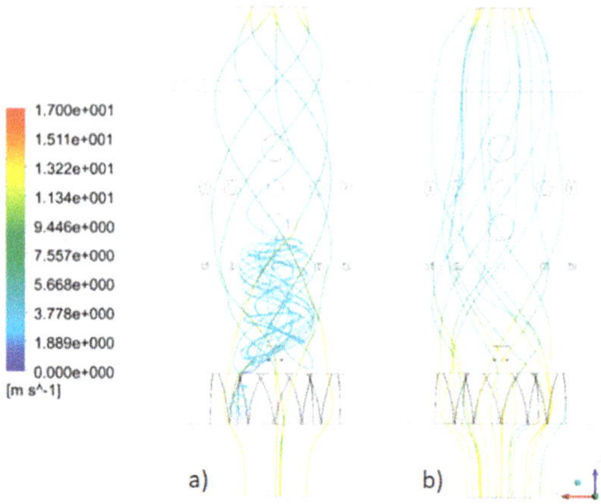

Figure 9. View of air velocity streamlines along *zx* plane of a CFD model of a combustion chamber.

Combustion

Figure 10 shows several *xy* planes along the combustion chamber used to show flow behavior during combustion conditions. The origin point is considered at the inlet of the combustion chamber, and each plane is located at the following distances: plane 1 (10.50 cm), plane 2 (14 cm), plane 3 (17.60 cm), plane 4 (23.60 cm), plane 5 (26.60 cm), and plane 6 (30 cm).

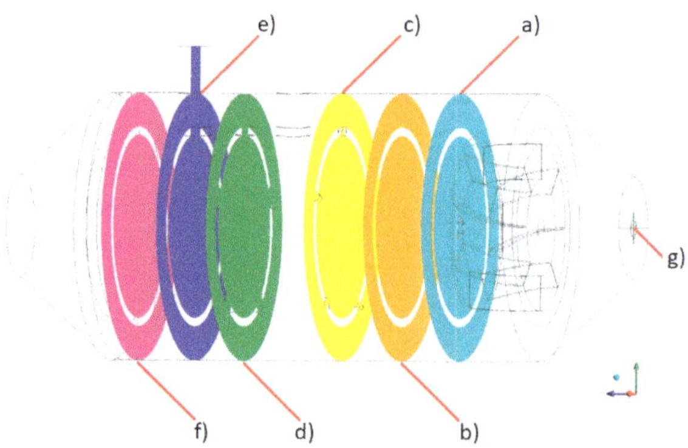

Figure 10. *xy* planes and origin point of the combustion chamber. (**a**) plane 1, (**b**) plane 2, (**c**) plane 3, (**d**) plane 4, (**e**) plane 5, (**f**) plane 6, and (**g**) origin point.

Figure 11a,b depict the optimized and baseline swirler velocity vectors in plane 1, respectively. Figure 11a shows that the slowest velocity zone is located at the center of the combustion chamber and has a hub-like shape. In this zone, average velocities of 3 m/s are obtained, and it is surrounded by a uniform high-velocity zone, with maximum velocities of 14 m/s. On the other hand, in Figure 11b, it is observed that the zone of low velocity begins at the center of the combustor and has a star-like shape, which expands from the center of the hub to the radially middle section of the swirler vanes. This low-velocity zone is surrounded by a high-velocity zone, with an average value of 9 m/s.

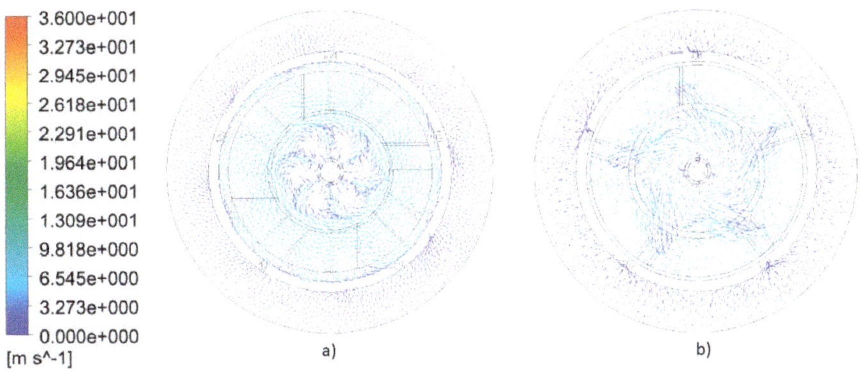

Figure 11. Velocity vectors during combustion in plane 1 of the combustion chamber. (**a**) Optimized swirler, (**b**) baseline swirler.

Figure 12a,b represent the velocity vectors in plane 2 of the combustion chamber with the optimized and baseline swirler, respectively. Figure 12a shows that the size of the low-velocity zone decreased and remains at the center of the combustion chamber. It is also observed that the vectors surrounding the low-velocity zone maintain a constant speed of 8 m/s and rotate inside the flame tube. On the other hand, in Figure 12b, it is observed that the velocity remains constant inside the flame tube with velocity magnitudes comparable to those of the optimized swirler; however, the direction of the vectors indicates that there is no a swirling pattern in the transverse plane to achieve a constant rotation. Most of the air flow passing through the combustion chamber is located in the annular area between the casing and the flame tube.

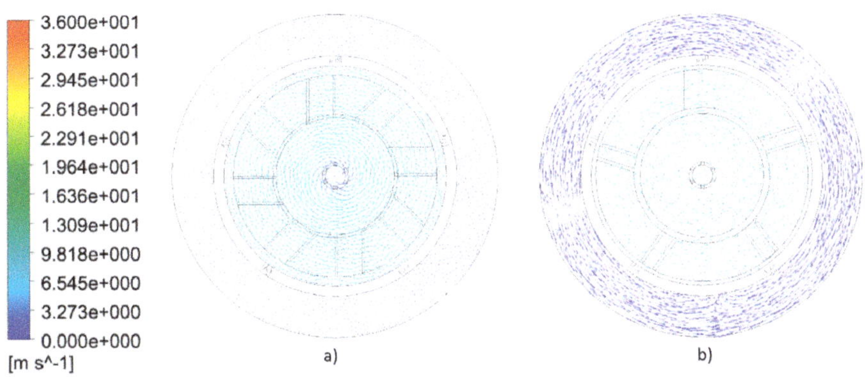

Figure 12. Velocity vectors during combustion in plane 2 of the combustion chamber. (**a**) Optimized swirler, (**b**) baseline swirler.

Figure 13a,b represent the velocity vectors in plane 3 of the combustion chamber with the optimized and baseline swirler, respectively. It is observed that in Figure 13a, high velocities are kept constant near the liner walls while in the central zone, a low-velocity profile is found; however, the flow is maintained rotating throughout the transverse plane of the flame tube. In addition, it is observed that fresh air enters the flame tube through the secondary holes, which will benefit the burning of unburned fuel in the primary zone. On the other hand, Figure 13b shows that a low-velocity zone is generated at the central portion of the combustion chamber, with an approximate velocity of 3 m/s, which is surrounded by a higher velocity zone. It is also observed that there is a rotation of the flow at the central zone of the combustor compared to the flow at the liner walls; moreover, the fresh air entering the flame tube has a higher velocity than the model with the optimized swirler.

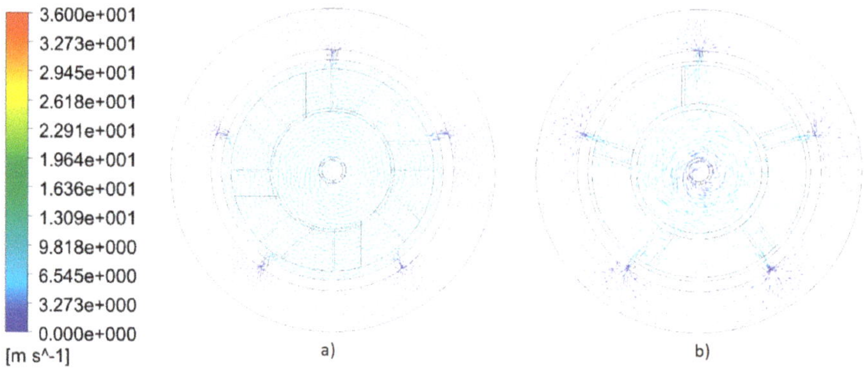

Figure 13. Velocity vectors during combustion in plane 3 of the combustion chamber. (**a**) Optimized swirler, (**b**) baseline swirler.

Figure 14, Figure 15, and Figure 16 show the vectors of planes 4, 5, and 6 of the combustion chamber; Figure 14a, Figure 15a, and Figure 16a represent the optimized swirler, and Figure 14b, Figure 15b, and Figure 16b correspond to the baseline swirler. Figure 14a, Figure 15a, and Figure 16a show that the velocity profile pattern in the transverse plane remains constant and rotating inside the flame tube; furthermore, it is observed that air flow enters through the tertiary holes of the combustor and peephole 2. On the other hand, in Figure 14b, Figure 15b, and Figure 16b, it is observed that the flow field does not have a constant velocity pattern along the flame tube; instead, low-velocity zones and a low-intensity rotating flow is given, even though flow also enters through the tertiary holes and peephole 2.

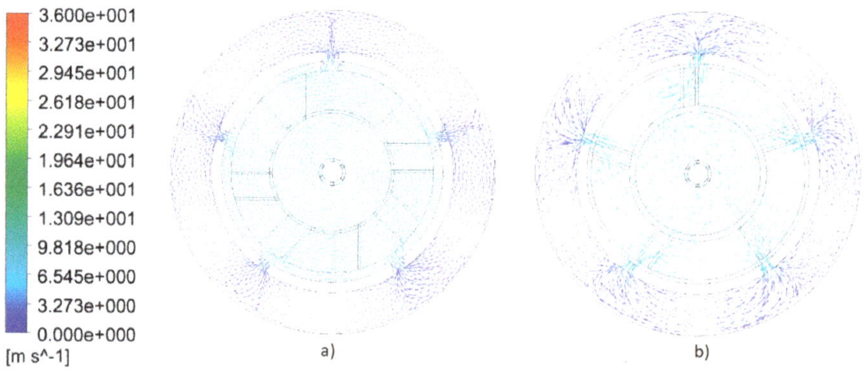

Figure 14. Velocity vectors during combustion in plane 4 of the combustion chamber. (**a**) Optimized swirler, (**b**) baseline swirler.

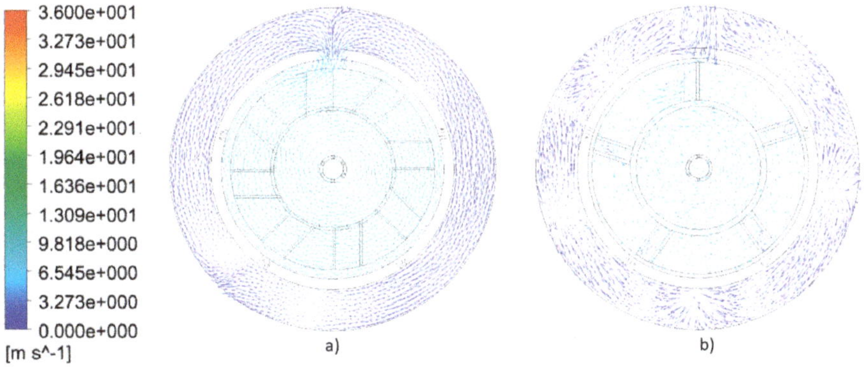

Figure 15. Velocity vectors during combustion in plane 5 of the combustion chamber. (**a**) Optimized swirler, (**b**) baseline swirler.

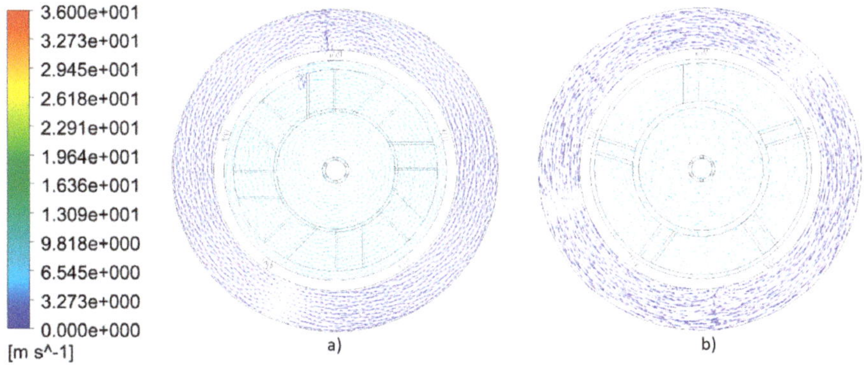

Figure 16. Velocity vectors during combustion in plane 6 of the combustion chamber. (**a**) Optimized swirler, (**b**) baseline swirler.

A pressure drops comparison of the air flow circulating through both optimized and baseline swirlers is shown in Figure 17. Figure 17a shows an air flow pressure at the optimized swirler inlet with an average value of 616 Pa; meanwhile, the same variable measured at the outlet has an average value of 587 Pa, as shown in Figure 17b. Pressure drop is created mainly on the swirler internal walls with an average value of 29 Pa; such value is compared with the reference value of Table 4 of 20 Pa. This represents an acceptable approximation to be used with the GA. On the other hand, from Figure 17c–d, it is observed that the baseline swirler has the following values: an inlet pressure of 616 Pa and an outlet pressure of 530 Pa. Therefore, the pressure drop is 86 Pa, which represents a higher value than the reference in Table 4.

The temperature profile in all six *xy* planes of the combustion chamber model using the optimized swirler are shown in Figure 18. Figure 18a shows the temperature distribution in the primary zone of the combustion chamber where an annular shaped high-temperature zone is developed around the fuel injector limited by the hub of the swirler. At plane 1, we observe a region with a local mixture with the highest fuel flow fraction, and it generates the highest temperature region matching the highest velocities, while at the lower velocity zones, the temperature decreases. Figure 18b shows that the highest local temperature is close to 1200 K and is located at the central low-velocity zone caused by a local air–fuel mixture with the highest fuel flow fraction in this plane. Figure 18c depicts that highest global temperature is close to 1790 K located at the center of the combustion chamber, where the local air–fuel mixture has the highest fuel flow fraction, matching in this case with the low-velocity zone.

In Figure 18d, a decrement in temperature at the combustor core is observed, which is caused mainly by fresh air entering the tertiary holes, obtaining a highest local temperature close to 1450 K. Figure 18e,f show the dilution zone with a reduction in the global temperature; comparison between planes 5 and 6 reveals a similar temperature profile corresponding with both planes having a similar velocity profile.

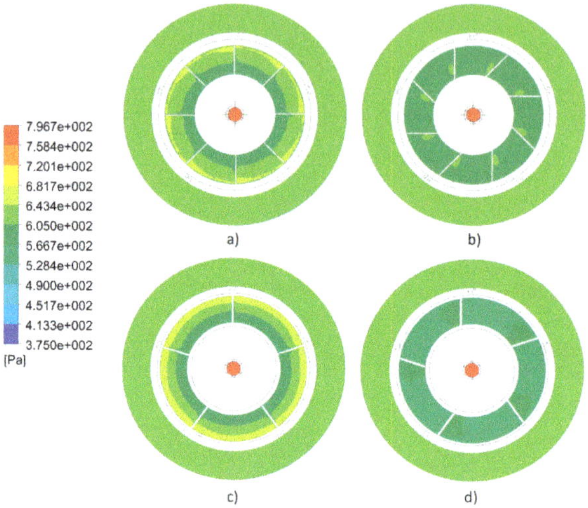

Figure 17. Pressure drop of the air flow generated by the swirler. (**a**) Pressure at the optimized swirler inlet, (**b**) pressure at the optimized swirler outlet, (**c**) pressure at the baseline swirler inlet, (**d**) pressure at the baseline swirler outlet.

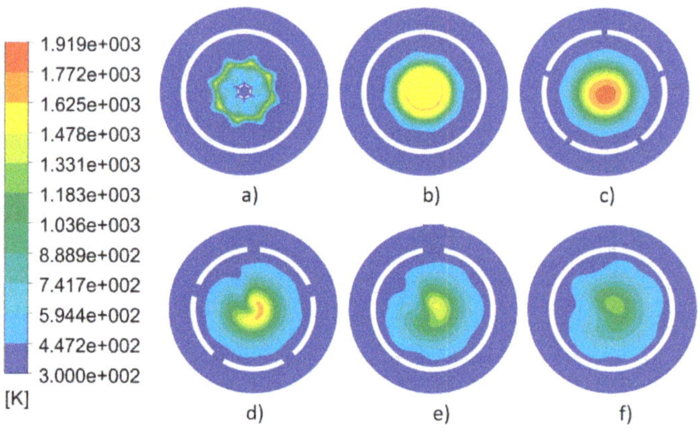

Figure 18. Results of temperature profile in different *xy* planes of a CFD model of a combustion chamber with optimized swirler. (**a**) Plane 1 (10.50 cm), (**b**) plane 2 (14 cm), (**c**) plane 3 (17.60 cm), (**d**) plane 4 (23.60 cm), (**e**) plane 5 (26.60 cm), and (**f**) plane 6 (30 cm).

Figure 19 shows the temperature profile in all six *xy* planes of the combustion chamber model with the baseline swirler. Figure 19a contains the highest global temperature with approximately 1900 K located at a low-velocity star-like shaped zone as in Figure 11b, while the lower local temperatures match the high-velocity zones; for this plane, a local air–fuel mixture with the highest fuel flow fraction is found at the aforementioned star-like shape. In Figure 19b, the global temperature decreases with values up to 1480 K corresponding to a decrease in the fuel flow fraction; this behavior matches the increment in flow velocity, keeping a similar shape as in plane 1. Figure 19c shows that the air flow coming from secondary holes makes the temperature decrease to an average value of 900 K, given that at this plane, a low-velocity zone is kept at the center of the flame tube. Figure 19d–f depict temperature profiles with approximate values of 740 K.

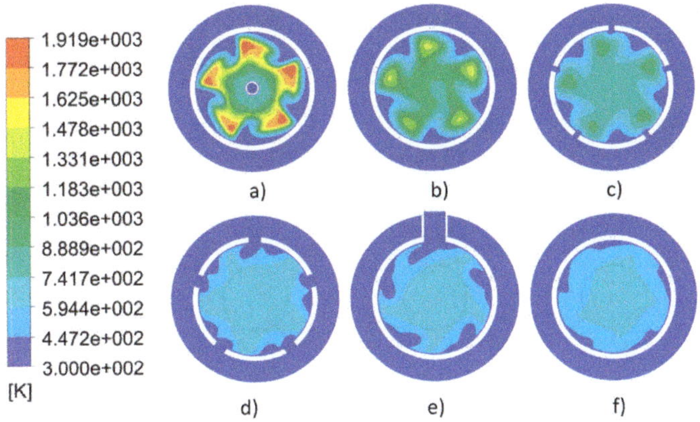

Figure 19. Results of temperature profile in different xy planes of a CFD model of a combustion chamber with a baseline swirler. (**a**) Plane 1 (10.50 cm), (**b**) plane 2 (14 cm), (**c**) plane 3 (17.60 cm), (**d**) plane 4 (23.60 cm), (**e**) plane 5 (26.60 cm), and (**f**) plane 6 (30 cm).

Figure 20 shows the mole fraction of the CO profile in all six xy planes of the combustion chamber model using the optimized swirler. Figure 20a shows a symmetric distribution of the mole fraction of CO at the center of the plane, where CO appears as a result of fuel burning. Figure 20b depicts plane 2 corresponding to the primary zone of the combustion chamber where the highest global mole fraction of CO is located at the combustion core. Figure 20c describes plane 3 containing mainly a cross-section of the secondary zone; it can be observed that the mole fraction of CO decreases as an effect of fresh air entering from the secondary holes, allowing CO to oxidize and form CO_2. Figure 20d shows a significant reduction in mole fraction of CO, while in Figure 20e,f, the formation of CO reaches its minimum.

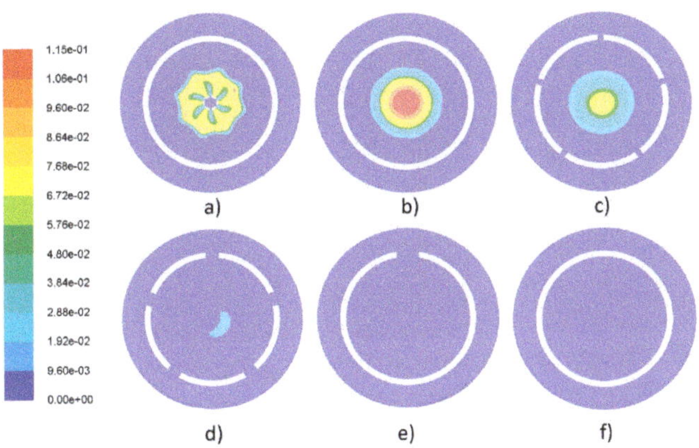

Figure 20. Results of mole fraction of CO profile in different xy planes of a CFD model of a combustion chamber with an optimized swirler. (**a**) Plane 1 (10.50 cm), (**b**) plane 2 (14 cm), (**c**) plane 3 (17.60 cm), (**d**) plane 4 (23.60 cm), (**e**) plane 5 (26.60 cm), and (**f**) plane 6 (30 cm).

In Figure 21, the mole fraction of CO generated during combustion with the baseline swirler is visualized. Figure 21a shows the mole fraction of CO in plane 1 keeping a star-like shape as in the low-velocity zone shown in Figure 11, indicating that combustion happens in this zone. The mole fraction of CO in planes 2–6 is minimal, as there is no CO generation as shown in Figure 21b–f.

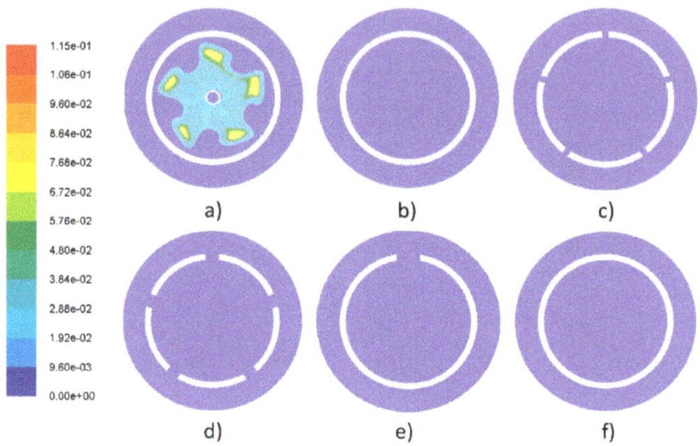

Figure 21. Results of mole fraction of CO profile in different *xy* planes of a CFD model of a combustion chamber with a baseline swirler. (**a**) Plane 1 (10.50 cm), (**b**) plane 2 (14 cm), (**c**) plane 3 (17.60 cm), (**d**) plane 4 (23.60 cm), (**e**) plane 5 (26.60 cm), and (**f**) plane 6 (30 cm).

Figure 22 shows the mole fraction of the CO_2 profile resulting from using the optimized swirler in the six reference planes used to describe the combustion chamber. Figure 22a shows the CO_2 generated when fuel starts to burn, getting a symmetric pattern around the fuel injector. A mole fraction of CO_2 with an annular shape is observed at the core area of plane 2 in Figure 22b. Figure 22c describes the behavior of the secondary zone in the combustion chamber where the highest global mole fraction of CO_2 is obtained matching the plane with the highest global temperature; this phenomenon is the result of CO reacting with oxygen entering from the secondary holes to produce CO_2. Figure 22d–f depict the dilution zone as the mole fraction of CO_2 decreases closer to the combustion chamber outlet caused by air flowing through tertiary holes and peephole 2; CO_2 has an uniform distribution across the planes.

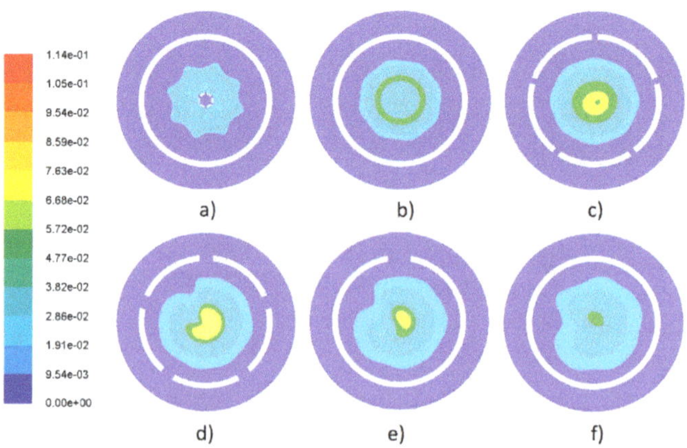

Figure 22. Results of mole fraction of a CO_2 profile in different *xy* planes of a CFD model of a combustion chamber with an optimized swirler. (**a**) Plane 1 (10.50 cm), (**b**) plane 2 (14 cm), (**c**) plane 3 (17.60 cm), (**d**) plane 4 (23.60 cm), (**e**) plane 5 (26.60 cm), and (**f**) plane 6 (30 cm).

The mole fraction of CO_2 distribution obtained with the baseline swirler is given in Figure 23. Figure 23a demonstrates that the highest global mole fraction of CO_2 has a star-like shape similar to the behavior of CO distribution across this plane, as shown in Figure 21a. Figure 23b shows how the star-like shape is preserved; however, the mole fraction of CO_2 decreases. In plane 3 (Figure 23c), the mole fraction of CO_2 decreases, given that fuel flow was burnt at the two previous planes. In Figure 23d–f, the mole fraction of CO_2 is uniform and keeps decreasing closer to the combustion chamber outlet.

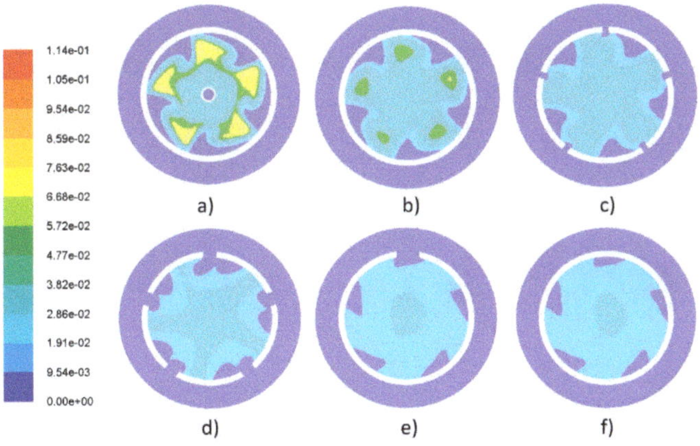

Figure 23. Results of mole fraction of the CO_2 profile in different xy planes of the CFD model of a combustion chamber with a baseline swirler. (**a**) Plane 1 (10.50 cm), (**b**) plane 2 (14 cm), (**c**) plane 3 (17.60 cm), (**d**) plane 4 (23.60 cm), (**e**) plane 5 (26.60 cm), and (**f**) plane 6 (30 cm).

5.2. Experimental Results

Once the objectives of the optimized swirler were verified with the numerical analysis, the combustion experimental tests were carried out to compare the numerical data with the experimental data and validate them.

A swirler prototype was used to measure the temperature behavior inside a combustion chamber. For these measurements, the following conditions were employed: atmospheric pressure of 101,325 Pa, ambient temperature of 303 K, air supply temperature of 305 K, fuel pressure of 2275 Pa, and fuel temperature of 300 K. Figure 24 shows the combustion chamber with its main components.

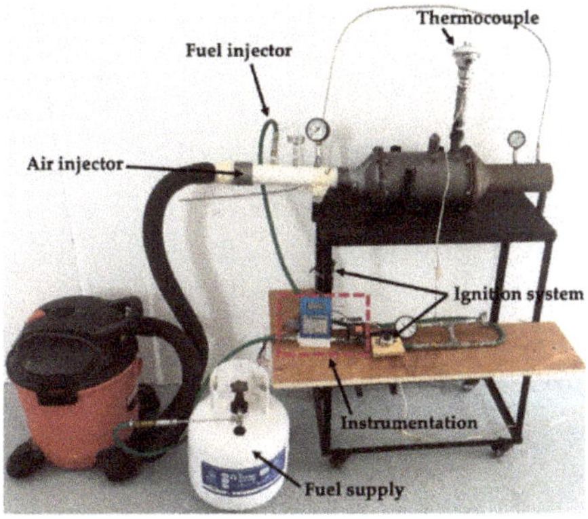

Figure 24. Prototype of combustion chamber used in experimental tests.

The temperature inside the combustion chamber was measured using a bimetallic thermocouple (platinum–rhodium 13%), which allows obtaining temperature measurements ranging from 0 to 1450 °C with an uncertainty of +/− 1.5 °C. It has a ceramic insulating bar of 30 cm length and a ceramic cover. The thermocouple was positioned in two peepholes of the combustion chamber at four different positions. The reference for placing the thermocouple was the central axis of the combustion chamber. Figure 25 depicts the thermocouple positions in both peepholes.

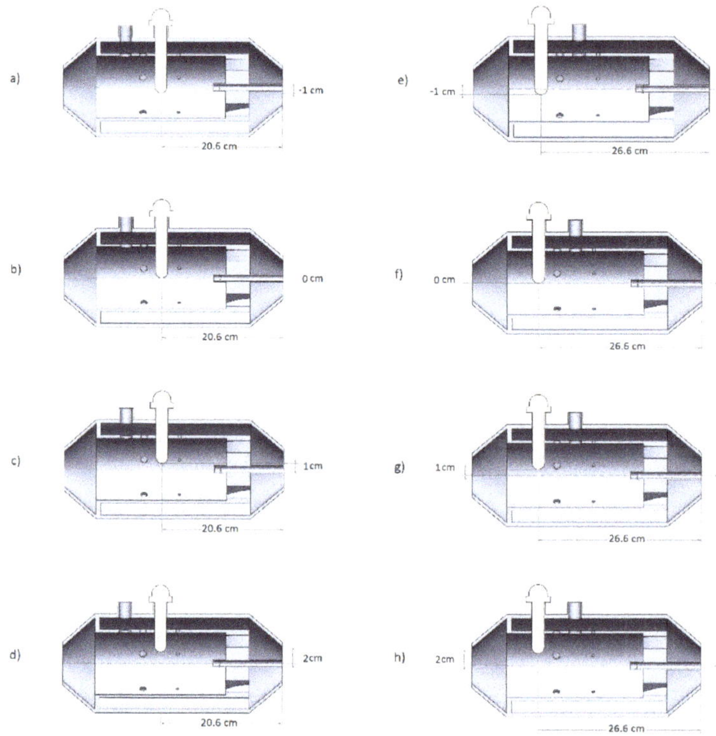

Figure 25. Schematic view of different positions of thermocouple in peephole 1 (left): (**a**) Position 1, (**b**) position 2, (**c**) position 3, and (**d**) position 4, and peephole 2 (right): (**e**) position 1, (**f**) position 2, (**g**) position 3, and (**h**) position 4.

Figure 26 shows the measurements of temperature in four different positions of the combustion chamber. For peephole 1, the highest temperature was measured in position 1 with average temperatures of 1002 K. On the contrary, the lowest temperature was registered in position 4 with average value of 438 K. For peephole 2, the highest temperature (1030 K) was achieved in position 1 and the lowest temperature (438 K) was measured in position 4.

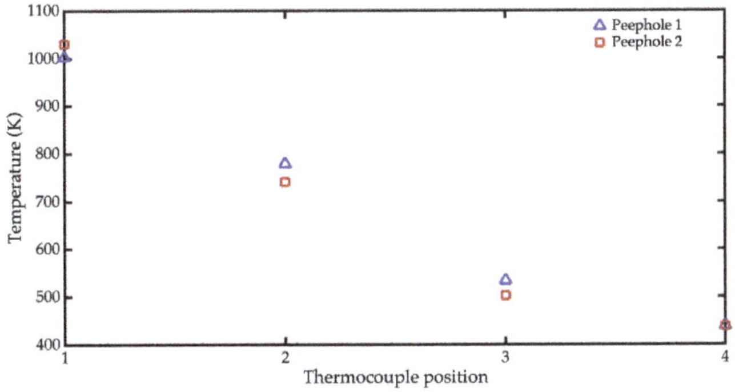

Figure 26. Measurements of temperatures in four different positions of the combustion chamber.

Validation of Numerical Modeling

Figure 27 shows the temperature results obtained using the CFD model and experimental tests in peephole 1 of the combustion chamber. The numerical results of the temperature agree well with the measurement results. Figure 28 depicts the numerical and experimental values of temperature in four different positions along peephole 2 of the combustion chamber. The numerical values of the temperature agree well in comparison with measurement results.

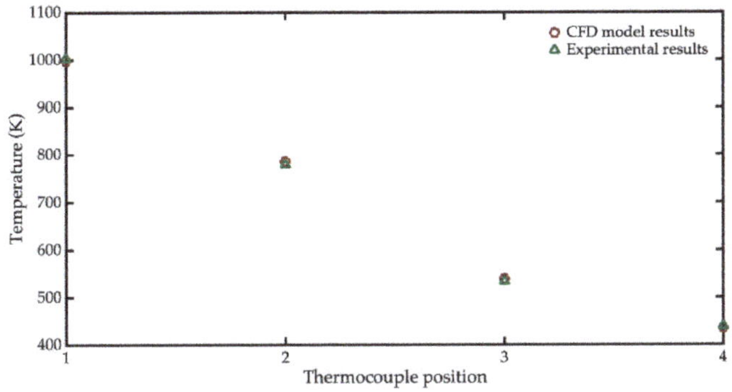

Figure 27. Temperature in peephole 1 obtained using a CFD model and experimental tests.

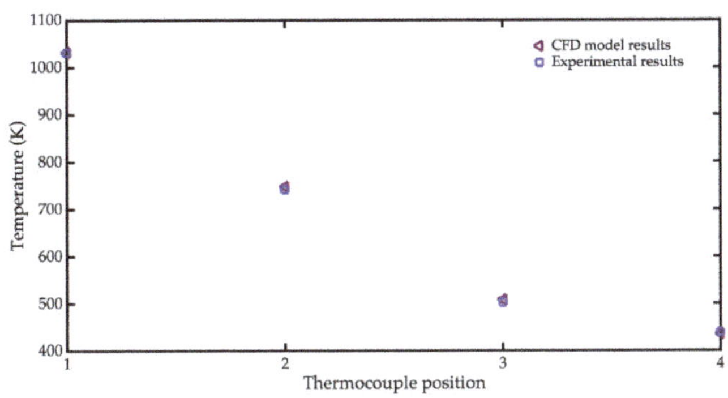

Figure 28. Temperature in peephole 2 obtained using a CFD model and experimental tests.

An infrared thermographic camera (Fluke) was used to visualize the distribution of temperature at the outlet of the combustion chamber. The thermographic camera has a spatial resolution of 320 × 240 and it was configured as follows: emissivity 0.95, bottom temperature of 308.15 K, and transmission of 100%. Figure 29 shows the measurement of the temperature distribution at the exterior environment of the combustion chamber. Low temperature is registered at the walls of the flame tube, while the area of the nozzle outlet reaches the maximum temperature of 557 K. In addition, Figure 30 depicts a thermal image of a cross-section at the combustion chamber outlet. It registers the temperature profile, in which the highest temperature (638 K) is located at the center of the combustion chamber. However, the internal walls of the combustion chamber have relatively low temperatures. Due to this, the generation of hot spots inside the combustor is decreased.

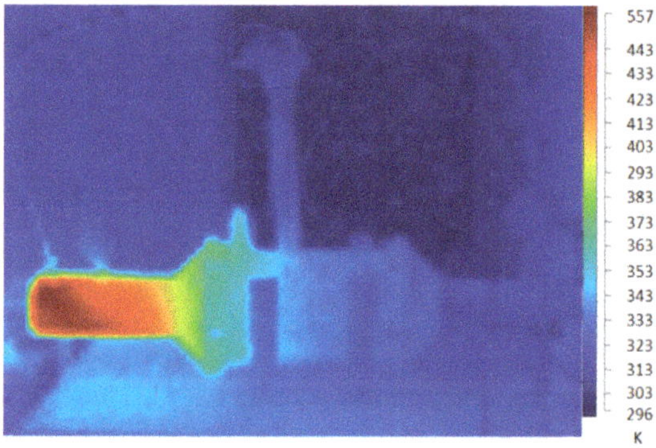

Figure 29. Lateral thermal image of the combustion chamber.

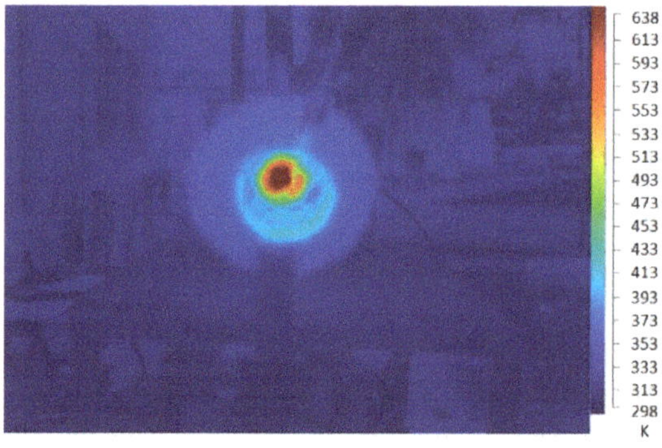

Figure 30. Cross thermal image of the combustion chamber outlet.

6. Conclusions

The optimized design of a swirler using genetic algorithms to generate a recirculation zone in a combustion chamber was presented. CFD models of the combustion chamber were developed using ANSYS software (16.2, ANSYS, Inc., Pittsburgh, USA) through the FLUENT module. In these models, the renormalization group (RNG) k-ε turbulence method was applied. The optimized design was compared to a baseline design to verify optimization. Results obtained numerically showed that the optimized swirler generates a recirculation zone with a longer length and less width. The flame obtained using the optimized swirler was located at the center of the combustion chamber, while using the baseline swirler, a flame was found upstream from the aforementioned location. The exhaust temperature using the optimized swirler was higher than its baseline counterpart, even though the same air–fuel ratio was used.

The CFD models of the combustion chamber using the optimized swirler showed that the stabilization was sensitive to the recirculation zone. In addition, the temperature behavior obtained using the CFD models agreed well in comparison with the experimental results. The Thermographic images showed that the external walls of the combustion chamber registered low-temperature values.

The prediction of the recirculation generated by the experimentally proposed swirler is beyond the scope of this investigation; however, the implementation of the use of particle image velocimetry (PIV) in future research is proposed to obtain parameters such as radial and axial velocities and the swirl number produced by the proposed swirler.

Author Contributions: CFD models of combustion chamber and experimental data, D.A.Z.-L. and M.E.T.-d.-C.; methodology, M.O.V.-Z.; writing—review and editing, A.L.H.-M.; materials, S.A.Z.-C. All authors have read and agreed to the published version of the manuscript.

Funding: The APC was funded by the Master in Applied Engineering of Veracruzana University.

Acknowledgments: Daniel Alejandro Zavaleta-Luna thanks to Consejo Nacional de Ciencia y Tecnología by the scholarship received during his master studies.

Conflicts of Interest: The authors declare no conflict of interest. The funders had no role in the design of the study; in the collection, analyses, or interpretation of data; in the writing of the manuscript, or in the decision to publish the results.

References

1. Hong, S.; Shanbhogue, S.J.; Ghoniem, A.F. Impact of Fuel Composition on the Recirculation Zone Structure and its Role in Lean Premixed Flame Anchoring. *Proc. Combust. Inst.* **2015**, *35*, 1493–1500. [CrossRef]
2. Weigand, P.; Meier, W.; Duan, X.R.; Stricker, W.; Aigner, M. Investigations of Swirl Flames in a Gas Turbine Model Combustor: I. Flow field, structures, temperature, and species distributions. *Combust. Flame* **2006**, *144*, 205–224. [CrossRef]
3. Durox, D. Flame Dynamics of a Variable Swirl Number System and Instability Control. *Combust. Flame* **2013**, *160*, 1729–1742. [CrossRef]
4. Krieger, G.C.; de Campos, A.P.V.; Filho, F.L.S.; de Souza, R.C. A Swirler Stabilized Combustion Chamber for a Micro-Gas Turbine Fuelled with Natural Gas. *J. Braz. Soc. Mech. Sci. Eng.* **2012**, *34*, 441–449. [CrossRef]
5. Zainul, M.N.M.; Jaafar, M.N.M.; Lazim, T.M. Effect of Varying the Double Radial Swirler Configuration on the Fluid Dynamic and Emissions Performances in a Can Combustor. *J. Teknol.* **2017**, *79*, 33–41. [CrossRef]
6. Tuma, J.; Kubata, J.; Betak, V.; Hybl, R. Experimental and Numerical Research of a Novel Combustion Chamber. *Eur. Phys. J. Conf.* **2013**, *45*, 1091. [CrossRef]
7. Lefebvre, A.H.; Ballal, D.R. *Gas Turbine Combustion Alternative Fuels and Emissions*, 3rd ed.; CRC Press: Boca Raton, FL, USA, 2010.
8. Fernández, P.D. Control y Eliminación de los NOx. *Cent. Térmicas* **2000**, *2*, 945–961.
9. Syred, N. A Review of Oscillation Mechanisms and the Role of the Precessing Vortex Core (PVC) in Swirl Combustion Systems. *Prog. Energy Combust. Sci.* **2006**, *32*, 93–161. [CrossRef]
10. Mundus, B.; Kremer, H. The Influence of Swirl Generator Characteristics on Flow and Combustion in Turbulent Diffusion Flames. *Aerothermodyn. Combustors* **1992**, 89–103. [CrossRef]
11. Kwark, J.H.; Jeong, Y.K.; Jeon, C.H.; Chang, Y.J. Effect of Swirl Intensity on the Flow and Combustion of a Turbulent Non-Premixed Flat Flame. *Flow Turbul. Combust.* **2005**, *73*, 231–257. [CrossRef]
12. Kumar, S.; Pandey, K.M. Computational Simulation and Effect of Swirl Angle on NOx Generation of 2D Swirl Burner in Gas Turbine. *IJERT* **2014**, *3*, 1243–1246.
13. Ruan, C.; Chen, F.; Yu, T.; Cai, W.; Li, X.; Lu, X. Experimental Study on Flame/Flow Dynamics in a Multi-Nozzle Gas Turbine Model Combustor Under Thermo-Acoustically Unstable Condition with Different Swirler Configurations. *Aerosp. Sci. Technol.* **2020**, *98*, 105692. [CrossRef]
14. Tanimura, S.; Nose, M.; Ishizaka, K.; Takiguchi, S.; Rodriguez, J. Advanced Dry Low NOx Combustor for Mitsubishi G Class Gas Turbines. *Combust. Fuels Emiss. Parts A B* **2008**, *3*, 607–615.
15. Rodríguez, N.P.; Tamariz, E.R.; Illescas, R.G. Emissions and Noise Pollution Diagnosis of a Turbogas Power Plant During Commissioning Service. *Combust. Fuels Emiss. Parts A B* **2010**, *2*, 1269–1276.
16. Júnior, J.A.C.; Lacava, P.T. *Emissões Em Processos de Combustão*, 1st ed.; Unesp: São Paulo, Brazil, 2003.

17. Cerutti, M. Dry Low NOx Emissions Operability Enhancement of a Heavy-Duty Gas Turbine by Means of Fuel Burner Design Development and Testing. In Proceedings of the ASME Turbo Expo 2018: Turbomachinery Technical Conference and Exposition, Oslo, Norway, 11–15 June 2018.
18. Tuttle, S.G. Lean Blowoff Behavior of Asymmetrically-Fueled Bluff Body-Stabilized Flames. *Combust. Flame* **2013**, *160*, 1677–1692. [CrossRef]
19. Tejeda-del-Cueto, M.E.; Vigueras, M.O.Z.; Welsh, C.M.R.; Ordoñez, C.L.R.; Marin, L.F.U.; Herrera, A.L.M. Alineamiento de Flama Usando Placas Perforadas Tándem en una Cámara de Combustión de Gas LP. *Ing. Mecánica. Tecnol. Desarro.* **2015**, *5*, 293–302.
20. Kilik, E. The Influence of Swirler Design Parameters on the Aerodynamics of Downstream Recirculation Region. Ph.D. Thesis, School of Mechanical Engineering, Cranfield Institute of Technology, Cranfield, UK, May 1976.
21. Valera, A.M. Ammonia–Methane Combustion in Tangential Swirl Burners for Gas Turbine Power Generation. *Appl. Energy* **2017**, *185*, 1362–1371. [CrossRef]
22. Khadase, R.D.; Bhele, S.K. Performance Evaluation of Combustor by Using Different Swirler. *Int. J. Res. Appl. Sci. Eng. Technol.* **2016**, *4*, 85–90.
23. Gupta, N.; Lilley, A.K.; Syred, D.G. *Swirl Flows*; Abacus Press: Turnbridge Wells, UK, 1984.
24. Archer, S.; Gupta, A.K. Swirl Effects on Global and Chemical Behavior of Flames. In Proceedings of the Intersociety Energy Conversion Engineering Conference, Washington, WA, USA, 29–31 July 2002; pp. 364–369.
25. Andrews, G.E.; Ahmad, N.T. Axial Swirler Design Influences on NOx Emissions for Premixed Combustion in Gas Turbine Combustors with All the Combustor Air Flow Passing Through the Swirler. *Proc. ASME Turbo Expo* **2011**, *2*, 449–460.
26. Sander, G.F. Axial Vane-Type Swirler Performance Characteristics Axial Vane-Type Swirler Performance Characteristics. Master's Thesis, Oklahoma State University, Stillwater, OK, USA, July 1983.
27. Thundil, R.K.R.; Ganesan, V. Study on the Effect of Various Parameters on Flow Development behind Vane Swirlers. *Int. J. Therm. Sci.* **2008**, *47*, 1204–1225. [CrossRef]
28. Mularz, E.J. Flow Characteristics of Various Swirl-Can Module Designs. NASA: Washington, DC, USA, 1975; Volume 3236, pp. 1–37.
29. Khandelwal, B.; Banjo, O.; Sethi, V. Design, Evaluation and Performance Analysis of Staged Low Emission Combustors. *ASME J. Eng. Gas Turbines Power* **2014**, *136*, 1–11. [CrossRef]
30. Bourgouin, J.F.; Moeck, J.; Durox, D.; Schuller, T.; Candel, S. Sensitivity of Swirling Flows to Small Changes in the Swirler Geometry. *Comptes Rendus Mec.* **2013**, *341*, 211–219. [CrossRef]
31. Gopakumar, R.; Mondal, S.; Paul, R.M.S.; Chaudhuri, S. Mitigating Instability by Actuating the Swirler in a Combustor. *Combust. Flame* **2016**, *165*, 361–363. [CrossRef]
32. Khanafer, K.; Aithal, S.M. Fluid-Dynamic and NOx Computation in Swirl Burners. *Int. J. Heat Mass Transf.* **2011**, *54*, 5030–5038. [CrossRef]
33. Eldrainy, Y.A.; Saqr, K.M.; Aly, H.S.; Lazim, T.M.; Jaafar, M.N.M. Large Eddy Simulation and Preliminary Modeling of the Flow Downstream a Variable Geometry Swirler for Gas Turbine Combustors. *Int. Commun. Heat Mass Transf.* **2011**, *38*, 1104–1109. [CrossRef]
34. Zhou, L.X. Comparison of Studies on Flow and Flame Structures in Different Swirl Combustors. *Aerosp. Sci. Technol.* **2018**, *80*, 29–37. [CrossRef]
35. Beer, J.M.; Chigier, N.A. *Combustion Aerodynamics*; Applied Science Publishers Ltd.: London, UK, 1972.
36. Bhuvana, R.G.; Srinivasan, S.A.; Thanikaivel, D.M. CFD Analysis on Swirl Angle Effect in Gas Turbine Combustion Chamber. *IOP Conf. Ser. Mater. Sci. Eng.* **2018**, *402*, 12206. [CrossRef]
37. Matlab, Genetic Algorithm, Help with Matlab Script. Available online: https://www.mathworks.com/discovery/genetic-algorithm.html (accessed on 9 April 2020).
38. Hiroyasu, H.; Miao, H.; Hiroyasu, T.; Miki, M.; Kamiura, J.; Watanabe, S. Genetic Algorithms Optimization of Diesel Engine Emissions and Fuel Efficiency with Air Swirl, EGR, Injection Timing and Multiple Injections. *SAE Tech. Pap. Ser.* **2003**, *1*, 1853.
39. Senecal, P.K.; Pomraning, E.; Richards, K.J. Multi-Mode Genetic Algorithm Optimization of Combustion Chamber Geometry for Low Emissions. *SAE Tech. Pap. Ser.* **2002**, *1*, 958.
40. Liu, J.; Wang, J.; Zhao, H. Optimization of the Injection Parameters and Combustion Chamber Geometries of a Diesel/Natural Gas RCCI Engine. *Energy* **2018**, *164*, 837–852. [CrossRef]
41. Brizuela, D.I.; Romano, D.; Loza, J.C. Combustión en Turbinas de Gas. *Combustión* **2007**, *8*, 132–145.

42. Halpin, J.L. Swirl Generation and Recirculation Using Radial Swirl Vanes G. E. Aircraft Engines ornw. *Am. Soc. Mech. Eng.* **1993**, *93*, 1–9.
43. Keshtkar, H. The Effect of Flow Pattern on Combustion and Pollution Generation in a Jet Engine. *J. Astrophys. Aerosp. Technol.* **2018**, *6*, 1–4. [CrossRef]
44. Martin, C.A. Aspects of the Design of Swirlers as Used in Fuel Injectors for Gas Turbine Combustors. *Am. Soc. Mech. Eng.* **1987**, *87*, 139.
45. Beer, M.; Syred, N. Combustion in Swirling Flows: A Review. *Combust. Flame* **1974**, *23*, 143–201.
46. Knight, H.A.; Walker, R.B. The Component Pressure Losses in Combustion Chambers. *Natl. Gas Turbine Establ.* **1953**, *143*, 65.
47. Khandelwal, B.; Lili, D.; Sethi, V. Design and study on performance of axial swirler for annular combustor by changing different design parameters. *J. Energy Inst.* **2014**, *87*, 372–382. [CrossRef]
48. Dina, A.; Danaila, S.; Pricop, M.V.; Bunescu, I. Using genetic algorithms to optimize airfoils in incompressible regime. *INCAS Bull.* **2019**, *11*, 79–90. [CrossRef]
49. How the Genetic Algorithm Works - MATLAB & Simulink - MathWorks América Latina. Available online: https://la.mathworks.com/help/gads/how-the-genetic-algorithm-works.html (accessed on 18 April 2020).
50. Bahr, D.W. *Combustion System Design*; Mello, A.M., Ed.; Academic Press: San Diego, CA, USA, 1990.
51. Martin, C.A. Air Flow Performance of Air Swirlers for Gas Turbine Fuel Nozzles. In Proceedings of the ASME 1988 International Gas Turbine and Aeroengine Congress and Exposition, Amsterdam, The Netherlands, 6–9 June 1988.
52. Yakhot, V.; Orszag, S.A. Renormalization group analysis of turbulence. I. Basic theory. *J. Sci. Comput.* **1986**, *1*, 3–51. [CrossRef]
53. ANSYS Inc. *ANSYS CFX-Solver Modeling Guide v14.0*; ANSYS Inc.: Canonsburg, PA, USA, 2011; Volume 15317.
54. Vondál, J.; Hájek, J. Prediction of Flow through Swirl Generator and Validation by Measured Data. *J. Phys. Conf. Ser.* **2011**, *318*, 022026. [CrossRef]
55. Liu, Y.L.; Tang, H. Numerical Study on the Interaction Mechanism between Swirl and Reverse Flow Rate in a Twin Swirl Combustor. *Adv. Mater. Res.* **2014**, *960–961*, 341–348. [CrossRef]
56. Breussin, F.; Pigari, F.; Weber, R. Predicting the near-burner-one flow field and chemistry of swirl-stabilized low-NOx flames of pulverized coal using the RNG-k-ε, RSM and k-ε turbulence models. *Symp. Combust.* **1996**, *26*, 211–217. [CrossRef]
57. Darmawan, S.; Budiarso, B.; Siswantara, A.I. CFD Investigation of Standard k-and RNG k-Turbulence Model in Compressor Discharge of Proto X-2 Bioenergy Micro Gas Turbine. In Proceedings of the 8th International Conference in Fluid Thermal and Energy Conversion, Semarang, Indonesia, 9–10 November 2013; pp. 4–8.
58. Mompean, G. Numerical simulation of a turbulent flow near a right-angled corner using the speziale non-linear model with RNG K-ε equations. *Comput. Fluids* **1998**, *27*, 847–859. [CrossRef]
59. Carbonell, D.; Oliva, A.; Perez, C.D. Implementation of two-equation soot flamelet models for laminar diffusion flames. *Combust. Flame* **2009**, *156*, 621–632. [CrossRef]
60. Peters, N. Multiscale Combustion and Turbulence. *Proc. Combust. Inst.* **2009**, *32*, 1–25. [CrossRef]
61. Peters, N. Laminar Diffusion Flamelet Models in Non-Premixed Turbulent Combustion. *Prog. Energy Combust. Sci.* **1984**, *10*, 319–339. [CrossRef]
62. Combustion Research Group at UC San Diego. Available online: http://web.eng.ucsd.edu/mae/groups/combustion/index.html (accessed on 4 February 2020).
63. Government of Mexico. Available online: https://www.gob.mx/cms/uploads/attachment/file/94616/gasLP.pdf (accessed on 4 February 2020).
64. ANSYS FLUENT 12.0 User's Guide - 26.2 Choosing the Spatial Discretization Scheme. Available online: https://www.afs.enea.it/project/neptunius/docs/fluent/html/ug/node778.htm (accessed on 9 April 2020).
65. ANSYS FLUENT 12.0 User's Guide - 26.2.1 First-Order Accuracy vs. Second-Order Accuracy. Available online: https://www.afs.enea.it/project/neptunius/docs/fluent/html/ug/node779.htm (accessed on 9 April 2020).

energies

MDPI

Article

Analysis of the Thermal Characteristics of Welding Spatters in SMAW Using Simplified Model in Fire Technology

Yeon Je Shin and Woo Jun You *

Department of Architecture & Fire Safety, Dongyang University, Yeongju 36040, Korea; jeje5842@naver.com
* Correspondence: wjyou@dyu.ac.kr; Tel.: +82-54-630-1297

Received: 16 February 2020; Accepted: 22 April 2020; Published: 4 May 2020

Abstract: In this study, the thermal characteristics of welding spatters were analyzed to predict the risk of fire spread in the shield metal arc welding. The mean diameters and the distribution shapes of the particles were investigated with the variation of the distribution coefficients using the modified gamma distribution function. To quantify the heat source of the welding spatters, the previous empirical equation for the heat transfer coefficient of Ranz and Marshall was analyzed regarding the particle velocity and surface temperature. The order of magnitude for the convection and the radiation were as performed to the particles and the base metal, respectively. The results, which are only valid for $T_{p,m}$ = 750 K and $T_{b,s}$ = 2300 K, show that the radiation term is only 10% for a particle but the convective term is only 6% for the base metal. Finally, the simplified model for the temperature of the welding spatters was obtained and the validation results were within ±13%. The variations of electrical power, droplet size, number of particles, and surface temperature were systematically analyzed with the prediction model. The importance of safety conditions to prevent the fires spread by welding spatter was suggested with electrical power, particle velocity, and numerical density of the particles.

Keywords: welding spatter; distribution; shield arc metal welding; particle heat transfer; fire risk

1. Introduction

Welding can be divided into fusion, pressure, and soldering by applying heat and pressure to the same or different types of metal materials to bond the solids together [1–3]. The fusion method uses an electrode and an electrical power source to create a heat source, resulting in the base metal reaching more than melting temperature and bonding together. Fusion types include gas metal arc welding (GMAW) and shield metal arc welding (SMAW) [4–6]. The GMAW is an arc welding process that uses a plasma arc between a continuous, consumable filler-metal electrode and the weld pool [7,8]. The SMAW, which is the most widely used in building construction, applies electrical power between the metals, such as carbon steels alloy steels nickel and copper alloys, and the electrode to generate arc heat, heating and binding the welds [9–11]. In this process, however, the unbound metals fall off and generate numerous or large particles due to the high temperature of the arc discharge, also known as welding spatters. The spread of these particles can make contact with a combustible material and, if the particles are above the ignition point of the material, can cause a fire. The fire hazards from the SMAW at building construction sites can occur when welding spatters make contact with the inward of a pipe or other enclosed space filled with the flammable vapor or liquid [9–13]. In addition, flammable materials, such as foam urethane or sandwich panels, can be ignited by welding spatters due to thermal storage [13–15]. In accordance with previous research, particles from 0.1 to 3 mm can be dispersed horizontally from the work from 11 to 15 m, igniting flammable materials [13–18]. Previous studies have reviewed major fire accidents caused by welding spatters [19,20] and the thermal characteristics of the particle distributions is one of the most important subjects in fire engineering.

Chol et al. numerically investigated the movement of the metal particles from the GMAW to be 150 A and 300 A [21]. The predictions using the VOF (Volume of Fluid) method for spray velocity, pressure and diameter of particles for the two values of current roughly agree with Kim's research [17]. However, the heat transfer between electrodes and base metal was not considered. Wang et al. [22] used high-speed photography, laser-shadow imaging, and metallographic analysis to validate the simulation results of the GMAW. The results were in broad agreement with experiments despite the assumptions that no drag effects on the droplet surface. Hu et al. analyzed the fluid flow and heat transfer in the GMAW with arc plasma [7]. The transient distributions of current density, arc temperature, and arc pressure were studied, with the assumption that Gaussian profiles were presented to explain the droplet transfer and deformed weld pool surface for GMAW. However, the thermal characteristics of the SMAW is still slow to progress since the distribution of the welding spatters are affected by welders' experiences or circumstances [12]. The electrode used in SMAW is thicker and the arc instability relies on the welders' sensory evaluations unlike the electrode used in GMAW [4]. This causes significant differences for the thermal–fluid characteristics of welding spatters in the fire technology. While the GMAW generates relatively small welding spatters since it is welded to thin electrodes continuously supplied by gas and heat sources such as plasma, the SMAW can generate large particles as it is welded with thick electrodes supplied with fluctuating heat sources [4–6]. The previous studies of the GMAW may predict the transfer phenomenon for welding spatters, but only complex computations can obtain numerical results regarding that the phenomenon. These issues would make it difficult to analyze the risk of fire spread. To overcome these difficulties, a simplified method for solving the heat transfer equation must be developed to explain, not only the thermal characteristics of welding spatters but also the distributions of particles in SMAW. In this study, the simplified prediction model of the surface temperature of welding spatters by SMAW is presented in accordance with electrical power, welding speed and a time, using energy conservation equation.

2. Materials and Methods

2.1. Heat Transfer Phenomenon of Welding Spatters

The thermo–fluid characteristics of the distribution particles, which are generated by welding spatters from the SMAW, depends on electronic power, base metal (material) and electrode (material, welding speed and contact angle between the base metal and electrode) as shown in Figure 1 [17,18]. The dependent variables are summarized as temperature, relative velocity, diameter, and numerical density of particles. Thus, the energy balance for the control volume of the base metal can be expressed as:

$$\Delta \dot{E}_{st} = \dot{E}_{in} + \dot{E}_{g} - \dot{E}_{out} \tag{1}$$

where $\Delta\dot{E}_{st}$, $\dot{E}_{g}$, $\dot{E}_{in}$ and $\dot{E}_{out}$ are the stored thermal and mechanical energy, thermal, and mechanical energy generation and thermal and mechanical energy transport across the control surfaces, i.e., the inflow and outflow terms, respectively. The flammable materials, which are continuously heated for a sufficient time as shown in Figure 1, could combust while the base metal maintains its melting temperature. Thus, the first term of Equation (1) can be simply assumed as Equation (2) for the steady state regarding the risk of fire spread.

$$\dot{E}_{st} = m_b c_{b,p} \frac{dT_b}{dt} \equiv 0 \tag{2}$$

where m_b, $c_{b,p}$, T_b, and t denote mass, specific heat, and temperature for the base metal and time, respectively. The thermal and mechanical energy transference to the base metal for the SMAW is mainly produced by electrical power. Thus, Equation (3) can be assumed as,

$$\dot{E}_{in} = \eta(V \times I) \tag{3}$$

where η, V, and I denote that efficiency of the electrical power, voltage, and current, respectively. When the heat source from the electrical power is transferred to the base metal, the surface temperature can be up to the melting temperature during the welding process [5,7]. The base metal can be thought of as at constant temperature since the equilibrium of the heat transfer by convection and radiation maintains the melting temperature of the base metal. The second assumption is very thin control volume of Figure 1 and the heat transfers to the base metal. Thus, the heat generation $\dot{E}_g$ for the base metal can be regarded as:

$$\dot{E}_g = -\left[h_b A_{b,s}\left(T_{b,s} - T_\infty\right) + \sigma\varepsilon A_{b,s}\left(T_{b,s}^4 - T_{sur}^4\right)\right] \tag{4}$$

where h_b, $A_{b,s}$, $T_{b,s}$, T_∞ and T_{sur} denote that heat transfer coefficient for the flat plate, surface area, surface temperature for base metal, free stream, and surrounding temperature, respectively. If the surface temperature of the base metal remains constant during the welding process, the heat transfer coefficient h_b can be approximated as 10 W/m^2-K [23]. The amount of energy released by the particle distribution can be expressed as:

$$\dot{E}_{out} = \sum_{i=1}^{\infty} E_{p,i} = \sum_{i=1}^{\infty} n_i q''_{p,i} A_i \tag{5}$$

where the subscript i, n_i, $q''_{p,i}$ and A_i mean that each size of the particles during the welding process, the number of particles, the heat flux, and the surface area for ith particle size. Therefore, Equation (1) can be rearranged as

$$\eta \times P_e = \left[h_b A_{b,s}\left(T_{b,s} - T_\infty\right) + \sigma\varepsilon\left(T_{b,s}^4 - T_{sur}^4\right)\right] + \sum_{i=1}^{\infty} n_i q_{p,i} A_i \tag{6}$$

where the left-hand side of Equation (6) explains the electrical power ($P_e = V \times I$) from the electrode to the base metal and the right-hand side explains the heat transfer of base metal and the particles, respectively.

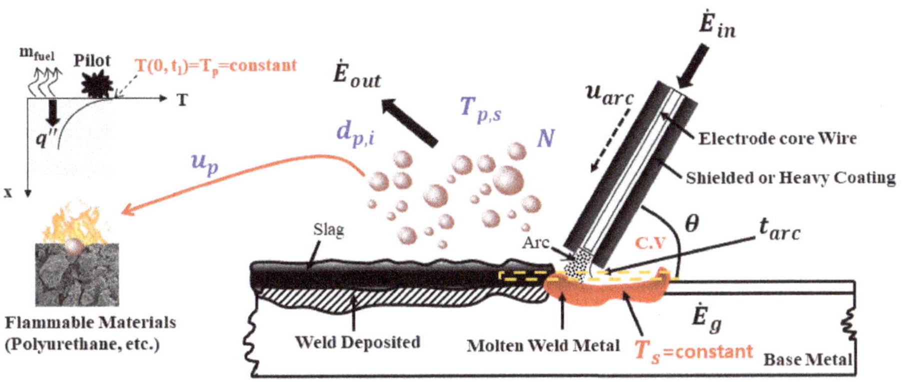

Figure 1. The schematic diagram of welding spatters by the SMAW and the assumptions for temperature of the particles and the base metal.

2.2. Momentum Equation for Simplification

The velocity distributions of the heated particles should consider the drag coefficient and the dynamic viscosity of air and density [7,8,14]. As temperature increases, so does the viscosity; meanwhile, the density decreases for the ideal gas. This means that the initial particle velocity decreases since the drag coefficient increases by the Reynolds number [5,14]. However, in the author's knowledge, the initial velocity depends on the welding conditions such as electrical power, the angle of contact between the electrode and base metal for the specified properties of each material and the thermal time constant of the base metal surface. In addition, the worst case of the fire spread by welding is the maximum velocity for steady state. In this study, the most important objective is the prediction of the fluid and the thermal characteristics of welding spatters which can ignite surrounding flammable materials. Thus, the transient and the velocity variations during the spray time are not considered.

2.3. Distribution Model of Welding Spatters

The size distribution of particles produced by the SMAW is approximated by a Deirmenjian modified gamma function distribution, which is the most widely used in poly dispersion systems as denoted below [24,25].

$$n(r) = ar^{\alpha}\exp(-br^{\gamma}) \quad (7)$$

where a, b, α and γ are distribution coefficients for the welding spatters. In general, α is an integer and the distribution coefficients are dependent on each other [24]. Since Equation (9) should have three zeros (at $r = 0$, $r = \infty$ and $r = r_m$), the constant b can be determined as $b = \alpha/\left(\gamma r_m^{\gamma}\right)$ at $r = r_m$ [24,25]. The relation of the numerical density (number of particles per unit volume) and the normalized distribution of $n(r)$ is defined as $n(r) = f_{exp}(r)\,N$ [26]. Thus, the coefficient a, the number density N and the normalized modified distribution $f_{exp}(r)$ can be obtained as Equations (8)–(10) [25]:

$$a = N \times \gamma b^{(\alpha+1)/\gamma}/\Gamma\left(\frac{\alpha+1}{\gamma}\right) \quad (8)$$

$$N = \int_0^{\infty} n(a)da = \int_0^{\infty} f(a)Nda = A\Gamma\left(\frac{\gamma+1}{\delta b(\gamma+1)}\right) \quad (9)$$

$$f_{\exp}(r) = a\prime C_0 \times r^{\alpha}\exp(-br^{\gamma}) \quad (10)$$

where Γ and N are gamma function and number of particles per unit volume. The symbol A in Equation (9) can be determined by the number density N for $a' = a/N$ and the experimental constant (C_0) and substitute $f(r)$ as denoted in Equation (10). Studies have examined the fitting process of the measured distributions [26]. The droplet mean diameter can be obtained from Equation (9) with various methods such as SMD (Sauter Mean Diameter) as shown in Figure 2 [27]. Therefore, as denoted in Equation (11), the summation of volume for each particle should be equal to the volume of particle mean diameter multiplied by the number.

$$V_{total} = \sum_{i=1}^{\infty} \frac{\pi}{6} d_{p,i}^3 \times n_i = \frac{\pi}{6}\overline{N}d_{p,m}^3 \quad (11)$$

where V_{total}, $d_{p,i}$, $d_{p,m}$ and $\overline{N}$ means the total volume, the diameter of each particle, the mean diameter of particles and the number of particle mean diameter, respectively. Thus, $d_{p,m}$ and $\overline{N}$ are related with:

$$\overline{N} = \frac{6V_{total}}{\pi d_{p,m}^3} \quad (12)$$

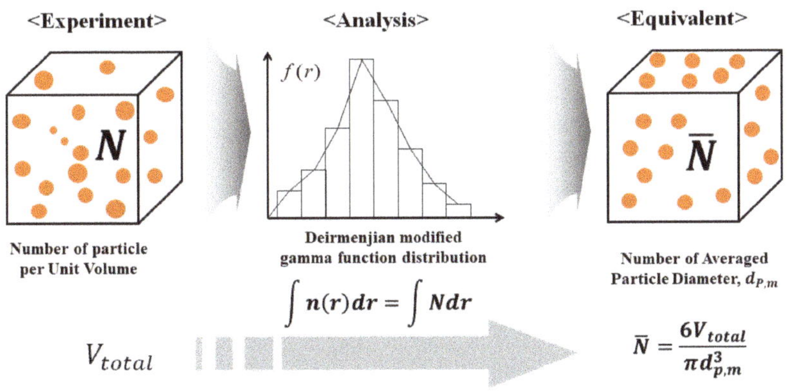

Figure 2. The schematic diagram of the analytic method for distribution of the particles.

Finally, $\dot{E}_{out}$ in Equation (5) can be replaced by:

$$\dot{E}_{out} = \sum_{i=1}^{\infty} n_i q''_{p,i} A_i = \overline{N} \times \overline{q}''_{p,d_m} A_{d_m} \quad (13)$$

where $\overline{q}''_{p,d_m}$ and A_{d_m} denote the heat flux and the surface area of the mean diameter, respectively.

2.4. Considering of Heat Transfer Coefficient

The heat transfer from the spray of spherical particles has the conservation of energy equation as denoted in Equation (14) by the convection for the free stream at the particle surface and the radiation with surrounding,

$$\overline{q}''_{p,d_m} = h(T_{P,s} - T_\infty) + \varepsilon\sigma(T^4_{P,s} - T^4_{sur}) \quad (14)$$

where h is the heat transfer coefficient. Several researchers have proposed the empirical equation of the convective heat transfer coefficient. In this study, Equation (15) is suggested by Ranz and Marshall [28], which is recently applied to investigation of Song et al. [14] and is analyzed with particle diameter and particle surface temperature as the Reynolds number increases. Table 1 denotes the thermal properties for determining the Reynolds number, the Prandtl number and the heat transfer coefficient when the particle diameters are $d_{p,m}$ = 0.001 m and $d_{p,m}$ = 0.003 m, respectively.

$$Nu_D = \frac{hd_{p,m}}{k} = 2 + 0.6\mathrm{Re}_D^{0.5}\mathrm{Pr}^{1/3} \quad (15)$$

where Nu_D, h, Re_D and Pr of Equation (15) denote the overall Nusselt number, the conductive heat transfer coefficient, Reynolds number, and Prandtl number. The properties of density (ρ), thermal conductivity (k), specific heat (C_p) and viscosity (μ) for Reynolds number and Prandtl number are considered at film temperature ($T_{ref} = (T_{p,s}+T_\infty)/2$, T_∞ = 273 K) [14,29].

Figure 3 denotes the previous investigations on the correlation between the critical temperature (for ignition of flammable materials) ($T_{p,c}$) and one particle size ($d_{p,c}$). As shown in the figure, Urban [30] et al. experimentally analyzed the fire spread in power grass blend according to particle size (marked with triangle). Song [14] and Hadden [31] applied the Frank–Kamenetskii equation on polyurethane foam and Powder cellulose to analyze the critical of fire spread, respectively (marked with rectangular and circle). In author's knowledge, the critical temperature of combustible materials may vary depending on the particle distribution of welding spatters. In addition, the ignition boundary could be affected by the particle velocity as well as the particle diameter due to the heat transfer of the particle [14,30,31]. Therefore, in this study, the lowest temperature boundary of J. Song et al. [14] is assumed with the critical temperature for fire spread as denoted in Equation (16), which is derived from curve-fitting, and the effects of particle diameter on the heat transfer are investigated.

$$d_{p,c} = 1.65 \times 10^{32} T_{p,c}^{-11.32} \quad (16)$$

Table 1. Results of the coefficients of the distribution function.

Fixed Value	a	b	α	γ	$d_{p,m}$
	0.0090	0.0030	2.0000	3.0000	6.0571
	0.0418	0.0030	2.0000	4.0000	3.5930
	0.1029	0.0030	2.0000	5.0000	2.6606
α = 2	0.1854	0.0030	2.0000	6.0000	2.1926
	0.2808	0.0030	2.0000	7.0000	1.9173
	0.3821	0.0030	2.0000	8.0000	1.7382
	0.4845	0.0030	2.0000	9.0000	1.6134
	0.0124	0.0030	1.0000	4.0000	3.0214
	0.0042	0.0030	2.0000	4.0000	3.5930
	0.0012	0.0030	3.0000	4.0000	3.9764
γ = 2	0.0003	0.0030	4.0000	4.0000	4.2729
	0.0001	0.0030	5.0000	4.0000	4.5180
	0.0000	0.0030	6.0000	4.0000	4.7287
	0.0000	0.0030	7.0000	4.0000	4.9145

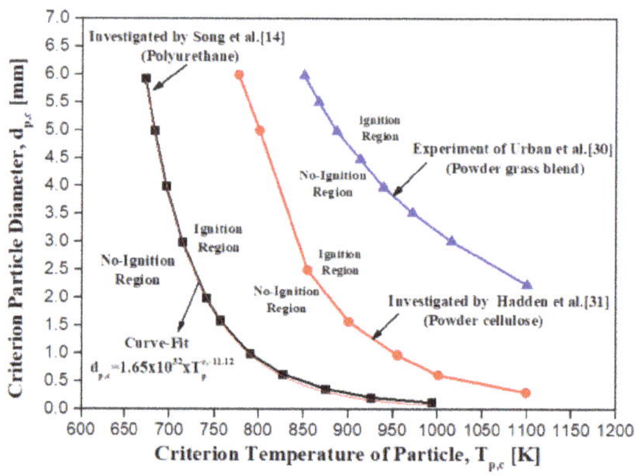

Figure 3. Investigation of the particle diameter and temperature criteria from previous results [14,30,31].

Figure 4 shows results of the overall Nusselt number and the heat transfer coefficient of particles as the Reynolds number increases when T_{ref} = 525 K and 1315 K, $d_{p,m}$ = 1 mm and 3 mm, respectively. As shown in the figure, the overall Nusselt number increases proportionally with the Reynolds number regardless of the particle diameter and temperature. However, for the fixed values of $d_{p,m}$ and the Reynolds number, as the particle temperature increases, the heat transfer coefficient increases due to the increase in viscosity and the decrease in density. In addition, for the fixed values of T_{ref} and the Reynolds number, as the particle diameter decreases, the heat transfer coefficient inversely increases proportionally to the particle diameter.

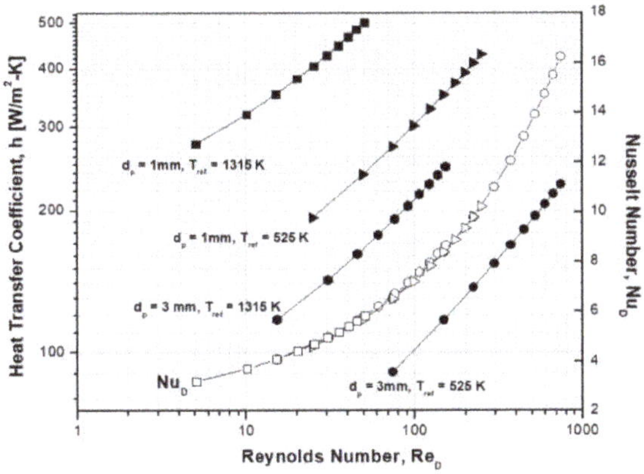

Figure 4. The Nusselt number and heat transfer coefficient with the particle diameter and temperature.

2.5. The Order of Magnitude for Particle and Base Metal

The governing equation of Equation (6) cannot be derived analytically due to the radiation term of particle as denoted in Equation (14). To predict the particle temperature with simple equation, the quantities of heat amount are compared with the method of the order of magnitude for the base metal and particle diameter, respectively [32]. Since the base metal can be welded when the melting temperature reaches about ~2327 K while the particle temperature can ignite the flammable materials more than ~750 K as referred to in [14]. The interest region of temperature can divide into ~750 K for particle and ~2327 K for base metal.

Figure 5 shows the comparative results of the radiative and the convective heat transfer for one particle at $d_{p,m}$ = 3 mm and $u_{p,m}$ = 10 m/s. Thus, the heat transfer coefficient, h can be obtained from the Reynolds number ($Re_D = \rho u_{p,m} d_{p,m} / \mu$) in Equation (15) [14,15]. The radiative term more rapidly increases than the convective term as particle temperature increases; the radiative term is higher than the convective term at 2240 K. The risk of fire spread by welding particles, however, is about $T_{p,s}$ = 750 K at $d_{p,m}$ = 3 mm. The radiative term is only 10% and the convective term is more dominant at $T_{p,s}$ = 750 K. Therefore, the assumption that the convective heat transfer for the particles are mainly an influence on the particle's temperature is a reasonable one. Thus, Equation (14) can be simplified as below,

$$\overline{q}''_{p,d_m} \equiv -h(T_{P,s} - T_\infty) \tag{17}$$

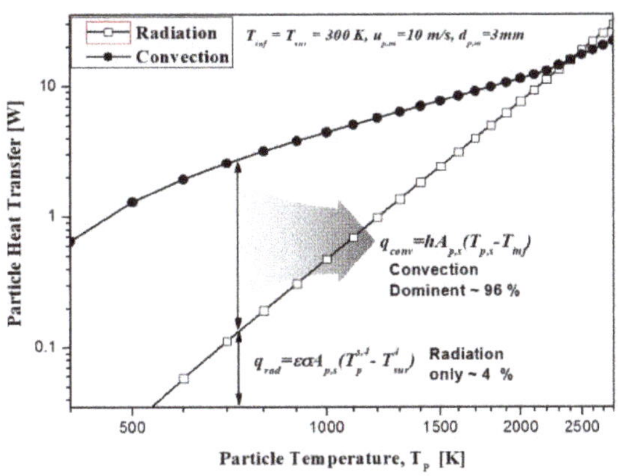

Figure 5. Comparison of the convection and the radiation of the particle.

Figure 6 shows the comparative results of the radiative and the convective heat transfer for the base metal at using ε = 0.3, σ = 5.67 × 10^{-3} W/m^2·K^4 and $d_{p,m}$ = 3 mm. Since the heat transfer coefficient is $h_{b,s}$ = 10 W/m^2-Ks for the plate [23], the convective term is less than the radiative term and more than $T_{p,s}$ = 700 K. Contrary to the heat transfer of one particle, the surface temperature of the base metal is more than 2300 K and the convective term is only 6%. Thus, the radiative term is mainly an influence on the surface temperature of the base metal. Equation (4) can therefore be simplified as follows,

$$\dot{q}_g'' = \frac{\dot{E}_g}{A_{b,s}} \equiv \sigma\varepsilon\left(T_{b,s}^4 - T_{sur}^4\right) \tag{18}$$

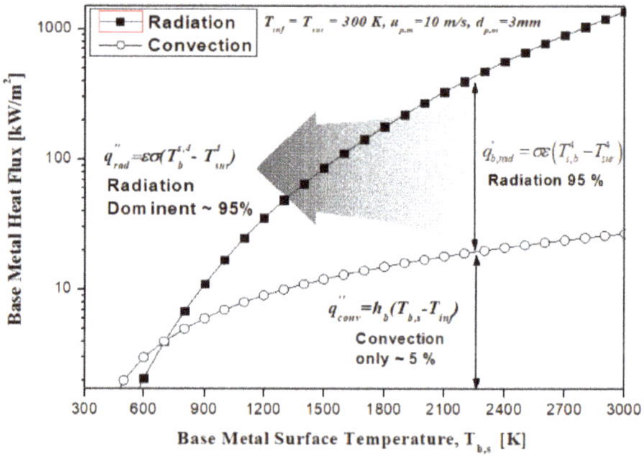

Figure 6. Comparison of the convection and the radiation of the base metal.

2.6. Model Equation for Surface Temperature of the Particles

From the analysis of the heat transfer for the particle and base metal, Equation (6) for the energy balance in the SMAW can be rearranged as Equation (19) using with Equations (3), (17) and (18) for the steady state condition.

$$\eta \times P_e = \sigma\varepsilon A_{b,s}\left(T_{b,s}^4 - T_{sur}^4\right) + \overline{N}hA_{P,s}(T_{P,s} - T_\infty) \tag{19}$$

Equation (19) explains that the electric power supplied during the welding is equal to the sum of the radiant heat from the base metal and the convection heat from the particles. Finally, the predictions of particles can be obtained as Equation (20) for $T_{sur} = T_\infty$ and $\eta = 1$:

$$T_{P,s} = T_\infty + \frac{P_e - \sigma\varepsilon A_{b,s}\left(T_{b,s}^4 - T_{sur}^4\right)}{\overline{N}hA_{P,s}} \tag{20}$$

From the results of the heat transfer with this distribution model of the particles, the overall calculation process can be presented as shown in Figure 7.

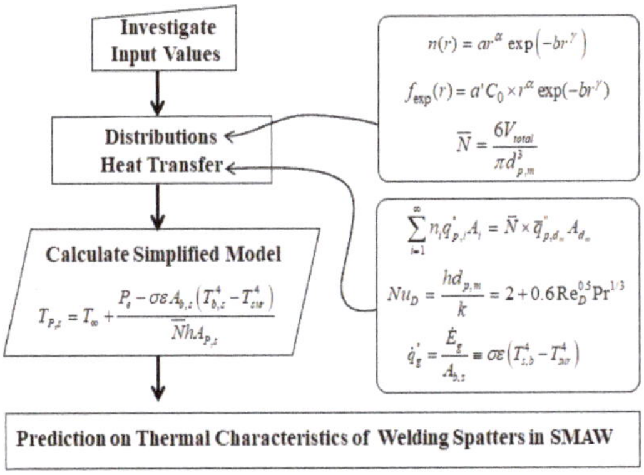

Figure 7. Calculation process of the prediction model.

3. Results

To obtain the particle temperature of Equation (20), the main parameters, such as particle mean diameter ($d_{p,m}$), the number of particle mean diameter ($\overline{N}$), the electric power (VI) and particle velocity (u_p), should be determined. However, only several simulation results have presented these parameters for a specified condition. Therefore, reliability of the predictions can be obtained from the correlations between the unknown terms, investigated by experiments.

3.1. Validation of the Presented Model Equation

Results of the particle temperature are compared to the previous simulation of Urban [5] at 22 kW $\varepsilon = 0.3$, $\sigma = 5.67 \times 10^{-3}$ W/m^2 K^4 and uarc = 0.0025 m/s to validate the accuracy of the prediction model in Equation (20). The surrounding and free stream temperature of 300 K, the particle velocity ($u_{p,m}$) of 17.4 m/s, the base metal area ($A_{b,s}$) of 0.04 m^2 and the total volume (V_{total}) of 2.5×10^{-5} m^3 are assumed. As particle diameter ($d_{p,m}$) increases, the convective heat transfer coefficient of particle (h) increases with root mean square of $d_{p,m}$ by Equation (15) while the number of particle mean diameter ($\overline{N}$) decreases with third square of $d_{p,m}$ by Equation (12). These induce the particle temperature ($T_{p,m}$) to increase and the comparisons between prediction and the previous result are in agreement within ±13% for $d_{p,m} = 1$, 4, and 7 mm as shown in Figure 8. However, to predict more accurate the particle temperature using the model presented in this study, the relationship between the electric power, the speed of the welding electrode and the surface area of the base metal should be determined in the steady state.

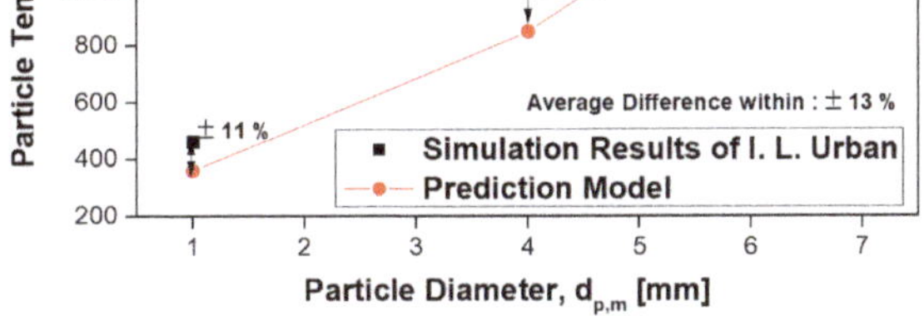

Figure 8. Comparison of the particle temperature of the prediction model with previous results.

3.2. Analysis on the Distribution Characteristics of Welding Spatters

Figure 9 presents that the effects of the normalized distribution $f(r)$ on the change of the distribution coefficient γ from 3 to 9 when fixed value of $\alpha = 2$. The mean diameters of the particles are in the range of 1.6–6.05 mm as γ decreases. The critical of the particle diameter($d_{p,c}$) which has the risk of igniting the flammable materials is assumed to be 3 mm by some studies [14], then the distributions from $\gamma = 3$ to $\gamma = 6$ can cause the fire spread. This result explains that the distribution coefficients are important engineering parameters and the distribution characteristics of the welding spatters significantly influence on the risk of the fire spread.

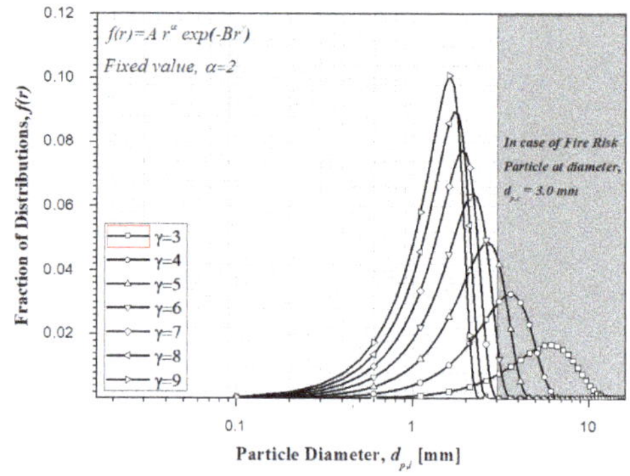

Figure 9. The results of the distribution of the particles for $\alpha = 2$ and $\gamma = 3, 4, 5, 6, 7, 8$, and 9.

Figure 10 denotes that the effects of normalized distribution $f(r)$ on change of the distribution coefficient α from 1 to 7 for $\gamma = 4$. The mean diameters of the particles are in the range of 3–4.9 mm, as denoted in Table 1 with the distribution coefficients. In particular, the distribution configurations are sharply shifted to the right when compared to Figure 9. It means that all the distribution features assumed to be greater than 3 mm in particle size could spread the fire. In this study, the distributions of particles are analyzed with the working conditions such as electrical power, particle velocity, and diameter.

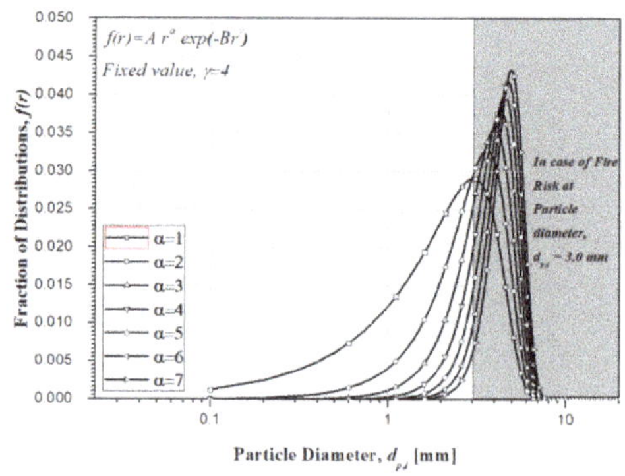

Figure 10. The results of the distribution of the particles for $\gamma = 4$ and $\alpha = 1, 2, 3, 4, 5, 6$, and 7.

3.3. Analysis on the Thermal Characteristics of Welding Spatters

Figure 11 shows that effects of the particle temperature ($T_{p,m}$), the number of particle mean diameter ($\overline{N}$) and the critical of particle temperature ($T_{p,c}$) on the particle mean diameter ($d_{p,m}$) change when reaching an electric power (P_e) of 22 kW, the particles velocities ($u_{p,m}$) of 17.4 m/s, the base metal temperature ($T_{b,s}$) of 2370 K (melting temperature of steel [14]), the base metal area ($A_{b,s}$) of 0.04 m^2 and the total volume of the electrode (V_{total}) of 2.5×10^{-5} m^3 for the 4 mm electrode's diameter. The $d_{p,m}$ is obtained from the droplet distribution coefficients of α and γ in Table 1. The other distribution coefficients a and b can be obtained from Equations (7) and (8). As shown in the table, the particle mean diameter ($d_{p,m}$) increases in accordance with α and γ. It causes the particle temperature ($T_{p,m}$) increases as presented in Figure 11 since the number of particle mean diameter ($\overline{N}$) of Equation (12) more rapidly decreases compared to the increase of the convective heat transfer coefficient (h) of Equation (15). In particular, if $T_{p,c}$, which has the risk of igniting the flammable materials assumed with the previous results using Equation (17) (red line in the figure), the fire spread can exist at $d_{p,m}$ > 3.4 mm. Moreover, the 'no ignition' of $d_{p,m}$ is 1.6–3.4 mm for $\alpha = 2$ and $\gamma = 3$–9 and 3–3.4 mm for $\gamma = 2$ and $\alpha = 1$–7. It means that the distribution characteristics is important to estimate the critical of the ignition by the welding spatters. It is a noticeable result that the 'no ignition' can be affected by the distribution coefficients

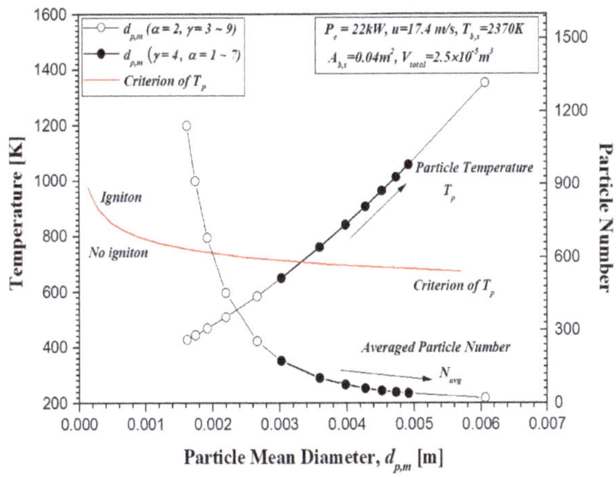

Figure 11. The effects of particle temperature, particle number, and the critical on the particle mean diameter (for ignition of flammable materials).

Figure 12 explains that the effects of the particle temperature ($T_{p,m}$) and the convective heat transfer coefficient of particle (h) on the particle velocity (u_p) change when reaching an electric power (P_e) 22 kW, the base metal temperature ($T_{b,s}$) 2370 K, the base metal area ($A_{b,s}$) 0.04 m^2 and the total volume of the electrode (V_{total}) 2.5×10^{-5} m^3. The particle temperature ($T_{p,m}$) of Equation (20) decreases as the particle velocity ($u_{p,m}$) increases since the convective heat transfer coefficient (h) increases in accordance with the Reynolds number ($Re_D = \rho u_{p,m}\, d_{p,m}/\mu$) of Equation (15) for each particle mean diameter ($d_{p,m}$ = 1 mm, 4 mm, and 7 mm). As shown in the figure, the $T_{p,m}$ is higher temperature as $d_{p,m}$ increases for the fixed value of and the total volume of the electrode (V_{total}) and the electric power (P_e) in order to converge on the energy equilibrium. Moreover, the critical temperature of $d_{p,m}$ = 1 mm, 4 mm, and 7 mm can be approximated with 700 K, 750 K and 800 K from Equation (17). Thus, it is possible to predict that $d_{p,m}$ = 1 mm satisfies 'no ignition' due to $T_{p,m}$ < 700 K, whereas $d_{p,m}$ = 4 mm and 7 mm are 'ignition' due to the minimum values of $T_{p,m}$ > 750 K. However, the maximum particle diameter can be determined by the distribution coefficients α and γ (as explained in Figures 4 and 5). Therefore, the fire risk of the welding spatters should consider the maximum welding particle diameter and particle temperature.

Figure 13 shows that effects of the electrical power (P_e) and the particle mean diameter ($d_{p,m}$) on the particle temperature ($T_{p,m}$) of Equation (20) change when reaching the particle velocity ($u_{p,m}$) of 17.4 m/s. Despite the dependence between $A_{b,s}$ and P_e, no correlation has been suggested. In this study, the linear fit result of $A_{b,s}$ from the comparison of Urban [30] are assumed with $A_{b,s} = 0.0004 \times P_e$. As shown in this figure, the number of particles ($\overline{N}$) decreases as particle mean diameter ($d_{p,m}$) increases since the total volume (V_{total}) of Equation (12) maintains constant. This causes the surface temperature of particles increase since the overall convective heat transfer coefficient of particles decrease. This means that as more particles are particulate, cooling effects by convection heat improves. Moreover, $d_{p,m}$, which satisfies the critical temperature of the particle ($T_{p,c}$), decreases as P_e increases. The critical of $d_{p,m}$ and $\overline{N}$ for preventing fire spread by welding spatters are investigated with $d_{p,m}$ < 13 mm and $\overline{N}$ > 3 at 2.2 kW, $d_{p,m}$ < 5.5 mm and $\overline{N}$ > 25 at 8.8 kW, $d_{p,m}$ < 4.0 mm and $\overline{N}$ > 70 at 15.4 kW, $d_{p,m}$ < 2.8 mm and $\overline{N}$ > 104 at 22 kW, respectively. However, the exact prediction requires the relations between the electrical power and base metal area as well as the particle mean diameter. Therefore, the results of the surface temperature are only for the analytical results of particle distributions.

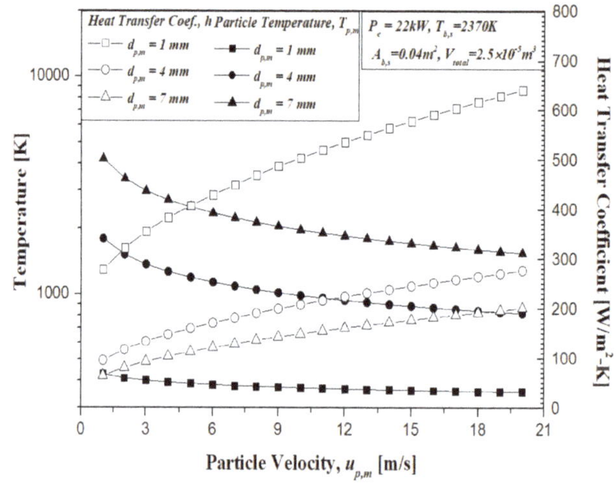

Figure 12. The effects of particle temperature and convective heat transfer coefficient on the particle velocity for $d_{p,m}$ = 1, 4, and 7 mm.

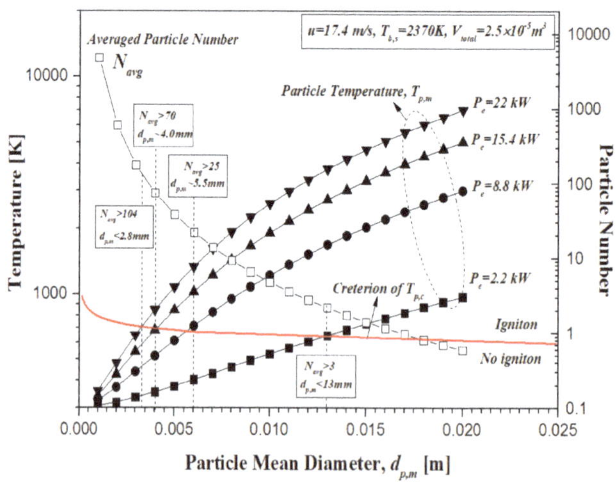

Figure 13. The effects of particle temperature and particle number on the particle mean diameter for Pe = 2.2 kW, 8.8 kW, 15.4 kW and 22 kW.

4. Conclusions

In this study, a theoretical investigation of the welding spatters by the SMAW is performed. To analyze the fire spread by welding spatters, the critical of particle temperature, which have the risk of igniting the flammable materials, is opted from the previous investigation of Song et al. [14]. The results of this study are summarized as:

Heat transfer equation of the welding spatters are derived from the energy conservation equation. The order of magnitude is performed for the base metal and the particles to present a simplified model. The simulation results of I. U. James are compared to the model equation to validate the accuracy. The prediction and the previous results agree within ±13%. The welding conditions for preventing the fire spread by welding spatters are analyzed with the prediction results with the electrode total volume of 2.5×10^{-5} m^3 and the base metal temperature of 2300 K.

The effects of a fire spread on the particles mean diameter in accordance with the distribution coefficients of the welding spatters are investigated with using by a Deirmenjian modified gamma function distribution. The prediction values of the particle temperature linearly increase as the particle mean diameter increases when reaching an electrical power of 22 kW, the particle velocity of 17.4 m/s and the base metal area of 0.04 m^2. The 'no ignition' region of particle mean diameter exists more widely when the distribution coefficients of $\alpha = 2$ and $\gamma = 3$–9 than $\gamma = 2$ and $\alpha = 1$–7. These results validate that not only the particle mean diameter, but the distribution coefficients with the number of particles are significantly influence on the fire spread by welding spatters.

Each particle temperature, which has the particle mean diameter of 1 mm, 4 mm, and 7 mm, is predicted in accordance with the variation of particle velocities when the base metal area is 0.04 m^2 and electrical power is 22 kW. The surface area increases as $d_{p,m}$ and the velocity increases, resulting in an increase in the convection heat transfer coefficient. In particular, $d_{p,m} = 4$ mm and $d_{p,m} = 7$ mm for the particle velocity from 1 m/s to 20 m/s were expected to reach an ignition temperature of more than 750 K.

Finally, the effects of the particle mean diameter ($d_{p,m}$) and electrical power (P_e) on the particle temperature are investigated with the assumption that the base metal temperature were linearly proportional to the size of the electric power. it was predicted that the particle temperature would increase proportionally to the size of the particle mean diameter ($d_{p,m}$) and the electric power and the number of particle mean diameter ($\overline{N}$) decrease. In addition, the critical of particle diameter ($d_{p,c}$) and the number of particle mean diameter ($\overline{N}$) for preventing the fire spread by welding spatters are investigated with $d_{p,m} < 13$ mm and $\overline{N} > 3$ at 2.2 kW, $d_{p,m} < 5.5$ mm and $\overline{N} > 25$ at 8.8 kW, $d_{p,m} < 4.0$ mm and $\overline{N} > 70$ at 15.4 kW, $d_{p,m} < 2.8$ mm and $\overline{N} > 104$ at 22 kW, respectively.

These results confirm that the simplified model presented in this study could predict the fire spread of the welding spatters depending on the distribution of particles, the particle mean diameter, the number of particles, the particle velocity, and the electrical power.

Author Contributions: Conceptualization, Y.J.S. and W.J.Y.; methodology, W.J.Y; software, Y.J.S.; validation, Y.J.S. and W.J.Y.; formal analysis, W.J.Y.; investigation, Y.J.S. resources, W.J.Y; data curation, W.J.Y.; writing—original draft preparation, W.J.Y.; writing—review and editing, W.J.Y.; visualization, Y.J.S.; supervision, Y.J.S.; project administration, W.J.Y.; funding acquisition, W.J.Y.; All authors have read and agreed to the published version of the manuscript.

Funding: This research was funded by Korea Agency for Infrastructure Technology Advancement, grant number (19CTAP-C151893-01).

Acknowledgments: This research was supported by Korea Agency for Infrastructure Technology Advancement, grant number (19CTAP-C151893-01) and we would like to thank Editage (www.editage.co.kr) for English language editing.

Conflicts of Interest: The authors declare no conflict of interest. The funders had no role in the design of the study; in the collection, analyses, or interpretation of data; in the writing of the manuscript, or in the decision to publish the results.

References

1. Hojjati, A.Y.M.; Immarigeon, J. Fusion Bonding/Welding of Thermoplastic Composites. *J. Thermoplast. Compos. Mater.* **2004**, *43*, 161–165.
2. Hashemabad, S.G.; Gu, Z.; Ando, T. Flux-less direct soldering of aluminum by ultrasonic surface activation. *J. Mater. Process. Technol.* **2016**, *233*, 135–141. [CrossRef]
3. Liu, Y.; Liang, C.; Wei, A.; Jiang, Y.; Tian, Q.; Wu, Y. Solder-free electrical Joule welding of macroscopic graphene assemblies. *J. Mater. Today Nano* **2018**, *3*, 1–8. [CrossRef]
4. Narasimhan, P.N.; Mehrotra, S.; Raja, A.R.; Vashista, M.; Yusufzai, M.Z.K. Development of hybrid welding processes incorporating GMAW and SMAW. *ICMPC* **2019**, *18*, 2924–2932. [CrossRef]
5. Urban, J.L. Spot Ignition of Natural Fuels by Hot Metal Particles. Ph.D. Thesis, University of California Berkeley, Berkeley, CA, USA, 2017.
6. Janusas, G.; Jutas, A.; Palevicius, A.; Zizys, D.; Barila, A. Static and vibrational analysis of the GMAW and SMAW joints quality. *J. Vibroeng.* **2012**, *14*, 1220–1226.
7. Hu, J.; Tsai, H.L. Metal Transfer and Arc Plasma in Gas Metal Arc Welding. *J. Heat Transf.* **2006**, *129*, 1025–1035. [CrossRef]
8. Hu, J.; Tsai, H.L. Heat and mass transfer in gas metal arc welding. Part I: The arc. *J. Heat Mass Transf.* **2007**, *50*, 833–846. [CrossRef]
9. Messler, R.W. *Principles of Welding: Processes, Physics, Chemistry, and Metallurgy*; Wiley: Hoboken, NJ, USA, 2008.
10. DuPont, J.N.; Lippold, J.C.; Kiser, S.D. *Welding Metallurgy and Weldability of Nickel-Base Alloys*; Wiley: Hoboken, NJ, USA, 2009.
11. Weman, K. *Welding Processes Handbook*; Cambridge Woodhead Publishing: Cambridge, UK, 2003.
12. Shigeta, M.; Peansukmanee, S.; Kunawong, N. Qualitative and quantitative analyses of arc characteristics in SMAW. *Weld. World* **2016**, *60*, 355–361. [CrossRef]
13. Mikkelsen, K. An Experimental Investigation of Ignition Propensity of Hot Work Processes in the Nuclear Industry. Master's Thesis, University of Waterloo, Waterloo, ON, Canada, 2014.
14. Song, J.; Wang, S.; Chen, H. Safety distance for preventing hot particle ignition of building insulation materials. *Theor. Appl. Mech. Lett.* **2014**, *4*, 034005. [CrossRef]
15. Hagimoto, N.; Kinoshita, K. Scattering and Igniting Properties of Sparks Generated in an Arc Welding. In Proceedings of the 6th Indo Pacific Congress on Legal Medicine and Forensic Sciences, Kobe, Japan, 26–30 July 1998.
16. NFPA. *NFPA Standard 51B: Standard for Fire Prevention During Welding, Cutting, and Other Hot Work*; Technical Report; National Fire Protection Association: Quincy, MA, USA, 2009.
17. Kim, Y.S.; Eagar, T.W. Analysis of metal Transfer in Gas Metal Arc Welding. *Weld. J.* **1991**, *72*, 269–278.
18. Lin, Q.; Li, X.; Simpson, S.W. Metal transfer measurements in gas metal arc welding. *J. Phys. D Appl. Phys.* **2001**, *34*, 347. [CrossRef]
19. Ahrens, M. *Brush, Grass, and Forest Fires*; Technical Report November; NFPA: Quincy, MA, USA, 2013.
20. Wakelin, H. Ignition Thresholds for Grassland Fuels and Implications for Activity Controls on Public Conservation Land in Canterbury. Ph.D. Thesis, University of Canterbury, Canterbury, UK, 2010.
21. Choi, S.K.; Yoo, C.D.; Kim, Y.S. Dynamic Simulation of Metal Transfer in GMAW, Part 1: Globular and Spray Transfer Modes. *Weld. J.* **1998**, *34*, 347.
22. Wang, F.; Hou, W.K.; Hu, S.J.; Asibu, E.K.; Schultz, W.W.; Wang, P.C. Modelling and analysis of metal transfer in gas metal arc welding. *J. Phys. D Appl. Phys.* **2003**, *36*, 1143. [CrossRef]
23. Ingason, H.; Wickstrom, U. Measuring incident radiant heat flux using the plate thermometer. *Fire Saf. J.* **2006**, *42*, 161–166. [CrossRef]
24. Viskanta, R. *Radiative Transfer in Combustion Systems: Fundamentals and Applications*; Begell House: New York, NY, USA, 2005.
25. Deirmendjian, D. *Electromagnetic Scattering on Spherical Polydispersions*; Elsevier: New York, NY, USA, 1969.
26. You, W.J.; Ryou, H.S. Development and application of a simplified radiative transport equation in water curtain systems. *Fire Saf. J.* **2018**, *96*, 124–133. [CrossRef]
27. Sauter, J. *Die Grössenbestimmung der in Gemischnebeln von Verbrennungskraftmaschinen vorhandenen Brennstoffteilchen. VDI-Forschungsheft Nr*; VDI-Verl: Berlin, Germany, 1926.

28. Ranz, W.E.; Marshall, W.R. Evaporation from drop, Part 1. *Chem. Eng. Prog.* **1952**, *48*, 141–146.
29. Incropera, F.P.; Lavine, A.S.; Bergman, T.L.; DeWitt, D.P. *Principles of Heat and Mass Transfer*; Wiley: Hoboken, NJ, USA, 2013.
30. Urban, J.L.; Zak, C.D.; Song, J. Smoldering spot ignition of natural fuels by a hot metal particle. *Proc. Combust. Inst.* **2017**, *36*, 3211–3218. [CrossRef]
31. Hadden, R.M.; Scott, S. Ignition of Combustible Fuel Beds by Hot Particles: An Experimental and Theoretical Study. *Fire Technol.* **2011**, *47*, 341–355. [CrossRef]
32. Bejan, A. *Adrian. Convection Heat Transfer*; John wiley & sons: Hoboken, NJ, USA, 2013.

energies

MDPI

Article

A Numerical Analysis of the Fire Characteristics after Sprinkler Activation in the Compartment Fire

Ho Trong Khoat [1], Ji Tea Kim [1], Tran Dang Quoc [2], Ji Hyun Kwark [3] and Hong Sun Ryou [1,*]

1 School of Mechanical Engineering, Chung-Ang University, Seoul 06974, Korea; hotrongkhoat93@cau.ac.kr (H.T.K.); sdd322@naver.com (J.T.K.)

2 School of Transportation Engineering, Hanoi University of Science and Technology, Hanoi 100000, Vietnam; trandangquoc@gmail.com

3 Fire Insurers Laboratories of Korea, Gyeonggi-do 469-881, Korea; kwark@kfpa.or.kr

* Correspondence: cfdmec@cau.ac.kr; Tel.: +82-10-3269-5280

Received: 27 April 2020; Accepted: 12 June 2020; Published: 15 June 2020

Abstract: Understanding fire characteristics under sprinkler spray is valuable for performance-based safety design. However, fire characteristics during fire suppression by sprinkler spray has seldom been studied in detail. In order to present a fire suppression model by sprinkler spray and determine the fire characteristics after sprinkler activation in a compartment, a numerical analysis was conducted using a fire dynamics simulator (FDS). A simple fire suppression model by sprinkler spray was calibrated by comparing ceiling temperatures from experimental data. An extinguishing coefficient of 3.0 was shown to be suitable for the fire suppression model. The effect of sprinkler spray on the smoke layer during fire suppression was explained, revealing a smoke logging phenomenon. In addition, the smoke, which spread under the influence of the sprinkler spray, was also investigated. The temperature, velocity, and mass flow rate of the smoke layer through the doorway was significantly reduced during fire suppression compared to a free burn case.

Keywords: sprinkler; fire dynamics simulator (FDS); fire suppression; extinguishing coefficient; smoke logging; smoke spread

1. Introduction

In Korean regulations, automatic sprinkler systems are required in buildings and factories to control a fire. Applications of sprinkler systems in the building are to suppress the fire and prevent the fire spread from one compartment to another. As a result, it can restrict the untenable conditions caused by smoke hazards and extend the available safe egress time when a fire occurs in the building. Extensive research has been conducted to improve the design and efficiency of sprinkler systems for many years, especially using computational fluid dynamics (CFD) modelling, which can simulate the interaction of water with fire environments and has been widely used and become a powerful tool with many advantages.

Computational fluid dynamics can provide details on an analysis of the interaction between the fire environment and water sprays. The numerical analysis can be completed with a less expensive cost and consumed time and can be used in combination with experimental work to improve technical design for water spray systems. Various numerical studies on the interaction between water spray and fire plume have been reported. The primary concerns are related to extinguishing times and the minimum water flow rate to extinguish the fire. Novozhilow et al. [1] developed a CFD model of burning rate and extinction in fires by water sprinklers. The extinguishing rates were modeled by calculating the heat transfer and temperature at the solid surface when the sprinklers were in operation. The extinction time was compared to experimental data and found to be in satisfactory agreement. They also developed a CFD model to predict extinguishment times of an array of wood slats by sprinkler water. The wood pyrolysis rate modeled by Arrhenius's reaction was fully coupled with the gas phase model, and the spray was treated in Lagrangian fashion. The extinguishment times were reasonably close to those measured from experiments [2]. Nam [3] investigated the trajectory capacity of water spray from fast response sprinklers into heptane fire via numerical simulations. The simulation was a combination of the water spray model and the free burn fire model. The fire was modeled to avoid complicated chemical combustion by defining prescribed fire source. The results from the simulations compared reasonably well with experimental measurements in terms of actual delivered densities. Hua et al. [4] introduced a numerical approach to investigate the effect of water spray characteristics on a fire plume. The chemical combustion used the Arrhenius reaction to present fire extinction by water. In the results, a solid cone pattern and a finer water droplet size of the water spray was more effective in extinguishing fires. The flow rate of the water spray had a certain critical value for suppressing a specific fire.

Other studies on the effects of water spray on fire environments without considering fire suppression have also been conducted. O'grady and Novozhilov [5] used the Fire Dynamic Simulator (FDS) version 4 software to present the effect of the water spray of a sprinkler on the ceiling jet. The numerical results of two different flow rates were validated to free burn fire tests including a 1.5 MW steady-state fire. The relative errors were 7–8% for velocity and 10–15% for the temperature [5]. Cunfeng Zhang and Wanki Chow [6] used the FDS code to investigate the interaction of the sprinkler spray and the smoke layer. The cooling effect of the spray and drag force of water particles was analyzed. As a result, the temperature of the smoke layer decrease was almost linear to the working pressure of the sprinkler systems. Chen et al. [7] discussed the cooling effect of water spray systems in experimental and numerical analyses. Although the flame continued after activating sprinkler due to the lack of the suppression model, the cooling effects by the water spray in a room test within 120 s were obvious. Zhi Tang et al. [8] presented the effect of water spray on the fire smoke layer inside a hood by FDS simulation. The influence of the water spray characteristics on the downward smoke displacement due to the fact of drag and cooling was analyzed in detail. Other applications of sprinkler modeling were also used in the suppression of rack storage [9] and conveyor belt fires [10].

When the water droplets encounter to a fire source, the water is not only cooling the burning surface and surrounding air, but it also impacts the chemical reaction rate of solid fuel. Simultaneously, water droplets also significantly affect the smoke movement by its trajectory. Meanwhile, the complex Arrhenius reaction model is usually used to present chemical combustion. Therefore, when adding fire extinction by a water spray, the numerical simulation is more complicated. Hence, it is necessary for a study that includes the phenomena above to better understand fire characteristics under water spray with a simple fire suppression model. In this present study, fire characteristics after sprinkler activation in a compartment were investigated by numerical analysis in FDS 6.7.0. The mathematical equation of the fire suppression model was introduced in FDS. Experimental work was conducted to collect data related to the heat release rate and ceiling temperature. Based on the experimental data, the numerical simulations proceeded to calibrate the fire suppression model by comparison to the ceiling temperature. Then, the calibrated model was used to investigate the characteristics of the smoke movement during fire suppression. In addition, the smoke spread was analyzed in terms of temperature, velocity, mass flow, and was compared to a case of free burn.

2. Fire Dynamics Simulator and Fire Suppression Model by Water

Fire Dynamics Simulator v. 6.7.0 (FDS) [11,12] was used to simulate the fire suppression in this study. The FDS is a computational fluid dynamics (CFD) model of fire-driven fluid flow. A form of the Navier–Stokes equation was applied for low-speed, thermally-driven flow such as smoke and heat transport from fires. The partial derivatives of the conservation equations of mass, momentum, and energy were approximated as finite differences, and the solution was updated in time on a three-dimensional, rectilinear grid. The combustion model was based on the mixing-limited, infinitely fast reaction of lumped species, called eddy dissipation concept (EDC) model. Thermal radiation was computed using a finite volume technique on the same grid as the flow solver. Water droplets can absorb and scatter thermal radiation. For calculation of absorption coefficients, a gray band model, RAD-CAL was implemented. Lagrangian particles were used to represent the liquid droplets. The mass, momentum, and energy transfer from the Lagrangian particles to gas flow were treated sufficiently by source term in the conservation equation. The model in FDS was successfully tested and validated for a variety of fire problems, natural convective flows, smoke movement [13,14], and water-based fire suppression [15].

The fire suppression by a water spray involving multiple physical phenomena adds complications to CFD. The fire suppression by water was first introduced by Yu et al. [16] and applied to the FDS simulation by Hamins and McGrattan [12] as a fire suppression model. When the water droplets drop into burning surfaces, the water is cooling the surface and the surrounding gas, and it is also changing the pyrolysis rate of the solid fuel. A liquid droplet hits a solid horizontal surface in FDS, it is assigned a random horizontal direction and moves at a fixed velocity until it reaches the edge, at which point it drops straight down at the same fixed velocity. While attached to the burning surface, the water droplet forms a thin film that transfers heat to the solid, and heat and mass to the gas by evaporating. In the FDS model, the cooling of unburned surfaces and the reduction in the Heat Release Rate (HRR) are computed locally. However, the exponential nature of fire suppression by water is observed both locally and globally; thus, the local heat release rate per unit area (kW/m^2) can be expressed in the following equation:

$$\dot{q}'(t) = \dot{q}_0{}'(t)e^{-\int k(t)dt} \tag{1}$$

where $\dot{q}''{}_0(t)$ is the user-specified heat release rate per unit area of fuel (kW/m^2) when no water droplets are applied, and $k(t)$ is a linear function of the local water mass per unit area, $m''{}_w(t)$, in units of kg/m^2, which is expressed as:

$$k(t) = \alpha m''_w(t) \qquad 1/s \tag{2}$$

where α is the extinguishing coefficient and is obtained experimentally in units of m^2/(kg·s). The extinguishing coefficient is dependent on the material properties of the solid fuel and its geometrical configuration. In order to obtain the optimal α value, simulations were performed for calibration by comparison of the ceiling temperature. The detail will be discussed in the next section.

3. Experiment Setup

In order to obtain data for calibrating the suppression model, a fire experiment was conducted in a compartment using a water sprinkler. The measurement used to collect ceiling temperature in a compartment. The compartment had dimensions of 3 m (W) × 7 m (L) × 3 m (H) coved by gypsum board in the ceiling, plywood walls, and a concrete floor. Two doorways on opposite walls provided for ventilation. The left doorway was 0.89 m (width) × 2.8 m (height), and the right doorway was 1.04 m (width) × 2.8 m (height) (Figure 1). A pendant water sprinkler was installed 76 mm below the ceiling. The manufacturer supplied the glass bulb sprinkler with an activation temperature of 68 °C and a K-factor of K = 50 Lpm/bar$^{0.5}$. The operation pressure was 1 bar. The spray angle of 10 to 80 degrees was estimated by image analysis. Two K-type thermocouples (range: −200 to +1000 °C, accuracy is ±1 °C) were installed 76 mm from the ceiling to measure the temperature. The ceiling temperature data were recorded by a portable data logger, a midi LOGGER GL240, every second (the measurement accuracy: 0.05% for the K-type thermocouple).

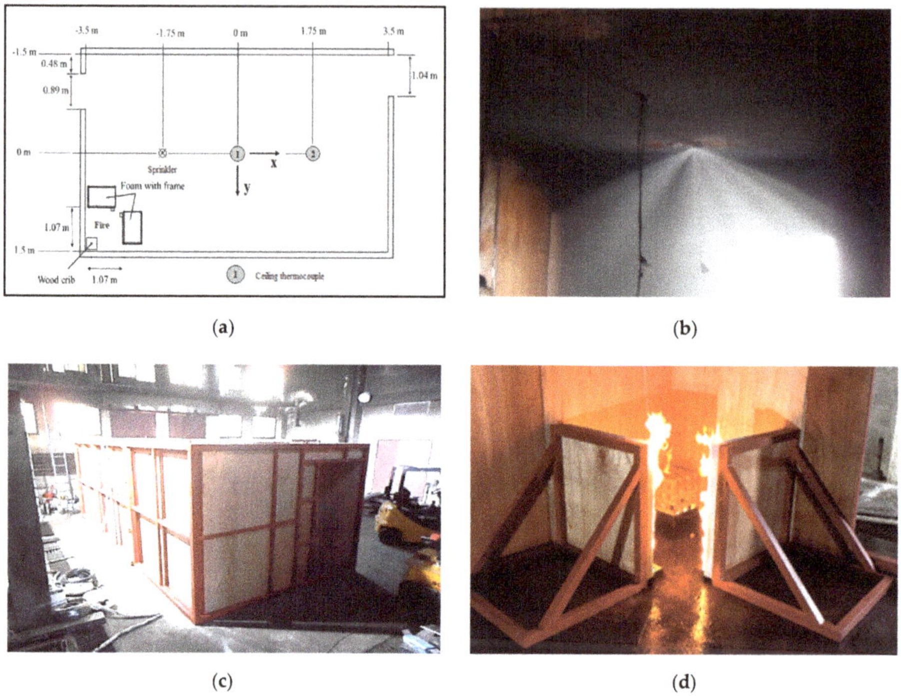

Figure 1. Detailed drawings of the compartment in the experiment: (**a**) test compartment layout; (**b**) sprinkler spray; (**c**) test compartment; and (**d**) fire source.

The fire source consisted of a wood crib and polyether foam, described in detail in ISO 6182-10 [17]. The wood crib, with dimensions of 305 mm × 305 mm × 152 mm, was made with fir lumber of 3 kg and ignited by a pan of heptane. The pure polyether foam of 810 mm × 760 mm × 76 mm was ignited by cotton wicks soaked in heptane. The polyether foam glued to a thick plywood backing and assembled to a steel frame for support. The fire source was placed in the corner of the compartment. The wood crib was positioned 5 mm from each wall (Figure 1a,d).

The heat release rate of the fire source could not be determined directly during the compartment fire experiment. Therefore, the HRR measurement was conducted separately without the effect of sprinkler spray. Figure 2 shows the wood crib and polyether foam were burnt individually under a large-scale calorimeter (LSC), applicable up to 3 MW (ISO 13784-1), to determine the unsteady heat release rate. The principle of LSC is that it is based on the amount of heat released from a burning sample and is proportional to the amount of oxygen consumed during the combustion. The dimensions and weight of the wood crib and polyether foam were the same in the compartment test.

(**a**) (**b**)

Figure 2. Independent measurement of heat release rate for (**a**) the wood crib and (**b**) polyether foam.

The experimental results of HRR in the free burn and ceiling temperature in the compartment fire equipped with a sprinkler are shown in Figure 3.

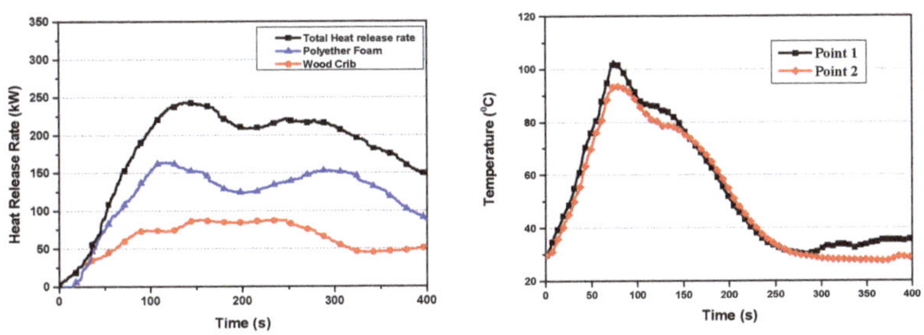

(**a**) (**b**)

Figure 3. Experiment results: (**a**) heat release rate in the free burn and (**b**) ceiling temperature in the compartment with sprinkler activation.

4. Numerical Detail

Figure 4 shows the computational domain coved the compartment test with dimensions of 3.4 m (W) × 11.4 m (L) × 4.4 m (H). The compartment geometry was made based on experiments. The wood crib and polyether foam modeled simple obstructions of the same dimension in the experiment. The fire source presented as burner with a specified heat release rate per unit area (HRRPUA) in units of kW/m^2. The HRR assigned to the top surface of the wood crib and interior surface of polyether foam (red surface, Figure 2). Sprinkler and ceiling thermocouples were installed in the same location in the experiment. The boundary conditions assigned "OPEN" for the computational domain, and "ADIABATIC" to the wall and floor. The ambient temperature was 29 °C, in agreement with the experiments. A Cartesian coordinate system indicated at the center of the compartment for convenience in the analysis.

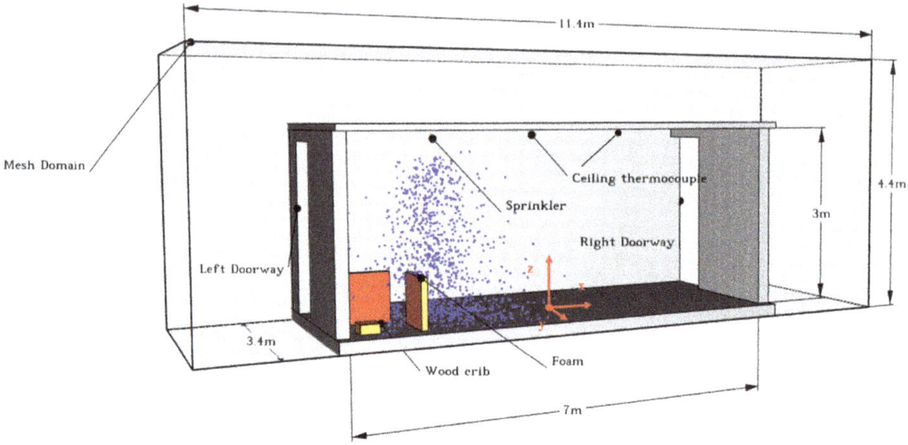

Figure 4. 3D schematic view of the compartment modeling.

In order to model a fire suppression correctly, numerical sprinkler spray parameters need to be specified in detail. Those parameters include the water flow rate, droplet diameter, initial droplet velocity, spray angle, and offset distance. Because of the limit of apparatus, there was a lack of data on sprinkler spray in the experimental report. Therefore, the necessary estimations have been made based on the experimental work and various studies reported in the literature [18,19].

The water flow rate of any sprinkler can be calculated based on the K-factor and pipe pressure by the equation:

$$\dot{m} = K\sqrt{p} \tag{3}$$

where $\dot{m}$ is the water flow rate in L/min, K is the K-factor for the sprinkler L/(min.bar$^{0.5}$) which can be obtained from the manufacturer and p is the pressure of the pipe in Bar.

Water spray usually includes various sizes of spherical droplets. The cumulative volume fraction (CVF) is specifies the size distribution of water droplets. This function indicates the fraction of the total mass carried by droplets less than the given diameter. The CVF for a sprinkler spray is represented by a combination of log-normal and Rosin–Rammer distributions [12]:

$$F(D) = \begin{cases} \frac{1}{\sqrt{2\pi}} \int_0^D \frac{1}{\sigma D'} \exp\left(-\frac{\left[ln\left(\frac{D'}{D_{v,0.5}} \right) \right]^2}{2\sigma^2} \right) dD' & (D \le D_{v,0.5}) \\ 1 - \exp\left[-0.693 \left(\frac{D}{D_{v,0.5}} \right)^{\gamma} \right] & (D > D_{v,0.5}) \end{cases} \tag{4}$$

where $D_{v,0.5}$ is the median volumetric droplet diameter, µm (i.e., half the mass is carried by droplets with diameters of $D_{v,0.5}$ or less), γ and σ are empirical constants equal to approximately 2.4 and 0.48, respectively [12]. The median droplet diameter, $D_{v,0.5}$, is estimated using the formula reported by Yu [19]:

$$\frac{D_{v,0.5}}{D} = C_{sp} W_e^{-1/3} \tag{5}$$

$$W_e = \frac{\rho_d u_d^2 D}{\sigma_d} \tag{6}$$

where D is the orifice diameter of the sprinkler. ρ_d is the density of water in kg/m^3, u_d is the initial droplet velocity in m/s, and σ_d is the water surface tension in N/m. Analysis of Sheppard's [18] data provided an average value of C_{sp} approximately 1.53 for sprinklers test.

Sheppard [18] used particle image velocimetry (PIV) and phase Doppler interferometry (PDI) to measure the initial droplet sizes and velocities for the various type of sprinklers. The initial droplet velocity had an average measurement at 0.6 of $\sqrt{p/\rho_d}$. The PIV measurements were taken from the spray atomization length of 0.2 m from the sprinkler orifice, the distance at which no more droplet breaks up downstream from this point.

The angles of the spray were estimated from experimental work providing a 10–80° spray angle. The activation temperature was 68 °C. All sprinkler parameters are summarized in Table 1.

Table 1. Sprinkler parameters and estimated sprinkler spray parameters.

Sprinkler Parameter	Value
Flow rate	50 L/min
Velocity	6.01 m/s
Droplet size	954 μm
Atomization distance	0.2 m
Angles	10–80°
Activation temperature	68 °C

Grid size is important to verify in numerical simulation, because it directly influences results. In theory, a very small grid can give the accuracy of the calculation, but the time may be wasted unnecessarily. When the deviation between the neighboring results is small enough, the grid can be considered as an independent grid. Before applying the sprinkler simulation, the independent grid tests for the free burn (without sprinkler) were conducted. Figure 5 shows the temperature at point 1 in the compartment with four different grids (15 cm, 10 cm, 7.5 cm, 5 cm). The result of 15 cm had larger deviations than the results of 10 cm, 7.5 cm, and 5 cm. Therefore, the grid size of 10 cm was chosen in this paper. The grid selection was the same in the simulation of sprinkler interaction with a fire ceiling jet by O'Grady and Novozhilov [5].

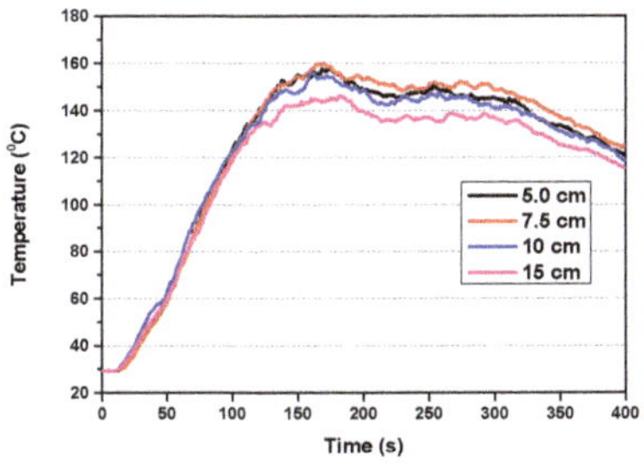

Figure 5. Ceiling temperature at point 1 with different grid size without sprinkler activation.

5. Result and Discussion

5.1. Optimal Extinguishing Coefficient

The extinguishing coefficient is not determined by experimental correlation, as mention by Hamins and McGrattan [12]. Instead, the suppression model is calibrated to find the optimal extinguishing coefficient (α value) in this study. When sprinkler activation, the water of sprinkler strikes to fire source and make a reduction on the burning rate. The HRR reduction with a certain water distribution depends on the α value in Equations (1) and (2). In order to define the α value, the comparison of ceiling temperature is conducted.

Figure 6 shows the ceiling temperature over time measured in simulation with various α values and experimental data at points 1 and 2. As α value varies between 0.5 $m^2/(kg{\cdot}s)$ and 4.0 $m^2/(kg.s)$, the ceiling temperature of numerical simulation decreases with an increasing extinguishing coefficient.

Comparing ceiling temperature, the value of the extinguishing coefficient $\alpha = 3.0\ m^2/(kg{\cdot}s)$ provides the best fit temperature between numerical and experimental data at both points. These comparisons demonstrate that the fire suppression model in FDS can capture the features of fire characteristics under the effect of sprinkler spray.

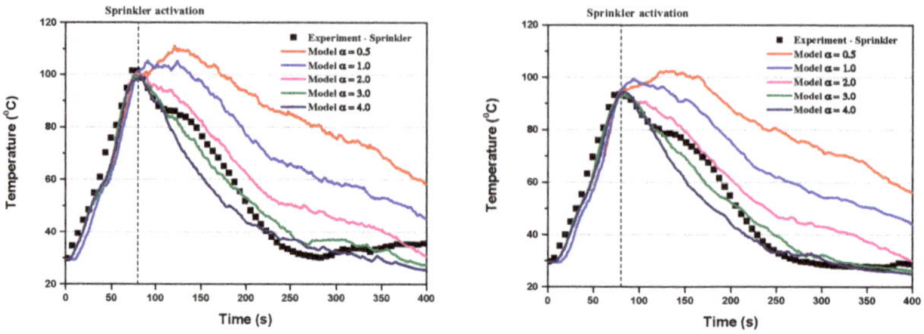

(**a**) (**b**)

Figure 6. Variation of ceiling temperature with different extinguishing coefficients: (**a**) Point 1 and (**b**) Point 2.

The effect of extinguishing coefficients on the HRR after activating sprinkler is shown in Figure 7. The HRR falls more instantly when sprinkler activates. The higher the alpha value, the greater the HRR decreases. As optimal $\alpha = 3.0$, it was observed that HRR stops growing to 165 kW at 70 s and then reduces rapidly to 10 kW at 400 s. The suppression model was calibrated in the prediction of the HRR affected by a water spray. Therefore, this model can be used to investigate the fire characteristics in the compartment.

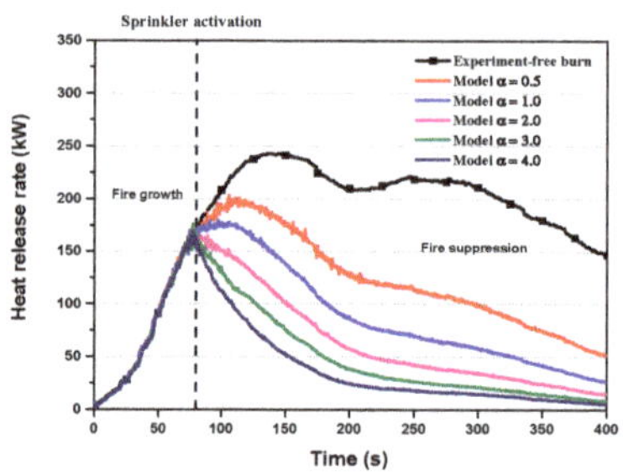

Figure 7. Variation of heat release rate for different extinguishing coefficients.

5.2. Temperature and Flow Field

Analysis of the temperature distribution and velocity field was necessary to understand the interaction between sprinkler spray and the smoke layer during fire suppression. The numerical view of the temperature and velocity field of the smoke layer during fire suppression is shown in Figure 8.

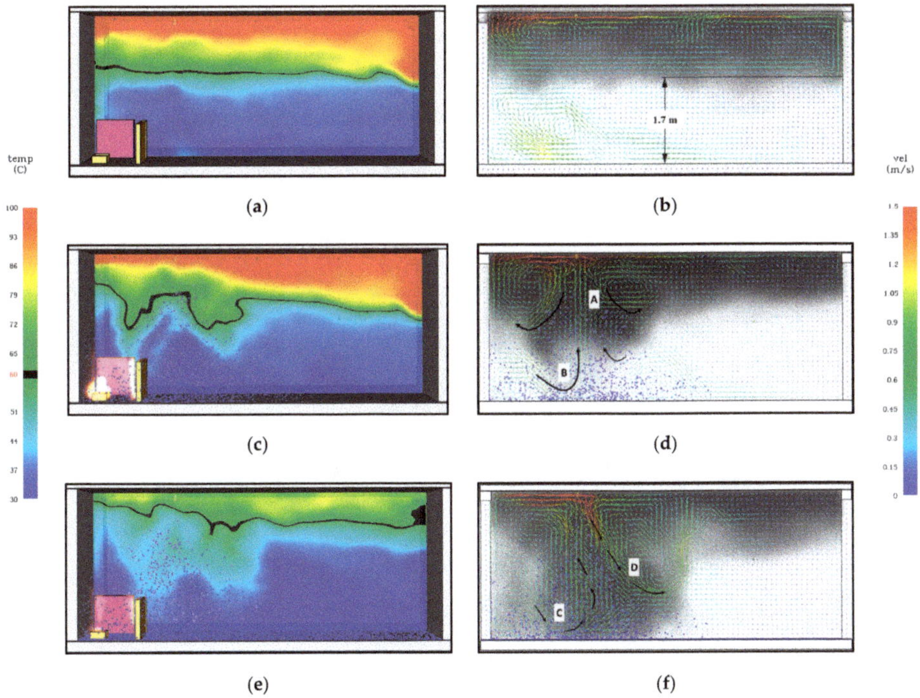

Figure 8. Temperature distribution and vector field of the smoke layer in vertical plane y = 0 m: (**a**) temperature distribution before sprinkler activation, 70 s; (**b**) velocity field before sprinkler activation, 70 s; (**c**) temperature distribution after sprinkler activation, 100 s; (**d**) velocity field after sprinkler activation, 100 s; (**e**) temperature distribution at 150 s; (**f**) velocity field at 150 s.

Figure 8a,c,e presents the temperature distribution with different times in the vertical plane y = 0 m. A smoke layer was formed under the ceiling and flowed out across the doors. Temperature contours showed the higher temperature distributed gradually from the bottom to the top and from left to right in the smoke layers (Figure 8a). There were great differences in temperature distribution when the sprinkler activated. Due to the reduction of HRR and the cooling effect on smoke, the temperature would decrease apparently over time. The hot layer rapidly reduced both the temperature value and the area as shown in Figure 8c,e.

The smoke movement and velocity vector with different times in the vertical plane y = 0 is shown in Figure 8b,d,f. There was a clear smoke layer at z = 1.7 m above the floor. Immediately, when the sprinkler was activated, the impact of the water spray on the smoke movement was observed. The smoke layer surrounding the sprinkler was entrained into the spray and flowed downward with the water spray due to the drag force (A). Smoke logging happened [20]. At that moment, fresh air entered into the inner water spray due to the buoyance force (B). The cone shape of the smoke logging was formed with symmetry right below the sprinkler location. The lower temperature due to the cooling and replacement of fresh air inside the smoke cone is clearly shown in Figure 8c.

After that, under the effect of the sprinkler spray, smoke logging combined with fresh air-entraining from the left doorway and swirled back inside the water spray (C). Simultaneously, it pushed the other side of the smoke logging to the right region, further away from the center sprinkler spray (D). The symmetrical cone of the smoke logging was broken; local smoke vortexes were formed. Smoke logging occurred more strongly, and it widely spreads towards the right doorway (Figure 8f).

5.3. Effect of Sprinkler Spray on Smoke Spread

In the event of a building fire, the smoke spread directly affects tenability criteria in the means of egress. The available safe egress time depends on how the smoke spreads from compartment to compartment more than considering the fire itself in many cases. Therefore, the ability to predict the impact of a sprinkler on the smoke spread from the compartment would be a valuable engineering tool for use with performance-based safety design. The smoke temperature, velocity, and mass flow out of the doorway are important factors in considering smoke spread during fire suppression.

The smoke movement has different behavior in different positions depending on the sprinkler spray region as well as the structure of the compartment. In order to investigate smoke spreading in detail, the numerical devices were measured inside the water spray region and the doorway, Figure 9. The numerical results of sprinkler case were compared to the case without the sprinkler, i.e., free burn.

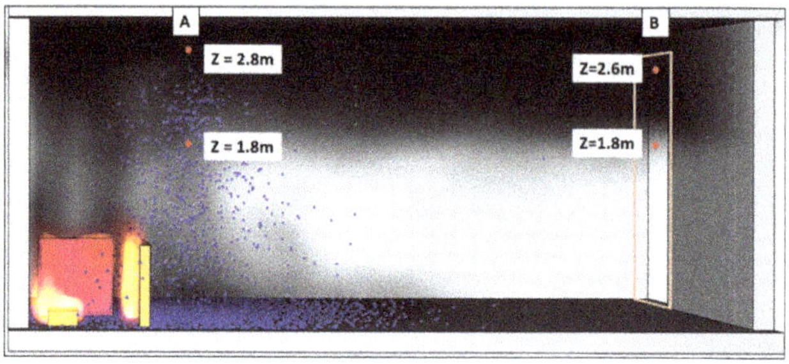

Figure 9. The location of numerical devices to investigate smoke spread in the compartment.

Figure 10 shows the change in temperature over time with different locations: inside spray with the height of 2.8 m, 1.8 m (location A), and a centerline of the right doorway (location B) with the height of 2.6 m and 1.8 m in two cases of sprinkler and free burn. In the sprinkler case, the temperature of the smoke layer varied from 50 °C to 110 °C just before sprinkler activation at both locations. Due to the influence of water spray, the temperature of the smoke layer reduced rapidly to an ambient temperature after 200 s of activating the sprinkler. The slopes of the temperature at the locations A and B were similar, involving the reduction of HRR and cooling by water spray, whereas the smoke temperature varied from 60 °C to 160 °C over time in the free burn case. It is much higher than the sprinkler case.

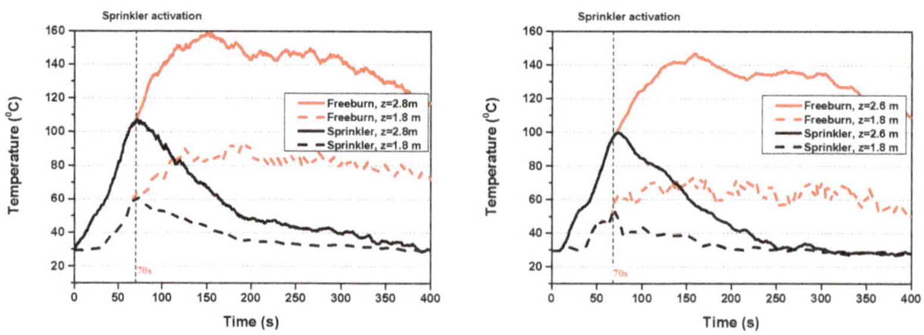

(**a**) (**b**)

Figure 10. Variation in the temperature at (**a**) location A; (**b**) location B.

Figure 11 illustrates the variation in terms of velocity at the same locations in temperature measurement. In location A, variation in smoke velocity became over 2 m/s with the sprinkler activation due to the direct effect of water spray. As observed, however, most smoke flows downward to the floor because of the smoke logging phenomenon. Therefore, smoke velocity reduction in the horizontal direction can be observed at the right doorway. The smoke velocity decreased immediately with sprinkler activation at location B, from 1.8 m/s to 0.5 m/s at 300 s. Whereas in the free burn case, the velocity of the smoke layer varied from 0.25 m/s to 1.3 m/s at location A. Due to the narrow geometry of the right doorway, the smoke velocity accelerated and varied from 0.5 m/s to over 2.0 m/s at location B (Figure 11b). It was much higher than the smoke velocity in the sprinkler case at the near ceiling.

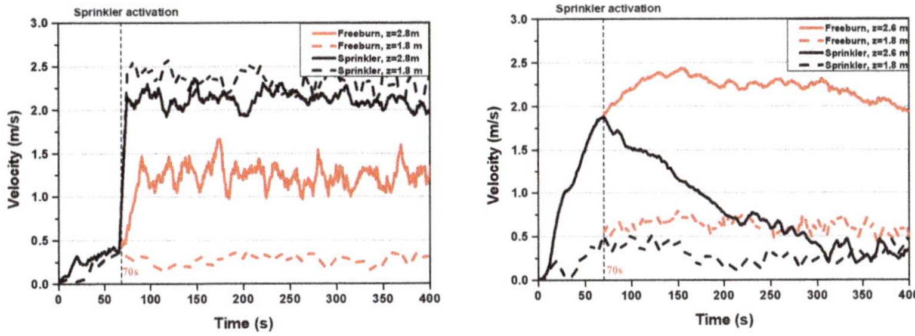

(**a**) (**b**)

Figure 11. Variation in the velocity at (**a**) location A and (**b**) location B.

The mass flow rate of the smoke layer flowing out was measured in the right doorway as shown in Figure 12. The sprinkler activation caused a significant reduction in the mass flow rate leaving the compartment. The mass flow rate reduced from 1.0 kg/s to 0.6 kg/s for 130 s after activating the sprinkler. The reasons were the reduction of HRR as well as smoke logging inside the compartment. Meanwhile, in the free burn case, the mass flow rate reached 1.2 kg/s at 100 s and kept almost constant afterward. It was more than twice that in the case of the sprinkler from 200 s.

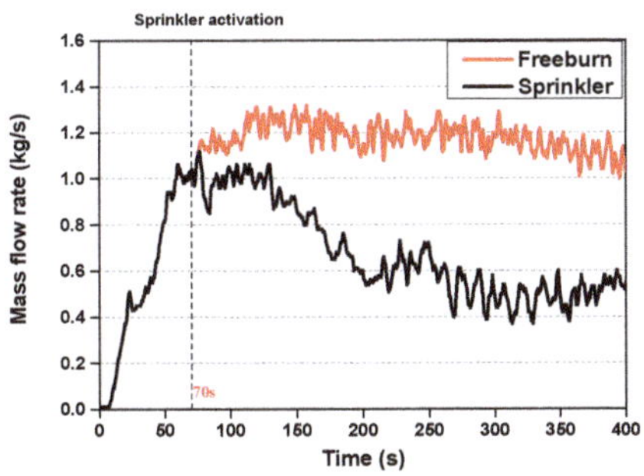

Figure 12. Variation in the mass flow rate of the smoke layer through the right doorway.

6. Conclusions

A numerical study on the fire characteristics during fire suppression by a sprinkler was conducted using FDS. The following conclusions were made:

- The extinguishing coefficient of 3.0 was chosen for the fire suppression model in this study. Under the effect of sprinkler spray, the HRR stopped growing at 165 kW at 70 s and then reduced rapidly to 10 kW at 400 s;
- The hot upper layer rapidly reduced both the temperature value and the area of the layer during fire suppression. The smoke layer was formed with a symmetrical cone right below the sprinkler, revealing the smoke logging phenomena. Under the influence of sprinkler spray, the combination

of smoke logging and fresh air-entraining from the left door pushed the smoke logging to occur strongly in the right region afterwards;

- Heat Release Rate reduction and smoke logging phenomena inside the compartment presented a significant effect of the sprinkler spray on the smoke spread in the doorway. The temperature of the smoke layer through the doorway reduced to an ambient temperature of 200 s after activating the sprinkler. The smoke velocity inside the sprinkler spray could reach over 2 m/s. At the doorway, however, the smoke velocity reduced to 0.5 m/s at 300 s, much lower than the 2 m/s in the free burn case. The mass flow rate through the doorway in the sprinkler case reduced to half compared to the free burn case at 200 s;
- The extinguishing coefficient in this study can be used as a first step trial for other researchers who want to apply the fire suppression model. However, the fire suppression model defined by the calibration method depends not only on the material properties and geometry of solid fuels but also on water spray distribution. Therefore, several important spray characteristics, such as water droplet size, spray angle, and initial velocity, need attention when referring to an extinguishing coefficient. Sensitive to these factors, we will further investigate.

Author Contributions: Conceptualization, H.T.K., J.T.K.; methodology, H.T.K., H.S.R.; software, H.T.K.; validation, H.T.K., J.T.K. and H.S.R.; formal analysis, H.T.K., J.H.K., H.S.R.; investigation, H.T.K., J.T.K.; resources, J.H.K.; data curation, H.T.K., J.H.K.; writing—original draft preparation, H.T.K., T.D.Q.; writing—review and editing, H.S.R.; supervision, H.S.R.; project administration, H.S.R. All authors have read and agreed to the published version of the manuscript.

Funding: This research was funded by the National Research Foundation of Korea (NRF) (NRF-2019R1F1A1061741), and Research and Development to Enhance Firefighting Response Ability (NFA002-007-01010002-2018).

Acknowledgments: This research was supported by Chung-Ang University Young Scientist Scholarship (CAYSS) Program.

Conflicts of Interest: The authors declare no conflict of interest.

References

1. Novozhilov, V.; Harvie, D.; Green, A.R.; Kent, J.H. A computational fluid dynamic model of fire burning rate and extinction by water sprinkler. *Combust. Sci. Technol.* **1997**, *123*, 227–245. [CrossRef]
2. Novozhilov, V.; Harvie, D.; Kent, J.; Apte, V.; Pearson, D. A computational fluid dynamics study of wood fire extinguishment by water sprinkler. *Fire Saf. J.* **1997**, *29*, 259–282. [CrossRef]
3. Nam, S. Development of a computational model simulating the interaction between a fire plume and a sprinkler spray. *Fire Saf. J.* **1996**, *26*, 1–33. [CrossRef]
4. Hua, J.; Kumar, K.; Khoo, B.C.; Xue, H. A numerical study of the interaction of water spray with a fire plume. *Fire Saf. J.* **2002**, *37*, 631–657. [CrossRef]
5. O'Grady, N.; Novozhilov, V. Large eddy simulation of sprinkler interaction with a fire ceiling jet. *Combust. Sci. Technol.* **2009**, *181*, 984–1006. [CrossRef]
6. Zhang, C.; Chow, W.K. Numerical studies on the interaction of sprinkler and smoke layer. *Procedia Eng.* **2013**, *62*, 453–462. [CrossRef]
7. Chen, Y.; Su, C.; Tseng, J.; Li, W. Experimental and numerical analysis of the cooling performance of water spraying systems during a fire. *PLoS ONE* **2015**, *10*, e0118306. [CrossRef] [PubMed]
8. Tang, Z.; Fang, Z.; Sun, J.; Beji, T.; Merci, B. Computational fluid dynamics simulations of the impact of a water spray on a fire-induced smoke layer inside a hood. *J. Fire Sci.* **2018**, *36*, 380–405. [CrossRef]
9. Wang, Y.; Meredith, K.; Zhou, X.; Chatterjee, P.; Xin, Y.; Chaos, M.; Ren, N.; Dorofeev, S. Numerical simulation of sprinkler suppression of rack storage fires. *Fire Saf. Sci.* **2014**, *11*, 1170–1183. [CrossRef]
10. Yuan, L.; Smith, A.C. Numerical modeling of water spray suppression of conveyor belt fires in a large-scale tunnel. *Process Saf. Environ. Prot.* **2015**, *95*, 93–101. [CrossRef] [PubMed]
11. McGrattan, K.B.; Forney, G.P. *Fire Dynamics Simulator User's Guide*, 6th ed.; NIST Special Publication 1019: Gaithersburg, MD, USA, 2018.
12. McGrattan, K.; Baum, H.R.; Rehm, R.G.; Hamins, A.; Forney, G.P. *Fire Dynamics Simulator-Technical Reference Guide*, 6th ed.; NIST Special Publication 1018-1: Gaithersburg, MD, USA, 2018.

13. Hietaniemi, J.; Hostikka, S.; Vaari, J. *FDS Simulation of Fire Spread—Comparison of Model Results with Experimental Data*; VTT Technical Research Centre of Finland: Espoo, Finland, 2004.
14. Rinne, T.; Hietaniemi, J.; Hostikka, S. *Experimental Validation of the FDS Simulations of Smoke and Toxic Gas Concentrations*; VTT Technical Research Centre of Finland: Espoo, Finland, 2007; ISBN 9789513866174.
15. Vaari, J.; Hostikka, S.; Sikanen, T.; Paajanen, A. *Numerical Simulations on the Performance of Waterbased Fire Suppressions Systems*; VTT Technical Research Centre of Finland: Espoo, Finland, 2012; ISBN 9789513878825.
16. Yu, H.; Lee, J.; Kung, H.; Brown, W. Suppression of rack-storage fires by water. *Fire Saf. Sci.* **1994**, *4*, 901–912. [CrossRef]
17. ISO. *ISO 6182-10:2006 Fire Protection-Automatic Sprinkler Systems—Part 10: Requirements and Test Methods for Domestic Sprinklers*; ISO: Geneva, Switzerland, 2006.
18. Sheppard, D. Spray Characteristics of Fire Sprinklers. Ph.D. Thesis, Northwestern University, Evanston, IL, USA, 2002.
19. You, H.Z. Investigation of spray patterns of selected sprinklers with the fmrc drop size measuring system. *Fire Saf. Sci.* **1986**, *1*, 1165–1176. [CrossRef]
20. Bullen, M.L. The effect of a sprinkler on the stability of a smoke layer beneath a ceiling. *Fire Technol.* **1977**, *13*, 21–34. [CrossRef]

energies

MDPI

Article

Analysis on the Fire Growth Rate Index Considering of Scale Factor, Volume Fraction, and Ignition Heat Source for Polyethylene Foam Pipe Insulation

Jung Wook Park [1], Ohk Kun Lim [2] and Woo Jun You [1,*]

1 Department of Architecture & Fire Safety, Dong Yang University, Yeongju 36040, Korea; kkisd0123@naver.com
2 R&D Laboratory, Korea Fire Institute, Yongin 17088, Korea; terence@kfi.or.kr
* Correspondence: wjyou@dyu.ac.kr; Tel.: +82-54-630-1297

Received: 17 June 2020; Accepted: 2 July 2020; Published: 15 July 2020

Abstract: The fire growth rate index (*FIGRA*), which is the ratio of the maximum value of the heat release rate (Q_{max}) and the time (t_{max}) to reach the maximum heat release rate, is a general method to evaluate a material in the fire-retardant performance in fire technology. The object of this study aims to predict *FIGRA* of the polyethylene foam pipe insulation in accordance with the scale factor (S_f), the volume fraction of the pipe insulation (*VF*) and the ignition heat source (Q_{ig}). The compartments made of fireboard have been mock-up with 1/3, 1/4, and 1/5 reduced scales of the compartment as specified in ISO 20632. The heat release rate data of the pipe insulation with the variation of S_f, *VF*, and Q_{ig} are measured from 33 experiments to correlate with *FIGRA*. Based on a critical analysis of the heat transfer phenomenon from previous research literature, the predictions of Q_{max} and t_{max} are presented. It is noticeable that the fire-retardant grade of the polyethylene foam pipe insulation could have Grade B, C, and D in accordance with the test conditions within ±15% deviation of the predicted *FIGRA*. In case of establishing the database of various types of insulation, the prediction models could apply to evaluate the fire-retardant performance.

Keywords: pipe insulation; fire growth rate index; scale factor; volume fraction; ignition heat source; maximum heat release rate; time to reach maximum HRR (heat release rate)

1. Introduction

Insulation is widely used in buildings as an important material to prevent energy loss of architecture [1–3]. Among them, polystyrene, poly-urethane, poly-ethylene, and elastomeric closed cell thermal insulation, which are made of organic substance, are mainly used as pipe insulation to prevent freezing and surface condensation by minimizing the heat loss [4–6]. However, pipe insulation could be ignited from the overheated hot wire or welding work for maintenance, and rapidly spread to the surrounding combustibles [7–9]. To fundamentally prevent the spread of fire occurred by pipe insulation, inorganic materials such as semi-combustible, which does not ignite at high temperature, is able to be applied. However, inorganic substance, especially molded stone wool, glass wool, etc., is not useful in the installation of piping compared to organic materials because of its highly absorbent feature as a mechanical weakness [10–12]. For these reasons, the specific material of pipe insulation is not regulated, and it is recommended to use materials that satisfy the fire retardants as an alternative [13]. The fire growth rate index (*FIGRA*) is a general method to evaluate the material in the fire-retardant performance in fire technology [14,15]. Therefore, the pipe insulation, which satisfies extremely low *FIGRA* values, should minimize the rapid spread of fire phenomenon even though the pipe insulation could not have the properties of the complete non-combustible. However, the regulated test conditions, such as the compartment size, the thermal properties of the wall, and the ignition heat source, are not equal for each test standard [16–19]. It means that *FIGRA* for evaluating the flame-retardant performance of the pipe insulation depends on the test conditions. Thus, the thermal characteristics inside the compartment in accordance with test conditions are important to analyze a fire risk of the pipe insulation.

There are several studies that the fire-retardant performance test result is changed by the test conditions in the use of the same pipe insulation material. The temperature inside the compartment increases in proportion to the thermal ratio, as resulted in the study of H. Pretrel et al., analyzed by the correlation under the ventilated pool fire phenomenon by the thermal ratio of dimensionless variables including thermal conductivity coefficient, the thickness of the wall, and the opening area of the compartment [20].

The fire growth rate decreases as the ignition source decreases in the same volume of pipe insulation, since the time to reach the maximum heat release is decreased proportional to the delay time of ignition, as investigation of N. Hernandez with the predictive model of the phenomenon of the penetration of radiation to the sold materials [21]. In addition, the fire growth rate decreases in proportion to the size of volume space, in the research of R.R. Leisted et al. with the analyze of the temperature distribution by the time in the case of polyisocyanurate or stone wool in the 1/5 reduced scale compartment of ISO 13784-1 [22].

In accordance with the previous research, it can be predicted that *FIGRA* would not be equal for the same material due to the thermal conditions. However, to the authors' knowledge, there have been no research to predict *FIGRA* since the values of Q_{max} and t_{max} cannot be closed with previous investigations [20–24]. Especially, if the value of *FIGRA* for a highly combustible material is evaluated too low at a specified test condition, it can cause a risk for the material to be used for building construction. From this point of view, the quantitative analysis on *FIGRA* in accordance with the experiment conditions can be considered as a significant object in terms of the evaluation for the risk of fire spread. The object of this study aims to predict *FIGRA* of polyethylene foam pipe insulation in accordance with the scale factor (S_f), the volume fraction of the pipe insulation (*VF*), and the ignition heat source (Q_{ig}).

2. Materials and Methods

2.1. Heat Transfer Phenomenon

Figure 1 explains that the heat transmission phenomenon of a polyethylene foam pipe insulation in the semi-closed compartment space. As shown in this figure, the surface temperature of the pipe insulation increases to reach the reference temperature by the ignition heat source. Therefore, the mass loss of the pipe insulation can be quantified as [25,26].

$$\dot{m}'''_{f_r} = \rho A_0 Y_f^n Y_{O_2}^m \mathrm{e}^{(-E/RT_\tau)} \tag{1}$$

where $\dot{m}'''_{f,r}$, ρ, A_0, Y, E, R, and T_r are the mass loss rate per unit volume, the density of pipe insulation, the pre-exponential factor, the mass fraction, the activation energy (kJ/mole), the gas constant (8.314 kJ/kmole), and the reference temperature, respectively. The heat release rate during the vaporization of the combustible (Q_f) can be expressed as,

$$Q_f = \eta \Delta h_c \dot{m}'''_{f,r} V_f \tag{2}$$

where η, Δh_c, and V_f are the combustion efficiency, the heat of combustion, and the ignited insulation volume during the combustion time, respectively [27]. When the ignition of the pipe insulation takes place inside the compartment, the surface area, which reaches the reference temperature (T_r)

by convective heat from the free stream and the radiative heat from the flame and the wall, could be time-variant. Thus, the total heat release rate can be denoted as,

$$Q_t(t) = Q_f(t) + Q_{ig} \tag{3}$$

where t denotes the combustion time, and the total heat release rate ($Q_t(t)$) and the heat release rate of the pipe insulation ($Q_f(t)$) in Equation (3) should be a time-variant function due to $A_f \rightarrow A_f(t)$. However, the heat of combustion (Δh_c, kJ/kg), which is one of the thermochemical properties, has a constant value for the pipe insulation as shown in Equation (4) [28–30].

$$\Delta h_c = \frac{\int_{t=0}^{t=t_{final}} \dot{Q}_f(t)dt}{\eta \Delta m_f} = const \tag{4}$$

where Δm_f means that the mass loss after the combustion of the pipe insulation. The fire growth rate index (*FIGRA*) to classify the fire-retardant grade is defined as the ratio of the maximum heat release rate (Q_{max}), which is the net heat of the pipe insulation, and the time (t_{max}) to reach the maximum heat release rate (Q_{max}) as denoted in Equation (5) [16].

$$FIGRA = \frac{Q_{\max}}{t_{\max}} \tag{5}$$

where *FIGRA* refers to a main parameter to evaluate the fire-retardant grade. As Q_{ig} increases, $\dot{Q}_f(t)$ is varied in proportion, since the size of the solid surface to reach the reference temperature (T_r) varies with the combustion time. However, when the heat of combustion (Δh_c) in Equation (4) maintains a constant value with a fixed value of Δm_f, the integral value of $Q_f(t)$ during the total combustion time should satisfy the first law of energy conservation. It means that Q_{max} increases as the total combustion time decreases, as shown in the lower right side in Figure 1. Especially, the convective and radiative heat can be mainly affected by the compartment space (V_M), the volume of the combustibles (V_f), and the geometrical shape of the opening area. Therefore, it is assumed that $\dot{Q}_{\max}$ and t_{max} can be functioned with the quantity of the volume of the compartment (V_M), the volume of the combustibles (V_f), and the ignition heat source (Q_{ig}) as denoted in Equation (6).

$$Q_{\max}, t_{\max} \sim f(Q_{ig}, V_M, V_f) \tag{6}$$

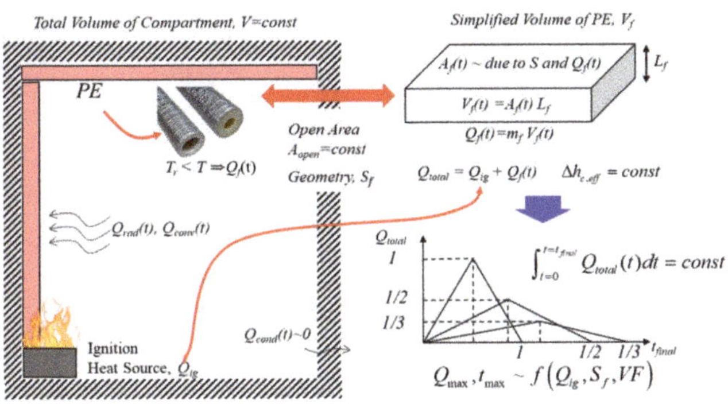

Figure 1. Schematic diagram of heat transfer and the heat release rate of the pipe insulation in a compartment fire.

2.2. Heat Transfer Phenomenon

Table 1 summarized the test conditions of ISO 20632 and NFPA 274. As shown in the table, the volume (V_M) of the compartment was reduced to 1/3, 1/4, and 1/5 of ISO 20632. In addition, the volume fractions of the pipe insulation (VF) consisted of a total 5 conditions, including those of ISO 20632 (VF = 0.024) and NFPA 274 (VF = 0.07). The experiment conditions of this study were opted with the definitions of the scale factor (S_f) and the volume fraction (VF) as denoted in Equation (6).

$$S_f = \frac{L_M}{L_{ISO}}, VF = \frac{V_f}{V_M} \tag{7}$$

where S_f, V_M, V_{ISO}, V_f, and VF are the scale factor, the volume of the compartment, the volume of the ISO 20632 compartment, and the volume fraction of the pipe insulation, respectively. The schematic diagram and pictures of the experiment conditions are explained in Figure 2.

Table 1. Test conditions investigated by ISO 20632 and NFPA 274.

Contents	ISO 20632	NFPA 274	Test Conditions
Volume of Compartment (m^3)	V_{ISO} = 20.76 m^3	$V_{M,NFPA}$ = 0.78 m^3	$V_{M,1/3}$ = 0.768 m^3 $V_{M,1/4}$ = 0.324 m^3 $V_{M,1/5}$ = 0.165 m^3
Scale Factor (-)	1	1/3 (= 0.33)	1/3 (= 0.33), $\frac{1}{4}$ (= 0.25), 1/5 (= 0.20)
Volume of Insulation (m^3)	$V_{f,iso}$ = 0.51 m^3	$V_{f,NFPA}$ = 0.055 m^3	$V_{f,1/5}$ = 0.004 m^3, 0.008 m^3, 0.019 m^3, 0.008 m^3 $V_{f,1/4}$ = 0.016 m^3, 0.038 m^3, 0.012 m^3, 0.023 m^3 $V_{f,1/3}$ = 0.053 m^3, 0.017 m^3, 0.032 m^3
Volume Fraction (-)	VF = 0.024	VF = 0.07	VF = 0.024, 0.05, 0.07, 0.1
Ignition (kW) and Time (s)	$Q_{ig,1}$ = 100 kW for 600 s $Q_{ig,2}$ = 300 kW for 600 s	$Q_{ig,1}$ = 20 kW for 180 s $Q_{ig,2}$ = 70 kW for 420 s	Q_{ig} = 10 kW, 15 kW, 20 kW for t_{final} = 600 s

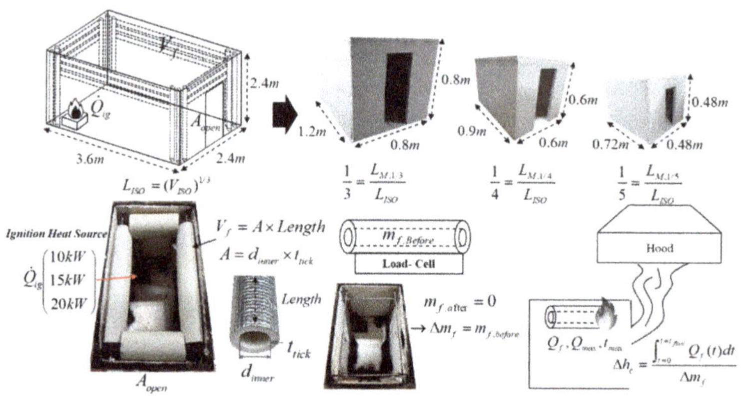

Figure 2. Schematic diagram and pictures of the test conditions and the experiment method.

2.3. Calibration of the Heat Release Rate

Figure 3 shows the calibration results of the heat release rate for a propane burner by the oxygen consumption method of the cone calorimeter. A mass flow controller (MFC, Model TSC-145) was used to control the flow rate of propane. In Figure 3a, the difference of maximum 50 s with the theoretical heat release rate was measured due to the increase of the response time of the MFC instrument. Thus, in the low flow range from 10 to 20 kW, the fluctuation of the mass flow rate and the delay time were minimized by the metering valve and the area flow meter. As a result of performing A-Type uncertainty under the repeated experiments, it is found that the cone-calorimeter apparatus used in this study had a reliability of ±5% when the coverage factor, k = 1.95 at 95% confidence level as displayed in Figure 3b. Table 2 shows the specifications of the experimental apparatus used in this study.

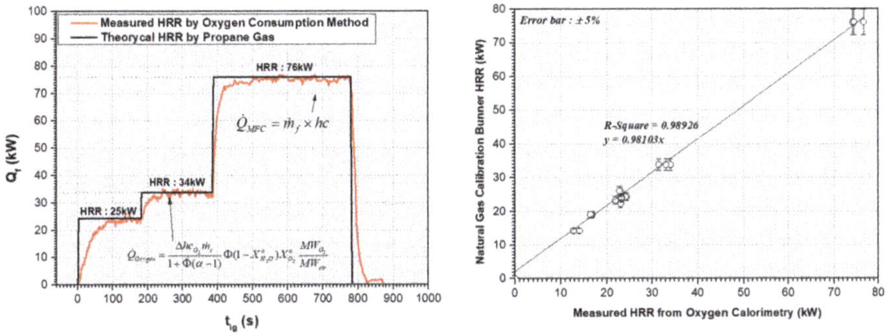

(a) (b)

Figure 3. (**a**) Calibration results of heat release rate (HRR) using propane burner and (**b**) comparison of the theoretical and measured HRR.

Table 2. Specification of the experiment apparatus of the cone-calorimeter.

Measurement	Specification
Duct Temperature	K-Type Wire, Range: −200–1000 °C
DAQ	Voltage: 20 mV to 100 V, 1–5 V F.S., 20 channels, Accuracy: ±0.1%
Duct Size & Blower	Length: 5 m, Diameter: 0.2 m, Fan capacity: 3 hp
O_2 Analyzer	Output: 4–20 mA, Range: 0.7–1.2 bar, Model: OXYMAT 61
Pressure Sensor	Output: 0–10 V, Range: 1250 Pa, Model: MS-311
Mass Flow Controller	Fuel: CH_4, C_3H_8, Output: 0–5 VDC, Range: 200 LPM, Model: TSC-145
Pressure Transmitter	Output: 0–20 mA, Range: 0–20 bar, Model: PSC-E-B-A-P-G

2.4. Material Properties of the Test Sample

Figure 4a shows the measurement of the mass loss rate of the standard specimen calcium oxalate to verify the accuracy of the apparatus of thermo gravimetric analysis (TGA, Model STA PT1000). The calcium oxalate changes in the symmetry of $CaC_2O_4{\cdot}H_2O$, CaC_2O_4, and $CaCO_3$ as mentioned in reference [31]. Thus, the reference temperature (T_r) was given as 190 °C, 450 °C, and 700 °C at ±30 °C, and the mass reduction rate was measured as −12.99%, −19.15%, and −29.98%, respectively. The measured values were in good agreement within about ±0.6% to the reference values.

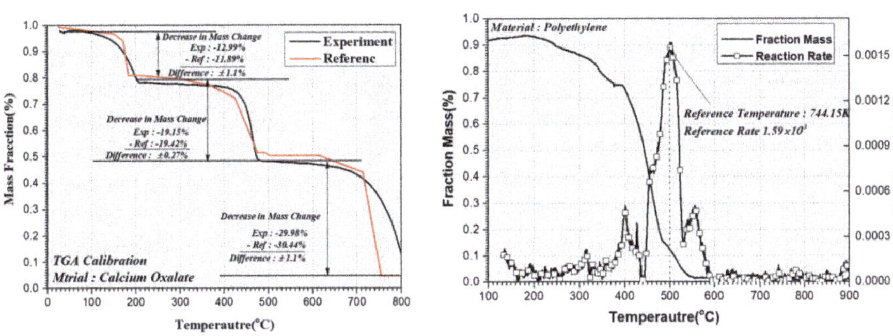

(a) (b)

Figure 4. (**a**) Calibration results of TGA using calcium oxalate and (**b**) the results of TGA for material properties of the test sample (polyethylene foam).

Figure 4b shows the result of calculating the reaction rate, pre-exponential factor, and reference temperature, which are the thermochemical properties of the polyethylene foam pipe insulation used in this study. The reaction and the heat rate mechanism of a solid fuel can be found in the combustion theory [32]. From T_r = 744 K, the pre-exponential factor (A_j) of Equation (8) and activation energy (E_j) of Equation (9), which are suggested by Lyon et al. [30,33], are arranged in Table 3.

$$A_j = \frac{er_{p,j}}{Y_{s,o}} e^{\left(\frac{E_j}{RT_{p,j}}\right)} \tag{8}$$

$$E_j = \frac{er_{p,j}}{Y_{s,o}} \frac{RT_{p,j}^2}{(dT/dt)} \tag{9}$$

where A_j, E_j, T_r, and $Y_{s,o}$ means the pre-exponential factor, activation energy, reference temperature, and mass fraction, respectively [33].

Table 3. Thermal properties of the polyethylene foam (PE) test sample.

Properties	Values	Properties	Values
Reference Temperature, (K)	744	Heat of Combustion, (kJ/kg)	42,660
Activation Energy (kJ/kmole)	1.19×10^{-5}	Density (kg/m^3)	26
Pre-exponential Factor (1/s)	1.05×10^{-6}	Specific Heat (kJ/kg-°C)	2.31

3. Results and Discussion

3.1. Experiment of Heat Release Rate

The effects of the ignition heat source on the heat release rate in the case of the fixed values of the scale factor (S_f = 1/3) and the volume fraction (VF = 0.024) are plotted in Figure 5. As denoted in this figure, the values of t_{max} were decreased as 589 s, 203 s, and 136 s in accordance with Q_{ig} = 12 kW, 16 kW, and 23 kW, respectively. While Q_{max} maintained a constant value at approximately 209 ± 10 kW. The results explained that the time for the pipe insulation to reach the reference temperature (T_r) of 744 K decreased as Q_{ig} increased. However, the values of Q_{max} maintained a constant value since the overall heat amount of the pipe insulation inside the compartment should be conserved. All results for S_f = 1/3, 1/4, and 1/5 and VF = 0.024, 0.05, 0.07, and 0.1 in accordance with the ignition heat sources are plotted in Figure 6. Test conditions and analysis based on the Figure 6 are summarized in Table 4. As shown in Table 4, t_{max} of Test #8, #16, and #31–#33, which have more than 20% deviation, were excluded due to the environment differences. However, in all the experimental result, the values of Q_{max} maintained a constant value within the range of ±3.48% average and ±12.26% maximum for the fixed volume fraction (VF), while the values of t_{max} were decreased, which were inversely proportional to the heat amount of ignition. From the results of Figure 6 and Table 4, the effective heat of combustion was investigated and Q_{max} and t_{max} were predicted with the effects of S_f, VF, and Q_{ig}.

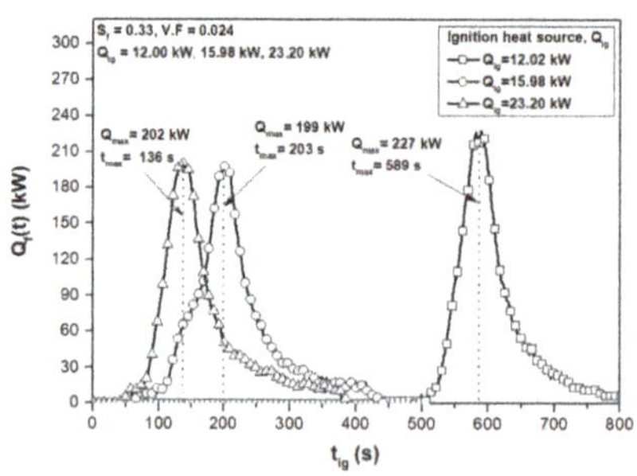

Figure 5. The results of the heat release rate with the variations of the ignition heat source for volume fraction 0.024 and scale factor 0.33.

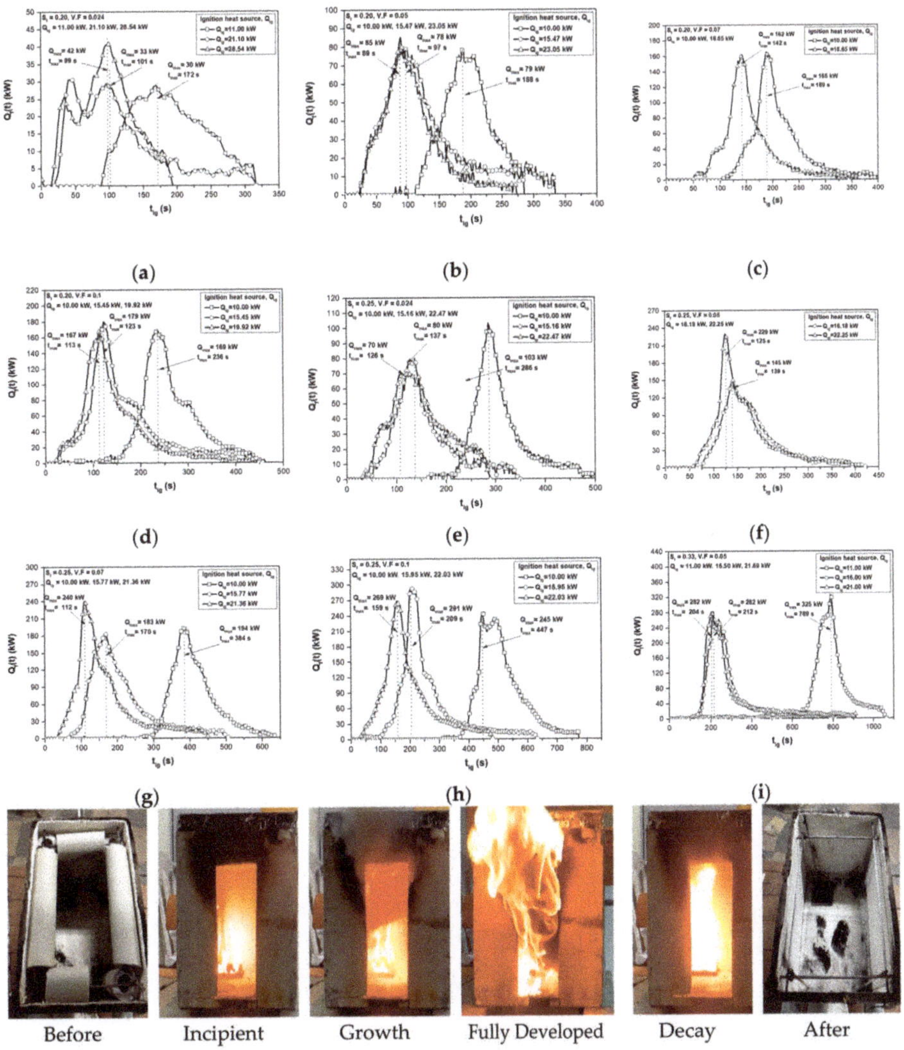

Figure 6. Experiment results of the heat release rate vs. ignition heat source for 0.33, 0.25, and 0.2 of the scale factor with a volume fraction of 0.024, 0.05, 0.07, and 0.1. (**a**) Test#1–Test#3. Heat release rate vs. time (S_f = 0.2, VF = 0.024, Q_{ig} = 11.00 kW, 21.10 kW, 28.54 kW); (**b**) Test#4–Test#6. Heat release rate vs. time (S_f = 0.2, VF = 0.05, Q_{ig} = 10.00 kW, 15.47 kW, 23.05 kW); (**c**) Test#7, Test#9. Heat release rate vs. time (S_f = 0.2, VF = 0.07, Q_{ig} = 10.00 kW, 18.65 kW); (**d**) Test#10–Test#12. Heat release rate vs. time (S_f = 0.2, VF = 0.1, Q_{ig} = 10.00 kW, 15.45 kW, 19.92 kW); (**e**) Test#13–Test#15. Heat release rate vs. time (S_f = 0.25, VF = 0.024, Q_{ig} = 10.00 kW, 15.16 kW, 22.47 kW); (**f**) Test#17–Test#18. Heat release rate vs. time (S_f = 0.25, VF = 0.05, Q_{ig} = 16.18 kW, 22.25 kW); (**g**) Test#19–Test#21. Heat release rate vs. time (S_f = 0.25, VF = 0.07, Q_{ig} = 10.00 kW, 15.77 kW, 21.36 kW); (**h**) Test#22–Test#24. Heat release rate vs. time (S_f = 0.25, VF = 0.1, Q_{ig} = 10.00 kW, 15.95 kW, 22.03 kW); and (**i**) Test#28–Test#30. Heat release rate vs. time (S_f = 0.33, VF = 0.05, Q_{ig} = 11.00 kW, 16.50 kW, 21.69 kW).

Table 4. Test conditions and summarized results of experiments.

Test Num.	Test Conditions					Results of Experiment				
	S_f	V_M	V_f	VF	Δm_f	$\dot{Q}_{ig}$	$\dot{Q}_f$	$\Delta m_{c,eff}$	$\dot{Q}_{max}$	t_{max}
	(-)	(m^3)	(m^3)	(-)	(kg)	(kW)	(kJ)	(kJ/kg)	(kW)	(s)
#1	0.2	0.165	0.004	0.024	0.109	11.00	3795	34,817	30.23	172
#2	0.2	0.165	0.004	0.024	0.109	21.10	3797	34,833	33.47	101
#3	0.2	0.165	0.004	0.024	0.109	28.54	3779	34,672	42.63	99
#4	0.2	0.165	0.008	0.050	0.227	10.00	7477	32,938	79.09	188
#5	0.2	0.165	0.008	0.050	0.227	15.47	8576	37,778	78.96	97
#6	0.2	0.165	0.008	0.050	0.227	23.05	6913	30,455	85.23	89
#7	0.2	0.165	0.012	0.070	0.312	10.00	11,847	37,972	165.25	189
#8	0.2	0.165	0.012	0.070	0.320	14.64	10,785	33,702	160.78	N/A
#9	0.2	0.165	0.012	0.070	0.318	18.65	11,282	35,479	162.03	142
#10	0.2	0.165	0.0165	0.1	0.450	10.00	17,344	38,541	169.34	236
#11	0.2	0.165	0.0165	0.1	0.454	15.45	18,482	40,708	179.98	123
#12	0.2	0.165	0.0165	0.1	0.452	19.92	16,436	36,362	167.22	113
#13	0.25	0.324	0.008	0.024	0.223	10.00	8260	37,042	103.18	286
#14	0.25	0.324	0.008	0.024	0.214	15.16	7660	35,797	80.16	137
#15	0.25	0.324	0.008	0.024	0.220	22.47	9101	41,366	70.60	126
#16	0.25	0.324	0.016	0.050	0.444	10.00	18,959	42,699	178.05	N/A
#17	0.25	0.324	0.016	0.050	0.444	16.18	14,961	33,696	145.56	139
#18	0.25	0.324	0.016	0.050	0.444	22.25	17,763	40,006	229.68	125
#19	0.25	0.324	0.023	0.070	0.626	10.00	22,223	35,500	194.57	384
#20	0.25	0.324	0.023	0.070	0.632	15.77	23,994	37,964	183.09	170
#21	0.25	0.324	0.023	0.070	0.620	21.36	24,962	40,261	240.65	112
#22	0.25	0.324	0.032	0.1	0.892	10.00	35,515	39,815	245.92	447
#23	0.25	0.324	0.032	0.1	0.892	15.95	32,812	36,785	291.99	209
#24	0.25	0.324	0.032	0.1	0.872	22.03	33,095	37,953	269.13	159
#25	0.33	0.768	0.019	0.024	0.495	12.00	19,455	39,303	227.52	589
#26	0.33	0.768	0.019	0.024	0.500	15.98	20,547	41,095	199.78	203
#27	0.33	0.768	0.019	0.024	0.503	23.20	18,974	37,722	202.00	136
#28	0.33	0.768	0.038	0.050	1.052	11.00	44,035	41,859	325.05	789
#29	0.33	0.768	0.038	0.050	1.046	16.50	37,700	36,042	282.23	212
#30	0.33	0.768	0.038	0.050	1.054	21.69	39,344	37,328	282.02	204
#31	0.33	0.768	0.053	0.070	1.448	13.57	49,677	34,307	382.10	N/A
#32	0.33	0.768	0.053	0.070	1.430	17.69	54,897	38,389	437.19	N/A
#33	0.33	0.768	0.053	0.070	1.451	21.83	50,609	34,878	370.69	N/A

3.2. Comparison of the Effective Heat of Combustion, $\Delta h_{c,eff}$

Regarding the previous studies, the heat of combustion (Δh_c) of the polyethylene foam pipe insulation is around 42,660 kJ/g [34]. In this study, the effective heat of combustion of Equation (10) as referred in [35] was compared with the heat of combustion by integrals on the combustion time under the measured heat release rate.

$$h_{c,eff} = \int_{t=0}^{t=t_{final}} Q_f(t)dt/\Delta m_f \tag{10}$$

where $\Delta h_{c,eff}$ and Δm_f are the effective heat of combustion and the mass loss of the pipe insulation after combustion, respectively. The measured effective heat of combustion for S_f = 1/3, 1/4, and 1/5, VF = 0.024, 0.05, 0.07, and 0.1 in accordance with the values of Q_{ig} are plotted in Figure 7. As shown in the figure, the average values of the effective heat of combustion was around 37,214 kJ/kg, which was only about 87% of the combustion efficiency compared to 42,660 kJ/kg. The main reason can be found that the incomplete combustion condition occurs due to the circumstance lack of ventilation regarding to the size of the opening area [34–37]. In addition, from R. N. Walters et al. [34], the combustion heat value could be changed by the composition ratio and the porosity of the molecule consisted of materials. Therefore, the correlation between the geometric shape of the pipe insulation and the thermochemical properties of the molecular structure of the pipe insulation should be analyzed with the combustion efficiency to obtain more accuracy reasons. However, the main purpose of this study was to predict the fire growth rate index related with the scale factor (S_f), the volume fraction (VF), and the ignition heat source (Q_{ig}). Therefore, Q_{max} and t_{max} were analyzed with the effective heat of combustion assumed to be the averaged value of 37,214 kJ/kg.

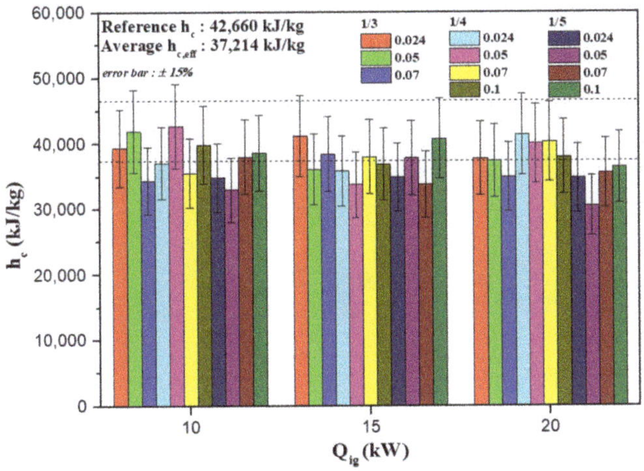

Figure 7. The results of the effective heat of combustions with volume fraction 0.024, 0.05, 0.07, and 0.1 for scale factor 1/3, 1/4, and 1/5.

3.3. Analysis of the Maximum Heat Release Rate, Q_{max}

The values of Q_{max} and Δm_f for S_f = 1/3, 1/4, and 1/5 and VF = 0.024, 0.05, 0.07, and 0.1 in Table 4 are plotted in Figure 8. The line marked in red can be obtain from the boundary condition, which is Q_{max} = 0 at Δm_f = 0, as shown in the Equation (11).

$$Q_{\max} = a_1 \times \Delta m_f^{b_1} \tag{11}$$

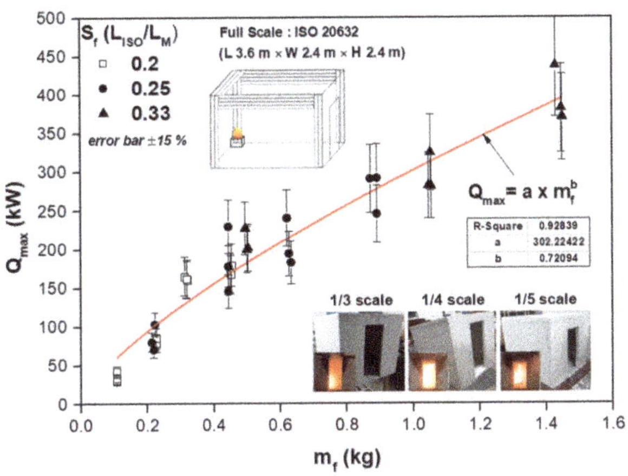

Figure 8. Experiment results of the maximum heat release rate vs. fuel mass loss with scale factors (S_f = 1/3, 1/4, and 1/5) and volume fraction (VF = 0.024, 0.05, 0.07, and 0.1).

The Δm_f can be assumed that the initial mass of the pipe insulation since all completely burned during the experiments for each condition in Table 4. The simple expression foam of Q_{max} can be curve-fitted with Δm_f, regardless of S_f, VF, and Q_{ig} in the case of $a_1 = 302.224$ and $b_1 = 0.721$ within ±15%. However, in the overall range of Δm_f, the deviations between Equation (11) and the experiments were higher in accordance with the volume of compartment and the pipe insulation. Thus, the effects of S_f and VF on Q_{max} were investigated to obtain more accurate prediction.

Figure 9a shows the correlations between Q_{max} and VF for S_f = 1/3, 1/4, and 1/5. As shown in the figure, Q_{max} intends to increase in proportional to VF as denoted in Equation (12).

$$Q_{\max,pre} = a_2(VF)^{b2} \tag{12}$$

where $Q_{max,pre}$ (kW), a_2(kW), and b_2 are the predictive value of the maximum heat release rate and the experimental constants, respectively. The mass loss is approximated in Equation (13).

$$\Delta m_f = VF \times V_M \times \rho_f \tag{13}$$

when a_2 is constant at 1766.78 of Equation (12), $b2$ decreases with S_f as shown in Figure 9b. Thus, the experiment constant, $b2$ can be curve-fitted as,

$$b_2 = 0.18308 \times S_f^{-1.04545} \tag{14}$$

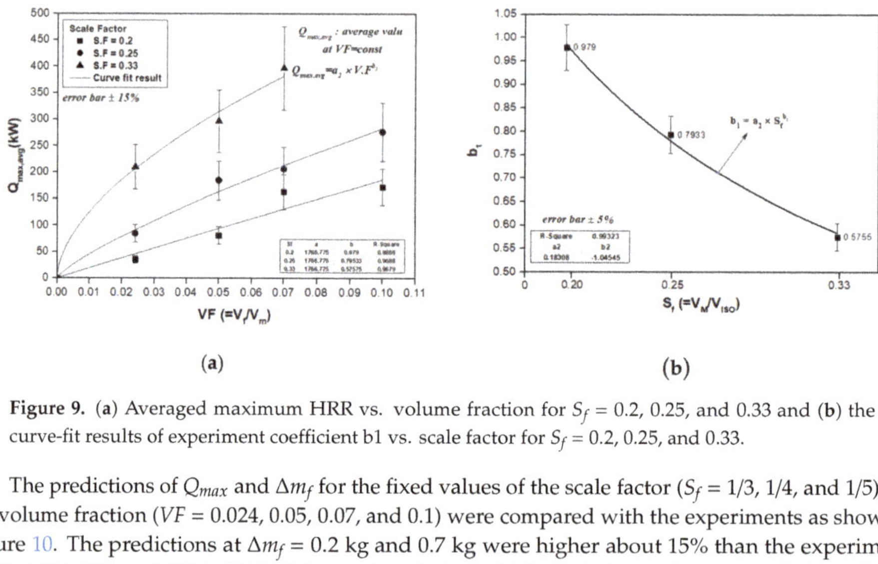

(a) (b)

Figure 9. (**a**) Averaged maximum HRR vs. volume fraction for S_f = 0.2, 0.25, and 0.33 and (**b**) the curve-fit results of experiment coefficient b1 vs. scale factor for S_f = 0.2, 0.25, and 0.33.

The predictions of Q_{max} and Δm_f for the fixed values of the scale factor (S_f = 1/3, 1/4, and 1/5) and the volume fraction (VF = 0.024, 0.05, 0.07, and 0.1) were compared with the experiments as shown in Figure 10. The predictions at Δm_f = 0.2 kg and 0.7 kg were higher about 15% than the experiments for Test #14, #15, and #18 in Table 4 due to the relatively high deviation of the averaged Q_{max}. On the other hand, the predictions at Δm_f = 0.2 kg were about 15% lower than the experiments for Test#7 and #9 in Table 4. The main reason would be expected to take place from the combustion efficiency in accordance with the opening area [20]. However, it is confirmed that the total of 27 experiments and the predicted values were in good agreement within ±5%. Thus, Equation (12) indicates that the improved accuracy approximately 10% or more compared to Equation (11) since S_f and VF were considered. The limitation of the prediction should consider the experiment constants. The prediction of Q_{max} can be applicable in the case of establishing the database of various types of insulation.

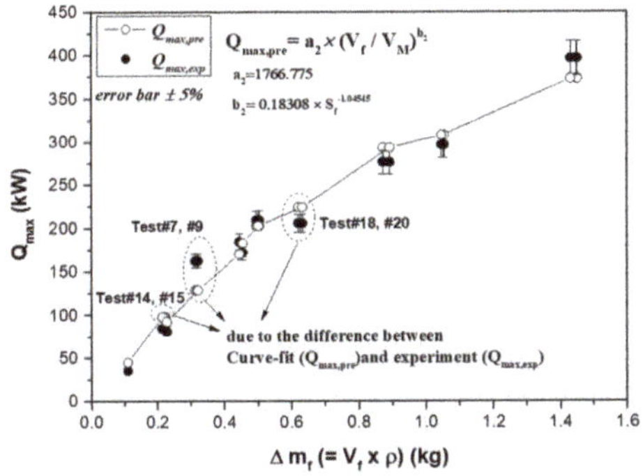

Figure 10. Comparison of the maximum heat release rate of experiments and predictions with mass loss.

3.4. Analysis of the Time to Reach the Maximum Heat Release Rate, t_{max}

From the results in Table 2, as the volume of the compartment increased, the heating time of the surface temperature for the pipe insulation by convection and radiation was proportionally increased in the case of the fixed values of Q_{ig} and VF. In addition, when Q_{ig} increased, t_{max} was decreased since the surface area of the pipe insulation to reach reference temperature (T_r) rapidly increased. These relations can be functioned as,

$$t_{\max} \sim f(S_f/Q_{ig}) \tag{15}$$

Figure 11a shows the relations between S_f/Q_{ig} and t_{max} in the case of VF = 0.024, 0.05, 0.07, and 0.1. The values of t_{max}, which was inversely proportional to Q_{ig} and proportional to S_f under the fixed value of VF, can be curve-fitted as,

$$t_{\max,pre} = c_1 + c_2 \times e^{(c_3 \times S_f/Q_{ig})} \tag{16}$$

where c_1, c_2, and c_3 represent the experimental constants for the polyethylene foam pipe insulation. The values of c_2 and c_3 were found to have the constant with 3.283 and 180.102. In addition, the values of the experimental constant c_1 were 74.93, 81.07, 83.35, and 122.64 when VF = 0.024, 0.05, 0.07, and 0.1 respectively, as shown in Figure 11b. Thus, it can be curve fitted as,

$$c_1 = d_1 + d_2 \times e^{(d_3 \times VF)} \tag{17}$$

where d_1, d_2, and d_3 have a constant value of 75.782, 0.167, and 56.36, respectively.

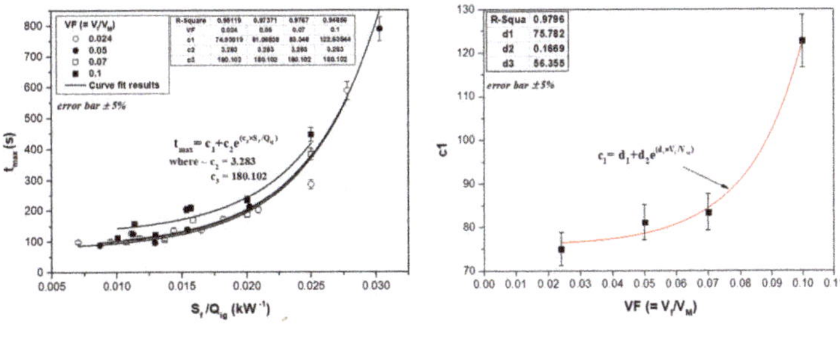

(a) (b)

Figure 11. (**a**) Maximum time vs. S_f/Q_{ig} for 0.024, 0.05, 0.07, and 0.1 of the volume fraction and (**b**) the curve-fit results of experiment coefficient c1 vs. volume Fraction.

The predictions of t_{max} and S_f/Q_{ig} for the fixed values of the scale factor (S_f = 1/3, 1/4, and 1/5) and the volume fraction (*VF* = 0.024, 0.05, 0.07, and 0.1) were compared with the experiments as shown in Figure 12. The predictions at S_f/Q_{ig} = 0.01, 0.015, and 0.024 kW^{-1} were about 20% deviation than the experiments for Test #5, #6, #11, #12, #13, #15, and #30 in Table 4 due to the heat loss by the leakage from the connection part in the compartment, the difference humidity or the relatively low surrounding temperature. The heat loss can cause the experiments of t_{max} that could be relatively delayed than the predictions of t_{max}. However, the total 23 of predictions were in good agreement with the experiments in the error range of ±5%. Therefore, as mentioned in Section 3.3, t_{max} can be applicable in the case of establishing the database of various types of insulation. As denoted in Equations (12) and (16), the predictions of Q_{max} and t_{max} were significantly correlated with S_f, *VF*, and Q_{ig}, which were the test conditions of fire resistance standard. It is noticeable that the values of *FIGRA* can be obtained without experiments for the polyethylene foam pipe insulation if the effects of the thermal properties of the compartment materials are determined.

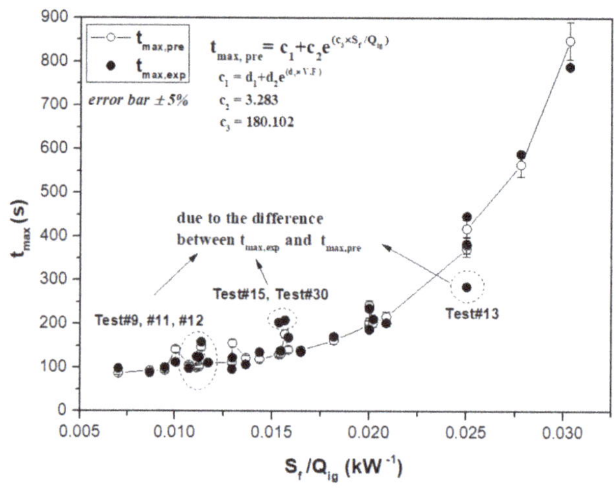

Figure 12. Comparison of the maximum time of experiments and predictions with S_f/Q_{ig}.

3.5. Estimation of the Fire Growth Rate Index, FIGRA

According to EN13501-1, the fire-retardant grade can be divided as Grade A_2 for $FIGRA \leq 0.16$ kW/s, Grade B for $0.16 \leq FIGRA \leq 0.6$ kW/s, Grade C for $0.6 \leq FIGRA \leq 1.5$ kW/s, the Grade D for $0.6 \leq FIGRA \leq 7.5$ kW/s, and Grade E for $FIGRA \geq 7.5$ kW/s. Therefore, Equations (12) and (16) are substituted into Equation (5), and the prediction of *FIGRA* can be arranged as,

$$FIGRA_{pre} = \frac{a_2(VF)^{b_2}}{c_1 + c_2 \times e^{(c_3 \times S_f/Q_{ig})}} \tag{18}$$

where $FIGRA_{pre}$ means the prediction value of the fire growth rate index (*FIGRA*) considering the scale factor (S_f), the volume ratio (*VF*), and the ignition heat source (Q_{ig}).

Figure 13 shows the results of the comparisons between the predictions by Equation (18) and the experiments in Table 4. The predictions of Q_{max} for Test #15, #20, and #30 in Table 4 were about 11% higher than the experiments, while the predictions of t_{max} were about 16% lower than the experiments at *FIGRA* = 0.975 kW/s, 1.58 kW/s, and 2.40 kW/s. On the other hand, the predictions of Q_{max} for Test #7 and #8 in Table 4 were about 19% lower than the experiments, while the predictions of t_{max} were about 10% lower than the experiments at *FIGRA* = 0.63 kW/s and 1.05 kW/s. These results caused more than a 30% deviation of $FIGRA_{pre}$ and *FIGRA*. It would be necessary to investigate the combustion efficiency and the surrounding temperature with environmental conditions to improve more accurate predictions. However, the final results shows that a total of 22 of the predictions were in good agreement with the experiments within ±15% since the predicted values of Q_{max} and t_{max} were satisfied with the experiments on the effects of S_f, *VF*, and Q_{ig} within ±5%. Especially, in the case of application of *FIGRA* of EN 13501-1, the polyethylene foam pipe insulation could have Grade B, C, or D in accordance with S_f, *VF*, and Q_{ig}.

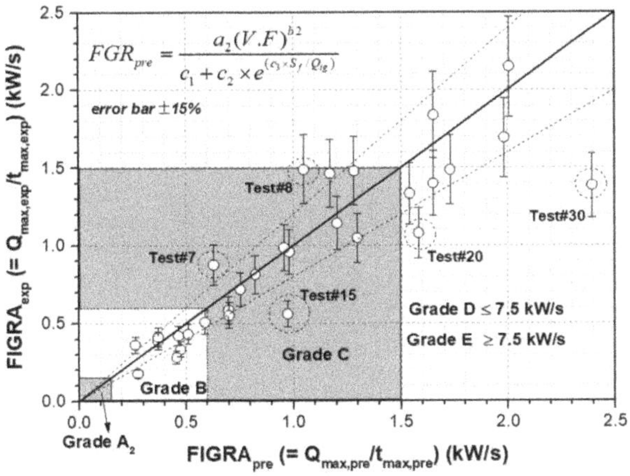

Figure 13. The comparison of the predictions and experiment of the fire growth rate index with scale factors (S_f = 1/3, 1/4, and 1/5), volume fraction (*VF* = 0.024, 0.05, 0.07, and 0.1), and ignition heat sources (Q_{ig}).

4. Conclusions

In this study, the effects of the scale factor (S_f), the volume fraction (*VF*), and the ignition heat source (Q_{ig}) on the fire growth rate index (*FIGRA*) of polyethylene foam pipe insulation were systematically investigated. From the 33 experiments of the heat release rate of the pipe insulation, the maximum heat release rate (Q_{max}), and the time (t_{max}) to reach the maximum heat release rate were analyzed with the effective heat of combustion assumed to be the averaged value of 37,214 kJ/kg. The results of this study can be summarized as follows,

First, the values of Q_{max} maintained a constant value within the range of ±3.48% average and ±12.26% maximum regardless of Q_{ig} when the value of *VF* was fixed. While t_{max} decreased, which was inversely proportional to Q_{ig}. These results explain that the heat amount of the pipe insulation should be conserved regardless of the ignition.

Second, the correlations between the values of Q_{max} and t_{max} in accordance with the variation of the scale factor (S_f = 1/3, 1/4, and 1/5), the volume fraction (*VF* = 0.024, 0.05. 0.07, and 0.1), and the ignition heat source (Q_{ig} = 10 kW, 15 kW, and 20 kW) were presented. It is possible to quantify that Q_{max} intended to increase in proportional to *VF* and S_f regardless of Q_{ig} while t_{max} increased in proportion to S_f/Q_{ig} and *VF*. However, the limitation of the predictions was that the experiment coefficients should be determined with the thermal properties of the wall and the type of the pipe insulation.

Finally, *FIGRA* as defined in EN 13501-1 was evaluated using the prediction models of the Q_{max} and t_{max}. It was verified that a total of 22 experiments in Table 4 were in good agreement with the predictive values of *FIGRA* within ±15%. Especially, the fire-retardant grade for the polyethylene foam pipe insulation could have a Grade B, C, and D in accordance with the scale factor (S_f = 1/3, 1/4, and 1/5), volume fraction (*VF* = 0.024, 0.05, 0.07, and 0.1), and the ignition heat sources (Q_{ig}). Therefore, in case of establishing the database of various types of insulation, it can be expected that the prediction models could apply to evaluate the fire-retardant performance with dimensionless methods for *FIGRA*.

Author Contributions: Conceptualization, J.W.P., O.K.L. and W.J.Y.; methodology, W.J.Y.; software, J.W.P.; validation, J.W.P. and W.J.Y.; formal analysis, W.J.Y.; investigation, J.W.P., O.K.L.; resources, W.J.Y.; data curation, W.J.Y.; writing—original draft preparation, W.J.Y.; writing—review and editing, W.J.Y.; visualization, J.W.P.; supervision, J.W.P., O.K.L.; project administration, W.J.Y.; funding acquisition, W.J.Y. Please turn to the CRediT taxonomy for the term explanation. Authorship must be limited to those who have contributed substantially to the work reported. All authors have read and agreed to the published version of the manuscript.

Funding: This research was supported by the Field-oriented Support of Fire Fighting Technology Research and Development Program funded by the Ministry of Public Safety and Security (2018-NFA002-008-01020000-2019).

Acknowledgments: This research was funded by the National Fire Agency (Republic of Korea), grant number 2018-NFA002-008-01020000-2019.

Conflicts of Interest: The authors declare no conflict of interest. The funders had no role in the design of the study; in the collection, analyses, or interpretation of data; in the writing of the manuscript, or in the decision to publish the results.

References

1. Kecebas, A. Determination of insulation thickness by means of exergy analysis in pipe insulation. *Energy Convers. Manag.* **2012**, *58*, 76–83. [CrossRef]
2. Kayfeci, M.; Yabanova, I.; Kecebas, A. The use of artificial neural network to evaluate insulation thickness and life cycle costs: Pipe insulation application. *Appl. Therm. Eng.* **2014**, *63*, 370–378. [CrossRef]
3. Kecebas, A.; Alkan, M.A.; Bayhan, M. Thermo-economic analysis of pipe insulation for district heating piping systems. *Appl. Therm. Eng.* **2011**, *31*, 17–18. [CrossRef]
4. Mangs, S. *Insulation Materials in District Heating Pipes: Environmental and Thermal Performance of Polyethylene Terephthalae and Polyurethane Foam*; Chalmers University of technology: Göteborg, Sweden, 2005; pp. 1–2.
5. Vega, A.; Yarahmadi, N.; Jakubowicz, I. Determination of the long-term performance of district heating pipes through accelerated ageing. *Polym. Degrad. Stab.* **2018**, *153*, 15–22. [CrossRef]
6. Winkler-Skalna, A.; Łoboda, B. Determination of the thermal insulation properties of cylindrical PUR foam products throughout the entire life cycle using accelerated aging procedures. *J. Build. Eng.* **2020**, *31*, 101348. [CrossRef]
7. Hilado, C.J.; Cumming, H.J. Fire Safety Aspects of Thermal Insulation. *J. Therm. Insul.* **1977**, *1*, 116–128. [CrossRef]
8. Goldfarb, I.; Zinoviev, A. A study of delayed spontaneous insulation fires. *Phys. Lett. A* **2003**, *311*, 491–500. [CrossRef]
9. Hidalgo, J.P.; Welch, S.; Torero, J.L. Performance criteria for the fire safe use of thermal insulation in buildings. *Constr. Build. Mater.* **2015**, *100*, 285–297. [CrossRef]
10. Ye, L.; Meng, X.Y.; Ji, X.; Li, Z.M.; Tang, J.H. Synthesis and characterization of expandable graphite–poly (methyl methacrylate) composite particles and their application to flame retardation of rigid polyurethane foams. *Polym. Degrad. Stab.* **2009**, *94*, 971–979. [CrossRef]
11. Xu, W.; Chen, R.; Du, Y.; Wang, G. Design water-soluble phenolic/zeolitic imidazolate framework-67 flame retardant coating via layer-by-layer assembly technology: Enhanced flame retardancy and smoke suppression of flexible polyurethane foam. *Polym. Degrad. Stab.* **2020**, *176*, 109152. [CrossRef]
12. D'Alessandro, F.; Baldinelli, G.; Sambuco, S.; Rufini, A. Experimental assessment of the water content influence on thermo-acoustic performance of building insulation materials. *Constr. Build. Mater.* **2018**, *158*, 264–274. [CrossRef]
13. Sundström, B. *The Development of a European Fire Classification System for Building Products*; Department of Fire Safety Engineering, Lund University: Lund, Sweden, 2007.
14. European Committee for Standardization. *EN ISO 11925-2 Reaction to Fire Tests—Ignitability of Building Products Subjected to Direct Impingement of Flame—Part 2: Single-Flame Source Test*; European Committee for Standardization: Brussels, Belgium, 2010.
15. European Committee for Standardization. *EN ISO 13823 Reaction to Fire Tests—Building Products-building Products Excluding Floorings Exposed to the Thermal Attack by a Single Burning Item*; European Committee for Standardization: Brussels, Belgium, 2002.

16. European Committee for Standardization. *EN ISO 13501-1 Reaction to Fire Tests—Fire Classification of Construction Products and Building Elements-Part 1: Classification using Test Data*; European Committee for Standardization: Brussels, Belgium, 2002.
17. NFPA. *NFPA Standard 274: Standard Test Method to Evaluate Fire Performance Characteristics of Pipe Insulation*; Technical Report; National Fire Protection Association: Quincy, MA, USA, 2013.
18. European Committee for Standardization. *EN ISO 13501-1 Reaction to Fire Tests—Small Room Test for Pipe Insulation Products or Systems*; European Committee for Standardization: Brussels, Belgium, 2008.
19. Janssens, M.L.; Huczek, J.; Sauceda, A. Development of a model of the ASTM E 84 Steiner tunnel test. *Fire Saf. Sci.* **2008**, *9*, 279–289. [CrossRef]
20. Pretrel, H.; Bouaza, L.; Suard, S. Multi-scale analysis of the under-ventilated combustion regime for the case of a fire event in a confined and mechanically ventilated compartment. *Fire Saf. J.* **2020**, *5*, 103069. [CrossRef]
21. Hernández, N.; Fuentes, A.; Reszka, P.; Fernández-Pello, A.C. Piloted ignition delay times on optically thin PMMA cylinders. *Proc. Combust. Inst.* **2019**, *37*, 3993–4000. [CrossRef]
22. Leisted, R.R.; Sørensen, M.X.; Jomaas, G. Experimental study on the influence of different thermal insulation materials on the fire dynamics in a reduced-scale enclosure. *Fire Saf. J.* **2017**, *93*, 114–125. [CrossRef]
23. You, W.J.; Park, J.W.; Sin, J.Y.; Park, H.G.; Ohk, K.L. Analysis of the Maximum Heat Release Rate in Accordance with the Test Method of the Flame Retardant Performance for Pipe Insulation. *Fire Sci. Eng.* **2020**, *34*, 18–25. [CrossRef]
24. Ohk, K.L.; Nam, D.G.; Jang, H.Y. Evaluation of the Reaction-to-fire Performance of Pipe Insulation Material using Small Room Test. *Fire Sci. Eng.* **2019**, *33*, 1–8.
25. McGrattan, K.; Klein, B.; Hostikka, S.; Floyd, J. *Fire Dynamics Simulator (Version 5), User's Guide*; NIST Special Publication; NIST: Gaithersburg, MD, USA, 2010; pp. 56–57.
26. Quintiere, J.G. *Fundamentals of Fire Phenomena*; John Wiley & Sons: West Sussex, UK, 2006; pp. 96–133.
27. Tewarson, A.; Jiang, F.H.; Morikawa, T. Ventilation controlled combustion of polymers. *Combust. Flame* **1993**, *95*, 151–169. [CrossRef]
28. Madrigal, J.; Guijarro, M.; Hernando, C.; Diez, C.; Marino, E. Effective heat of combustion for flaming combustion of Mediterranean forest fuels. *Fire Technol.* **2011**, *47*, 461–474. [CrossRef]
29. Al-Ghouti, M.; Al-Degs, Y.; Mustafa, F. Determination of hydrogen content, gross heat of combustion, and net heat of combustion of diesel fuel using FTIR spectroscopy and multivariate calibration. *Fuel* **2010**, *89*, 193–201. [CrossRef]
30. Lyon, R.E.; Walters, R.N. Pyrolysis combustion flow calorimetry. *Anal. Appl. Pyrolysis* **2004**, *71*, 27–46. [CrossRef]
31. Minakshi, M.; Visbal, H.; Mitchell, D.R.; Fichtner, M. Bio-waste chicken eggshells to store energy. *Dalton Trans.* **2018**, *47*, 16828–16834. [CrossRef] [PubMed]
32. Minakshi, M.; Barmi, M.; Mitchell, D.R.; Barlow, A.J.; Fichtner, M. Effect of oxidizer in the synthesis of NiO anchored nanostructure nickel molybdate for sodium-ion battery. *Mater. Today Energy* **2018**, *10*, 1–14. [CrossRef]
33. Lyon, R.E. An integral method of nonisothermal kinetic analysis. *Thermochim. Acta* **1997**, *297*, 117–124. [CrossRef]
34. Walters, R.N.; Hackett, S.M.; Lyon, R.E. Heats of combustion of high temperature polymers. *Fire Mater.* **2000**, *24*, 245–252. [CrossRef]
35. Bundy, M.F.; Hamins, A.P.; Johnsson, E.L.; Kim, S.C.; Ko, G.; Lenhert, D.B. *Measurements of Heat and Combustion Products in Reduced-Scale Ventilation-Limited Compartment Fires*; Technical Note (NIST TN) -1483; National Institute of Standards and Technology: Gaithersburg, MD, USA, 2007; pp. 114–116.
36. Van Krevelen, D.W. Some basic aspects of flame resistance of polymeric materials. *Polymer* **1975**, *16*, 615–620. [CrossRef]
37. Filipczak, R.; Crowley, S.; Lyon, R.E. Heat release rate measurements of thin samples in the OSU apparatus and the cone calorimeter. *Fire Saf. J.* **2005**, *40*, 628–645. [CrossRef]

www.ingramcontent.com/pod-product-compliance
Lightning Source LLC
Chambersburg PA
CBHW061553220926
51850CB00005B/322

MDPI
St. Alban-Anlage 66
4052 Basel
Switzerland
Tel. +41 61 683 77 34
Fax +41 61 302 89 18
www.mdpi.com

Energies Editorial Office
E-mail: energies@mdpi.com
www.mdpi.com/journal/energies